Calculus
A Problem-Solving Approach

Calculus
A Problem-Solving Approach

Neal Reid – *Principal Author*
Parkside High School

Douglas Mark
Parkside High School

Phil Feldman
Northview Heights Secondary School

Russell Garrett
Hillcrest High School

Joseph Geiser
Cardinal Leger Secondary School

David Handley
Belleville Collegiate Institute
and Vocational School

John Wiley & Sons
TORONTO   NEW YORK   CHICHESTER   BRISBANE   SINGAPORE

Canadian Cataloguing in Publication Data

Main entry under title:

Calculus: a problem solving approach

For use in high schools.
ISBN 0–471–79690–5

1. Calculus.   I. Reid, Neal E.,
    1938–   .

QA303.C347 1988   515   C88-093272-4

Cover and text design: JACK STEINER GRAPHIC DESIGN
Technical illustration: ESTEBAN & JORGE O. GUASTAVINO/GRAFIKA ART STUDIOS

Printed and bound in Canada
1 2 3 4 5 ML 98 97 96 95 94

Table of Contents

PART 2 Integration

Foreword

We believe that it is our job as teachers not only to teach calculus, but also to teach students how to learn calculus. We hope that when students leave classrooms, they leave with the tools they need to learn on their own. *Calculus: A Problem-Solving Approach* has been developed to encourage independent learning in two ways:

1. Beyond describing solutions to particular problems, this textbook shows how one thinks of such solutions, i.e., what encourages particular problem solving methods?

2. Asking the right questions about a problem is often the key to a solution. *Calculus: A Problem-Solving Approach* expressly illustrates and encourages this as a problem-solving tool.

Features

Calculus: A Problem-Solving Approach is intended for use in introductory calculus courses.

Organization
The textbook is organized into short, manageable chapters on closely defined topics, making reinforcement of difficult concepts easier.

Illustrations
You will see that our textbook is liberally illustrated. Beginners in mathematics frequently have difficulty extracting concepts from mathematical formulas; yet when the concepts are properly illustrated, they become clear immediately.

Extension Material
This textbook contains more material than can be covered in a first course in calculus. The key to using this textbook for teachers and students alike is to be selective. Optional sections and chapters permit teachers to be flexible in designing a course that extends beyond the basic core material. Students will find here ample material for term papers or independent study projects.

Special Sections
Special sections designated by the letters A through K are found between the regular chapters. Most of these sections are reviews of topics from pre-calculus mathematics courses. They provide, at strategic points, a convenient reference to formulas and methods.

Computer Applications

Other lettered sections describe computer applications, such as computer graphing. These may be omitted unless there is a need or an interest. There are also computer-related sections within chapters, which may be used to supplement your course of study.

Applications

Many applications to physics, engineering, biology, chemistry, and economics appear throughout the textbook.

Exercises

Exercises appear at the end of most sections. Each exercise set begins with routine drill problems and progresses toward problems of greater difficulty.

Chapter Summary

Review sections at the end of each chapter summarize the essential points of the chapter and provide suggestions on how the subject material can be mastered. Additional exercises will also be found here.

Answers

Answers to all numerical end-of-section exercises and end-of-chapter exercises are included at the end of the textbook. The *Solutions Manual* for *Calculus: A Problem-Solving Approach* contains complete solutions for *all* questions.

Acknowledgements

We gratefully acknowledge the fine work of Craig Fraser, Assistant Professor, Institute for the History and Philosophy of Science and Technology, University of Toronto, who prepared the historical profiles of mathematicians included in this textbook.

We would like to express our appreciation to the following reviewers, who, as teachers of calculus, provided us with many useful comments on the manuscript at various stages of its development: Professor Hugh Allen, Queen's University; Bill Haehnel, Bendale Secondary School; John LeBrun, Hillcrest High School; Paul Macallum, Banting Memorial High School; Sebastian Reisch, Clarke High School; and Steve Spiro, St. Clair Secondary School. Many of their recommendations were followed as we developed this textbook, for whose final content we alone are responsible.

The Authors

Differentiation

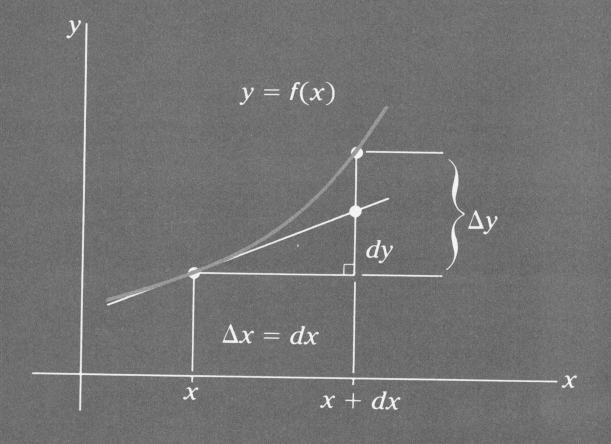

3

What is Calculus? 1

Credit for the invention of calculus is generally ascribed to Sir Isaac Newton in the latter part of the seventeenth century. In the intervening years, a great many mathematicians have worked to clarify the concepts and sharpen the logic of the theorems, until today, it has grown to become the cornerstone of modern mathematics.

The purpose of this first chapter is to help you to gain an overview of what calculus is about. In the following paragraphs, we will attempt to answer the question posed in the title of the chapter by discussing the principal ideas upon which calculus rests. We will briefly introduce the operations of calculus and the nature of the mechanical skills you will need to acquire. Lastly, we will point out some of the kinds of problems that can be solved using calculus and suggest why it would be difficult, if not impossible, to solve them by more elementary methods.

In keeping with our practice throughout the text, we shall depend on intuition, common sense, and simple algebraic demonstrations. Once you have become acquainted with the basic ideas and methods developed in this text, you will be ready to study the theorems and the rigorous proofs that justify these methods.

Basic Concepts

In essence, calculus deals with two fundamental problems: (a) finding the **tangent to a curve**, and (b) finding the **area under a curve**. Underlying the solution to both of these problems is the concept of a **limit**.

To develop the notion of a tangent to a curve, consider Figure 1.1a in which a curve C is cut by a straight line l in two points. This line is called a **secant** to the curve. By sliding or rotating the secant in the proper direction as in Figures 1.1b or 1.1c, it is possible to cause the two points of intersection to move toward each other. At the moment they coincide, the secant becomes tangent to the curve.

Figure 1.1

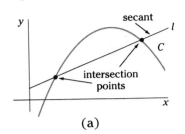
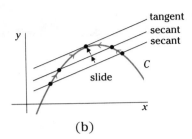
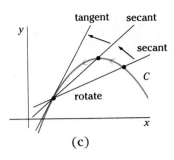

(a)   (b)   (c)

The resulting single point of intersection is the point of tangency. In the vicinity of this point, the tangent closely approximates the position and orientation of the curve, which makes it valuable in all types of applications. The following example shows how it is possible to discover the tangent to a curve by sliding a secant.

EXAMPLE A

Find the equation of the line having a slope of 1 that is tangent to the curve

$$y = 4 - (x - 2)^2$$

SOLUTION

Figure 1.2

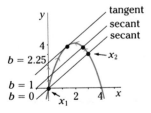

The curve is a parabola opening down with vertex at $(2, 4)$, as shown in Figure 1.2. The equation $y = x + b$ is a line with slope 1 and y-intercept b. The line $y = x + b$ and the curve $y = 4 - (x - 2)^2$ intersect when

$$x + b = 4 - (x - 2)^2$$

$$x + b = 4 - x^2 + 4x - 4$$

$$x^2 - 3x + b = 0$$

The x-coordinates of the points of intersection, x_1, and x_2, can be found by solving this equation for x using the quadratic formula:

$$x = \frac{3 \pm \sqrt{(-3)^2 - 4b}}{2}$$

$$x_1 = \frac{3 - \sqrt{9 - 4b}}{2}$$

$$x_2 = \frac{3 + \sqrt{9 - 4b}}{2}$$

As expected, the positions of the points of intersection depend on b, the value of which determines where the line is located.

Now move the line by varying b. Table 1.1 shows that as the value of b increases, the points of intersection get closer together. When b reaches the value 2.25, the expression in the radical is zero, and x_1 and x_2 both equal 1.5. At this moment when the points coincide, the secant has become the tangent to the curve. The equation of the tangent is therefore

$$y = x + 2.25$$

The act of moving the secant in such a way as to bring the two points of intersection into coincidence is an example of a **limit process**. Limits and tangents will be thoroughly investigated in Chapters 2 and 3. It is difficult, in general, to determine the slope of a tangent to a curve by elementary methods. It is simple, however, to calculate the slope of a secant at any position. By moving the secant and following a sequence of values of its slope, you can discover that the slope approaches a specific value, a *limit*, which is the slope of the tangent.

The expression "area under a curve" refers to the area between the curve and the x-axis over some interval from a to b, as indicated in Figure 1.3a. There is no general formula for areas of this sort, because of the curved upper boundary. However, the measure of the area can be approximated by summing the areas of a series of narrow rectangular strips, as indicated in Figure 1.3b.

Table 1.1

b	x_1	x_2
0	0	3
1	0.38	2.62
2	1	2
2.2	1.28	1.72
2.25	1.50	1.50

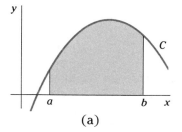

(a)

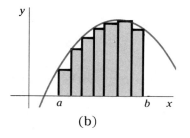

(b)

Figure 1.3

This diagram suggests that the narrower the rectangles, the more nearly exact is the value of the area so obtained. This act of reducing the size of the rectangles to obtain a better approximation is another **limit process**. The exact value of the area is found "in the limit" as the width of the rectangles approaches zero and their number becomes exceedingly large.

EXAMPLE B

Determine an approximate value for the area under the curve

$$f(x) = 4 - (x - 2)^2$$

in the interval $x = 0$ to $x = 4$ shown in Figure 1.4a.

Figure 1.4

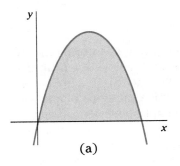

(a)

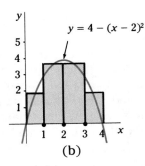

(b)

SOLUTION

Cover the area with rectangles having a width arbitrarily chosen to be one unit and a height equal to the value of f at the midpoint of the base as in Figure 1.4b. The area is then approximately equal to the sum

$$A = h_1 b_1 + h_2 b_2 + h_3 b_3 + h_4 b_4$$

$$= [f(0.5)](1) + [f(1.5)](1) + [f(2.5)](1) + [f(3.5)](1)$$

$$= 1.75 + 3.75 + 3.75 + 1.75$$

$$= 11$$

Using narrower rectangles would of course give a more accurate result. A calculation similar to the one above using rectangles having a width of 1/2 unit, for instance, gives an approximate value of 10.75 square units for the area. The consequence of using narrower rectangles, however, is a lengthier calculation since there are more rectangular areas to compute and sum. As you will see in Chapter 16, calculus lets you deal with an infinite number of infinitely narrow rectangles, and yields an exact answer of 32/3 or $10.\overline{6}$ square units for the area in this particular example.

The Operations of Calculus

Calculus breaks down naturally into two major subdivisions: **differential calculus** and **integral calculus**. Central to differential calculus is a mathematical operation called **differentiation**. This operation changes any given mathematical function f into a related function f' (read: "f prime") called the **derivative** of f. The **value of the derivative** at a particular point is the slope of the tangent to the graph of f at that point. The formal definition of a derivative will be given in Chapter 3. In that and the next several chapters, considerable work will be done to develop the rules for differention and to study the applications of the derivative to a wide variety of problems.

EXAMPLE C

Given that

$$f'(x) = 6x$$

is the derivative of the function

$$f(x) = 3x^2 + 5$$

determine the equation of the line tangent to f at $x = 3$.

SOLUTION

At $x = 3$, the value of the function is

$$f(3) = 3(3)^2 + 5$$

$$= 32$$

This point $(3, 32)$ is the point of tangency. The slope of the tangent at that point is found by evaluating the derivative at $x = 3$:

$$f'(3) = 6(3)$$

$$= 18$$

The equation of the line through the point $(3, 32)$ with slope 18 is therefore

$$\frac{y - 32}{x - 3} = 18$$

$$y - 32 = 18(x - 3)$$

or

$$y = 18x - 22$$

This is the equation of the tangent to the curve at $x = 3$. ❏

EXAMPLE D

According to one of the rules of differentiation, the derivative of the function $f(x) = x^n$ is the function $f'(x) = nx^{n-1}$, where n is any real number. Use this rule to determine the derivative of

(a) $f(x) = x^5$ and   (b) $f(x) = x^{-5}$

SOLUTION

(a) In the function $f(x) = x^5$, the exponent is 5. This is the value to use for n in the differentiation rule. Therefore, $n - 1 = 4$, and the rule states that the derivative of f is

$$f'(x) = 5x^4$$

(b) In the case of the function $f(x) = x^{-5}$, the exponent is -5, which means that $n - 1 = -6$, and the derivative is

$$f'(x) = -5x^{-6}$$ ❏

Integral calculus deals with another mathematical operation called **integration**. When this operation is applied to a given function f, the result is a new function called the **indefinite integral** of f. What is most remarkable is that the operations of differentiation and integration are inverse operations in somewhat the same sense that multiplication and division are inverse operations.

Closely related to the indefinite integral of f is the **definite integral**. It is a number, which can be interpreted as the measure of the area under the graph of f over a specific interval. The subject of integration occupies the latter half of this text. In those chapters, you will learn a number of methods of integration and study a variety of applications.

EXAMPLE E

A definite integral of the function $f(x) = 6x^2$, which represents the area under the curve $y = 6x^2$ from the origin up to some value $x = a$, is $F(a) = 2a^3$. What is the area under the curve from $x = 3$ to $x = 5$, shown in Figure 1.5?

SOLUTION

The area under the curve from 3 to 5 is the area from 0 to 5 less the area from 0 to 3.

Figure 1.5

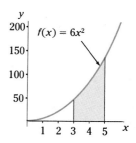

Since

$$F(5) = 2(5^3)$$

$$= 250$$

and

$$F(3) = 2(3^3)$$

$$= 54$$

therefore

$$F(5) - F(3) = 250 - 54$$

$$= 196$$

Thus, the area in question is 196 square units.

Applications

The application of mathematics to real situations requires assumptions that oversimplify and idealize the real, physical world. The earth is not a perfect sphere, yet it is convenient to assume so in computing the basic parameters of a satellite orbit. To say that a projectile follows a parabolic path is to assume that the earth is a perfectly flat, airless world. To say that an object moves at a constant speed is to neglect gravity, friction, and other forces present in the real world.

Calculus, like elementary mathematics, describes the world only imprecisely. However, by virtue of the limit processes, it goes beyond elementary mathematics in two important respects: first, it is able to analyze how small variations in one quantity affect another (the tangent problem) and, second, it is able to sum a (very great) number of (incredibly) small values (the area problem). The next examples give some indication how these new capabilities extend the range of problems that can be covered.

EXAMPLE F

The position of an object falling under the force of gravity is given by

$$s = \frac{1}{2}gt^2 + v_0 t + s_0$$

where v_0 is the initial velocity,
  s_0 is the initial position,
and g is the acceleration due to gravity, $-9.8\,\text{m/s}^2$.

If a stone falls from rest from the top of a cliff 200 m high, with what velocity does it strike the ground at the base of the cliff?

Figure 1.6

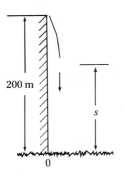

200 m

s

0

SOLUTION

It makes sense to use the ground at the base of the cliff as the point of reference. Then s is the height of the stone above the ground at any time t after it starts to fall (Figure 1.6). At the start of the fall (at $t = 0$), the stone is 200 m above the ground. This is its initial position, $s_0 = 200$. To fall from rest means that the initial velocity v_0 of the stone is zero. The equation which describes the motion of the stone is, therefore,

$$s = -4.9t^2 + 200$$

The velocity of the stone is the rate at which its position changes. It is given by the derivative of s with respect to t, which is found using the rules of differentiation to be

$$v = s'$$

$$= -9.8t$$

You can determine when the stone hits the ground by setting $s = 0$:

$$0 = -4.9t^2 + 200$$

$$\therefore \quad t \doteq 6.39$$

The stone hits the ground approximately 6.39 s after it starts to fall. At this moment, the velocity is

$$v = -9.8(6.39)$$

$$\doteq -62.6$$

The velocity with which the stone hits the ground is, therefore, approximately 62.6 m/s. ❑

EXAMPLE G

Determine the amount of glass required to make a circular lens having a radius of 5 cm and a thickness of 2 cm at its centre. One surface is flat, and the other is parabolic in shape.

SOLUTION

The lens is shown in Figure 1.7 with its vertex at the origin. Imagine that the lens is sliced like a loaf of bread perpendicular to the x-axis. The volume of glass in the lens is equal to the sum of the volumes of all the slices:

$$V = \Delta V_1 + \Delta V_2 + \Delta V_3 + \cdots + \Delta V_n$$

$$= \sum_{k=1}^{n} \Delta V_k$$

Figure 1.7

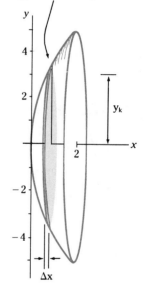

a typical disk-shaped slice

Here the symbol Σ is a shorthand way of denoting the sum of n similar terms, and ΔV_k is the volume of a typical slice. Each slice has approximately the shape of a thin circular disk. The volume of a slice, therefore, is approximately equal to

$$\Delta V_k \doteq \text{circular area} \times \text{thickness}$$
$$\doteq \pi y_k^2 \cdot \Delta x$$

It follows that the volume of the lens is approximately equal to

$$V \doteq \sum_{k=1}^{n} \pi y_k^2 \cdot \Delta x$$

This approximation improves as the slices are made thinner. The exact value of the volume is found in the limit as the slice thickness Δx approaches zero. It is expressed as the integral

$$V = \int_0^2 \pi y^2 \cdot dx$$

The integral is evaluated using methods you will learn about in later chapters. The result is that 25π cm^3 of glass are required to make the lens. ❑

Exercise 1.1

1. The line $y = 2x - 5$ is tangent to the graph of the function $y = (x - 2)^2$ (Figure 1.8). What are the coordinates of the point of tangency? What is the slope of the tangent line?

Figure 1.8

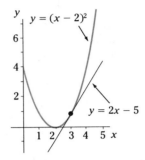

2. What is the slope of the tangent to the parabola $y = (x - 3)^2 + 4$ at its vertex?

3. By sliding the secant $y = 2x + b$ until it is tangent to the curve $y = x^2 - 4$ shown in Figure 1.9, determine the equation of the tangent with slope 2.

Figure 1.9

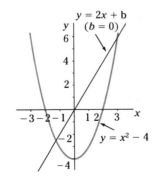

4. Determine an approximate value for the area under the curve $y = x^2 + 2$, for $0 \le x \le 3$ in two ways:

(a) using three rectangles, the upper *left* corners of which lie on the curve, as in Figure 1.10a;

(b) using three rectangles, the upper *right* corners of which lie on the curve, as in Figure 1.10b.

Figure 1.10

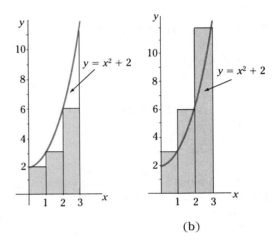

(b)

How is the exact value of the area related to the answers in (a) and (b)? Estimate its value.

5. Determine an approximate value for the area in Figure 1.11 under the curve $y = 21/x$ for $1 \le x \le 4$ using

(a) three rectangles

(b) six rectangles.

Measure the heights of the rectangles at the midpoints of the intervals.

Figure 1.11

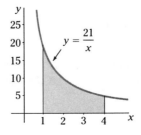

6. It takes an automobile 12 min to travel from exit 15 to exit 16 on the expressway, a distance of 21 km. What is its average speed in km/h? Does the automobile actually travel the entire distance at exactly this speed? Explain.

7. According to one of the rules of differentiation, the derivative of the function $f(x) = \sqrt{ax}$ is the function

$$f'(x) = \frac{\sqrt{a}}{2\sqrt{x}}$$

where a is a real number. Using this rule, determine the derivative of

(a) $f(x) = \sqrt{2x}$

(b) $f(x) = \sqrt{4x}$

8. According to another one of the rules of differentiation, the derivative of the function $f(x) = \tan(ax)$ is the function $f'(x) = a\sec^2(ax)$, where a is a real number. Using this rule, determine the derivative of

(a) $f(x) = \tan(5x)$

(b) $f(x) = \tan\left(\dfrac{x}{2}\right)$

9. According to an integration formula, the indefinite integral of the function $f(x) = x^n$ is the function

$$F(x) = \frac{x^{n+1}}{n+1} + C$$

where n is a real number, $n \neq -1$, and C is an arbitrary constant. Using this formula, determine the indefinite integral of

(a) $f(x) = -x^{-2}$

(b) $f(x) = x^{1/2}$

(c) $f(x) = x^{2.5}$

10. According to another integration formula, the indefinite integral of the function $f(x) = \cos(ax)$ is the function

$$F(x) = \frac{1}{a}\sin(ax) + C$$

where a is a real number, and C is an arbitrary constant. Using this formula, determine the indefinite integral of

(a) $f(x) = \cos(4x)$

(b) $f(x) = \cos\left(-\dfrac{1}{3}x\right)$

(c) $f(x) = \cos(0.5x)$

11. The graph in Figure 1.12 shows the height of an individual each year from birth to age 21. Copy the graph and draw the tangent to the graph at age 6. What meaning can be given to the slope of the tangent? Would you expect the slope to be different at age 13? At age 20? Explain.

Figure 1.12

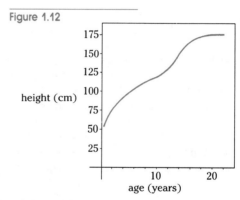

12. On graph paper draw a series of rectangles each having the same perimeter, 20 units, but different widths: $1, 2, 3, \ldots, 9$ units. As the width increases, how does the length change? How does the area change? Which of these rectangles has the largest area?

13. The perimeter p of a regular polygon is given by the formula

$$p = 2n\sin\left(\frac{180°}{n}\right)$$

where n is the number of sides, and the distance from the centre to any vertex is one unit (Figure 1.13). Determine the perimeter of polygons having 4, 6, 12, 20, and 36 sides. As the value of n increases, the shape of the polygon approaches what geometrical figure? What numerical value does the perimeter approach?

Figure 1.13

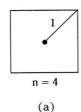

n = 4

(a)

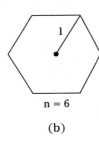

n = 6

(b)

14. (a) Determine the approximate volume of a cone of height 10 cm and base radius 10 cm shown in Figure 1.14a as follows:

    i. Approximate the volume using a stack of five cylindrical disks, as in Figure 1.14b, each of which has a volume of $V = \pi r^2 h$.

    ii. Find the sum of the volumes of the disks.

  (b) Does this approximation overestimate or underestimate the volume?

  (c) What could be done to improve the approximation?

  (d) Compare the result of part (a) to the exact value, which according to calculus is given by the formula $V = \pi r^2 h/3$.

Figure 1.14

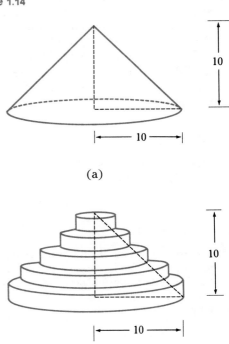

10

10

(a)

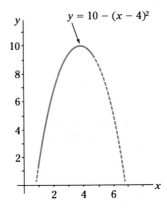

10

10

(b)

15. (a) Determine approximately the length of the arc of the parabola $y = 10 - (x - 4)^2$ shown in Figure 1.15 from $x = 1$ to $x = 5$ as follows:

Figure 1.15

$y = 10 - (x - 4)^2$

i. Approximate the arc of the parabola using four straight line segments drawn between points on the arc with x-coordinates 1, 2, 3, 4, and 5.

ii. Find the sum of the lengths of these line segments.

(b) Does this approximation overestimate or underestimate the length of the arc?

(c) What could be done to improve the approximation?

(d) Compare the result of part (a) to the exact value 11.226023, found using calculus.

16. A frog sets out to cross a lily pond 10 m in diameter. The first jump takes it to the centre of the pond. Because the frog grows tired, each jump after the first is half the length of the previous one.

(a) How long is the third jump? The fourth jump? The fifth jump?

(b) How far does the frog travel in three jumps? Four jumps? Five jumps?

(c) How many jumps are required to cross the pond? How long is the last jump? Explain.

A.   Preparation for Calculus

On beginning the study of calculus, you are expected to have a certain amount of skill in manipulating algebraic expressions and solving equations. You should also be familiar with the analytical geometry of the straight line and be able to graph simple functions. In addition, you will find it helpful to be acquainted with the factor theorem, the binomial theorem, and sequences and series. These and certain other topics are outlined in this section and Section B. Thumb through these sections briefly. You may skip them both, if a review of these subjects is not needed, but you may want to refer back to them later.

Laws of Exponents

The symbol a^n for positive, integer values of n is defined as the product:

$$a^n = \underbrace{a \cdot a \cdot a \cdot \ldots \cdot a}_{n \text{ factors}}$$

The following laws of exponents arise directly out of this definition:

(a) $a^m \cdot a^n = a^{m+n}$

(b) $\dfrac{a^m}{a^n} = a^{m-n} \qquad (m > n, a \neq 0)$

(c) $(a^m)^n = a^{mn}$

(d) $(a \cdot b)^m = a^m \cdot b^m$

(e) $\left(\dfrac{a}{b}\right)^m = \dfrac{a^m}{b^m} \qquad (b \neq 0)$

Even when n is not a positive integer, it is required that these laws still be satisfied in all situations. Extending the definition of a^n in this way gives some additional rules for zero, negative, and fractional exponents:

(f) $a^0 = 1 \qquad (a \neq 0$, the symbol 0^0 is not defined.$)$

(g) $a^{-n} = \dfrac{1}{a^n}$

(h) $a^{1/2} = \sqrt{a}$

In general, $a^{1/q}$ is defined to be the positive qth root of a, and $a^{p/q}$ is the pth power of this root.

Radicals

The use of good mathematical form includes the simplification of expressions which contain radicals. To reduce a radical to lowest terms, remove all common factors that are perfect squares from under the radical sign. For example,

$$\sqrt{12 - 16x^2} = \sqrt{4(3 - 4x^2)}$$

$$= 2\sqrt{3 - 4x^2}$$

Sometimes a situation calls for a radical to be rationalized, for instance, one in the denominator of a fraction. There are two ways to do this. If the radical expression is simply $\sqrt{b}$, then multiply by $\sqrt{b}$:

$$\frac{10}{3\sqrt{5}} = \frac{10}{3\sqrt{5}} \cdot \frac{\sqrt{5}}{\sqrt{5}}$$

$$= \frac{10\sqrt{5}}{3 \cdot 5}$$

$$= \frac{2\sqrt{5}}{3}$$

On the other hand, if the radical expression is of the form $(a + \sqrt{b})$, then multiply by the conjugate $(a - \sqrt{b})$ For example,

$$\frac{1}{x + \sqrt{x^2 - 4}} = \frac{1}{(x + \sqrt{x^2 - 4})} \cdot \frac{(x - \sqrt{x^2 - 4})}{(x - \sqrt{x^2 - 4})}$$

$$= \frac{x - \sqrt{x^2 - 4}}{x^2 - (x^2 - 4)}$$

$$= \frac{x - \sqrt{x^2 - 4}}{4}$$

Factoring

The recognition of common factors is essential in simplifying expressions. A typical example is the common factor

$$2x^3(x + 3) + 3x^2(x + 3)^2 = x^2(x + 3)[2x + 3(x + 3)]$$
$$= x^2(x + 3)(5x + 9)$$

Other common patterns for factoring are

the difference of squares:

$$a^2 - b^2 = (a + b)(a - b)$$

the sum of cubes:

$$a^3 + b^3 = (a + b)(a^2 - ab + b^2)$$

the difference of cubes:

$$a^3 - b^3 = (a - b)(a^2 + ab + b^2)$$

EXAMPLE A

Factor the trinomial $6x^2 + 11x - 10$.

SOLUTION

$$6x^2 + 11x - 10 = \frac{6(6x^2 + 11x - 10)}{6}$$

$$= \frac{(6x)^2 + 11(6x) - 60}{6}$$

$$= \frac{(6x + 15)(6x - 4)}{6}$$

$$= \frac{3(2x + 5) \cdot 2(3x - 2)}{6}$$

$$= (2x + 5)(3x - 2)$$

The Factor Theorem

The Factor Theorem is a useful tool for finding linear factors of polynomials and thus for solving polynomial equations.
If $P(x)$ is a polynomial in x, then

(a) $P(a)$ is the remainder when $P(x)$ is divided by $(x - a)$;

(b) $(x - a)$ is a factor of $P(x)$, if and only if $P(a) = 0$.

EXAMPLE B

Factor the polynomial $P(x) = 3x^3 + 14x^2 + 13x - 6$.

SOLUTION

The value of a in $(x - a)$ must be a number of the form p/q, where p is a factor of the constant term, -6, and q is a factor of the coefficient, 3, of the leading power of x. Thus it is necessary to test only the values $\pm 1, \pm 2, \pm 3, \pm 6, \pm 1/3, \pm 2/3$ in this case.

If

$$P(x) = 3x^3 + 14x^2 + 13x - 6$$

then

$$P(1) = 3 + 14 + 13 - 6$$

$$\neq 0$$

and $(x - 1)$ is *not* a factor.

But

$$P(-2) = -24 + 56 - 26 - 6$$

$$= 0$$

therefore $(x + 2)$ *is* a factor.

The other factor is obtained by dividing $P(x)$ by $(x - a)$ using long division. Fully factored, the polynomial is

$$P(x) = (x + 2)(3x^2 + 8x - 3)$$

$$= (x + 2)(3x - 1)(x + 3)$$

Solution of Quadratic Equations

Besides the usual algebraic processes, factoring is helpful in solving equations. For example, by factoring,

$$2x^2 - 7x - 15 = 0$$

$$(2x + 3)(x - 5) = 0$$

The product is zero, only when the factors are zero. Therefore,

$$2x + 3 = 0 \quad \text{or} \quad x - 5 = 0$$

$$x = -\frac{3}{2} \quad \text{or} \quad x = 5$$

When factors cannot be found, a quadratic equation can nevertheless be solved using the **quadratic formula**.

If

$$ax^2 + bx + c = 0$$

then

$$x = \frac{-b \pm \sqrt{b^2 - 4ac}}{2a}$$

The solutions are the **roots** of the equation. The nature of the roots depends on the **discriminant** $b^2 - 4ac$:
If $b^2 - 4ac > 0$, there are two unequal real roots.
If $b^2 - 4ac = 0$, there are two equal real roots.
If $b^2 - 4ac < 0$, there are no real roots.

Solution of Inequations

The solution of an inequation is a range of values of x, rather than a set of single values.

EXAMPLE C

Solve for x:
(a) $-2x > 10$ 　　　　　　　　　(b) $x^2 - x - 6 < 0$

SOLUTION

(a) If 　　$-2x > 10$ 　←————————————————
　　 then 　　$x < -5$

when multiplying or dividing by a negative number, the direction of the inequality is reversed

(b) 　　$x^2 - x - 6 < 0$

　　$(x - 3)(x + 2) < 0$

It follows that the factors must have opposite signs. Graphing the signs of the factors as in Figure A.1 makes the solution evident: x must lie in the interval $-2 < x < 3$. ❏

Figure A.1

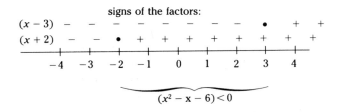

signs of the factors:

$(x^2 - x - 6) < 0$

Absolute Value

The absolute value of a quantity is defined by

$$|x| = \quad x \quad \text{if} \quad x \geq 0$$

$$= -x \quad \text{if} \quad x < 0$$

Therefore, depending on whether x is negative or positive, the statement $|x| \leq a$ is expressed as

either	$x \geq 0$ and $x \leq a$	or	$x < 0$ and $-x \leq a$
that is,	$0 \leq x \leq a$	or	$-a \leq x < 0$
which means		$-a \leq x \leq a$	

In other words, x lies in the interval $[-a, a]$. Likewise, the statement $|x| > a$ means that x lies outside the interval $[-a, a]$.

Similarly, the statement $|x - k| \leq a$ means that

$$(x - k) \geq 0 \text{ and } (x - k) \leq a \quad \text{or} \quad (x - k) < 0 \text{ and } -(x - k) \leq a$$
$$0 \leq (x - k) \leq a \quad \text{or} \quad -a \leq (x - k) < 0$$
$$-a \leq (x - k) \leq a$$
$$k - a \leq x \leq k + a$$

In this case, x lies in an interval centered on k, extending a units to the left and to the right (see Figure A.2). Notation like this is used to describe errors in measured quantities. For instance, the statement $|x - 5| \leq 0.01$ means that x lies within ± 0.01 units of 5, that is, $4.99 \leq x \leq 5.01$.

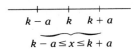

Figure A.2

Factorials

The symbol $n!$ (read: "n factorial") is used to represent the product of the integers 1 to n:

$$n! = n \cdot (n - 1) \ldots 3 \cdot 2 \cdot 1$$

For instance $5! = 5 \cdot 4 \cdot 3 \cdot 2 \cdot 1 = 120$.
By definition $0! = 1$

The Binomial Theorem

When n is a natural number, the expansion of $(a + b)^n$ is given by

$$(a + b)^n = a^n + na^{n-1}b + \frac{n(n - 1)}{2 \cdot 1}a^{n-2}b^2$$
$$+ \frac{n(n - 1)(n - 2)}{3 \cdot 2 \cdot 1}a^{n-3}b^3 + \cdots + nab^{n-1} + b^n$$

where the kth term is

$$\frac{n(n - 1)(n - 2)\ldots(n - k + 1)}{k(k - 1)(k - 2)\ldots 3 \cdot 2 \cdot 1}a^{n-k}b^k$$

Upon multiplying numerator and denominator by $(n - k)!$, the kth term can be written more simply as

$$\frac{n!}{(n - k)!\, k!}a^{n-k}b^k.$$

When n is a small integer, the coefficients of the terms in this expansion can easily be found from Pascal's Triangle. This is an array of numbers (see Figure A.3) in which each number is the

Figure A.3

$n = 0$					1			
$n = 1$				1		1		
$n = 2$			1		2		1	
$n = 3$		1		3		3		1
$n = 4$	1		4		6		4		1
$n = 5$	1		5		10		10		5		1

Pascal's Triangle

sum of the two numbers diagonally above it. To expand $(x - 4)^3$, for instance, use the coefficients 1 3 3 1 in the row for $n = 3$:

$$(x - 4)^3 = (1)x^3 + (3)(-4)x^2 + (3)(-4)^2x + (1)(-4)^3$$
$$= x^3 - 12x^2 + 48x - 64$$

Function Notation

A relation is a set of ordered pairs (x, y) with a rule for determining which pairs belong to the set. A function is a relation with the additional requirement that there is never more than one value of y corresponding to each value of x. A function is written

$$f = \{(x, y) \mid y = f(x)\}$$

or simply

$$y = f(x)$$

where f is the name of the function, x is the independent variable, and $f(x)$ (read: "f of x") is the value of the function at x.

EXAMPLE D

Evaluate $f(x) = 3x^2 - x + 5$ at
(a) -1 (b) k (c) $x + h$

SOLUTION

If

$$f(x) = 3x^2 - x + 5$$

then

(a) $\quad f(-1) = 3(-1)^2 - (-1) + 5$

$\qquad\qquad = 9$

(b) $\quad f(k) = 3k^2 - k + 5$

(c) $\quad f(x + h) = 3(x + h)^2 - (x + h) + 5$

$\qquad\qquad = 3(x^2 + 2hx + h^2) - x - h + 5$

$\qquad\qquad = 3x^2 + (6h - 1)x + (3h^2 - h + 5)$ ❑

Linear Functions

A **linear function** is defined by an equation of the form

$$f(x) = mx + b$$

where m and b are constants. The graph of a linear function is a straight line. The **slope m** of the line is defined as

$$m = \frac{y_2 - y_1}{x_2 - x_1}$$

where (x_1, y_1) and (x_2, y_2) are any pair of points that satisfy the function. b is the **y-intercept**. When the slope m and one point (x_1, y_1) on the line are known, the equation of the line can be found from

$$y - y_1 = m(x - x_1)$$

Quadratic Functions

A **quadratic function** is defined by an equation of the form

$$f(x) = ax^2 + bx + c$$

where a, b, and c are constants, $a \neq 0$. The graph of a quadratic function is a parabola, which opens up if a is positive and down if a is negative. You can determine the points at which $f(x) = 0$ by factoring or by using the quadratic formula. These are the points where the graph crosses the x-axis.

Another form of a quadratic function can be found by "completing the square." The result is

$$f(x) = a(x - p)^2 + q$$

where $x - p = 0$ is the equation of the axis of symmetry and the point (p, q) is the vertex of the parabola.

EXAMPLE E

Determine the position of the vertex of $f(x) = 2x^2 + 5x + 5$ by completing the square.

SOLUTION

$$f(x) = 2x^2 + 5x + 5$$

$$= 2\left[x^2 + \frac{5}{2}x\right] + 5 \qquad \text{———— half the coefficient of } x, \text{ squared}$$

$$= 2\left[x^2 + \frac{5}{2}x + \frac{25}{16} - \frac{25}{16}\right] + 5 \qquad \longleftarrow \text{ write as a square}$$

$$= 2\left[\left(x + \frac{5}{4}\right)^2 - \frac{25}{16}\right] + 5$$

$$= 2\left(x + \frac{5}{4}\right)^2 - \frac{25}{8} + 5$$

$$= 2\left(x + \frac{5}{4}\right)^2 + \frac{15}{8}$$

The vertex of the parabola is at $\left(-\frac{5}{4}, \frac{15}{8}\right)$.   ❑

Sequences and Series

A **sequence** is a function f defined on the natural numbers N. The values of the function are called the terms of the sequence and denoted by $t_1, t_2, t_3, \ldots, t_n, \ldots$, where $t_n = f(n)$ is the general term or nth term of the sequence. A sequence may have a finite or an infinite number of terms. Two sequences that are of particular interest are the arithmetic and the geometric sequences.

An **arithmetic sequence** is one in which consecutive terms differ by a constant, d:

$$d = t_2 - t_1 = t_3 - t_2 = \cdots = t_n - t_{n-1} = d$$

The general term is

$$t_n = t_1 + (n - 1)d$$

A **geometric sequence** is one in which the ratio of two consecutive terms is a constant, r:

$$r = \frac{t_2}{t_1} = \frac{t_3}{t_2} = \cdots = \frac{t_n}{t_{n-1}} = r$$

The general term is

$$t_n = t_1 r^{n-1}$$

A **series** is the sum of the terms of a sequence:

$$S_n = t_1 + t_2 + t_3 + \cdots + t_n$$

The series

$$S_n = t_1 + (t_1 + d) + (t_1 + 2d) + \cdots + [t_1 + (n-1)d]$$

is a finite **arithmetic series** with n terms. Its sum is given by

$$S_n = \frac{n}{2}(t_1 + t_n)$$

For example, the arithmetic sequence $2, 5, 8, 11, \ldots$ has a common difference of 3. The nth term of the sequence is

$$t_n = 2 + (n-1)3$$

$$= 3n - 1$$

The twenty-fifth term of the sequence is $t_{25} = 74$. The sum of the first 25 terms of this sequence is the series

$$S_{25} = 2 + 5 + 8 + 11 + \cdots + 74$$

This sum is

$$S_{25} = \frac{25}{2}(2 + 74)$$

$$= 950$$

The sum of the terms of a geometric sequence is a **geometric series**. The series

$$S_n = t_1 + t_1 r + t_1 r^2 + \cdots + t_1 r^{n-1}$$

is a finite geometric series with n terms. Its sum is given by

$$S_n = \frac{t_1(1 - r^n)}{1 - r}$$

For instance, the geometric sequence

$$\frac{3}{2}, \frac{3}{4}, \frac{3}{8}, \dots, \frac{3}{2}\left(\frac{1}{2}\right)^{n-1}$$

has a common ratio $r = 1/2$. The sum of the first 10 terms of this sequence is

$$S_{10} = \frac{\frac{3}{2}\left(1 - \left(\frac{1}{2}\right)^{10}\right)}{\left(1 - \frac{1}{2}\right)}$$

$$= 3\left(1 - \frac{1}{1024}\right)$$

$$= \frac{3069}{1024}$$

An **infinite geometric series** has the sum

$$S = \frac{t_1}{1 - r}, \quad \text{if} \quad |r| < 1.$$

Exercise A

1. Simplify:

(a) $(4x^{-2}) \cdot (8x^{-2})$

(b) $(4x^{-2}) - (8x^{-2})$

(c) $(4x^{-2}) \div (8x^{-2})$

(d) $(4x^{-2}) + (8x^{-2})$

2. Factor fully:

(a) $x^2 - 10x - 24$

(b) $x^2 - 64$

(c) $x^4 - 13x^2 + 36$

(d) $6x^2 - 41x + 30$

(e) $4x^3 - 32x^2 - x + 8$

(f) $x^3 - 10x^2 + 19x + 30$

(g) $56x^3 - 7$

3. Rationalize the denominator and simplify:

(a) $\dfrac{4x^2}{\sqrt{2x}}$

(b) $\dfrac{x^2 - 4}{\sqrt{x - 2}}$

(c) $\dfrac{8}{\sqrt{4 + x} - \sqrt{x}}$

4. Solve the following equations by factoring:

(a) $3x^2 + 9x - 30 = 0$

(b) $6x^2 + 11x - 10 = 0$

(c) $4x^3 + 3x^2 - 16x - 12 = 0$

(d) $2x^3 - x^2 - 13x - 6 = 0$

5. Solve the following equations using the quadratic formula:

(a) $x^2 + 6x - 12 = 0$

(b) $5x^2 - 10x + 4 = 0$

6. Simplify by extracting a common factor:

(a) $x^{3/2} + \dfrac{3}{2}x^{1/2}(x - 3)$

(b) $6x^2(x^2 - 4)^2 - (x^2 - 4)^3$

(c) $5x(x - 6)^{3/2} + 2(x - 6)^{5/2}$

7. Find a common denominator and simplify:

(a) $\dfrac{4}{x - 2} + \dfrac{3}{x + 6}$

(b) $\dfrac{-4x}{(2x + 3)^3} + \dfrac{1}{(2x + 3)^2}$

(c) $\dfrac{x^2}{2\sqrt{x - 5}} + 2x\sqrt{x - 5}$

8. Determine the missing factor in each of the following:

(a) $6 + \sqrt{4x + 16} = 2(\ldots)$

(b) $\dfrac{3}{x^2} - \dfrac{1}{x} + 5 = x^{-2}(\ldots)$

(c) $\sqrt{4x + 1} = \sqrt{x}(\ldots)$

(d) $(x^2 - 3) = (x - \sqrt{3})(\ldots)$

(e) $x + 8 = (\sqrt[3]{x} + 2)(\ldots)$

(f) $\dfrac{2}{3}x(2x - 1)^{-2/3} + (2x - 1)^{1/3} =$

$\dfrac{1}{3}(2x - 1)^{-2/3}(\ldots)$

9. Solve:

(a) $(3x - 2)(x + 3) \le 0$

(b) $\dfrac{2x - 3}{x + 4} \le -1$

(c) $\dfrac{3}{x + 2} - \dfrac{2}{x - 1} \ge 0$

10. On what interval(s) are the following functions positive?

(a) $f(x) = x^2 - 7x + 12$

(b) $f(x) = (x - 4)(x - 1)(x + 5)$

(c) $f(x) = \dfrac{x^2 - 4}{x^2}$

11. Solve for x:

(a) $|3x - 4| = 5$

(b) $|2x - 1| = x + 3$

12. Find the equation of the line:
(a) with slope $-2/3$ and y-intercept 4.
(b) with slope $-2/3$ and x-intercept 4.
(c) that passes through the point $(4, -5)$ with slope 3/4.
(d) that passes through the points $(-6, 2)$ and $(-3, 4)$

13. Express each of the following in the form $a(x - p)^2 + q$ by completing the square.

(a) $x^2 + 8x + 20$

(b) $x^2 - 5x + 8$

(c) $-2x^2 + 12x - 9$

14. Evaluate the following at 2 and at $2 + h$:

(a) $f(x) = -3x + 5$

(b) $f(x) = 2x^2 - 4x + 7$

(c) $f(x) = \dfrac{6}{x - 5}$

15. In each of the following cases, determine if the sequence is arithmetic or geometric and find the eighth term and the nth term:

(a) $-3, -12, -21, -30, \ldots$

(b) $3, 12, 48, 192, \ldots$

(c) $18, 6, 2, 2/3, \ldots$

16. In each of the following cases, determine if the series is arithmetic or geometric and find the sum:

(a) $1 - 2 + 4 - 8 + \ldots,$ to 12 terms

(b) $1 + 3 + 5 + 7 + \ldots,$ to 25 terms

(c) $37 + 34 + 31 + 28 + \ldots,$ to 15 terms

17. Consider the geometric series

$$1 + \frac{1}{2} + \frac{1}{4} + \frac{1}{8} + \cdots$$

(a) Find the sum of the first 4 terms.

(b) Find the sum of the first 8 terms.

(c) Show that the sum of the first n terms is

$$S_n = \frac{2^n - 1}{2^{n-1}}$$

(d) What is the sum of an infinite number of terms?

B.   Graph Sketching

The overall behaviour of a function can be seen at a glance in a graph of the function. This section discusses transformations, symmetries, asymptotes, and other tools that enable you to sketch a graph quickly and easily. In the study of calculus, this skill is essential. You may skip this section if a review of these topics is not needed.

In sketching a graph, it is always possible to create a table of values and plot individual points. However, this is a tedious process and is generally done as a last resort or when great precision is required. In that case, it might be more productive to use a computer, rather than do it by hand. (See C. Computer Applications and D. Graphing by Computer.) Usually, a rough sketch made using other methods is sufficient.

Intercepts

The y-intercept of a function f is the value of f at the point where the graph of the function crosses the y-axis. It is found by setting $x = 0$. The x-intercepts, known as the **zeros** of a function, are the solutions or **roots** of the equation $f(x) = 0$. Zeros of polynomial functions can be found by factoring.

EXAMPLE A

Sketch the following functions defined by

(a) $f(x) = \dfrac{2}{3}x + 4$

(b) $f(x) = x^2 - 10x + 21$

(c) $f(x) = x^3 + 2x^2 - 5x - 6$

(d) $f(x) = x(x + 2)(x - 1)^2$

SOLUTION

(a) The graph of this linear function is a straight line with y-intercept 4. Start at 4 on the y-axis and follow the slope of the line, moving right 3 and up 2, as in Figure B.1.

(b) The quadratic expression factors

$$f(x) = (x - 3)(x - 7)$$

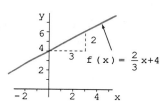

Therefore, $f(x) = 0$ when $x = 3$ or $x = 7$. These are the x-intercepts. The graph is a parabola that crosses the x-axis at 3 and 7, opening up (Figure B.2).

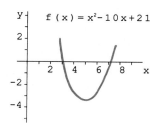

(c) Using the factor theorem (See A. Preparation for Calculus.)

$$f(x) = (x - 2)(x + 1)(x + 3)$$

Therefore, $f(x) = 0$ when $x = 2$, $x = -1$, or $x = -3$. When $x > 2$, all three factors are positive, so the graph must look like that in Figure B.3.

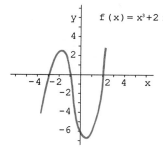

(d) By inspection, the x-intercepts of the graph of

$$f(x) = x(x + 2)(x - 1)^2$$

are 0, -2, and 1. The factor $x - 1$ occurs twice, so 1 is a double root. At such a point, the graph touches the axis but does not cross it, as shown in Figure B.4.

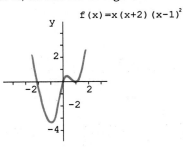

$f(x) = x(x+2)(x-1)^2$

Translations

In many cases, the graph of a function can be obtained by translating the graph of a more elementary function. A translation of a function is made by changing each point (x, y) into a new point $(x + h, y + k)$. The effect is to slide the graph of the function to a new position in the xy-plane.

EXAMPLE B

Sketch the graph of

(a) $y = (x - 2)^2 - 5$ (b) $y = \sin(x - \pi/3)$

SOLUTION

(a) The graph of $y = (x - 2)^2 - 5$ can be viewed as a translation of the graph of the parabola $y = x^2$, the vertex of which has been moved from the origin to the point $(2, -5)$, as in Figure B.5.

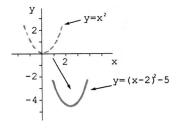

$y = x^2$

$y = (x-2)^2 - 5$

(b) The graph of $y = \sin(x - \pi/3)$ is obtained by shifting the graph of $y = \sin x$ to the right $\pi/3$ units, as in Figure B.6.

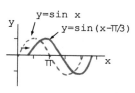

Graphs of Inverse Functions

From the graph of the function $y = f(x)$, the graph of the inverse function $y = f^{-1}(x)$ is obtained by a reflection in the line $y = x$.

EXAMPLE C

Sketch the graph of

$$y = \sqrt{x - 3}$$

SOLUTION

The function $y = \sqrt{x}$ is the inverse of the function $y = x^2$, $x \geq 0$. Its graph is half of a parabola with vertex at the origin, opening right as in Figure B.7. The graph of $y = \sqrt{x - 3}$ is obtained by translating the graph of $y = \sqrt{x}$ three units to the right.

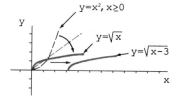

Asymptotes

A **vertical asymptote** occurs at $x = a$ if the value of a function exceeds all bounds as x approaches a. If, for instance, the denominator of a rational function is zero when $x = a$, then $x = a$ is a vertical asymptote. A **horizontal asymptote** occurs when the value of a function approaches a constant as the value of x grows very large. The subject of asymptotes will be discussed more fully in Section 5.1 in connection with limits.

EXAMPLE D

Sketch the graphs of

(a) $y = \dfrac{1}{x + 3}$ (b) $y = \dfrac{x^2}{(x - 1)(x - 2)}$

SOLUTION

(a) As x approaches -3, the value of $y = 1/(x + 3)$ exceeds all bounds. At $x = -3$, the function is undefined. The line $x = -3$ is a vertical asymptote. When the value of x becomes very large in magnitude, the fraction $1/(x + 3)$ approaches zero. Thus, the line $y = 0$ is a horizontal asymptote. The resulting graph is shown in Figure B.8.

Figure B.8

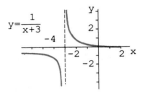

(b) You can see that the graph of $y = x^2/(x - 1)(x - 2)$ goes through the origin, and that there are vertical asymptotes at $x = 1$ and $x = 2$. When x becomes very large in magnitude, the value of the function approaches 1. Therefore, the line $y = 1$ is a horizontal asymptote. The value of the function is negative only when x is between 1 and 2. The graph of the function must be like that in Figure B.9.

Figure B.9

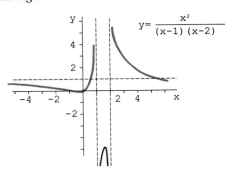

Even/Odd Symmetry

A function is called an **even** function, if $f(-x) = f(x)$. An even function is symmetric about the y-axis. A function is **odd**, if $f(-x) = -f(x)$. An odd function is symmetric about the origin. If you discover that a function is not even, you cannot conclude that it is odd. Many functions are neither even nor odd.

EXAMPLE E

Sketch the graphs of

(a) $y = \dfrac{1}{x^2}$ (b) $y = \dfrac{8x}{x^2 + 4}$

SOLUTION

(a) The x- and y-axes are asymptotes of $y = 1/x^2$. Since the function is an even function, the left half is a mirror image of the right half (see Figure B.10).

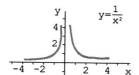

Figure B.10

(b) Since the function $y = 8x/(x^2 + 4)$ is odd, the graph of the function is symmetric about the origin. For large values of x the function approaches zero, so the x-axis is a horizontal asymptote. Its graph goes through the origin, and then somewhere bends back toward the x-axis again as in Figure B.11. You can discover exactly where such bends occur using methods of calculus developed in Chapter 5.

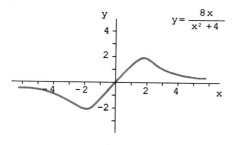

Figure B.11

Exercise B

1. Sketch the graphs of the following linear functions:

   (a) $f(x) = x - 2$

   (b) $f(x) = 2x$

   (c) $f(x) = -x + 2$

   (d) $f(x) = \dfrac{x + 2}{2}$

2. Sketch the graphs of the following quadratic functions:

   (a) $f(x) = (x - 2)(x - 4)$

   (b) $f(x) = x(x - 6)$

   (c) $f(x) = x^2 - 6$

3. State the zeros and sketch the graph of each of the following functions. For what values of x is $f(x)$ positive? Negative?

   (a) $f(x) = x^2 - 7x + 12$

   (b) $f(x) = x^3 - 1$

   (c) $f(x) = (x - 2)(x + 3)^2(x - 4)$

4. Sketch the graphs of the following functions. For what values of x are these functions undefined?

   (a) $f(x) = \sqrt{2x - 5}$

   (b) $f(x) = \sqrt{1 - x^2}$

5. (a) $P(x)$ is a polynomial, the equation $P(x) = 0$ has roots at -2, 1, and 3, and the graph of $y = P(x)$ has a y-intercept of 4. Find an expression for $P(x)$.

   (b) Find an expression for $P(x)$ if -2 in part (a) is a double root.

6. Are the following even or odd functions? Sketch the graph of each function, taking its symmetry into account.

   (a) $y = -2x^2 + 1$

   (b) $y = x^3 + 3x$

   (c) $y = \dfrac{-2x}{x^2 + 1}$

   (d) $y = \dfrac{3}{x^2 + 3}$

7. How are the following functions related? Sketch their graphs.

   (a) $y = \dfrac{1}{x}$ and $y = \dfrac{1}{x - 2}$

   (b) $y = \dfrac{1}{x}$ and $y = \dfrac{1}{x} - 2$

8. Find the asymptotes and sketch the graphs of:

   (a) $y = \dfrac{1}{(2x - 1)^2}$

   (b) $y = \dfrac{4}{(x - 2)(x + 1)}$

   (c) $y = \dfrac{x - 1}{(x + 2)(x - 3)}$

   (d) $y = \dfrac{2x^2 - 1}{x^2}$

9. Sketch the following functions:

   (a) $y = x + \dfrac{1}{x}$

   (b) $y = \dfrac{4}{2x + 3}$

   (c) $y = \dfrac{2x^2}{x^2 + 3}$

   (d) $y = x^2 - \dfrac{1}{x}$

   (e) $y = \dfrac{x}{(x - 2)^2}$

   (f) $y = \dfrac{x}{\sqrt{x^2 + 9}}$

   (g) $y = \dfrac{x^2}{x^2 - 1}$

C. Computer Applications

Throughout this text you will find examples of ways that a computer can be of help to you in your study of calculus. These consist for the most part of short programs written in BASIC. They require little prior experience to use and should run on any machine.

Each program has been written in a straightforward manner to make its connection with the mathematics as clear as possible. This means that there are no safeguards against nonsensical input. It is up to you to enter reasonable values and judge the validity of the output.

Two small programs for finding values of functions are given here. You will find them useful in preparing tables of values and graphs. The first, shown in Figure C.1, prints out the value y of a function for any input value of x. The second, in Figure C.2, prints out a range of values of a function in equal steps from $x = a$ to $x = b$.

```
10    REM VALUES OF A FUNCTION
100   :
110   DEF FN F(X)=X^2+3*X+2
200   :
210   REM LOOP
220   INPUT "X = ";X
230   Y=FNF(X)
240   PRINT TAB(20);Y
250   GOTO 210
300   :
310   END
```

Figure C.1

```
10    REM TABLE OF VALUES OF A FUNCTION
100   :
110   DEF FN F(X)=(X+5)*(X-2)
120   :
130   INPUT "FROM X = ";A
140   INPUT "   TO X = ";B
150   INPUT "IN STEPS OF ";DX
200   :
210   FOR X=A TO B STEP DX
220      Y=FNF(X)
230      PRINT X,Y
240   NEXT X
300   :
310   END
```

Figure C.2

Here, as in subsequent programs, it is necessary to key the function of interest in the DEF FN (define function) statement on line 110. Use ^ for exponents, * for multiplication, / for division, and use parentheses to group terms as in algebra. For example, x^2 becomes X^2 and $3x$ becomes 3*X. When the program has been keyed in, key RUN to execute it.

Do not be afraid to experiment and explore. Nothing you key at the keyboard can damage the computer.

Exercise C

1. Using the program in Figure C.1, find the values of the following functions at the points indicated. (Each function is followed by the equivalent expression in BASIC.)

(a) $y = x^2$, $x = -5, 0, 3$

   Y = X^2

(b) $y = (x - 2)(x + 3)$, $x = -2, 1, 4$

   Y = (X − 2)*(X + 3)

(c) $y = \dfrac{2x}{x^2 + 1}$, $x = -3, 0.5, 10$

   Y = 2*X/(X^2 + 1)

2. Certain mathematical functions are "built in" as a part of BASIC. Using the program in Figure C.1, evaluate the following functions at the points indicated:

(a) $y = \sqrt{x + 1}$, $x = 0, 3, 9, 48$

   Y = SQR(X + 1)

(b) $y = \sin x$, $x = 0, \pi/6, \pi/4,$
                                 $\pi/3$ radians
   Y = SIN(X)   Note: input x as a
                                 decimal

(c) $y = |x - 2|$, $x = -2, 0, 2, 4$

   Y = ABS(X − 2)

3. Using a computer to perform a calculation

has its limitations. Explain why the calculation fails when you attempt to evaluate each of the following functions at the given value of x. Note the error message in each case.

(a) $y = \dfrac{5}{x - 3}$, $x = 3$

   Y = 5/(X − 3)

(b) $y = \sqrt{x - 4}$ $x = 2$

   Y = SQR(X − 4)

(c) $y = 10^x$, $x = 100$

   Y = 10^X

4. Using the program in Figure C.2, print a table of several values of the following functions over the given range.

(a) $y = x^3 - 2x^2 - 5x + 6$, $x = -3$ to 4

   Y = X^3 − 2*X^2 − 5*X + 6

(b) $y = \dfrac{x}{2^x}$, $x = -4$ to 4

   Y = X/2^X

(c) $y = x - \tan x$, $x = 0$ to $\pi/2$

   Y = X − TAN(X)

5. Experiment with functions of your own choosing.

Gottfried Wilhelm Leibniz

THE FIRST PUBLICATION of the differential calculus took place in 1684 in the *Acta Eruditorum*, a European scientific journal that had been founded in 1682. The author of the article, Gottfried Wilhelm Leibniz, was a thirty-eight-year-old German philosopher, mathematician, and diplomat. Leibniz had duplicated the achievement of his English contemporary, Isaac Newton, who several years earlier had communicated privately to friends his results in the calculus. Although controversy over priority developed later, it is clear that both men worked independently and pursued different courses in their development of the new analysis.

Throughout the seventeenth century, mathematicians had accumulated, in their study of the geometrical curve, methods to solve two kinds of problems—the determination of tangents and the calculation of areas and pathlengths. In the 1670s Leibniz studied the algebraic techniques René Descartes had introduced to analyze curves and invented a calculus that provided a general algorithm for the solution of tangent and area problems. In 1693 he published the fundamental theorem of the calculus, demonstrating the inverse character of differentiation and integration. The fundamental theorem resolved the basic problem of the calculus, to describe mathematically change along a curve, and to connect change at a point (slope) to change over the entire curve (area, path-length).

In arriving at the idea for the calculus Leibniz was guided by his study of numerical sequences.

Beginning with a given sequence, one forms the sum and difference sequences, whose general terms are given respectively as the sum of the first n members and the difference of the nth and $(n - 1)$st members. The particular examples that Leibniz investigated were generated from the arithmetic and harmonic sequences, 1, 2, 3, . . ., and 1/1, 1/2, 1/3, . . . He examined arrays of sequences and used their sum and difference properties to determine relations among them. He connected this interest in combinatorial analysis to his investigation of the geometry of curves to arrive at his fundamental insight concerning the inverse character of differentiation and integration.

Although Leibniz shared credit for the invention of the calculus with Newton, his own work was historically much more influential. Newton was slow to publish and preferred to work alone. Leibniz, by contrast, established a vigorous school of mathematicians to promote his methods, among whom were l'Hôpital in Paris and the Bernoulli brothers in Basel. The notation and algorithms of the Leibnizian calculus were more effective and easier to use than Newton's awkward fluxional calculus. Throughout his life Leibniz was concerned with the philosophical problem of language, and he consciously emphasized the importance of notation in his new calculus. The decline of English mathematics following Newton's death may be explained in part by the failure of English researchers to appreciate the power and superiority of Leibniz's methods.

Limits

2

A function in mathematics describes how two quantities are related. Given such a relationship, calculus is concerned with how a *change* in one quantity is related to or dependent upon a *change* in the other.

In this chapter, we shall investigate this question from a particular standpoint: What happens to the value of a function *f*, as *x* gets closer and closer to a particular value *a*? Does *f(x)* tend to "home in" on some specific value, that is, does it have a **limit**?

The concept of a limit developed here underlies everything that follows in subsequent chapters. In particular, you will see (a) that the slope of the tangent to a curve at a point can be obtained as the limit of a sequence of values; and (b) that quantities such as the area under a curve can be expressed as the limit of a sum of terms. Solutions to these two basic problems have far-reaching repercussions throughout mathematics and science.

The Limit of a Function   2.1

To illustrate the concept of a limit, let us start with a particular example. Take the function $f(x) = x^2 + 1$ shown in Figure 2.1, and consider what happens to the value of *f* as *x* takes on a sequence of values that get closer and closer to 2. As *x* approaches 2, you can see by direct calculation that the corresponding values of *f* appear to approach 5 (Table 2.1). To state this result verbally, you would say that "as *x* approaches 2, the **limit** of $(x^2 + 1)$ is 5." The proper mathematical notation for writing this limit is

$$\lim_{x \to 2} (x^2 + 1) = 5$$

Observe that in this example, it happens to be true that the limit of the function as *x approaches* 2 is the same as the value of the function when *x equals* 2. However, this is not always the case. It is important to understand that the value of a limit depends only on

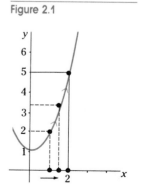

Figure 2.1

Table 2.1

x	$f(x)$
1.5	3.25
1.9	4.61
1.99	4.9601
1.999	4.996001
1.9999	4.99960001

Table 2.2

x	$f(x)$
2.5	7.25
2.1	5.41
2.01	5.0401
2.001	5.004001
2.0001	5.00040001

how a function behaves *near* the point in question, and not on its value *at* that point.

In the above example, the values of x approached 2 through a sequence of numbers less than 2, that is, from the left on the graph. This **left limit** is denoted by placing a small *negative sign* above and to the right of 2:

$$\lim_{x \to 2^-} (x^2 + 1) = 5$$

The limit is the same if x approaches 2 from the right, as Table 2.2 shows. **Right limits** are denoted by a small *positive sign* placed above and to the right of 2:

$$\lim_{x \to 2^+} (x^2 + 1) = 5$$

A limit exists only if the left and right limits both exist and are equal, in which case the small signs are dropped, and the limit is written simply

$$\lim_{x \to 2} (x^2 + 1) = 5$$

The following examples will illustrate some of the properties of limits. Many limits can be found by **inspection**, that is, by direct replacement of x. Cases in which the evaluation of a limit is not so straightforward will be discussed in Section 2.2.

EXAMPLE A

Find $\lim_{x \to 2} f(x)$ where $f(x) = 3x^2 - x + 4$.

SOLUTION

$$\lim_{x \to 2} f(x) = \lim_{x \to 2} (3x^2 - x + 4)$$

$$= 3(2)^2 - 2 + 4$$

$$= 14$$

Observe that as x approaches 2,
x^2 approaches 4,
$3x^2$ approaches 12,
$-x$ approaches -2,
and since 4 is a constant, independent of x,
4 remains 4

$$\therefore \quad \lim_{x \to 2} 3x^2 + \lim_{x \to 2} (-x) + \lim_{x \to 2} 4 = 12 + (-2) + 4$$

$$= 14$$

which equals the limit of $f(x)$ as $x \to 2$ found above.
Therefore, the limit of a sum of terms can be written as the sum of their separate limits. ❏

EXAMPLE B

Find $\lim_{x \to 0} (3x - 2)(5 - 4x^2)$.

SOLUTION

$$\lim_{x \to 0} (3x - 2)(5 - 4x^2)$$

$$= [3(0) - 2][5 - 4(0)^2]$$

$$= (-2) \cdot (5)$$

$$= -10$$

In this case,

as $x \to 0, (3x - 2)$ approaches -2
and $(5 - 4x^2)$ approaches 5

therefore, the limit of a product is equal to the product of limits of the separate factors. ❏

EXAMPLE C

Find the limit of the function $y = \sqrt{x - 3}$ as $x \to 3$.

SOLUTION

Unlike the previous examples, this function is defined only for values of $x \geq 3$, so the left limit

$$\lim_{x \to 3^-} \sqrt{x - 3} \quad \text{does not exist.}$$

By inspection, the right limit is

$$\lim_{x \to 3^+} \sqrt{x - 3} = 0$$

Nevertheless, with no left limit, it must be concluded that

$$\lim_{x \to 3} \sqrt{x - 3} \text{ does not exist.}$$

The basic properties of limits illustrated by these examples are given here without proof:

1. The limit of a constant times a function is the constant times the limit of the function:

$$\lim_{x \to a} Cf(x) = C \lim_{x \to a} f(x)$$

2. The limit of a sum is the sum of the limits:

$$\lim_{x \to a} [f(x) + g(x)] = \lim_{x \to a} f(x) + \lim_{x \to a} g(x)$$

3. The limit of a product is the product of the limits:

$$\lim_{x \to a} \left(f(x) \cdot g(x) \right) = \left(\lim_{x \to a} f(x) \right) \left(\lim_{x \to a} g(x) \right)$$

4. The limit of a quotient is the quotient of the limits:

$$\lim_{x \to a} \frac{f(x)}{g(x)} = \frac{\lim\limits_{x \to a} f(x)}{\lim\limits_{x \to a} g(x)} \text{ (provided } \lim_{x \to a} g(x) \neq 0)$$

It was emphasized in the foregoing discussion regarding the function $f(x) = x^2 + 1$ that the limit has a lot to do with what happens

Figure 2.2

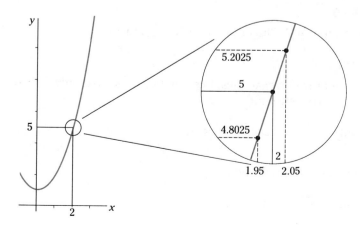

as x gets close to 2. This idea of "closeness" is illustrated in Figure 2.2. If the value of x lies for instance within ± 0.05 of 2, that is, in the interval $1.95 \leq x \leq 2.05$, then the value of this function lies between $f(1.95)$ and $f(2.05)$, that is, between 4.8025 and 5.2025. If you require the value of the function to lie in a narrower interval around 5, you can achieve this by restricting x to a smaller interval around 2.

The essence of the limit process rests in the observation that the interval around 5 to which $f(x)$ is confined can be made as small as you please, by restricting x to a sufficiently small interval about 2. In general, if the value of a function cannot be confined to a small interval, when x is restricted to a small interval about a, then the limit of the function as $x \to a$ does not exist.

EXAMPLE D

(a) State the range of values of the function $f(x) = 5x - 3$ that correspond to values of x in the intervals

   i. $3.95 \leq x \leq 4.05$

   ii. $3.995 \leq x \leq 4.005$

(b) Find $\lim_{x \to 4} f(x)$.

SOLUTION

(a) $f(3.95) = 5(3.95) - 3 = 16.75$

   $f(4.05) = 5(4.05) - 3 = 17.25$

   $f(3.995) = 5(3.995) - 3 = 16.975$

   $f(4.005) = 5(4.005) - 3 = 17.025$

The function f is increasing everywhere in these intervals so that

   if $3.95 \leq x \leq 4.05$ then $16.75 \leq f(x) \leq 17.25$

  and if $3.995 \leq x \leq 4.005$ then $16.975 \leq f(x) \leq 17.025$.

(b) $\lim_{x \to 4} f(x) = \lim_{x \to 4} (5x - 3)$

           $= 5(4) - 3$

           $= 17$

EXAMPLE E

Consider the function $f(x) = 2x^2 - 4$.

(a) What range of values of $f(x)$ corresponds to the values of x in the interval $-2.01 \leq x \leq -1.99$?

(b) For what *positive* values of x is $|f(x) - 14| \leq 0.02$?

SOLUTION

(a) If $x = -2.01$, then $f(-2.01) = 2(-2.01)^2 - 4 = 4.0802$.

   If $x = -1.99$, then $f(-1.99) = 2(-1.99)^2 - 4 = 3.9202$.

   The function is decreasing everywhere in this interval. Therefore, if x is within ± 0.01 of -2, $f(x)$ will lie in the interval $3.9202 \leq f(x) \leq 4.0802$.

(b) There are two cases to consider, depending on whether the quantity in absolute values is positive or negative. When it is positive,

$$f(x) - 14 \geq 0 \qquad \text{and} \qquad f(x) - 14 \leq 0.02$$

$$
\begin{array}{rcl}
0 \leq & f(x) - 14 & \leq 0.02 \\
14 \leq & f(x) & \leq 14.02 \\
14 \leq & 2x^2 - 4 & \leq 14.02 \\
18 \leq & 2x^2 & \leq 18.02 \\
9 \leq & x^2 & \leq 9.01 \\
3 \leq & x & \leq 3.0017
\end{array}
$$

On the other hand, when it is negative,

$$f(x) - 14 < 0 \qquad \text{and} \qquad -[f(x) - 14] \leq 0.02$$

$$
\begin{array}{rcl}
-0.02 \leq & f(x) - 14 & < 0 \\
13.98 \leq & f(x) & < 14 \\
13.98 \leq & 2x^2 - 4 & < 14 \\
17.98 \leq & 2x^2 & < 18 \\
8.99 \leq & x^2 & < 9 \\
2.9983 \leq & x & < 3
\end{array}
$$

Together, these results mean that the value of the function will lie within ± 0.02 of 14, if x is restricted to the interval $2.9983 \le x \le 3.0017$. ◻

Exercise 2.1

1. (a) Calculate the values of the function $f(x) = x^2 - 10x + 12$ at the following values of x:

x	$f(x)$
5.1	
5.01	
5.001	
5.0001	

(b) As x approaches 5 from the right, what does the limit of the function appear to be?

(c) In a similar fashion, calculate the limit of $f(x)$ as x approaches 5 from the left.

2. Does the limit of the function shown in Figure 2.3 exist as $x \to 2$? Explain.

Figure 2.3

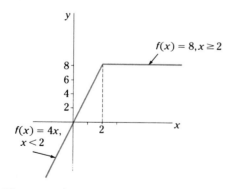

$f(x) = 8, x \ge 2$

$f(x) = 4x,$
$x < 2$

3. The function f is defined by the graph in Figure 2.4. State the value of the following limits, if they exist:

(a) $\lim\limits_{x \to -3} f(x)$

(b) $\lim\limits_{x \to 0} f(x)$

(c) $\lim\limits_{x \to 2} f(x)$

(d) $\lim\limits_{x \to 5} f(x)$

(e) $\lim\limits_{x \to 7} f(x)$

(f) $\lim\limits_{x \to 9} f(x)$

Figure 2.4

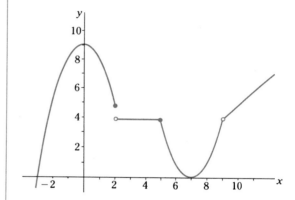

4. (a) Sketch the graph of f, where

$$f(x) = \begin{cases} 2x + 5, & x < 1 \\ (x - 1)^2 + 3, & x \ge 1 \end{cases}$$

(b) Find $\lim\limits_{x \to 1^+} f(x)$ and $\lim\limits_{x \to 1^-} f(x)$.

(c) Does the limit of $f(x)$ as $x \to 1$ exist? Explain.

5. Given that

$$g(x) = \begin{cases} 3x^2 - 4, & x \le 2 \\ 12x - 16, & x > 2 \end{cases}$$

find the left and right limits of g as $x \to 2$. Does the limit of $g(x)$ exist as $x \to 2$? Explain, with the help of a graph.

6. Given that $f(x) = 3 - 4x + 5x^2$, evaluate the following limits:

(a) $\lim\limits_{x \to 0} f(x)$

(b) $\lim\limits_{x \to -1} f(x)$

(c) $\lim\limits_{x \to 10} f(x)$

(d) $\lim\limits_{x \to k} f(x), k \in R$

7. Find the limit of each of the following functions as $x \to 2$:

(a) $f(x) = 16 - 2x^3$

(b) $f(x) = \dfrac{x^2 + 5}{x^2 - 3}$

(c) $f(x) = \dfrac{4}{x} + \dfrac{2}{x - 4}$

(d) $f(x) = \sqrt{5x^2 - 2x}$

8. Evaluate the following limits:

(a) $\lim\limits_{x \to 0} (3x^2 - 1)(5x + 1)$

(b) $\lim\limits_{x \to 1.4} (4x - 1)(8x^2 - 2x + 3)$

(c) $\lim\limits_{x \to -4} \dfrac{9}{x + 3}$

(d) $\lim\limits_{x \to -2} \dfrac{x + 2}{x - 10}$

(e) $\lim\limits_{x \to 0} \dfrac{(x - 3)(4 - x)}{x - 2}$

(f) $\lim\limits_{x \to -1} \dfrac{(x - 3)(4 - x)}{x - 2}$

9. Determine the following limits:

(a) $\lim\limits_{x \to \sqrt{2}} (8x^3 - 1)$

(b) $\lim\limits_{x \to 0} \dfrac{1}{\sqrt{7x + 4}}$

(c) $\lim\limits_{x \to 4} \sqrt{\dfrac{x}{5x^2 - 7x + 12}}$

(d) $\lim\limits_{x \to 2} (\sqrt{x^2 + 5} - x)$

10. If x lies in the given interval, in what interval is the value of each of the following functions found?

(a) $f(x) = 3x^3 + 6x, \quad 2.9 \le x \le 3.1$

(b) $f(x) = \sqrt{3x - 2}, \quad x = 9 \pm 0.02$

(c) $f(x) = \dfrac{x}{x^2 + 4}, \quad |x + 2| \le 0.01$

11. Given $f(x) = \dfrac{3x}{x - 2}$

(a) for what value of x is $f(x) = 4$?

(b) for what values of x is $|f(x) - 4| \le 0.05$?

(c) for what values of x is $|f(x) - 4| \le 0.005$?

2.2 Evaluation of Limits: Limits Involving 0/0

In the preceding section, the evaluation of the limit $\lim\limits_{x \to a} f(x)$ depended largely on being able to find the value of the function $f(x)$ at $x = a$. However, this is not always possible to do. Consider for instance the function

$$f(x) = \frac{4 - x^2}{2 - x}$$

At $x = 2$, direct substitution gives $f(2) = 0/0$. How is the quotient $0/0$ to be interpreted? How can you find the limit of f as $x \to 2$ in this situation?

The quotient a/b of two real numbers is defined in terms of the product:

$$\frac{a}{b} = c \qquad \text{if and only if} \qquad a = b \cdot c$$

Consider the possibilities when a or b or both are zero:

1. $a = 0$ but $b \neq 0$:

   If $\dfrac{0}{b} = c$, then $0 = b \cdot c$

   which is true only if $c = 0$.

   Therefore, the quotient $0/b$ is zero, as expected.

2. $a \neq 0$ but $b = 0$:

   If $\dfrac{a}{0} = c$, then $a = 0 \cdot c$

   but there is no real number c for which $0 \cdot c \neq 0$.

   A quotient of the form $a/0$ is said to be **undefined**. Limits involving such quotients will be dealt with in Section 5.1 in connection with finding asymptotes to graphs of functions.

3. Both $a = 0$ and $b = 0$:

   If $\dfrac{0}{0} = c$, then $0 = 0 \cdot c$

   which is true for any value of c.

   Therefore, a quotient of the form $0/0$ is said to be **indeterminate**; it may have any value, depending on the limit process and the function in question.

   To evaluate the limit of a function when $0/0$ occurs, the principal method is to discover what factors give rise to $0/0$ and divide them out before taking the limit. This is illustrated by the following examples.

EXAMPLE A

Evaluate $\displaystyle\lim_{x \to 2} \frac{4 - x^2}{2 - x}$

SOLUTION

Substituting $x = 2$ yields the indeterminate form $0/0$. Proceed, therefore, by first factoring:

$$\lim_{x \to 2} \frac{4 - x^2}{2 - x} = \lim_{x \to 2} \frac{(2 - x)(2 + x)}{(2 - x)}$$

$$= \lim_{x \to 2}(2 + x)$$

$$= 2 + 2$$

$$= 4$$

In evaluating the above limit, the only concern is how the function behaves as x approaches 2. In this process, x is never exactly equal to 2; therefore, the factor $(2 - x)$ is never exactly zero, and it may be divided out. □

EXAMPLE B

Evaluate $\lim\limits_{x \to 0} \dfrac{1 - \sqrt{1 + 3x}}{x}$

SOLUTION

Direct replacement of x by 0 yields 0/0. In this case, proceed by first rationalizing the numerator:

$$\lim_{x \to 0} \frac{1 - \sqrt{1 + 3x}}{x} = \lim_{x \to 0} \frac{(1 - \sqrt{1 + 3x})}{x} \cdot \frac{(1 - \sqrt{1 + 3x})}{(1 + \sqrt{1 + 3x})}$$

$$= \lim_{x \to 0} \frac{1 - (1 + 3x)}{x(1 + \sqrt{1 + 3x})}$$

$$= \lim_{x \to 0} \frac{-3}{(1 + \sqrt{1 + 3x})}$$

$$= \frac{-3}{1 + \sqrt{1}}$$

$$= -\frac{3}{2}$$

□

EXAMPLE C

Evaluate $\lim\limits_{x \to \sqrt{2}} \dfrac{x^2 - 2}{3 - \sqrt{7 + x^2}}$

SOLUTION

As $x \to \sqrt{2}$, the numerator and denominator both approach 0. Proceed by rationalizing the denominator:

$$\lim_{x \to \sqrt{2}} \frac{x^2 - 2}{3 - \sqrt{7 + x^2}} = \lim_{x \to \sqrt{2}} \frac{(x^2 - 2)}{(3 - \sqrt{7 + x^2})} \cdot \frac{(3 + \sqrt{7 + x^2})}{(3 + \sqrt{7 + x^2})}$$

$$= \lim_{x \to \sqrt{2}} \frac{(x^2 - 2)(3 + \sqrt{7 + x^2})}{9 - (7 + x^2)} \qquad \leftarrow \text{\textit{do not} expand the numerator}$$

$$= \lim_{x \to \sqrt{2}} \frac{(x^2 - 2)(3 + \sqrt{7 + x^2})}{(2 - x^2)}$$

$$= \lim_{x \to \sqrt{2}} (-1)(3 + \sqrt{7 + x^2})$$

$$= (-1)(3 + \sqrt{7 + 2})$$

$$= -6 \qquad \qquad \square$$

Exercise 2.2

1. Evaluate the limits:

(a) $\lim\limits_{x \to 0} \dfrac{x + x^2}{3x}$

(b) $\lim\limits_{x \to 4} \dfrac{x^2 - 16}{3x - 12}$

(c) $\lim\limits_{x \to 5} \dfrac{5x(5 - x)}{3x - 15}$

(d) $\lim\limits_{x \to 0} \dfrac{4x - \sqrt{x}}{3\sqrt{x}}$

2. Evaluate the limits:

(a) $\lim\limits_{x \to 0} \dfrac{x^2 + x - 6}{x - 2}$

(b) $\lim\limits_{x \to -2} \dfrac{x^2 + x - 6}{x - 2}$

(c) $\lim\limits_{x \to 2} \dfrac{x^2 + x - 6}{x - 2}$

(d) $\lim\limits_{x \to -3} \dfrac{x^2 + x - 6}{x + 3}$

(e) $\lim\limits_{x \to 3} \dfrac{x^2 + x - 6}{x + 3}$

3. Evaluate the limits:

(a) $\lim\limits_{x \to 0} \dfrac{x^3 + 8}{x + 2}$

(b) $\lim\limits_{x \to 1/2} \dfrac{8x^3 - 1}{1 - 5x + 6x^2}$

(c) $\lim\limits_{x \to -2} \dfrac{2x^4 - 32}{x + 2}$

(d) $\lim\limits_{x \to 1} \dfrac{x^2 - 2x + 1}{x^3 - x}$

4. Evaluate the limits:

(a) $\lim\limits_{x \to 2} \dfrac{\dfrac{1}{x} - \dfrac{1}{2}}{x - 2}$

(b) $\lim\limits_{x \to 3} \dfrac{\dfrac{1}{x - 4} + 1}{x - 3}$

(c) $\lim\limits_{x \to 0} \dfrac{\dfrac{1}{a} - \dfrac{1}{x + a}}{x}, a \in R$

5. Evaluate the limits, if they exist:

(a) $\lim\limits_{x \to 0} \dfrac{\sqrt{x}}{\sqrt{x} + x}$

(b) $\lim\limits_{x \to 0} \dfrac{x}{1 + \sqrt{1 + x^2}}$

(c) $\lim\limits_{x \to 1} \dfrac{\sqrt{x + 8} - 3}{x - 1}$

(d) $\lim\limits_{x \to 0} \dfrac{\sqrt{x + 3} - \sqrt{3}}{x}$

(e) $\lim\limits_{x \to 0} \dfrac{3 - \sqrt{3x}}{3 - x}$

(f) $\lim\limits_{x \to 0} \dfrac{\sqrt{x} - x}{\sqrt{x}}$

(g) $\lim\limits_{x \to \sqrt{5}} \dfrac{x^2 - 5}{x - \sqrt{5}}$

(h) $\lim\limits_{x \to 2} \dfrac{\dfrac{1}{3} - \dfrac{1}{\sqrt{x}}}{x - 9}$

6. If $f(x) = 3x^2 - 2x + 4$, find

(a) $\lim\limits_{h \to 0} \dfrac{f(h) - f(0)}{h}$

(b) $\lim\limits_{h \to 1} \dfrac{f(h) - f(1)}{h - 1}$

(c) $\lim\limits_{h \to 0} \dfrac{f(1 + h) - f(1)}{h}$

7. For each of the following functions, construct the expression

$$\frac{f(3 + h) - f(3)}{h}$$

Then find the limit of this expression as $h \to 0$:

(a) $f(x) = 6x$

(b) $f(x) = 2 - x$

(c) $f(x) = 7x^2$

(d) $f(x) = \dfrac{3}{\sqrt{x}}$

8. Given $f(x) = ax^2 + bx + c$, show that

$$\lim\limits_{\Delta x \to 0} \frac{f(x + \Delta x) - f(x)}{\Delta x} = 2ax + b$$

2.3 Continuity

In mathematics, continuity has to do with the "smoothness" of the graph of a function. A continuous function is commonly understood to be one whose graph has no gaps or abrupt jumps or breaks. It is one whose graph can be drawn without taking the pen off the paper.

As an example of a function, which is *not* continuous, consider how the postage required on a letter depends on its mass (Figure 2.5). If a letter has a mass equal to 30 g or less, the postage is 37¢. But if the mass of a letter exceeds 30 g by any amount, however small, the cost of postage increases abruptly to 57¢. There is *no* mass for which the required postage is 48¢, for instance. Discontinuities of this sort are typical in functions determined by legislation.

Contrast this with the graph in Figure 2.6, which describes the current through a lamp as a function of time. When the lamp is switched on, there is a brief interval of time during which the current rises rapidly but smoothly from 0 to its maximum value, 0.5 amps. The current takes on every value between 0 and 0.5 amps at some time during the change. Physical processes generally vary smoothly, that is continuously, in this fashion.

Figure 2.5

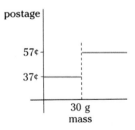

Figure 2.6

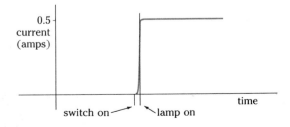

These intuitive notions of continuity are captured in the following definition. A function is continuous at a point $x = a$ if and only if

$$\lim_{x \to a} f(x) = f(a)$$

In essence, a function is continuous at a point if the limit of the function as you *approach* the point is equal to the value of the function *at* that point.

This definition implies that the following three specific conditions must be satisfied:

1. The function must have a value at $x = a$;

2. The limit of the function as $x \to a$ must exist;

3. The two quantities in (1) and (2) above must be equal.

If these conditions are satisfied at every point of a certain interval, then the function is said to be continuous over the interval.

The following examples show various ways in which a function can fail to be continuous.

EXAMPLE A

Describe the discontinuities in each of the following functions.

(a) $f(x) = \dfrac{x^2 - 16}{x - 4}$ (b) $f(x) = \sqrt{x^2 - 1}$ (c) $f(x) = \dfrac{1}{(x - 2)^2}$

(d) $f(x) = 4^{1/x}$ (e) $f(x) = \begin{cases} 1, x > 0 \\ 0, x \le 0 \end{cases}$ (f) $f(x) = \begin{cases} 1, x \in I \\ 0, x \notin I \end{cases}$

SOLUTION

(a) $f(x) = \dfrac{x^2 - 16}{x - 4}$

$\qquad = \dfrac{(x - 4)(x + 4)}{x - 4}$

$\qquad = x + 4$, if $x \ne 4$

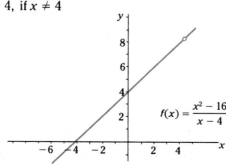

$f(x) = \dfrac{x^2 - 16}{x - 4}$

Figure 2.7

Figure 2.8

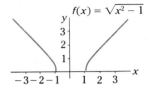

The graph of f is therefore the same as the graph of $x + 4$, except at $x = 4$, where it is not defined (see Figure 2.7). Hence, this function has a discontinuity at $x = 4$; condition (1) is not satisfied.

(b) $f(x) = \sqrt{x^2 - 1}$

This function is not defined when x is between -1 and $+1$, so it is not continuous in that interval (Figure 2.8). Condition (1) is not satisfied.

(c) $f(x) = \dfrac{1}{(x - 2)^2}$

Figure 2.9

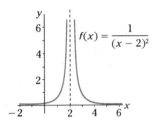

This function has a vertical asymptote at $x = 2$ (Figure 2.9). The function is not defined at $x = 2$, and the limit does not exist as $x \to 2$, so neither condition (1) nor (2) are satisfied.

(d) $f(x) = 4^{1/x}$

This function has a discontinuity at $x = 0$ because as $x \to 0$ the left and right limits are not equal (Figure 2.10). Neither condition (1) nor condition (2) are satisfied.

Figure 2.10

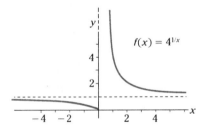

Figure 2.11

(e) $f(x) = \begin{cases} 1, x > 0 \\ 0, x \le 0 \end{cases}$

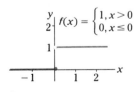

This function is like the "postage" function; it has an abrupt jump at $x = 0$ (Figure 2.11). Condition (2) is not satisfied since the left and right limits as $x \to 0$ are not equal. This function is an example of a **step function**.

(f) $f(x) = \begin{cases} 1, x \in I \\ 0, x \notin I \end{cases}$

This function is defined for all values of x; however it has a discontinuity whenever x is an integer (Figure 2.12). Condition (3) is not satisfied at any integer n, because

$$\lim_{x \to n} f(x) = 0,$$

but

$$f(n) = 1$$

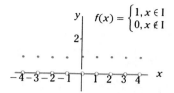

Figure 2.12

Exercise 2.3

1. Describe the function that determines the fine when the driver of an automobile exceeds the speed limit. Is this function continuous?

2. Describe the function that gives the speed as a function of time when an automobile hits a barrier during a crash test. Is this function continuous?

3. Discuss whether or not the cost of filling the gas tank of a car with gasoline is a continuous function of the amount of gasoline purchased.

4. Each of the following functions has a discontinuity at $x = 3$. In each case state which of the continuity conditions is not satisfied.

(a) $f(x) = \begin{cases} 0, & x \leq 3 \\ 0.001, & x > 3 \end{cases}$

(b) $f(x) = \dfrac{x^2 - x - 6}{x - 3}$

(c) $f(x) = \begin{cases} \dfrac{x^2 - x - 6}{x - 3}, & x \neq 3 \\ 3, & x = 3 \end{cases}$

(d) $f(x) = \dfrac{x^2 - x + 6}{x - 3}$

(e) $f(x) = \sqrt{3 - x}$

(f) $f(x) = \begin{cases} x - 1, & x \leq 2 \\ x - 3, & x \geq 4 \end{cases}$

(g) $f(x) = \begin{cases} 1 - x, & x \leq 3 \\ 3 - x, & x > 3 \end{cases}$

5. Sketch the graph of each of the following functions, and determine if there are any discontinuities.

(a) $f(x) = |x - 2|$

(b) $f(x) = \dfrac{3}{x - 2}$

(c) $f(x) = x - \dfrac{1}{x}$

(d) $f(x) = \dfrac{2 + (3/x)}{3 + (1/x)}$

(e) $f(x) = \dfrac{x^2 - 1}{x + 1}$

(f) $f(x) = \dfrac{x^2 - x^4}{x^2 - 1}$

(g) $f(x) = \dfrac{2x}{|x|}$

(h) $f(x) = \dfrac{x - 1}{x^2 + 2x - 3}$

6. Given $f(x) = \dfrac{x + 1}{x^2 + 1}$

(a) Find $f(a)$ and $\lim\limits_{x \to a} f(x)$.

(b) Show that f is continuous everywhere.

(c) Sketch the graph.

7. Show that the function $f(x) = \dfrac{x^2 - 2x - 15}{x + 3}$ has a discontinuity at $x = -3$. How must $f(-3)$ be defined to make the function continuous?

8. Discuss the continuity of the function

$$f(x) = \begin{cases} 1, & x \text{ rational} \\ 0, & x \text{ irrational.} \end{cases}$$

2.4 Computer Application: Roots by the Bisection Method

One of the most important problems in mathematics is that of solving equations. In other words, given a function $y = f(x)$, find the real numbers r (if any) such that $f(r) = 0$. Such a number r is called a **zero** of f or a **root** of the equation $f(x) = 0$.

The bisection method of finding roots rests on the observation that there must be a root of the equation $f(x) = 0$ in the interval $[a, b]$, if $f(a)$ and $f(b)$ have opposite signs as in Figure 2.13, that is, if

$$f(a) \cdot f(b) < 0$$

assuming, of course, that f is continuous on the interval. The algorithm consists of the following steps:

1. Select an interval $[a, b]$ for which $f(a) \cdot f(b) < 0$.
2. Find the midpoint m of the interval.
3. Evaluate $f(m)$.
4. If $f(m)$ is "close enough" to 0, then quit.
5. Otherwise, determine which half of the interval the root is in:
   (a) if $f(a) \cdot f(m) < 0$, then the root is in $[a, m]$.
   (b) if $f(m) \cdot f(b) < 0$, then the root is in $[m, b]$.
6. Repeat the process, with either $[a, m]$ or $[m, b]$.

This algorithm repeatedly cuts in half the interval in which the root is located, thus converging on the position of the root. This process is a limit process, which in principal can go on forever, giving unlimited accuracy. In practice, it must be terminated at some point when $f(m)$ is "close enough" to 0, that is, when $|f(m)|$ is less than some small value such as 0.000001. The size of this margin of error depends both on the capabilities of the computer used and on the requirements of the problem in question, and no general rule can be given.

A BASIC program using this algorithm is given in Figure 2.14. Before running this program, you should have some idea about where

Figure 2.13

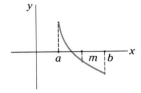

```
10   REM ROOTS BY THE BISECTION METHOD
100  :
110  DEF FNF(X)=TAN(X) − 2*X
120  INPUT "LOW END OF INTERVAL: X = ";A
130  INPUT "HIGH END OF INTERVAL: X = ";B
200  :
210  REM LOOP
220     M=(A+B)/2
230     IF ABS(FNF(M))<.000001 THEN GOTO 300 QUIT
240     IF FNF(A)*FNF(M)<0 THEN B=M
250     IF FNF(M)*FNF(B)<0 THEN A=M
260  GOTO 210 REPEAT
300  :
310  PRINT "THE ROOT IS AT X = ";M
320  END
```

Figure 2.14

the roots lie. There is no guarantee of finding a root if you input arbitrary numbers for the endpoints of the interval. There must be just one root in the interval. If there is more than one, the program may never terminate. This algorithm will not find a double root. Why not?

Exercise 2.4

1. Carry out the bisection method by hand to find the root of $y = x^2 − 7x + 6$ that lies beween 4 and 7.

2. Find the roots of the following functions by factoring:

(a) $y = 3x^2 + 13x − 10$

(b) $y = x^3 − 2x^2 − 6x + 12$

3. Find the roots of the functions given in Question 2 using the bisection algorithm given in Figure 2.14.

4. Find the roots of $y = 4x^2 − 12x + 1$ using the bisection algorithm. Check using the quadratic formula.

5. Find the roots of the following functions using the bisection algorithm:

(a) $y = x^3 + 6x^2 − 2x + 3$

(b) $y = x − \cos x$

(c) $y = 2^x − 5x$

6. Find the point of intersection of the functions $y = 2x$ and $y = \tan x$ that lies in the interval $0 < x < \pi/2$.

2.5 Summary and Review

The limit of a function has to do with how a function behaves as x takes on a sequence of values. A limit at $x = a$ exists, if the value of a function approaches some unique, finite value L, as $x \to a$ from either side.

From a practical standpoint, you can usually determine a limit simply by examining values of a function at a few specific values of x. In the event that the value of a function is indeterminate at a particular value of x, it may be helpful to change the form of the function by

1. dividing out common factors in the numerator and denominator

2. rationalizing the numerator or denominator of a fraction

As you work out problems involving limits, you should keep in mind the distinction between the value of a function at a point and the limit of the function as you approach that point. These two quantities will be equal only if the function is continuous:

$$\lim_{x \to a} f(x) = f(a)$$

Exercise 2.5

1. Evaluate the following limits:

(a) $\lim_{x \to -2} (x^2 - 4)$

(b) $\lim_{x \to 1} \dfrac{x^2 - 3x + 5}{x^2 + 9}$

(c) $\lim_{x \to 0} \dfrac{x^3 - 7x}{5x}$

(d) $\lim_{x \to 6} \dfrac{x^2 - x - 30}{x - 6}$

(e) $\lim_{x \to -2} \left(x^3 - \dfrac{x^2}{2x + 1} \right)$

(f) $\lim_{x \to 0} \dfrac{4x^3 + x - 1}{7x^3 + x^2 + 5}$

2. Evaluate the following limits:

(a) $\lim_{x \to 3} \left(\dfrac{5}{x - 5} - \sqrt{12x} \right)$

(b) $\lim_{x \to -5} \left(\dfrac{2x}{x + 5} + \dfrac{10}{x + 5} \right)$

(c) $\lim_{x \to 1} \dfrac{x^3 - 1}{x - 1}$

(d) $\lim_{x \to 1} \dfrac{x - 1}{\sqrt[3]{x} - 1}$

(e) $\lim_{x \to 6} \dfrac{x^2 - 4}{x^2 - 3x + 2}$

(f) $\lim_{x \to 1} \dfrac{x - 1}{x^2 - 1}$

3. Find the following limits if they exist:

(a) $\lim\limits_{x \to 0} \dfrac{\sqrt{4 + x^2} - 2}{x}$

(b) $\lim\limits_{x \to 0} \dfrac{\sqrt{1 + x} - 1}{x}$

(c) $\lim\limits_{x \to 10} \dfrac{\sqrt{x - 1} - 3}{x/5 - 2}$

(d) $\lim\limits_{x \to 1} \dfrac{x^2 - \sqrt{x}}{1 - \sqrt{x}}$

(e) $\lim\limits_{x \to 3^-} \dfrac{\sqrt{3x + 7} - 4}{\sqrt{3 - x}}$

4. Show that the expression $(\sqrt[3]{a} - \sqrt[3]{b})$ is rationalized by multiplying by $(\sqrt[3]{a^2} + \sqrt[3]{ab} + \sqrt[3]{b^2})$. Use this result to find the following limits:

(a) $\lim\limits_{x \to 0} \dfrac{\sqrt[3]{1 + x^2} - 1}{x^2}$

(b) $\lim\limits_{x \to 0} \dfrac{\sqrt[3]{27 + x} - \sqrt[3]{27}}{x}$

(c) $\lim\limits_{x \to 0} \dfrac{\sqrt[3]{1 + x} - \sqrt[3]{1 - x}}{x}$

5. Given $f(x) = \sqrt{25 - x^2}$, find

$$\lim\limits_{x \to 4} \dfrac{f(x) - f(4)}{x - 4}$$

6. Given $f(x) = 2x + 5$ when $x \neq 1$, and $f(1) = 6$.

(a) Sketch the graph of f.

(b) For what values of x is it true that $6 \leq f(x) \leq 8$?

7. Given $f(x) = 2x - 3$. How close to 4 must x be, if the values of the functon are required to lie within ± 0.01 of 5?

8. Are the following functions continuous at every point? Explain.

(a) $f(x) = \begin{cases} (x - 2)^2 + 1, & x < 6 \\ 8x - 31, & x \geq 6 \end{cases}$

(b) $f(x) = |x^2 - 2|$

9. Determine the intervals on which each of the following functions is continuous.

(a) $f(x) = \begin{cases} 2x + 3, & x \neq 4 \\ 5, & x = 4 \end{cases}$

(b) $f(x) = \dfrac{x^2 + x - 12}{x^2 - 9}$

(c) $f(x) = \dfrac{1 - \sqrt{x}}{x} + \sqrt{x}$

10. Due to depreciation, the value of an automobile t years after purchase is

$$f(t) = \left(\dfrac{9000 + 6t}{1 + 0.5t} + 50 \right) \text{ dollars}$$

(a) What is the value of the automobile when new?

(b) As t increases, what happens to the value of the automobile?

(c) What is the value of the automobile after 6 years?

(d) What is the scrap value?

(e) Sketch the graph of f.

11. What happens to the roots of the quadratic equation

$$ax^2 + bx + c = 0 \text{ as } a \to 0?$$

12. Fold and crease a narrow strip of paper along some transversal AB as in Figure 2.15. Lay AB along the edge BY and fold and crease again along BC. Observe that BC bisects the angle at B. Fold and crease along lines $CD, DE, EF, \ldots$ in a similar manner. Show that the triangles so formed tend to approach an equilateral triangle.

Figure 2.15

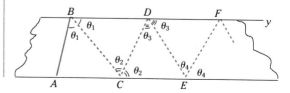

13. Two circles, $(x - 1)^2 + y^2 = 1$ and $x^2 + y^2 = r^2$, intersect at point B as shown in Figure 2.16. The second circle also intersects the positive y-axis at A. A line from A through B is extended to intersect the x-axis at C. Find the position of C in the limit as $r \to 0$.

Figure 2.16

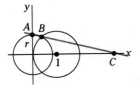

The Derivative

3

The derivative of a function describes how changes in one variable are related to changes in another. One representation of this concept in geometry is in the slope of the tangent to a curve. In later chapters, other interpretations and applications of the derivative are investigated.

Limits play a fundamental role in the process of finding the derivative of a function. Starting from the definition in terms of limits, certain basic rules for finding derivatives will be developed.

Slopes of Secants and Tangents   3.1

Let P be a point on the graph of a function f as in Figure 3.1. Consider a line through P that cuts the graph of the function at some other point Q. A line such as this, which intersects the graph of a function in at least two points, is called a **secant**. A line such as PR, which just grazes the graph, "touching" it only at the point P, is called the **tangent** at P.

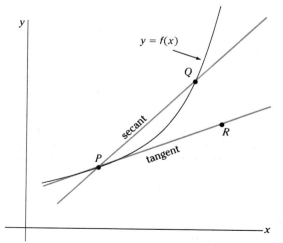

Figure 3.1

In order to find the slope of a secant, it is necessary to know the coordinates of points P and Q on the graph of f. We shall express these coordinates in a way that is particularly convenient for the work to follow. Let x denote the x-coordinate of P. The y-coordinate is the value of the function f at x, namely, $f(x)$. So the point P is $(x, f(x))$.

Let the x-coordinate of Q be $x + \Delta x$, where Δx is the amount by which the x-coordinate changes in going from P to Q (Figure 3.2). The y-coordinate of Q is the value of the function f at $x + \Delta x$, namely, $f(x + \Delta x)$. Thus, the point Q is $(x + \Delta x, f(x + \Delta x))$. The difference in the y-coordinates of P and Q can be written

Figure 3.2

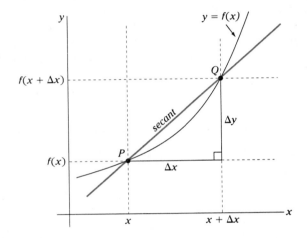

$$\Delta y = f(x + \Delta x) - f(x)$$

The slope of the secant PQ is, therefore,

$$\frac{\Delta y}{\Delta x} = \frac{f(x + \Delta x) - f(x)}{\Delta x}$$

EXAMPLE A

Find the slope of the secant PQ to the graph of the function

$$f(x) = (x - 3)^2$$

where $\Delta x = 4$ and the x-coordinate of P is 2.

SOLUTION

The x-coordinate of Q is

$$x + \Delta x = 2 + 4$$
$$= 6$$

The values of the function $f(x) = (x - 3)^2$ at $x = 2$ and 6 are

$$f(2) = (2 - 3)^2$$
$$= (-1)^2$$
$$= 1$$

and

$$f(6) = (6 - 3)^2$$
$$= (3)^2$$
$$= 9$$

The two points on the graph of f (Figure 3.3) are thus $P(2, 1)$ and $Q(6, 9)$. The slope of the secant is, therefore,

$$\frac{\Delta y}{\Delta x} = \frac{f(6) - f(2)}{4}$$

$$= \frac{9 - 1}{4}$$

$$= \frac{8}{4}$$

$$= 2$$

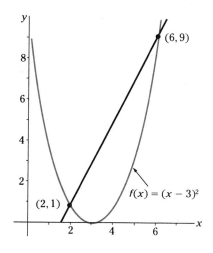

Figure 3.3

EXAMPLE B

(a) Show that the slope of a secant PQ to the graph of the function $f(x) = ax^2$ is $a(x_1 + x_2)$, where x_1 and x_2 are the x-coordinates of P and Q, respectively.

(b) Using the result of (a), find the slope of the secant PQ of the function $f(x) = x^2/4$, if P is the point $P(2, 1)$ and Q has various x-coordinates: 1.9, 1.99, 1.999, 2.001, 2.01, 2.1.

SOLUTION

(a) $y_1 = f(x_1)$ $y_2 = f(x_2)$

$\quad = ax_1{}^2$ $\quad = ax_2{}^2$

The slope of secant PQ is

$$\frac{\Delta y}{\Delta x} = \frac{y_2 - y_1}{x_2 - x_1}$$

$$= \frac{ax_2{}^2 - ax_1{}^2}{x_2 - x_1}$$

$$= \frac{a(x_2 - x_1)(x_2 + x_1)}{x_2 - x_1} \qquad \longleftarrow \qquad \text{divide out the common factor } (x_2 - x_1)$$

$$= a(x_1 + x_2)$$

Table 3.1

x-coordinate of Q	slope of secant PQ
1.9	$\frac{1}{4}(2 + 1.9) = 0.975$
1.99	$\frac{1}{4}(2 + 1.99) = 0.9975$
1.999	$\frac{1}{4}(2 + 1.999) = 0.99975$
2.001	$\frac{1}{4}(2 + 2.001) = 1.00025$
2.01	$\frac{1}{4}(2 + 2.01) = 1.0025$
2.1	$\frac{1}{4}(2 + 2.1) = 1.025$

(b) When $x_1 = 2$, the secant to $f(x) = x^2/4$ through $P(2, 1)$ and $Q(x_2, y_2)$ has the slope $(2 + x_2)/4$. Table 3.1 shows the value of the slope for several values of x_2 near 2. Notice that as Q approaches P from either side, the slope of the secant appears to approach the value 1. Furthermore, in Figure 3.4 you can see that

slope of		slope of		slope of
secant PQ	$<$	tangent	$<$	secant PQ
(Q left of P)		at P		(Q right of P)

and that as Q approaches P from either side, the slope of the secant approaches the slope of the tangent. Thus, it is reasonable to conjecture that the slope of the tangent at P is equal to 1.

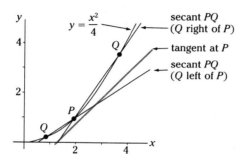

Figure 3.4

In general, to find the slope of the tangent to the curve $y = f(x)$ at P, assume P to be a fixed point, and let Q move along the curve from Q_1 to $Q_2, Q_3, Q_4, \ldots$, approaching P as in Figure 3.5. In the limit, as Q approaches P, Δx approaches 0, and the secant PQ becomes the

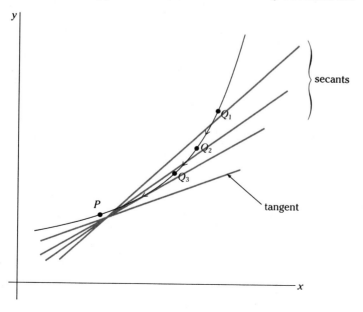

Figure 3.5

tangent at P. Hence, we define

$$\left(\begin{array}{c}\text{slope of}\\\text{tangent at } P\end{array}\right) = \lim_{\Delta x \to 0} \left(\begin{array}{c}\text{slope of}\\\text{secant } PQ\end{array}\right) = \lim_{\Delta x \to 0} \frac{\Delta y}{\Delta x}$$

EXAMPLE C

Find the slope of the tangent to the graph of $f(x) = (x - 3)^2$ at the point where $x = 2$.

SOLUTION

Let P be the point with x-coordinate 2.

Since
$$f(x) = (x - 3)^2$$

then

$$f(2) = (2 - 3)^2$$
$$= (-1)^2$$
$$= 1$$

Q is a point on the graph in the neighbourhood of P with x-coordinate $(2 + \Delta x)$ and y-coordinate $f(2 + \Delta x)$, where

$$f(2 + \Delta x) = ((2 + \Delta x) - 3)^2$$
$$= (\Delta x - 1)^2$$
$$= (\Delta x)^2 - 2(\Delta x) + 1$$

Thus,

$$\Delta y = f(2 + \Delta x) - f(2)$$
$$= (\Delta x)^2 - 2(\Delta x) + 1 - 1$$
$$= (\Delta x)^2 - 2(\Delta x)$$

and the slope of the secant PQ is

$$\frac{\Delta y}{\Delta x} = \frac{(\Delta x)^2 - 2(\Delta x)}{\Delta x} \qquad \longleftarrow \qquad \text{divide out the common factor } \Delta x$$

$$= \Delta x - 2$$

The slope of the tangent at P (Figure 3.6) is the limit of $\Delta x - 2$ as $\Delta x \to 0$:

Figure 3.6

$$\lim_{\Delta x \to 0} (\Delta x - 2) = -2$$

The slope of the tangent at P is -2.

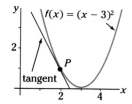

EXAMPLE D

Find the slope of the tangent to the graph of

$$f(x) = \frac{x}{2 - 3x}$$

at $x = 2$.

SOLUTION

Since

$$f(x) = \frac{x}{2 - 3x}$$

then

$$f(2) = \frac{2}{2 - 3 \cdot 2}$$

$$= -\frac{1}{2}$$

and

$$f(2 + \Delta x) = \frac{2 + \Delta x}{2 - 3(2 + \Delta x)}$$

$$= \frac{2 + \Delta x}{-4 - 3(\Delta x)}$$

The slope of the tangent at $x = 2$ is, therefore,

$$\begin{aligned}
\text{slope of} \atop \text{tangent} &= \lim_{\Delta x \to 0} \frac{f(2 + \Delta x) - f(2)}{\Delta x} \\
&= \lim_{\Delta x \to 0} \frac{1}{\Delta x} \left[\left(\frac{2 + \Delta x}{-4 - 3(\Delta x)} \right) - \left(-\frac{1}{2} \right) \right] \qquad \text{obtain a common denominator} \\
&= \lim_{\Delta x \to 0} \frac{1}{\Delta x} \left[\frac{2(2 + \Delta x) + (-4 - 3(\Delta x))}{2(-4 - 3(\Delta x))} \right]
\end{aligned}$$

$$= \lim_{\Delta x \to 0} \frac{1}{\Delta x} \left[\frac{-\Delta x}{2(-4 - 3(\Delta x))} \right]$$ ⟵ divide out the common factor Δx

$$= \lim_{\Delta x \to 0} \frac{-1}{2(-4 - 3(\Delta x))}$$ ⟵ evaluate the limit

$$= \frac{1}{8}$$

The slope of the tangent at $x = 2$ is $1/8$.   ❏

You should be aware of the possibility that a graph may not have a tangent at a particular value of x. For different reasons, for instance, the tangent does not exist at $x = -2$ in any of the following cases (see Figure 3.7).

(a) $f(x) = \dfrac{1}{x + 2}$

The tangent does not exist because there is an asymptote at $x = -2$.

(b) $f(x) = \sqrt{2x + 4}$

The tangent does not exist because the left limit as $x \to -2$ does not exist.

(c) $f(x) = |x + 2|$

The tangent does not exist because the left and right limits of the slope of the secant as $x \to -2$ are not equal, although the function itself is continuous.

Figure 3.7

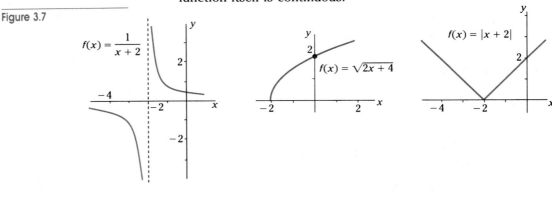

(a)   (b)   (c)

Exercise 3.1

1. If $f(x) = 3x + 1$, find $f(3)$, $f(3 + \Delta x)$, $f(-1)$, and $f(-1 + \Delta x)$.

2. Evaluate $f(a)$ and $f(a + \Delta x)$, and then express Δy in terms of Δx:

   (a) $f(x) = 1 - 2x$, $\quad a = -2$

   (b) $f(x) = x^2 - 4x + 3$, $a = 2$

   (c) $f(x) = \dfrac{2x}{3x + 5}$, $\quad a = 0$

   (d) $f(x) = (5 - x)^{1/2}$, $\quad a = -3$

3. In each of the following, find the slope of the secant to the curve through the points at which $x = -2$ and $x = 3$. Sketch the curve and the secant.

   (a) $y = x^2 - 6$

   (b) $y = -(x + 2)^2 + 5$

   (c) $y = 2x^3 - 16$

   (d) $y = \dfrac{4}{x^2 + 1}$

4. Find the slope of the secant through the points with x-coordinates 3 and $3 + \Delta x$. Express it in terms of Δx.

   (a) $f(x) = 3x - 2$

   (b) $f(x) = -7x^2$

   (c) $f(x) = 5 - 2x + x^2$

   (d) $f(x) = \dfrac{-3}{x + 1}$

5. Sketch graphs of each of the following functions and describe the slope of the tangent. Are there points where the tangent does not exist? Explain.

   (a) $y = x - 2$

   (b) $y = (x - 2)^2$

   (c) $y = |x - 2|$

(d) $y = \begin{cases} x, & x < 2 \\ 2, & x \geq 2 \end{cases}$

(e) $y = \begin{cases} 2, & x < 2 \\ -2, & x \geq 2 \end{cases}$

6. Sketch the graph of $y = 6/x$, and draw the secant through the points with x-coordinates -2 and 3. With the help of this graph, explain why secants formed by points on different branches of the same curve cannot be used to develop the slope of a tangent.

7. For each of the following functions, form the expression

$$\frac{\Delta y}{\Delta x} = \frac{f(a + \Delta x) - f(a)}{\Delta x}$$

for the given value of a, then find the slope of the tangent at a.

   (a) $y = x^2 - 1$, $\qquad a = 1$

   (b) $y = \dfrac{1}{2x}$, $\qquad a = -2$

   (c) $y = \dfrac{1}{x^2 - 4}$, $\qquad a = 1$

   (d) $y = \sqrt{x^2 - 9}$, $\qquad a = 5$

8. Find the slope of the tangent to each of the following functions at the point indicated:

   (a) $f(x) = 3x^2 - 5$, $\qquad$ at $(2, 7)$

   (b) $f(x) = x^2 - 3x$, $\qquad$ at $(0, 0)$

   (c) $f(x) = 2x^2 + 3x - 4$, $\quad$ at $(1, 1)$

   (d) $f(x) = (2x - 1)^2$, $\qquad$ at $(-1, 9)$

9. Find the slope of the tangent at the given value of x.

   (a) $y = \dfrac{2}{x - 1}$, $\qquad$ at $x = 7$

   (b) $y = \dfrac{x - 3}{x}$, $\qquad$ at $x = 2$

(c) $y = \sqrt{3x - 5}$,   at $x = 3$

(d) $\dfrac{1}{\sqrt{2x - 1}}$,   at $x = 1$

(e) $y = \dfrac{x + 3}{x - 1}$,   at $x = -3$

3.2 The Derivative of a Function

The preceding section explained how to find the slope of the tangent to the graph of $y = f(x)$ at a particular point. Rather than repeat this calculation for each different point at which a tangent is required, it is more practical to determine the slope of the tangent once and for all at some arbitrary point $(x, f(x))$. The result of this exercise will, in general, be some new *function* of x, which is called the **derivative** of f. The slope of the tangent to the graph of f at a particular point is then the *value* of the derivative of f at that point.

The derivative of the function $y = f(x)$ is denoted by a number of different symbols:

$$\frac{dy}{dx}, \qquad \frac{d}{dx}y, \qquad y', \qquad \frac{d}{dx}f(x), \qquad f'(x)$$

These symbols are read: "the derivative of y with respect to x" or "the derivative of f with respect to x." They represent the *value* of the derivative function at x. Take particular note, that dy/dx is to be viewed as a single symbol rather than as a fraction. Another symbol for the derivative is $D_x y$ or $D_x f(x)$, but this notation will not be used in this text.

The derivative of a function $y = f(x)$ is defined as

$$\frac{d}{dx}f(x) = \lim_{\Delta x \to 0} \frac{f(x + \Delta x) - f(x)}{\Delta x}$$

This quantity, as we have stated, is the slope of the tangent to f at a general point (x, y). This, in turn, is the limit of the slope of the secant. Finding the derivative of a function from this definition, is known as finding a derivative from **first principles**. In the examples which follow, you will see how this definition is applied to a variety of functions.

EXAMPLE A

Find the derivative of $f(x) = x^2$ from first principles.

SOLUTION

From the definition,

$$\frac{d}{dx}f(x) = \lim_{\Delta x \to 0} \frac{f(x + \Delta x) - f(x)}{\Delta x}$$

you can see that first it is necessary to find an expression for $f(x + \Delta x)$. In this example,

since
$$f(x) = x^2$$

then
$$f(x + \Delta x) = (x + \Delta x)^2$$

This gives two points on the graph: $P(x, f(x))$ and $Q(x + \Delta x, f(x + \Delta x))$. Next, you must form the expression for the slope of the secant PQ:

$$\frac{\Delta y}{\Delta x} = \frac{f(x + \Delta x) - f(x)}{\Delta x}$$

$$= \frac{(x + \Delta x)^2 - x^2}{\Delta x}$$

Evaluating the limit as $\Delta x \to 0$ at this stage results in the indeterminate form: 0/0. To avoid this, the expression for the slope of the secant must be simplified:

$$\frac{\Delta y}{\Delta x} = \frac{x^2 + 2x(\Delta x) + (\Delta x)^2 - x^2}{\Delta x}$$

$$= \frac{2x(\Delta x) + (\Delta x)^2}{\Delta x} \qquad \longleftarrow \qquad \text{divide out the common factor } \Delta x$$

$$= 2x + \Delta x$$

Now the limit can be evaluated:

$$\frac{d}{dx}f(x) = \lim_{\Delta x \to 0} (2x + \Delta x)$$

$$= 2x$$

Thus, the derivative with respect to x of the function

$$f(x) = x^2$$

Figure 3.8

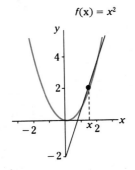

$f(x) = x^2$

is
$$\frac{d}{dx}f(x) = 2x$$

This is the slope of the tangent at any point x on the graph of f (Figure 3.8).

EXAMPLE B

From first principles, find the slope of the tangent to the graph of

$$f(x) = \frac{5}{x}$$

at $x = 4$

SOLUTION

Since

$$f(x) = \frac{5}{x}$$

then

$$f(x + \Delta x) = \frac{5}{x + \Delta x}$$

The slope of the secant is

$$\frac{\Delta y}{\Delta x} = \frac{\dfrac{5}{x + \Delta x} - \dfrac{5}{x}}{\Delta x} \qquad \longleftarrow \quad \text{obtain a common denominator}$$

$$= \left(\frac{5x - 5(x + \Delta x)}{x(x + \Delta x)}\right) \cdot \frac{1}{\Delta x} \qquad \longleftarrow \quad \begin{array}{l}\text{dividing by } \Delta x \text{ is} \\ \text{equivalent to} \\ \text{multiplying by } 1/\Delta x\end{array}$$

$$= \left(\frac{-5(\Delta x)}{x(x + \Delta x)}\right) \cdot \frac{1}{\Delta x} \qquad \longleftarrow \quad \begin{array}{l}\text{divide out the} \\ \text{common factor } \Delta x\end{array}$$

$$= \frac{-5}{x(x + \Delta x)}$$

Evaluating the limit as $\Delta x \to 0$ produces the derivative of the function, which is the slope of the tangent at the point (x, y):

$$\frac{dy}{dx} = \lim_{\Delta x \to 0} \frac{-5}{x(x + \Delta x)}$$

$$= \frac{-5}{x^2}$$

The slope of the tangent at $x = 4$ is found by substituting the value 4 into the expression for the derivative (see Figure 3.9). At $x = 4$,

$$\frac{dy}{dx} = \frac{-5}{(4)^2}$$

$$= -\frac{5}{16}$$

Figure 3.9

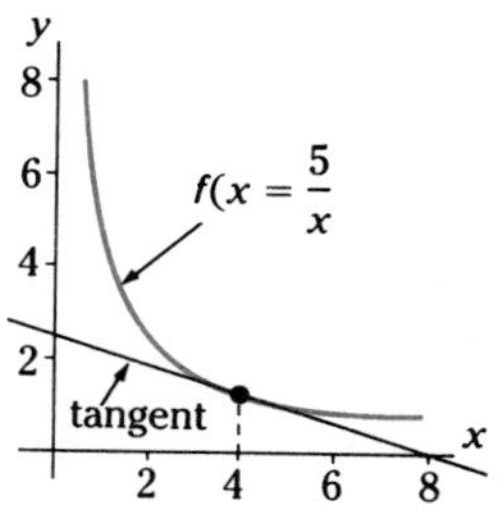

EXAMPLE C

Find the equation of the line tangent to the graph of

$$f(x) = \sqrt{3x}$$

at $x = 4$.

SOLUTION

The slope of the tangent and the coordinates of the point of tangency are both required.

Since

$$f(x) = \sqrt{3x}$$

then

$$f(x + \Delta x) = \sqrt{3(x + \Delta x)}$$

The slope of the secant is

$$\frac{\Delta y}{\Delta x} = \frac{\sqrt{3(x + \Delta x)} - \sqrt{3x}}{\Delta x}$$

Again, it is necessary to simplify this expression because evaluating the limit at this stage produces the indeterminate form: 0/0. The simplification is carried out by rationalizing the numerator:

$$\frac{\Delta y}{\Delta x} = \frac{(\sqrt{3(x + \Delta x)} - \sqrt{3x})}{\Delta x} \cdot \frac{(\sqrt{3(x + \Delta x)} + \sqrt{3x})}{(\sqrt{3(x + \Delta x)} + \sqrt{3x})} \quad \longleftarrow \text{conjugate radicals}$$

$$= \frac{3(x + \Delta x) - 3x}{\Delta x(\sqrt{3(x + \Delta x)} + \sqrt{3x})} \quad \longleftarrow \begin{array}{l} \text{the expression is} \\ \text{simplified more} \\ \text{easily if the} \\ \text{denominator is } not \\ \text{expanded} \end{array}$$

$$= \frac{3(\Delta x)}{\Delta x(\sqrt{3(x + \Delta x)} + \sqrt{3x})}$$

            ⟵   divide out the common factor Δx

$$= \frac{3}{\sqrt{3(x + \Delta x)} + \sqrt{3x}}$$

Now evaluate the limit as $\Delta x \to 0$:

$$\frac{dy}{dx} = \lim_{\Delta x \to 0} \frac{3}{\sqrt{3(x + \Delta x)} + \sqrt{3x}}$$

$$= \frac{3}{\sqrt{3x} + \sqrt{3x}}$$

$$= \frac{3}{2\sqrt{3x}}$$

$$= \frac{\sqrt{3}}{2\sqrt{x}}$$

If the slope m, and a point (x_1, y_1) on the line are known, it is possible to find the equation of the tangent at $x = 4$ from the point-slope formula for a straight line:

$$y - y_1 = m(x - x_1)$$

The slope at $x = 4$ is

$$m = \frac{\sqrt{3}}{2\sqrt{4}}$$

$$= \frac{\sqrt{3}}{4}$$

while the value of the function at $x = 4$ is

Figure 3.10

$$f(4) = \sqrt{3(4)}$$

$$= 2\sqrt{3}$$

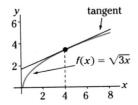

so that the point of tangency is $P(4, 2\sqrt{3})$. Therefore the required tangent line is (Figure 3.10)

$$y - 2\sqrt{3} = \frac{\sqrt{3}}{4}(x - 4)$$

or

$$y = \frac{\sqrt{3}}{4}x + \sqrt{3}$$

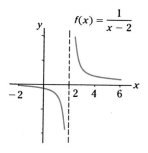

You have seen in these examples how to find the derivative of any function *f* from first principles. The procedure can be summarized in three steps:

1. Find $f(x + \Delta x)$.
2. Form the expression for the slope of the secant:

$$\frac{\Delta y}{\Delta x} = \frac{f(x + \Delta x) - f(x)}{\Delta x}$$

and simplify this algebraically.

3. Evaluate the limit

$$\lim_{\Delta x \to 0} \frac{\Delta y}{\Delta x} = \lim_{\Delta x \to 0} \frac{f(x + \Delta x) - f(x)}{\Delta x}$$

Differentiable functions

The process of finding the derivative of a function is called **differentiation**. The symbol $f'(a)$ is used to represent the derivative $f'(x)$ evaluated at $x = a$. If $f'(a)$ exists then the function *f* is said to be **differentiable** at $x = a$. Generally speaking, a function is differentiable if it is continuous and if it has no vertical tangents or abrupt, discontinuous changes in slope. The following example will illustrate several common ways in which a function can fail to be differentiable at a point.

EXAMPLE D

Explain why the following functions are not differentiable at $x = 2$:

(a) $f(x) = \dfrac{1}{x - 2}$ (b) $f(x) = \sqrt[3]{x - 2}$ (c) $f(x) = \dfrac{x^2 - 4}{x - 2}$

(d) $f(x) = \sqrt{x - 2}$ (e) $f(x) = |x - 2|$ (f) $f(x) = 16^{1/x - 2}$

SOLUTION

(a) $f(x) = \dfrac{1}{x - 2}$

This function is not defined at $x = 2$. It has a vertical asymptote there (Figure 3.11).

(b) $f(x) = \sqrt[3]{x - 2}$

Figure 3.11

Figure 3.12

Although this function has a tangent at $x = 2$, its slope is infinite (Figure 3.12). Therefore $f'(2)$ does not exist, and the function is not differentiable at that point.

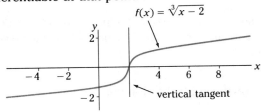

Figure 3.13

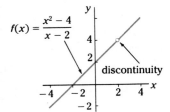

(c) $f(x) = \dfrac{x^2 - 4}{x - 2}$

This function is not defined at $x = 2$. The graph is a straight line with one point missing at $x = 2$ (Figure 3.13). There can be no tangent at the gap or break where f is discontinuous.

(d) $f(x) = \sqrt{x - 2}$

This function is not defined for $x < 2$ (Figure 3.14). Therefore, the limit as $x \to 2$ from the left does not exist.

Figure 3.14

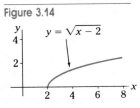

Figure 3.15

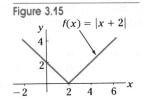

(e) $f(x) = |x - 2|$

This function has a corner at $x = 2$, where the slope of f changes abruptly from -1 to $+1$ (Figure 3.15).

(f) $f(x) = 16^{1/x - 2}$

The left and right limits of this function as $x \to 2$ are different (Figure 3.16).

Figure 3.16

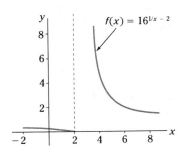

Exercise 3.2

1. State the function whose derivative is given as follows:

(a) $\dfrac{dy}{dx} = \lim\limits_{\Delta x \to 0} \dfrac{2(x + \Delta x)^3 - 2x^3}{\Delta x}$

(b) $\dfrac{dy}{dx} = \lim\limits_{\Delta x \to 0} \dfrac{5\sqrt{x + \Delta x} - 5\sqrt{x}}{\Delta x}$

2. Write down the expressions

i. $f(x + \Delta x)$

ii. $\dfrac{f(x + \Delta x) - f(x)}{\Delta x}$

for each of the following functions:

(a) $f(x) = 3x^2$

(b) $f(x) = \dfrac{1}{2}x^3 - 2$

(c) $f(x) = \dfrac{-2}{x + 4}$

(d) $f(x) = \sqrt{2x - 1}$

(e) $f(x) = \cos\dfrac{\pi x}{2}$

(f) $f(x) = 7^{3x}$

(g) $f(x) = \log_a x$

3. From first principles, find the derivative of each of the following functions:

(a) $y = 4x^2 + 5$

(b) $y = \dfrac{3}{x}$

(c) $y = \sqrt{x - 1}$

(d) $y = \dfrac{x - 1}{2 + x}$

(e) $y = \dfrac{1}{\sqrt{2 - 3x}}$

(f) $y = \dfrac{2x^2 - 3}{x}$

4. Find the slope of the tangent to the graph of

$$y = 5 - \dfrac{2}{x}$$

at $x = 2$.

5. Find the equation of the tangent to the graph of

$$y = \dfrac{4}{2 - x}$$

at the point where $x = 0$. Sketch the graph of the function and the tangent.

6. Find the equation of the tangent to the graph of

$$y = \dfrac{x}{2 - x}$$

at the point $(1, 1)$. Sketch the graph of the function and the tangent.

7. Find the equation of the tangent to the graph of $y = x^{1/3}$ at $(0, 0)$. Sketch the graph and the tangent.

8. (a) Find the derivative of $f(x) = 2x^2 - 7x + 9$.

(b) At what value of x is the slope of the tangent to the graph of f equal to 1?

9. (a) At what points on the graph of

$$y = \dfrac{x^3}{12} - 4x$$

is the slope of the tangent zero?

(b) Find the equations of the tangents in (a).

10. Find the equation of the tangent to $y = 3x^2 - x$ that has slope 5.

11. Explain why the following functions are *not* differentiable at the points indicated:

(a) $y = x^{2/3}$,   at $x = 0$

(b) $y = \begin{cases} 2x - 1, & x \le 3 \\ 0, & x > 3 \end{cases}$ at $x = 3$

(c) $y = \sqrt{3 - 2x}$ at $x = 3/2$

(d) $y = \dfrac{1}{x^2 - 5}$ at $x = \pm\sqrt{5}$

12. At what values of x are the following functions *not* differentiable? Explain.

(a) $y = \dfrac{4}{x - x^2}$

(b) $y = \dfrac{x^2 - x - 6}{x^2 - 4}$

(c) $y = \sqrt{x^2 - 6x + 5}$

13. $A(x_1, y_1)$ and $B(x_2, y_2)$ are two points on the parabola $y = ax^2 + bx + c$. At what point is the tangent to the parabola parallel to the secant AB?

3.3 Differentiation Rules

It is possible to find the derivative of any function from first principles. It is impractical, however, to have to return to first principles in every instance. It would be more useful to have general formulas or rules, which give the derivative of a function directly and by-pass the limit process entirely. In this section, rules for differentiating powers of x and polynomial functions will be developed.

The derivative of a constant

EXAMPLE A

Find the derivative of $f(x) = C$, where C is a constant, by means of the first principles method.

SOLUTION

Since
$$f(x) = C$$

then

$$f(x + \Delta x) = C$$

because C is a constant and its value does not change when x changes. Therefore,

$$\frac{\Delta y}{\Delta x} = \frac{f(x + \Delta x) - f(x)}{\Delta x}$$

$$= \frac{C - C}{\Delta x}$$

$$= 0$$

$$\frac{d}{dx}f(x) = \lim_{\Delta x \to 0} \frac{\Delta y}{\Delta x}$$

$$= \lim_{\Delta x \to 0} 0$$

$$= 0$$

Therefore,

$$\boxed{\frac{d}{dx}C = 0}$$

Figure 3.17

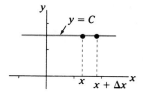

Thus, the derivative of a constant is zero. The graph of the function $y = C$ is a line parallel to the x-axis with slope zero (Figure 3.17).☐

The derivative of x^n—The power rule

To discover what the power rule might be like, let us find the derivative of some specific powers of x.

EXAMPLE B

Find the derivative of $f(x) = x$.

SOLUTION

Since

$$f(x) = x,$$

then

$$f(x + \Delta x) = x + \Delta x$$

$$\frac{\Delta y}{\Delta x} = \frac{f(x + \Delta x) - f(x)}{\Delta x}$$

$$= \frac{(x + \Delta x) - x}{\Delta x}$$

$$= \frac{\Delta x}{\Delta x}$$

$$= 1$$

$$\frac{d}{dx}f(x) = \lim_{\Delta x \to 0} \frac{\Delta y}{\Delta x}$$

$$= \lim_{\Delta x \to 0} 1$$

$$= 1$$

Figure 3.18

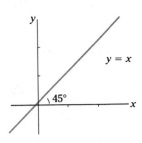

Therefore,

$$\frac{d}{dx} x = 1$$

The graph of $y = x$ (Figure 3.18) is a line through the origin making an angle of $45°$ with the positive x-axis. It has a slope of 1 everywhere.

☐

EXAMPLE C

Find the derivative of $f(x) = x^3$.

SOLUTION

Since
$$f(x) = x^3$$
then
$$f(x + \Delta x) = (x + \Delta x)^3$$

$$\frac{\Delta y}{\Delta x} = \frac{(x + \Delta x)^3 - x^3}{\Delta x}$$

$$= \frac{[x^3 + 3x^2(\Delta x) + 3x(\Delta x)^2 + (\Delta x)^3] - x^3}{\Delta x}$$

$$= \frac{3x^2(\Delta x) + 3x(\Delta x)^2 + (\Delta x)^3}{\Delta x}$$

$$= 3x^2 + 3x(\Delta x) + (\Delta x)^2$$

$$\frac{d}{dx} f(x) = \lim_{\Delta x \to 0} \frac{\Delta y}{\Delta x}$$

$$= \lim_{\Delta x \to 0} [3x^2 + 3x(\Delta x) + (\Delta x)^2]$$

$$= 3x^2$$

Therefore,

$$\frac{d}{dx} x^3 = 3x^2$$

☐

The results of these examples, including that of Example A in the preceding section, suggest a possible general rule for the derivative of any power x^n, where n is a natural number. Observe the pattern:

Function: $y = x$ $y = x^2$ $y = x^3$ ... $y = x^n$
Derivative: $y' = 1$ $y' = 2x$ $y' = 3x^2$ ... $y' = nx^{n-1}$

It is possible to arrive at this rule from first principles, following the same process used in previous examples. Starting with the function

$$f(x) = x^n$$

then

$$f(x + \Delta x) = (x + \Delta x)^n$$

and

$$\frac{\Delta y}{\Delta x} = \frac{(x + \Delta x)^n - x^n}{\Delta x}$$

The result of evaluating the limit at this stage is the indeterminate form 0/0. It first necessary to remove a common factor of Δx. To accomplish this, the binomial theorem is used to expand the term $(x + \Delta x)^n$. (You may wish to review A. Preparation for Calculus, pages 16-29, at this time.) According to the binomial theorem,

$$(a + b)^n = a^n + na^{n-1}b + \frac{n(n-1)}{2}a^{n-2}b^2 + \cdots + b^n$$

Replacing a by x, and b by Δx, then

$$\frac{\Delta y}{\Delta x} = \frac{\left[x^n + nx^{n-1}(\Delta x) + \frac{n(n-1)}{2}x^{n-2}(\Delta x)^2 + \cdots + (\Delta x)^n \right] - x^n}{\Delta x}$$

$$= \frac{nx^{n-1}(\Delta x) + \frac{n(n-1)}{2}x^{n-2}(\Delta x)^2 + \cdots + (\Delta x)^n}{\Delta x} \quad \leftarrow \text{divide out the common factor } \Delta x$$

$$= nx^{n-1} + \frac{n(n-1)}{2}x^{n-2}(\Delta x) + \cdots + (\Delta x)^{n-1}$$

$$\frac{d}{dx}f(x) = \lim_{\Delta x \to 0} \frac{\Delta y}{\Delta x}$$

$$= \lim_{x \to 0} \left[nx^{n-1} + \frac{n(n-1)}{2}x^{n-2}(\Delta x) + \cdots + (\Delta x)^{n-1} \right]$$

$$= nx^{n-1}$$

Therefore,

$$\boxed{\frac{d}{dx}x^n = nx^{n-1}}$$

This is the **power rule**. In the preceding derivation, n has been taken to be a positive integer. However, the binomial expansion is valid when n is any real number. In the general case, the expansion has an infinite number of terms. Each one after the first, however, contains a power of Δx. They all therefore vanish when the limit is evaluated. The conclusion is that the power rule is valid for all real values of the exponent n.

EXAMPLE D

Find the derivative of the following functions using the power rule:

$$\frac{d}{dx}x^n = nx^{n-1}$$

(a) $y = x^{12}$ (c) $y = \dfrac{1}{x^4}$

(b) $y = \sqrt{x}$ (d) $y = \dfrac{1}{\sqrt{x^3}}$

SOLUTION

(a) $y = x^{12}$

$$\frac{dy}{dx} = 12x^{11}$$

(b) $y = \sqrt{x}$

$$= x^{1/2}$$

$$\frac{dy}{dx} = \left(\frac{1}{2}\right)x^{-1/2}$$

$$= \frac{1}{2\sqrt{x}}$$

(c) $y = \dfrac{1}{x^4}$

$$= x^{-4}$$

$$\frac{dy}{dx} = (-4)x^{-5}$$

$$= -\frac{4}{x^5}$$

(d) $y = \dfrac{1}{\sqrt{x^3}}$

$\quad\quad = x^{-3/2}$

$\dfrac{dy}{dx} = \left(-\dfrac{3}{2}\right)x^{-5/2}$

$\quad\quad = -\dfrac{3}{2\sqrt{x^5}}$ ❑

The derivative of a constant times a function

The next task is to find the derivative of $f(x) = Cg(x)$ from first principles, where g is any differentiable function, and C is a constant.

Since

$$f(x) = Cg(x)$$

then

$$f(x + \Delta x) = Cg(x + \Delta x)$$

$$\dfrac{\Delta y}{\Delta x} = \dfrac{Cg(x + \Delta x) - Cg(x)}{\Delta x}$$

$$= C \cdot \dfrac{g(x + \Delta x) - g(x)}{\Delta x}$$

$$\dfrac{d}{dx}f(x) = \lim_{\Delta x \to 0}\dfrac{\Delta y}{\Delta x}$$

$$= \lim_{\Delta x \to 0} C \cdot \dfrac{g(x + \Delta x) - g(x)}{\Delta x}$$

$$= \lim_{\Delta x \to 0} C \cdot \lim_{\Delta x \to 0}\dfrac{g(x + \Delta x) - g(x)}{\Delta x} \quad\longleftarrow \text{since the limit of a} \atop \text{product is the} \atop \text{product of the limits}$$

$$= C\dfrac{d}{dx}g(x)$$

Therefore,

$$\boxed{\dfrac{d}{dx}Cg(x) = C \cdot \dfrac{d}{dx}g(x)}$$

In practical terms, this rule means that it is permissible to take a constant factor out of the derivative, as shown in the following example.

EXAMPLE E

Differentiate:

(a) $y = -5x^4$ (b) $y = \dfrac{15}{4}x^{2/3}$

SOLUTION

(a) $y = -5x^4$

$$\frac{dy}{dx} = \frac{d}{dx}(-5 \cdot x^4)$$

$$= -5 \cdot \frac{d}{dx}x^4$$

$$= -5 \cdot 4x^3$$

$$= -20x^3$$

(b) $y = \dfrac{15}{4}x^{2/3}$

$$\frac{dy}{dx} = \frac{d}{dx}\left[\frac{15}{4} \cdot x^{2/3}\right]$$

$$= \frac{15}{4} \cdot \frac{d}{dx}x^{2/3}$$

$$= \frac{15}{4} \cdot \frac{2}{3}x^{-1/3}$$

$$= \frac{5}{2}x^{-1/3}$$

The derivative of the sum of two functions

The derivative of $f(x) = g(x) + h(x)$, where g and h are any differentiable functions is found as follows:

Since

$$f(x) = g(x) + h(x)$$

then

$$f(x + \Delta x) = g(x + \Delta x) + h(x + \Delta x)$$

$$\frac{\Delta y}{\Delta x} = \frac{[g(x + \Delta x) + h(x + \Delta x)] - [g(x) + h(x)]}{\Delta x} \quad \text{regroup the terms}$$

$$= \frac{g(x + \Delta x) - g(x)}{\Delta x} + \frac{h(x + \Delta x) - h(x)}{\Delta x}$$

$$\frac{d}{dx}f(x) = \lim_{\Delta x \to 0} \frac{\Delta y}{\Delta x}$$

$$= \lim_{\Delta x \to 0} \frac{g(x + \Delta x) - g(x)}{\Delta x}$$

$$+ \lim_{\Delta x \to 0} \frac{h(x + \Delta x) - h(x)}{\Delta x}$$

since the limit of the sum is the sum of the limits

$$= \frac{d}{dx}g(x) + \frac{d}{dx}h(x)$$

Therefore,

$$\frac{d}{dx}[g(x) + h(x)] = \frac{d}{dx}g(x) + \frac{d}{dx}h(x).$$

In practice, this means that if a function is made up of a sum of differentiable functions, the derivative of the sum is the sum of the derivatives.

EXAMPLE F

Differentiate:

$$y = 8x^2 - 7x + 12$$

SOLUTION

$$y = 8x^2 - 7x + 12$$

$$\frac{dy}{dx} = \frac{d}{dx}(8x^2 - 7x + 12)$$

the derivative of a sum

$$= \frac{d}{dx}8x^2 + \frac{d}{dx}(-7x) + \frac{d}{dx}12$$

the derivative of a constant times a function

$$= 8\frac{d}{dx}x^2 - 7\frac{d}{dx}x + \frac{d}{dx}12$$

the power rule and the derivative of a constant

$$= 8(2x) - 7(1) + 0$$

$$= 16x - 7$$

You can see in this example the part played by each of the differentiation rules. With practice, you will be able to do these steps automatically.

EXAMPLE G

Differentiate:

$$y = (1 - 2x)^3$$

SOLUTION

$$y = (1 - 2x)^3$$

$$\frac{dy}{dx} = \frac{d}{dx}(1 - 2x)^3$$

It is important to realize that none of the differentiation rules introduced so far can be applied to this function as it stands. In particular, it is *incorrect* to think that you can use the power rule in this situation and write

$$\frac{d}{dx}(1 - 2x)^3 = 3(1 - 2x)^2 \qquad \text{incorrect!}$$

The power rule is concerned with powers of x itself and *not* with powers of functions. The way to proceed in this example is to expand the term $(1 - 2x)^3$ first, before differentiating.

$$\frac{dy}{dx} = \frac{d}{dx}(1 - 2x)^3$$

$$= \frac{d}{dx}(1 - 6x + 12x^2 - 8x^3)$$

$$= -6 + 24x - 24x^2 \qquad \square$$

If the function in this example had carried an exponent other than a small integer, it would be impractical if not impossible to expand it. In the next section, you will see how to apply another differentiation rule known as the **chain rule** to such cases.

EXAMPLE H

Differentiate:

$$y = (x + 2)(x^2 - 1)$$

SOLUTION

In the next chapter, you will learn a differentiation rule to apply to products of functions. For now it is necessary to express y as a polynomial:

$$y = (x + 2)(x^2 - 1)$$

$$= x^3 + 2x^2 - x - 2$$

$$\frac{dy}{dx} = \frac{d}{dx}(x^3 + 2x^2 - x - 2)$$

$$= 3x^2 + 4x - 1 \qquad \square$$

EXAMPLE I

Differentiate:

$$s = \frac{3t^5 - 2t^2 + 5t^{-1}}{t^2}$$

SOLUTION

Write each term as a constant times a power of t:

$$s = \frac{3t^5 - 2t^2 + 5t^{-1}}{t^2}$$

$$= 3t^3 - 2 + 5t^{-3}$$

$$\frac{ds}{dt} = 9t^2 - 15t^{-4} \qquad \square$$

EXAMPLE J

Differentiate:

$$u = \sqrt{\frac{2}{x}} + \sqrt{\frac{x}{3}}$$

SOLUTION

Again, write each term as a constant times a power of x:

$$u = \sqrt{\frac{2}{x}} + \sqrt{\frac{x}{3}}$$

$$= \frac{\sqrt{2}}{\sqrt{x}} + \frac{\sqrt{x}}{\sqrt{3}}$$

$$= \sqrt{2}\, x^{-1/2} + \frac{1}{\sqrt{3}} x^{1/2}$$

$$\frac{du}{dx} = \sqrt{2}\left(-\frac{1}{2}\right)x^{-3/2} + \frac{1}{\sqrt{3}}\left(\frac{1}{2}\right)x^{-1/2}$$

$$= -\frac{\sqrt{2}}{2}x^{-3/2} + \frac{1}{2\sqrt{3}}x^{-1/2} \qquad \square$$

EXAMPLE K

Given the parabola $y = 10x - x^2 - 16$

(a) Find the equation of the tangent at the point $(4, 8)$

(b) Find the equations of the two tangents that pass through the point $(4, 12)$.

SOLUTION

Figure 3.19

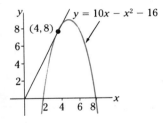

(a) The point $(4, 8)$ lies on the parabola as shown in Figure 3.19. The slope m of the tangent at this point is the value of the derivative of y with respect to x at $x = 4$:

$$y = 10x - x^2 - 16$$

$$\frac{dy}{dx} = 10 - 2x$$

$$\therefore \quad m = 10 - 2(4)$$

$$= 2$$

The equation of the tangent at $(4, 8)$ is therefore

$$y - 8 = 2(x - 4)$$

or

$$y = 2x$$

(b) The point $(4, 12)$ does *not* lie on the parabola. From this point, two lines can be drawn tangent to the curve (see Figure 3.20). Let $(a, f(a))$ be one of the points of tangency. The slope of the line through $(4, 12)$ and $(a, f(a))$ is

$$\frac{f(a) - 12}{a - 4} = \frac{(10a - a^2 - 16) - 12}{a - 4}$$

$$= \frac{10a - a^2 - 28}{a - 4}$$

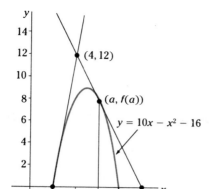

Figure 3.20

This must equal the slope m of the tangent at $x = a$:

$$\frac{dy}{dx} = 10 - 2x$$

$$\therefore \quad m = 10 - 2a$$

Equating these two expressions for the slope of the tangent determines the value of a:

$$\frac{10a - a^2 - 28}{a - 4} = 10 - 2a$$

$$10a - a^2 - 28 = (10 - 2a)(a - 4)$$

$$10a - a^2 - 28 = 10a - 40 - 2a^2 + 8a$$

$$a^2 - 8a + 12 = 0$$

$$(a - 2)(a - 6) = 0$$

$$\therefore \quad a = 2 \quad \text{or} \quad a = 6$$

Thus, one line from $(4, 12)$ is tangent at $x = 2$. The other is tangent at $x = 6$.

At $x = 2$,

$$y = 10(2) - (2)^2 - 16$$

$$= 0$$

At $x = 6$

$$y = 10(6) - (6)^2 - 16$$

$$= 8$$

$$m = 10 - 2(2)$$

$$= 6$$

$$y - 0 = 6(x - 2)$$

$$y = 6x - 12$$

$$m = 10 - 2(6)$$

$$= -2$$

$$y - 8 = -2(x - 6)$$

$$y = -2x + 20$$

The required tangents are $y = 6x - 12$ and $y = -2x + 20$. ❑

Exercise 3.3

1. Differentiate each of the following functions:

   (a) $y = 10 - x^3$

   (b) $y = 3x^2 - 7x$

   (c) $y = ax^2 + bx + c$

   (d) $y = 6x^{2/3}$

   (e) $y = \dfrac{4}{x}$

   (f) $y = \sqrt{6x}$

2. Find the derivative:

   (a) $y = 2x(3 - x^2)$

   (b) $y = 2(5 - x)^2$

   (c) $y = \sqrt[3]{24x}$

   (d) $y = \dfrac{-6}{x^3} - x$

   (e) $y = \dfrac{3 - 2x + 7x^3}{x^2}$

   (f) $y = (1 - x^3)(1 + x)$

   (g) $y = x(4 - 3x)(x + 1)$

   (h) $y = \dfrac{(2x - 1)^2}{4x}$

3. Find $f'(x)$ in each of the following cases:

   (a) $f(x) = \sqrt{\dfrac{x}{2}} + \sqrt{2x}$

   (b) $f(x) = 2\sqrt{x} + \sqrt[4]{6x}$

   (c) $f(x) = 1.6\sqrt[4]{x} - \dfrac{x^3}{0.9} + \dfrac{1}{5x^2}$

   (d) $f(x) = \dfrac{2}{\sqrt{x}} - \sqrt{4x}$

   (e) $f(x) = \dfrac{x^2 - 8}{\sqrt{8x}}$

4. Differentiate the following functions with respect to the independent variable.

   (a) $g(u) = 0.8 - 3(2 - u)^3$

   (b) $h(w) = (w - 1)(w^2 + 1)$

   (c) $k(r) = 0.6r^{-2/3} - \dfrac{5.2}{\sqrt[3]{r}} + \dfrac{2.5}{\sqrt[5]{r}}$

5. Differentiate each of the following with respect to t:

   (a) $f(t) = \dfrac{mt^2}{\sqrt{t}} + \dfrac{nt\sqrt{t}}{\sqrt[3]{t}} - \dfrac{p\sqrt{t}}{t}$

   where m, n, and p are constants.

   (b) $y = \dfrac{at^3 + bt^2 + c}{(a - b)t}$

   where a, b, and c are constants.

6. Evaluate the derivatives at the indicated values:

   (a) $f(x) = x^3 - 5x$,　　　　$f'(0), f'(a)$

   (b) $g(x) = 3x - 2\sqrt{x}$,　　　$g'(1), g'(4)$

   (c) $h(\theta) = \dfrac{2\theta^3 - 3\theta + \sqrt{\theta} - 1}{\theta}$,　$h'(1), h'(1/4)$

7. Find the equation of the tangent to the graph of $y = x^2 - 3x + 2$ at the point where $x = 1$.

8. Find the equation of the tangent to the curve defined by

$$f(x) = \frac{6}{x^2} - 2x$$

at the point where $x = -1$.

9. Find the equations of the tangents to the curve

$$f(x) = 3 - \frac{12}{x^2}$$

at the points where the curve crosses the x-axis. Sketch the graph and the tangent lines.

10. Given the graph defined by $y = x^3$.

   (a) Find the equation of the tangent at the point $(1, 1)$.

   (b) Find where the tangent of part (a) intersects the graph at a second point.

   (c) Find the equation of the tangent to the graph at the second point.

11. From the fact that at the vertex of a parabola the slope of the tangent is zero, find the vertex of $y = -5x^2 + 2x + 1$.

12. Find the values for a and b in $f(x) = ax^2 + bx$, given that $f(-1) = 5$ and the slope of the tangent at $x = 1$ is 7.

13. Find the second degree polynomial which passes through the point $(2, -1)$ and which has slope 0 at $x = 1$ and slope -4 at $x = -1$.

14. At what values of x does the tangent to the curve $y = x^3 - 3x$ have a slope of zero?

15. Find the values of x, for which the slopes of the tangents to the graphs $y = x^2 + 3x - 5$ and $y = x^3 + x^2 + 4$ are equal.

16. Find the point on the graph of $y = -3x^2 + 2x$ at which the tangent makes an angle of $45°$ with the positive x-axis. Illustrate the result with a sketch of the graph and the tangent line.

17. Find the equations of the two tangents to the graph of $f(x) = x^2 - 1$ that pass through the point $(1, -4)$. Sketch the graph and the tangent lines.

18. The **normal** to the graph of the function $y = f(x)$ at the point (x, y) is the line that is perpendicular to the tangent at that point. Find the equation of the normal to:

   (a) $f(x) = 3x^2 - 5x + 7$, at $x = 1$

   (b) $f(x) = \sqrt{x} - 4$,   at $x = 4$

In each case sketch the graph of the function, the tangent, and the normal.

The Chain Rule   3.4

Differentiation rules are relationshps between functions, which do not depend on the actual names of the variables involved. Using u instead of x, for instance, as the independent variable, the rule for differentiating a power could be written:

$$\text{if } y = u^n \quad \text{then} \quad \frac{dy}{du} = nu^{n-1}$$

As you will see in what follows, the secret to using a differentiation rule correctly lies in identifying the independent variable and recognizing what function is involved.

   In the preceding section, in Example G, you saw that the power rule could not be directly applied to

$$y = (1 - 2x)^3$$

because the expression on the right is a power of a function, not a simple power of x. In that example, it was necessary to expand $(1 - 2x)^3$, before taking the derivative. Observe, however, that y can be written

$$y = u^3$$

where

$$u = (1 - 2x)$$

In other words, you can regard y as a function of u, where u in turn is a function of x. Introduction of the variable u makes it possible to write a complicated function in terms of two simpler ones, each of which can be differentiated directly:

since $y = u^3$ and $u = (1 - 2x)$

then $\dfrac{dy}{du} = 3u^2$ and $\dfrac{du}{dx} = -2$

Are these derivatives related in a simple way to dy/dx? Yes! In this case, it is not hard to verify that dy/dx is their product:

$$\frac{dy}{dx} = \frac{dy}{du} \cdot \frac{du}{dx}$$

$$= 3u^2 \cdot (-2)$$

$$= 3(1 - 2x)^2 \cdot (-2)$$

$$= (-6)(1 - 4x + 4x^2)$$

$$= -6 + 24x - 24x^2 \qquad \text{(compare to Example G, Section 3.3)}$$

The connection between the three derivatives illustrated above is known as the **chain rule**:

$$\boxed{\frac{dy}{dx} = \frac{dy}{du} \cdot \frac{du}{dx}}$$

To establish that this formula is true in general, consider the case

where $y = f(u)$ (y is a function of u)

and $u = g(x)$ (u is a function of x).

In other words, y is the composition of two functions:
$y = f(g(x))$. Any change Δx in x causes a change Δu in u, which in turn causes a change Δy in y. Observe that (provided that $\Delta u \neq 0$), you can write

$$\frac{\Delta y}{\Delta x} = \frac{\Delta y}{\Delta u} \cdot \frac{\Delta u}{\Delta x}$$

Evaluating the limit of both sides gives

$$\lim_{\Delta x \to 0} \frac{\Delta y}{\Delta x} = \lim_{\Delta x \to 0} \left[\frac{\Delta y}{\Delta u} \cdot \frac{\Delta u}{\Delta x} \right]$$

Since u is a function of x, $\Delta u \to 0$ as $\Delta x \to 0$

$$\lim_{\Delta x \to 0} \frac{\Delta y}{\Delta x} = \lim_{\Delta x \to 0} \frac{\Delta y}{\Delta u} \cdot \lim_{\Delta x \to 0} \frac{\Delta u}{\Delta x} \qquad \longleftarrow \qquad \text{the limit of the product is the product of the limits}$$

If the limits exist, the result is the chain rule:

$$\frac{dy}{dx} = \frac{dy}{du} \cdot \frac{du}{dx}$$

A particular consequence of the chain rule is a more general form of the power rule. It now can be applied to powers of functions:

if
$$y = u^n \quad \text{where } u = g(x)$$

then

$$\frac{dy}{dx} = \frac{du^n}{du} \cdot \frac{du}{dx}$$

or

$$\boxed{\frac{d}{dx} u^n = nu^{n-1} \cdot \frac{du}{dx}}$$

The chain rule is a very powerful tool. As you will see, it arises frequently in the differentiation of functions of all sorts, including trigonometric functions, logarithmic functions, and exponential functions.

EXAMPLE A

Differentiate:

$$y = (2 - x^3)^{50}$$

SOLUTION

To start with, write

$$y = u^{50}$$

where $u = 2 - x^3$

The chain rule is applied in two steps. First differentiate the power u^{50} with respect to u:

$$\frac{dy}{dx} = \frac{d}{dx}(u^{50})$$

$$= \frac{d}{du}(u^{50})\frac{du}{dx}$$

$$= 50u^{49}\frac{du}{dx}$$

The variable u is introduced only to facilitate the application of the chain rule. Now that it has served its purpose, it can be replaced by $(2 - x^3)$.

$$\frac{dy}{dx} = 50(2 - x^3)^{49}\frac{d}{dx}(2 - x^3)$$

The second step is to differentiate u, that is, $(2 - x^3)$ with respect to x:

$$\frac{dy}{dx} = 50(2 - x^3)^{49}(-3x^2)$$

$$= -150x^2(2 - x^3)^{49} \qquad \square$$

EXAMPLE B

Find the derivative of:

(a) $y = (1 - x^2)^6$ and   (b) $y = (x^3 + 5)^6$

SOLUTION

In both cases, the function can be written

$$y = u^6$$

where in (a) $u = 1 - x^2$ but in   (b) $u = x^3 + 5$

Using the chain rule in either case gives

$$\frac{dy}{dx} = 6u^5 \cdot \frac{du}{dx}$$

Observe that this first step does not depend at all on the form of the function u; however, the next step does:

(a)
replace u by $(1 - x^2)$:

$$\frac{dy}{dx} = 6(1 - x^2)^5 \cdot \frac{d}{dx}(1 - x^2)$$

$$= 6(1 - x^2)^5 \cdot (-2x)$$

$$= -12x(1 - x^2)^5$$

(b)
replace u by $(x^3 + 5)$:

$$\frac{dy}{dx} = 6(x^3 + 5)^5 \cdot \frac{d}{dx}(x^3 + 5)$$

$$= 6(x^3 + 5)^5 \cdot (3x^2)$$

$$= 18x^2(x^3 + 5)^5 \qquad \square$$

With practice, you will be able to work out the derivatives using the chain rule without explicitly introducing the variable u, as in the following example.

EXAMPLE C

Differentiate:

$$y = \sqrt{6x - 8x^2}$$

SOLUTION

First write the radical as a power:

$$y = (6x - 8x^2)^{1/2}$$

In applying the chain rule, view this function as $y = u^{1/2}$, and differentiate first with respect to u using the power rule:

$$\frac{dy}{dx} = \frac{d}{dx}(6x - 8x^2)^{1/2}$$

$$= \frac{1}{2}(6x - 8x^2)^{-1/2} \cdot \frac{d}{dx}(6x - 8x^2)$$

Now carry out the differentiation of u with respect to x and simplify:

$$\frac{dy}{dx} = \frac{1}{2}(6x - 8x^2)^{-1/2} \cdot (6 - 16x)$$

$$= \frac{3 - 8x}{\sqrt{6x - 8x^2}}$$

❑

EXAMPLE D

Differentiate:

$$f(x) = \frac{5}{2 + \sqrt{3x^2 - 4}}$$

SOLUTION

When it is necessary to apply the chain rule more than once to find a derivative, as in this example, the best method is to carry out each differentiation in a separate step.

$$\frac{d}{dx}f(x) = \frac{d}{dx}\left(\frac{5}{2 + \sqrt{3x^2 - 4}}\right)$$

$$= \frac{d}{dx}5(2 + \sqrt{3x^2 - 4})^{-1}$$

$$= -5(2 + \sqrt{3x^2 - 4})^{-2} \cdot \frac{d}{dx}(2 + \sqrt{3x^2 - 4})$$

$$= -5(2 + \sqrt{3x^2 - 4})^{-2} \cdot \frac{1}{2}(3x^2 - 4)^{-1/2} \cdot \frac{d}{dx}(3x^2 - 4)$$

$$= -5(2 + \sqrt{3x^2 - 4})^{-2} \cdot \frac{1}{2}(3x^2 - 4)^{-1/2} \cdot (6x)$$

$$= \frac{-15x}{(2 + \sqrt{3x^2 - 4})^2 \sqrt{3x^2 - 4}}$$

❑

Exercise 3.4

1. Determine the value of k:

(a) $\dfrac{d}{dx}(7x^2 - 1)^5 = kx(7x^2 - 1)^4$

(b) $\dfrac{d}{dx}(8 - 2x^3)^2 = kx^2(8 - 2x^3)$

(c) $\dfrac{d}{dx}(5x^4 - 2x - 1)^k = \dfrac{10x^3 - 1}{(5x^4 - 2x - 1)^{1/2}}$

2. Find the derivative:

(a) $y = (3 + x^3)^5$

(b) $y = \left(1 - \dfrac{x}{6}\right)^7$

(c) $y = \dfrac{1}{4}(3x^2 - 2x + 8)^{24}$

(d) $y = (2 - 3x + 9x^2)^{-3}$

(e) $y = \dfrac{1}{(3 - 4x)^2}$

(f) $y = \dfrac{1}{4x - x^2}$

(g) $y = \sqrt{2 - 3x}$

(h) $y = 6\sqrt{14x^2 - 12x + 9}$

(i) $y = \sqrt[3]{x^2 - m^2}$, m is a constant.

3. Differentiate the following functions:

(a) $f(x) = (ax^2 + bx + c)^m$

(b) $f(x) = \dfrac{4}{\sqrt{7x^2 - 4x + 1}}$

(c) $f(x) = (6 - 0.5x^4)^{1.5}$

(d) $f(x) = \dfrac{-2}{\sqrt[4]{1 - 9x - 8x^2}}$

4. Differentiate with respect to the independent variable:

(a) $g(x) = (2x^{-2} - 3x^{-1})^{-2}$

(b) $h(u) = \dfrac{1}{\sqrt{9 - u^2} - u}$

(c) $f(t) = \sqrt{\sqrt{t^6 - 1} - (1/t^2)}$

(d) $g(s) = \sqrt{s - \sqrt{s + \sqrt{s}}}$

Computer Application: Newton's Method of Finding Roots   3.5

Consider the function $y = f(x)$ shown in Figure 3.21. The x-coordinate of the point R where the function crosses the x-axis is a root of the equation $f(x) = 0$. Its exact value is found algebraically by setting y equal to zero and solving the equation $f(x) = 0$, a task which may not always be possible. Newton's method, on the other hand, is an approximation method for finding the root, which works even when the algebraic problem is not solvable. It consists of finding a sequence of tangents to the curve whose x-intercepts approach the root.

Figure 3.21

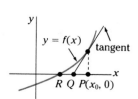

To develop the equations needed for Newton's method, let P be a point at $x = x_0$ on the axis near R. The slope m of the tangent to the function through P has the value $f'(x_0)$. The equation of the tangent at (x_0, y_0) is

$$y - y_0 = m(x - x_0)$$

or

$$y = mx + (y_0 - mx_0)$$

Point Q is the x-intercept of the tangent. The x-coordinate of Q is found by setting $y = 0$ and solving for x:

$$0 = mx + (y_0 - mx_0)$$

$$x = x_0 - y_0/m$$

It is evident from the graph that Q is closer to the root than P and thus is a better approximation to the root. Now the process can be repeated, using Q as the starting point. The iterative character of this solution makes it an impractical process to carry out by hand; it is, however, a process ideally suited to a computer.

The algorithm for finding roots by Newton's method thus consists of the following steps:

1. Choose a starting value of x. Call it x_0.

2. Calculate the value of f at x_0. Call it y_0.

3. If y_0 is "sufficiently close" to 0, then quit.

4. Otherwise calculate:
   the slope of the tangent: $m = f'(x_0)$
   the intercept of the tangent: $x = x_0 - y_0/m$.
   Repeat using the intercept as the new starting value of x.

A BASIC program for Newton's method is given in Figure 3.22. Before running it, you must type in the function of interest and its derivative in lines 110 and 120, respectively, using correct BASIC syntax. The number of significant figures in the answer can be adjusted in line 230. The program gives 2.61803399 and 0.381966011 as the roots of the equation $x^2 - 3x + 1 = 0$. You can double check these results using the quadratic formula.

When a function has more than one root, the root to which the calculation converges depends on the initial value chosen for x_0. Furthermore, in certain circumstances, the intercepts of the tangents may recede from rather than approach the root. There are also other situations in which the calculation does not converge. Consequently,

```
10      REM ROOTS BY NEWTON'S METHOD
100     :
110     DEF FNF(X)=X^2−3*X+1   : REM THE FUNCTION
120     DEF FND(X)=2*X−3       : REM ITS DERIVATIVE
130     INPUT "INITIAL VALUE OF X = ";X0
200     :
210     REM START LOOP
220     Y0 = FNF(X0)
230     IF ABS(Y0)<.000001 GOTO 300 QUIT
240     M = FND(X0)
250     X0 = X0 − Y0/M
260     GOTO 200 REPEAT
300     :
310     PRINT "THE ROOT IS AT X = ";X0
320     END
```

Figure 3.22

you should have at the outset a rough idea of the shape of the graph and the expected position of the roots. As a general rule, any computed results should be viewed with considerable skepticism until thoroughly checked.

Exercise 3.5

1. Find an approximation for $\sqrt{2}$ by carrying out three iterations (that is, three repetitions of the algorithm for Newton's method) by hand on the function $f(x) = x^2 − 2$. Start at $x = 1$.

2. Find an approximation for the cube root of 20. Use the function $f(x) = x^3 − 20$ and carry out two iterations by hand, starting at $x = 3$.

3. Using the program in Figure 3.22, find the roots of the following quadratic equations. Check using the quadratic formula.

   (a) $x^2 − 10x + 17 = 0$

   (b) $5 − x − 4x^2 = 0$

   (c) $16x^2 − 18x + 63 = 0$

4. Find all the real roots of

   (a) $x^3 − 2x^2 + 5x + 3 = 0$

   (b) $5x^3 + 9x^2 − 24x − 44 = 0$

   (c) $x^4 + 3x^3 + 15x^2 − 19x − 20 = 0$

   (d) $3x − \dfrac{4}{x^2} = 0$

3.6 Summary and Review

For a continuous function, $y = f(x)$, the quantity

$$\frac{f(x + \Delta x) - f(x)}{\Delta x}$$

is the slope of the secant through the points $(x, f(x))$ and $(x + \Delta x, f(x + \Delta x))$. The limit of this quantity as $\Delta x \to 0$ is the slope of the tangent line at the point $(x, f(x))$, which by definition is the derivative of the function f:

$$\frac{d}{dx} f(x) = \lim_{\Delta x \to 0} \frac{f(x + \Delta x) - f(x)}{\Delta x}$$

From this basic definition, it is possible to work out the derivative of any function.

The following differentiation rules derived from first principles facilitate the process of finding a derivative:

The derivative of a constant:

$$\frac{d}{dx} C = 0$$

The derivative of a power of x:

$$\frac{d}{dx} x^n = n x^{n-1}$$

The derivative of a constant times a function:

$$\frac{d}{dx} C f(x) = C \frac{d}{dx} f(x)$$

The derivative of a sum (or difference) of functions:

$$\frac{d}{dx} [f(x) + g(x)] = \frac{d}{dx} f(x) + \frac{d}{dx} g(x)$$

The chain rule:

$$\frac{dy}{dx} = \frac{dy}{du} \cdot \frac{du}{dx}$$

The derivative of a power of a function:

$$\frac{d}{dx}u^n = nu^{n-1} \cdot \frac{du}{dx}$$

In this chapter, these rules have been applied to algebraic functions, but they are valid for functions of all types.

In order to master this material, you should

1. become thoroughly familiar with the use of the differentiation rules by working a variety of problems;

2. draw graphs of functions and their tangents, and look for graphical interpretations of the algebraic results;

3. get used to simplifying algebraic expressions both before and after differentiation, with the object of finding the most direct route to the solution of a problem.

Exercise 3.6

1. (a) Given $f(x) = 2x^2 - 3x + 1$, find

$$f(3), \ f(3 + \Delta x), \ \frac{f(3 + \Delta x) - f(3)}{\Delta x}$$

(b) Given $g(x) = \sqrt{3 - 2x}$, find

$$g(1/2), \ g(1/2 + \Delta x),$$

$$\frac{g(1/2 + \Delta x) - g(1/2)}{\Delta x}$$

(c) Given $h(x) = \dfrac{3 - x}{3 + x}$, find

$$h(0), \ h(0 + \Delta x), \ \frac{h(0 + \Delta x) - h(0)}{\Delta x}$$

2. Sketch the graph of

$$f(x) = \begin{cases} x^2 - 9, & x \le 4 \\ 8x - 25, & x > 4 \end{cases}$$

Does the tangent exist at $x = 4$? Explain.

3. Starting from first principles, find the derivative of

(a) $y = 7 - 4x$

(b) $y = 3x^2 - 1$

(c) $y = \sqrt{2x - 1}$

(d) $y = \dfrac{1}{3x + 1}$

(e) $y = 3x - \dfrac{4}{x^2}$

(f) $y = \dfrac{x^2}{x - 4}$

4. Find the derivative:

(a) $y = 3x^2 - 4x + 7$

(b) $y = 8x^3 + 12x^5$

(c) $y = (4x - 1)^3$

(d) $y = \dfrac{2x^3 + 5x - 7}{2x}$

(e) $y = x^{4/3} - \dfrac{1}{x^{2/3}} + 3$

(f) $y = \dfrac{12}{5x\sqrt{x}}$

5. Find $f'(x)$ with the help of the chain rule:

(a) $f(x) = \sqrt{x^2 - 4x}$

(b) $f(x) = \dfrac{4}{\sqrt{1 - 3x}}$

(c) $f(x) = 6x\sqrt{x - 5}$

(d) $f(x) = \dfrac{x^3}{2}\sqrt{2 - x}$

(e) $f(x) = \dfrac{1}{2 - \sqrt{2 - x}}$

(f) $f(x) = \left(\dfrac{x}{3} - \dfrac{x^2}{4} + \dfrac{x^3}{6}\right)^{-2}$

(g) $f(x) = \left(3x - \dfrac{1}{3x}\right)^6$

(h) $f(x) = \left(\dfrac{1}{x - 2} + \dfrac{1}{x + 2}\right)^8$

(i) $f(x) = \sqrt{\dfrac{1 + x}{x^3}}$

6. Differentiate:

$$f(x) = \dfrac{2x^2 - 2x + 7}{x - 3}$$

using long division first to write the function in the form

$$f(x) = Ax + B + \dfrac{C}{x - 3}$$

7. Differentiate, following the method of problem 6:

(a) $f(x) = \dfrac{x^2 - 14x - 7}{x + 5}$

(b) $f(x) = \dfrac{x^3 - 6x}{x - 2}$

(c) $f(x) = \dfrac{4x + 6}{2x - 8}$

8. Find the equation of the tangent line at the point $(a, f(a))$. Sketch the graph and the tangent line.

(a) $f(x) = -3x^2 + 12, \quad a = -2$

(b) $f(x) = 2x - \dfrac{1}{x}, \quad a = 2$

(c) $f(x) = -\sqrt{2x - 5}, \quad a = 7$

(d) $f(x) = \dfrac{4}{1 + x^2}, \quad a = -1$

9. If $f(x) = \sqrt{x + 3}$, find the slope of the tangent at $x = a$. For what values of a does the tangent exist?

10. Find the equations of the tangent and the normal to the graph of $y = \sqrt{1 - 4x}$ at $(-2, 3)$.

11. Find the equations of the tangents to the curve $y = -(5 - 2x)^2$ at the points of intersection with the x-axis.

12. Find two points on the graph of $y = 2 - 4x^3$ at which the slope of the normal is 3.

13. Find the equations of the tangents to the graph of $y = x^3/3 - 3x$ that are parallel to the line $6x - y + 2 = 0$. Sketch the graph and the tangent lines.

14. Find the equations of the tangents to the graph of $xy = 9$ that are perpendicular to the line $y - 9x = 0$. Sketch the graph of the function, the given line, and the tangent lines.

15. Find the equations of the tangents to the graph of $y = x^3 - 4x$ at the points of intersection of the graph and the x-axis. Sketch the graph and the tangent lines.

16. The tangent to the graph of $y = 2x^2 - x$ intersects the x-axis at an angle of $45°$. Find the x-intercept of the tangent.

17. The line $2x + 3y = 1$ is tangent to the graph of $y = 2/x + k$. Find the value of k and the coordinates of the point of tangency.

18. Show that the graphs of $y = 2x^3 + 1$ and $y = 3x^2$ have a common tangent at the point $(1, 3)$. Sketch the graphs and their common tangent.

19. Find the equations of the lines through the origin that are tangent to the graph of $y = (x - 1)^3$.

20. Show that the line tangent to the graph of $f(x) = -x^4 + 2x^2 + x$ at $x = 1$ is also tangent to the graph at a second point. Find the two points of tangency and the equation of the tangent line.

Differentiation Rules and Methods

4

This chapter continues the work begun in the previous chapter on the differentiation of algebraic functions. Of principal interest are the rules for differentiating products and quotients of functions, and how these rules are used in conjunction with the power rule and the chain rule.

The Product Rule 4.1

When a function f consists of the product of two other functions g and h, it may be complicated or even impossible to find its derivative using the differentiation rules of the preceding chapter. What is required is a way to express the derivative of the product $g(x) \cdot h(x)$ in terms of the separate derivatives of the two functions g and h.

The derivative of the product of two functions g and h is

$$\frac{d}{dx}(g \cdot h) = g \cdot \frac{dh}{dx} + h \cdot \frac{dg}{dx}$$

This formula, known as the **product rule**, is derived from first principles as follows:

$$f(x) = g(x) \cdot h(x)$$

$$\frac{d}{dx}f(x) = \frac{d}{dx}g(x) \cdot h(x)$$

$$= \lim_{\Delta x \to 0} \frac{g(x + \Delta x) \cdot h(x + \Delta x) - g(x) \cdot h(x)}{\Delta x}$$

By adding and subtracting the quantity $g(x) \cdot h(x + \Delta x)$ in the numerator, it is possible to rewrite this expression as follows:

$$\frac{d}{dx} f(x) = \lim_{\Delta x \to 0} \frac{g(x + \Delta x) \cdot h(x + \Delta x) - g(x) \cdot h(x + \Delta x) + g(x) \cdot h(x + \Delta x) - g(x) \cdot h(x)}{\Delta x} \quad \longleftarrow$$

common
factoring

$$= \lim_{\Delta x \to 0} \frac{[g(x + \Delta x) - g(x)] \cdot h(x + \Delta x) + g(x) \cdot [h(x + \Delta x) - h(x)]}{\Delta x}$$

$$= \lim_{\Delta x \to 0} \left\{ \frac{[g(x + \Delta x) - g(x)]}{\Delta x} \cdot h(x + \Delta x) + g(x) \cdot \frac{[h(x + \Delta x) - h(x)]}{\Delta x} \right\}$$

$$= \left(\frac{dg(x)}{dx} \right) \cdot h(x) + g(x) \cdot \left(\frac{dh(x)}{dx} \right)$$

In the last step, as $\Delta x \to 0$, $h(x + \Delta x) \to h(x)$ and the two fractions become the derivatives of g and h, respectively. The result is the product rule:

$$\frac{d}{dx}(g \cdot h) = g \cdot \frac{dh}{dx} + h \cdot \frac{dg}{dx}$$

EXAMPLE A

Find $\dfrac{dy}{dx}$, where $y = (x - 6)(3x + 2)$

SOLUTION

y is the product of two functions g and h, where

$$g(x) = (x - 6) \qquad \text{and} \qquad h(x) = (x + 2)$$

According to the product rule,

$$\frac{d}{dx}(g \cdot h) = g \frac{dh}{dx} + h \frac{dg}{dx}$$

Therefore,

$$\frac{dy}{dx} = \frac{d}{dx}(x - 6) \cdot (3x + 2)$$

$$= (x - 6) \cdot \frac{d}{dx}(3x + 2) + (3x + 2) \cdot \frac{d}{dx}(x - 6)$$

$$= (x - 6) \cdot 3 + (3x + 2) \cdot 1$$

$$= 6x - 16 \qquad \qquad \square$$

Instead of using the product rule to find the derivative in this example, you could simply multiply out the two factors, and then differentiate. In a given problem, you must decide which approach is the most appropriate.

EXAMPLE B

Differentiate:

$$y = (7x - 2)^4(2x + 3)^5$$

SOLUTION

In this problem, it is inadvisable to try to multiply out the factors. Differentiating immediately,

$$\frac{dy}{dx} = \frac{d}{dx}(7x - 2)^4(2x + 3)^5 \qquad \longleftarrow \text{ first use the product rule}$$

$$= (7x - 2)^4 \frac{d}{dx}(2x + 3)^5 + (2x + 3)^5 \frac{d}{dx}(7x - 2)^4 \qquad \text{then use the chain rule}$$

$$= (7x + 2)^4 \cdot 5(2x + 3)^4 \cdot 2 + (2x + 3)^5 \cdot 4(7x - 2)^3 \cdot 7$$

$$= 10(7x - 2)^4(2x + 3)^4 + 28(7x - 2)^3(2x + 3)^5$$

The differentiation is now complete. It is possible to simplify the expression for the derivative by extracting common factors:

$$y' = (7x - 2)^3(2x + 3)^4 \cdot [10(7x - 2) + 28(2x + 3)]$$

$$= (7x - 2)^3(2x + 3)^4(126x + 64)$$

It is not always necessary to simplify the derivative in this manner. However, in certain applications, for instance in solving $dy/dx = 0$, it is essential. ◻

EXAMPLE C

Find $\dfrac{dy}{dx}$, where $y = 3x^4\sqrt{x + 5}$

SOLUTION

First, write the radical as a power:

$$y = 3x^4 \cdot (x + 5)^{1/2}$$

According to the product rule,

$$\frac{dy}{dx} = \frac{d}{dx} 3x^4 \cdot (x + 5)^{1/2}$$

$$= 3x^4 \frac{d}{dx}(x + 5)^{1/2} + (x + 5)^{1/2} \frac{d}{dx} 3x^4$$

$$= 3x^4 \cdot \frac{1}{2}(x + 5)^{-1/2} + (x + 5)^{1/2} \cdot 12x^3$$

$$= \frac{3x^4}{2\sqrt{x + 5}} + 12x^3\sqrt{x + 5}$$

As in Example B, it is possible to simplify the expression for the derivative, if desired. If you multiply and divide the second term by $2\sqrt{x + 5}$, the two terms will have a common denominator:

$$y' = \frac{3x^4}{2\sqrt{x + 5}} + \frac{24x^3(x + 5)}{2\sqrt{x + 5}}$$

$$y' = \frac{3x^3(9x + 40)}{2\sqrt{x + 5}}$$ ❏

Exercise 4.1

1. Find $\dfrac{dy}{dx}$:

(a) $y = (x - 3)(2x + 1)$

(b) $y = (6x^7 - 4x^5)(1 - x^2)$

(c) $y = (3x^2 - 1)(x^2 - 2x + 2)$

(d) $y = (4 + 5x)^2(8 - x)^6$

(e) $y = (x^3 - 5x^2)^4(2 - 3x^2)^2$

(f) $y = 6x^8(3x^2 - 4x + 9)^3$

2. Differentiate with respect to the independent variable:

(a) $f(t) = 5t(t^2 - 4)^3$

(b) $f(r) = (r + 5)^2(r - 5)^7$

(c) $f(z) = (z^2 - 3z + 6)^2(z^2 + 4z - 2)^2$

3. Differentiate with respect to x:

(a) $f(x) = (x - 2)\sqrt{1 - x^2}$

(b) $f(x) = 4\sqrt{x}(9 - 4x)^2$

(c) $f(x) = \sqrt{1 - x^2}\sqrt{1 + x^2}$

(d) $f(x) = \sqrt{(5 - 2x)(3x - 1)^3}$

(e) $f(x) = (x + 1)^2\sqrt{x^2 + 1}$

(f) $f(x) = (x^2 + 1)^2\sqrt{x + 1}$

4. Differentiate with respect to the independent variable:

(a) $x = (3t - 7)\sqrt{t + 2}$

(b) $v = 4r^{3/2}(r^2 + 8)^3$

(c) $y = \sqrt{3 - 2z}\sqrt{1 + 2z}$

5. Differentiate:

(a) $y = x^2(x - 1)(x + 2)$

(b) $y = x^2(1 - x^2)(1 - x^4)$

(c) $y = (x - 2)(x - 1)(x + 1)(x + 2)$

6. Differentiate:

(a) $f(x) = 5x^{10}(6x^2 - 1)^{2/3}$

(b) $f(x) = (x - 1)(2 - 9x)^{1/3}$

(c) $f(x) = (3x + 6)^{-3}(2x + 8)^{-2}$

7. Express as a product and differentiate:

(a) $y = \dfrac{(x - 3)^2}{(x + 9)}$

(b) $y = \dfrac{2x}{\sqrt{2 - x}}$

8. Find the equations of the tangent lines at the points where the graph of $y = (x - 2)(3 + x)$ intersects the x-axis. Sketch the graph and the tangent lines.

9. Find to the nearest tenth the x-coordinates of the points on the graph of $f(x) = (4 - x^2)(x + 5)$ at which the slope of the tangent is equal to zero. Sketch the graph and the tangents at these points.

10. Find the value(s) of a, if the tangent to $f(x) = -(x + 2)(x + 3)$ at $(a, f(a))$ passes through the origin. Sketch the graph of f and the tangent(s) at a.

The Quotient Rule   4.2

The **quotient rule** is a general purpose differentiation rule like the product rule. It is used for differentiating the quotient of two functions g and h:

$$\boxed{\dfrac{d}{dx}\left(\dfrac{g}{h}\right) = \dfrac{h\dfrac{dg}{dx} - g\dfrac{dh}{dx}}{h^2}}$$

Although the quotient rule can be derived from first principles, it is simpler to obtain it as a consequence of the product rule and the chain rule:

$$f(x) = \dfrac{g(x)}{h(x)}$$

$$\dfrac{d}{dx}f(x) = \dfrac{d}{dx}\left(\dfrac{g(x)}{h(x)}\right)$$

$$= \dfrac{d}{dx}g(x) \cdot [h(x)]^{-1} \qquad\longleftarrow \text{first use the product rule}$$

$$= g(x) \cdot \dfrac{d}{dx}[h(x)]^{-1} + [h(x)]^{-1} \cdot \dfrac{d}{dx}g(x) \qquad\longleftarrow \text{then use the chain rule}$$

$$= g(x) \cdot (-1)[h(x)]^{-2} \cdot \frac{dh}{dx} + [h(x)]^{-1} \cdot \frac{dg}{dx}$$

$$= \frac{1}{h} \cdot \frac{dg}{dx} - \frac{g}{h^2} \cdot \frac{dh}{dx}$$

$$= \frac{h \dfrac{dg}{dx} - g \dfrac{dh}{dx}}{h^2}$$

EXAMPLE A

Differentiate:

$$y = \frac{2x}{3 - x^2}$$

SOLUTION

y is the quotient of two functions g and h, where

$$g(x) = 2x \qquad \text{and} \qquad h(x) = (3 - x^2)$$

According to the quotient rule,

$$\frac{d}{dx}\left(\frac{g}{h}\right) = \frac{h \cdot \dfrac{dg}{dx} - g \cdot \dfrac{dh}{dx}}{h^2}$$

Therefore,

$$\frac{dy}{dx} = \frac{d}{dx}\left(\frac{2x}{3 - x^2}\right)$$

$$= \frac{(3 - x^2)\dfrac{d}{dx}(2x) - (2x)\dfrac{d}{dx}(3 - x^2)}{(3 - x^2)^2}$$

$$= \frac{(3 - x^2) \cdot 2 - (2x) \cdot (-2x)}{(3 - x^2)^2}$$

$$= \frac{6 + 2x^2}{(3 - x^2)^2} \qquad \qquad \square$$

Instead of using the quotient rule, you can find the derivative of the quotient by writing the function first as $y = (2x)(3 - x^2)^{-1}$ and then using the product rule. Try this approach in some of the exercises. In some cases you may find it to be simpler than the direct use of the quotient rule.

EXAMPLE B

Find the derivative of y with respect to x:

$$y = \left(\frac{3x + 1}{2x - 1}\right)^4$$

SOLUTION

Apply the chain rule first, then the quotient rule:

$$\frac{dy}{dx} = \frac{d}{dx}\left(\frac{3x + 1}{2x - 1}\right)^4$$

$$= 4\left(\frac{3x + 1}{2x - 1}\right)^3 \frac{d}{dx}\left(\frac{3x + 1}{2x - 1}\right)$$

$$= 4\left(\frac{3x + 1}{2x - 1}\right)^3 \cdot \frac{(2x - 1)\dfrac{d}{dx}(3x + 1) - (3x + 1)\dfrac{d}{dx}(2x - 1)}{(2x - 1)^2}$$

$$= 4\left(\frac{3x + 1}{2x - 1}\right)^3 \cdot \frac{(2x - 1)\cdot 3 - (3x + 1)\cdot 2}{(2x - 1)^2}$$

$$= 4\left(\frac{3x + 1}{2x - 1}\right)^3 \cdot \frac{6x - 3 - 6x - 2}{(2x - 1)^2}$$

$$= \frac{-20(3x + 1)^3}{(2x - 1)^5}$$

EXAMPLE C

Differentiate:

$$y = \frac{5x + 1}{\sqrt{3 - 4x}}$$

SOLUTION

Direct application of the quotient rule gives

$$\frac{dy}{dx} = \frac{d}{dx}\frac{5x + 1}{\sqrt{3 - 4x}}$$

$$= \frac{d}{dx}\frac{5x + 1}{(3 - 4x)^{1/2}}$$

$$= \frac{(3 - 4x)^{1/2}\frac{d}{dx}(5x + 1) - (5x + 1)\frac{d}{dx}(3 - 4x)^{1/2}}{[(3 - 4x)^{1/2}]^2}$$

$$= \frac{(3 - 4x)^{1/2} \cdot 5 - (5x + 1) \cdot \left(\frac{1}{2}\right)(3 - 4x)^{-1/2} \cdot (-4)}{(3 - 4x)}$$

$$= \frac{5(3 - 4x)^{1/2} + 2(5x + 1)(3 - 4x)^{-1/2}}{(3 - 4x)}$$

The differentiation of the function is now completed. To simplify this expression further, remove a common factor of $(3 - 4x)^{-1/2}$ from the numerator:

$$\frac{dy}{dx} = \frac{(3 - 4x)^{-1/2}[5(3 - 4x) + 2(5x + 1)]}{(3 - 4x)}$$

$$= \frac{15 - 20x + 10x + 2}{(3 - 4x)^{3/2}}$$

$$= \frac{17 - 10x}{(3 - 4x)^{3/2}} \qquad \square$$

Exercise 4.2

1. Differentiate each of the following functions:

(a) $y = \dfrac{3x^2}{4 - x}$

(b) $y = \dfrac{5x^4}{x^2 - 10}$

(c) $y = \dfrac{x - 1}{x + 1}$

(d) $y = \dfrac{2x^2 - 1}{x + 2}$

(e) $y = \dfrac{4x^3 - 3x}{2x^2}$

(f) $y = \dfrac{4 - (1/x)}{x - 4}$

2. Find f':

(a) $f(x) = \dfrac{x}{(x^2 + 1)^2}$

(b) $f(t) = \dfrac{3t^2 + 1}{(1 - t)^3}$

(c) $f(u) = \dfrac{u^3 - 2u}{u^2 + u + 1}$

(d) $f(v) = \dfrac{v^2 + 5v - 6}{(v^2 - 4)}$

(e) $f(r) = \dfrac{(r + 2)^2}{(r^2 - 16)^2}$

(f) $f(z) = \dfrac{(z + 3)(z - 3)}{3z(z + 2)}$

3. Find $\dfrac{dy}{dx}$:

(a) $y = \dfrac{x + 5}{\sqrt{2 - x}}$

(b) $y = \dfrac{3x^3}{4\sqrt{2x - 1}}$

(c) $y = \dfrac{\sqrt{1 + x^2}}{\sqrt{1 - x^2}}$

(d) $y = \dfrac{\sqrt[3]{x}}{\sqrt[3]{x - 1}}$

4. Find $\dfrac{dy}{dx}$. a, b, c, and d are constants.

(a) $y = \dfrac{ax + b}{cx + d}$

(b) $y = \dfrac{\sqrt{a} + \sqrt{x}}{\sqrt{a} - \sqrt{x}}$

(c) $y = \dfrac{ax^2 + bx + c}{ax + b}$

(d) $y = \dfrac{x}{\sqrt{a^2 - x^2}}$

5. Differentiate:

(a) $y = \dfrac{(x + 1)(x - 2)}{(x + 2)}$

(b) $y = \dfrac{(x - 5)}{(2x + 1)(2x - 3)}$

6. Find the equations of the tangents to the graph of

$$y = \dfrac{x - 3}{x + 5}$$

at the points where it intersects the coordinate axes. Sketch the graph and the tangent lines.

7. Find the equations of the tangents from the origin to the graph of

$$y = \dfrac{x + 9}{x + 8}$$

Sketch the graph and the tangent lines.

8. Let f be the quotient of two functions g and h

$$f(x) = \dfrac{g(x)}{h(x)}$$

From first principles, derive the quotient rule

$$\frac{d}{dx}\left(\frac{g}{h}\right) = \frac{h \cdot \dfrac{dg}{dx} - g\dfrac{dh}{dx}}{h^2}$$

Implicit Differentiation   4.3

In the preceding sections, the relationship between the variables x and y has always been written in the form $y = f(x)$. Consider, however, the equation $xy = y + 5$, in which the variables are "mixed up." In a case like this, you may still consider y to be a function of x, but since it is not written as $y = f(x)$, the exact form of the function is not immediately evident. y, therefore, is said to be an **implicit function** of x.

One way to differentiate such a function is to solve first for y, expressing it explicitly as a function of x:

$$xy = y + 5$$

$$xy - y = 5$$

$$y(x - 1) = 5$$

$$\therefore \qquad y = \frac{5}{x - 1}$$

Then

$$y' = -\frac{5}{(x - 1)^2}$$

But how can one find the derivative dy/dx when it is inconvenient or impossible to solve for y in terms of x?

The technique of finding dy/dx without first solving for y is called **implicit differentiation**. It depends upon treating y as an unspecified, but differentiable function of x, and upon a careful application of the product rule, the quotient rule, and/or the chain rule.

EXAMPLE A

Find dy/dx if $xy = y + 5$.

SOLUTION

Differentiate both sides of the equation with respect to x:

$$xy = y + 5$$

$$\frac{d}{dx}(xy) = \frac{d}{dx}(y + 5)$$

In differentiating, keep in mind that y is a function of x. This means in particular that xy is the product of two functions and you must use the product rule to differentiate it. It follows that

$$x\frac{d}{dx}y + y\frac{d}{dx}x = \frac{d}{dx}y$$

or

$$xy' + y = y'$$

Now solve for y':

$$xy' - y' = -y$$

$$y'(x - 1) = -y$$

$$y' = -\frac{y}{x - 1}$$

The final expression for y' is different from what you are accustomed to seeing, in that it contains the variable y as well as x. Since it is possible in this particular example to solve the original equation for y, you can replace y, obtaining

$$y' = \frac{y}{x - 1}$$

$$= -\frac{\dfrac{5}{x - 1}}{x - 1}$$

$$= -\frac{5}{(x - 1)^2}$$ ❑

EXAMPLE B

Find $\dfrac{dy}{dx}$, where $3x^2y - xy^2 = 6$

SOLUTION

$$\frac{d}{dx}\left(3x^2y - xy^2\right) = \frac{d}{dx}6$$

$$\frac{d}{dx}\left(3x^2y\right) - \frac{d}{dx}\left(xy^2\right) = 0 \quad \longleftarrow \quad \text{use the product rule}$$

$$3\left(x^2\frac{d}{dx}y + y\frac{d}{dx}x^2\right) - \left(x\frac{d}{dx}y^2 + y^2\frac{d}{dx}x\right) = 0 \quad \longleftarrow \quad \text{use the chain rule here}$$

$$3(x^2y' + 2xy) - (x \cdot 2yy' + y^2) = 0 \quad \longleftarrow \quad \text{solve for } y'$$

$$3x^2y' + 6xy - 2xyy' - y^2 = 0$$

$$(3x^2 - 2xy)y' = y^2 - 6xy$$

$$y' = \frac{y^2 - 6xy}{3x^2 - 2xy}$$ ❑

EXAMPLE C

Find the slope of the tangent to the curve

$$\sqrt{y} = \sqrt{xy} - 2\sqrt{x}$$

at $x = 4$.

SOLUTION

In this example, it is helpful to use a form of the power rule specifically designed for the square root function:

$$\frac{d}{dx}\sqrt{v} = \frac{1}{2\sqrt{v}}\frac{dv}{dx}$$

The differentiation now proceeds as follows:

$$\frac{d}{dx}\sqrt{y} = \frac{d}{dx}\sqrt{xy} - \frac{d}{dx}2\sqrt{x}$$

$$\frac{1}{2\sqrt{y}}y' = \frac{1}{2\sqrt{xy}}\frac{d}{dx}(xy) - \frac{2}{2\sqrt{x}}$$

$$\frac{1}{2\sqrt{y}}y' = \frac{1}{2\sqrt{xy}}(xy' + y) - \frac{2}{2\sqrt{x}} \qquad \longleftarrow \text{multiply by } 2\sqrt{xy}$$

$$y'\sqrt{x} = xy' + y - 2\sqrt{y} \qquad \longleftarrow \text{solve for } y'$$

$$(\sqrt{x} - x)y' = y - 2\sqrt{y}$$

$$y' = \frac{y - 2\sqrt{y}}{\sqrt{x} - x}$$

To find the numerical value of the slope at $x = 4$, it is necessary first to determine the y-coordinate of the point of tangency. The equation of the curve at $x = 4$ gives

$$\sqrt{y} = \sqrt{4y} - 2\sqrt{4}$$

$$\sqrt{y} = 2\sqrt{y} - 4$$

$$\sqrt{y} = 4$$

$$y = 16$$

The values of both x and y are now substituted into the expression for y':

$$y' = \frac{16 - 2\sqrt{16}}{\sqrt{4} - 4}$$

$$= \frac{16 - 2\cdot 4}{2 - 4}$$

$$= -4$$

Thus, the slope of the tangent line at $x = 4$ is -4.   □

Exercise 4.3

1. Find $\dfrac{dy}{dx}$ by implicit differentiation:

(a) $x^2y + y^2 = 8$

(b) $x - 2xy + y - 1 = 0$

(c) $x^3y^2 = 12$

(d) $6x^2 + 2xy - xy^3 = 5$

(e) $(1 - x^5)^{72} = (1 - y^{72})^5$

(f) $(2x - 3)^4 = 3y^4$

2. Find $\dfrac{dy}{dx}$, where a is a constant:

(a) $x^{1/2} + y^{1/2} = a^{1/2}$

(b) $\sqrt{x} + xy = a$

3. Find $\dfrac{dy}{dx}$ by implicit differentiation:

(a) $y^2 = \dfrac{x^2}{4 - x}$

(b) $\dfrac{1}{x} + \dfrac{1}{y} = 5$

4. Find the slope of the tangent(s) to each of the following graphs at the indicated position:

(a) $3xy - 5y = 4$ at $x = -2$

(b) $y^2 = 3x - 14$ at $x = 6$

(c) $x^2 + xy - 2y^2 + 8 = 0$ at $(0, -2)$

(d) $(x - 1)^2 + (y + 3)^2 = 17$ at $(2, 1)$

(e) $x^3 - kxy + 3ky^2 - 3k^3 = 0$ at $(k, k), k \neq 0$

5. Given $2x^3y - 5y + 4x = 7$, find y' by
(a) implicit differentiation
(b) solving for y, then differentiating y explicitly.
Show that the results of parts (a) and (b) are equivalent.

6. Find the equation of the tangent to the graph of $y^3 - 3xy - 5 = 0$ at the point $(2, -1)$.

7. Show that the normal to the graph of $x^2 - y^2 - 2y + 2 = 0$ at $(1, 1)$ is also tangent to the graph of $y = x^2 - 6x + 7$. Find the point of tangency.

8. Find the equation of the tangent to the ellipse $\dfrac{x^2}{8} + \dfrac{y^2}{18} = 1$ at the point $(2, -3)$.

9. Show that the equation of the tangent to the circle $x^2 + y^2 = r^2$ at the point (x_1, y_1) can be written in the form $x_1x + y_1y = r^2$.

10. Show that the equation of the tangent to the hyperbola

$$\frac{x^2}{a^2} - \frac{y^2}{b^2} = 1 \text{ at the point } (x_1, y_1)$$

can be written in the form

$$\frac{x_1x}{a^2} - \frac{y_1y}{b^2} = 1$$

11. Show that the equation of the tangent to the parabola $y^2 = 4px$ at (x_1, y_1) is $y_1 y = 2p(x + x_1)$.

12. There are two lines that are perpendicular to $2x + 4y - 3 = 0$ and tangent to the hyperbola $7x^2 - 2y^2 = 14$. Find their equations.

13. Show that at the points where the hyperbola $x^2 - y^2 = 5$ and the ellipse $4x^2 + 9y^2 = 72$ intersect, their tangents are perpendicular.

14. Given that the power rule,

$$\frac{dy}{dx} = ax^{a-1} \text{ where } y = x^a$$

is valid when a is an integer, show by implicit differentiation that it is also valid when a is a rational number: $a = m/n, m, n \in I, n \neq 0$.

15. By differentiating $y = f(x)$ both with respect to x and with respect to y, show that

$$\frac{dx}{dy} = \frac{1}{\dfrac{dy}{dx}}$$

4.4 Derivatives of Higher Order

If $y = f(x)$ is a differentiable function, then differentiation produces a new function $y' = f'(x)$ called the **first derivative** of y with respect to x. If $y' = f'(x)$ is in turn a differentiable function, then its derivative, dy'/dx, is called the **second derivative** of y with respect to x.

Symbols for the second derivative are

$$y'', \qquad \frac{d^2y}{dx^2}, \qquad f''(x), \qquad \frac{d^2}{dx^2}f(x)$$

In d^2y/dx^2, the 2 is *not* to be interpreted as an exponent. It stands for two successive differentiations. Likewise, the **third derivative** is the derivative of the second derivative. The third and **higher order** derivatives are denoted by

$$y''', \qquad \frac{d^3y}{dx^3}, \qquad f'''(x), \qquad \frac{d^3}{dx^3}f(x)$$

$$y^4, \qquad \frac{d^4y}{dx^4}, \qquad f^{(4)}(x), \qquad \frac{d^4}{dx^4}f(x)$$

$$y^{(n)}, \qquad \frac{d^ny}{dx^n}, \qquad f^{(n)}(x), \qquad \frac{d^n}{dx^n}f(x)$$

The second derivative, like the first, has a graphical interpretation. It tells something about the way a graph curves, or in other words, how its slope changes with x. The relationship between derivatives and graphs of functions will be thoroughly explored in the next chapter.

EXAMPLE A

Find the second derivative of y with respect to x, where $y = -7x^4 + 5x^3$.

SOLUTION

The first derivative is

$$y' = \frac{d}{dx}(-7x^4 + 5x^3)$$

$$= -28x^3 + 15x^2$$

The second derivative is the derivative of the first derivative:

$$y'' = \frac{d}{dx}(-28x^3 + 15x^2)$$

$$= -84x^2 + 30x$$

EXAMPLE B

Find the second derivative of y with respect to x, where

$$y = \frac{x^2}{x - 2}$$

SOLUTION

The first derivative is

$$y' = \frac{d}{dx}\frac{x^2}{x - 2}$$

$$= \frac{(x - 2)2x - x^2 \cdot 1}{(x - 2)^2}$$

$$= \frac{x^2 - 4x}{(x - 2)^2}$$

To find the second derivative, you must differentiate a second time using the quotient rule:

$$y'' = \frac{d}{dx}\frac{x^2 - 4x}{(x-2)^2}$$

$$= \frac{(x-2)^2 \cdot (2x-4) - (x^2 - 4x) \cdot 2(x-2)}{(x-2)^4}$$

Divide out a factor of $(x-2)$, then expand and simplify:

$$y'' = \frac{(x-2)(2x-4) - 2(x^2 - 4x)}{(x-2)^3}$$

$$= \frac{2x^2 - 4x - 4x + 8 - 2x^2 + 8x}{(x-2)^3}$$

$$= \frac{8}{(x-2)^3}$$

❏

EXAMPLE C

Find y'' if $2y - y^2 = 4x$, and evaluate y'' at $(-2, 4)$.

SOLUTION

If $2y - y^2 = 4x$ then by implicit differentiation,

$$2y' - 2yy' = 4$$

$$2y'(1 - y) = 4$$

$$y' = \frac{2}{1 - y}$$

or

$$y' = 2(1 - y)^{-1}$$

Differentiating again,

$$y'' = 2(-1)(1 - y)^{-2}(-y')$$

$$= \frac{-2}{(1 - y)^2}\left(-\frac{2}{1 - y}\right)$$

$$= \frac{4}{(1 - y)^3}$$

Thus at $(-2, 4)$,

$$y'' = \frac{4}{(1-4)^3}$$

$$= -\frac{4}{27}$$

EXAMPLE D

Find all the derivatives of

$$f(x) = 5x^3 - 4x^2 + 7x - 3$$

with respect to x.

SOLUTION

$$f(x) \;\; = 5x^3 - 4x^2 + 7x - 3$$
$$f'(x) \;\; = 15x^2 - 8x + 7$$
$$f''(x) \;\; = 30x - 8$$
$$f'''(x) \;\; = 30$$
$$f^{(4)}(x) = 0$$

It follows that derivatives of all orders higher than 4 are also zero for this function.

EXAMPLE E

Determine an expression for the nth derivative with respect to x of $y = 1/x$.

SOLUTION

$$y = \frac{1}{x}$$

or

$$y = x^{-1}$$
$$y' = (-1)x^{-2}$$
$$y'' = (-1)(-2)x^{-3}$$
$$y''' = (-1)(-2)(-3)x^{-4}$$
$$y^{(4)} = (-1)(-2)(-3)(-4)x^{-5}$$

etc.

$$y^{(n)} = (-1)(-2)(-3)\ldots(-n)x^{-n-1}$$

or

$$y^{(n)} = \frac{(-1)^n n!}{x^{n+1}}$$

Exercise 4.4

1. Find the first and second derivatives of y with respect to x:

(a) $y = x^5 - 3x^3 + 4x$

(b) $y = (2x - 3)(x + 5)$

(c) $f(x) = x^2\sqrt{x} - 7x$

(d) $f(x) = \sqrt{x} - \dfrac{1}{\sqrt{x}}$

(e) $y = \dfrac{2x^2 - 3}{5x}$

(f) $y = \left(\dfrac{1}{x^2} - x^2\right)^2$

2. Determine $f'(x)$ and $f''(x)$:

(a) $f(x) = (x + 3)^2(x + 2)^3$

(b) $f(x) = x^6(1 - x^6)$

(c) $f(x) = (5x - 4)^2(3x + 1)$

(d) $f(x) = \dfrac{x^2 + 4}{x^2 - 4}$

(e) $f(x) = \dfrac{x}{(x - 4)^3}$

(f) $f(x) = \dfrac{2x + 1}{(5x - 4)^2}$

3. Find the second derivative:

(a) $f(x) = \sqrt{4 - x^2}$

(b) $f(x) = x\sqrt{4 - x^2}$

(c) $f(x) = \dfrac{4}{\sqrt{2x^2 + 6}}$

(d) $f(x) = \dfrac{4x}{\sqrt{2x^2 + 6}}$

4. Find the second derivative of y with respect to x for the following implicit functions:

(a) $x^2 + y^2 = 25$

(b) $9x^2 + 16(y - 4)^2 = 144$

(c) $\sqrt{y} - \sqrt{x} = \sqrt{a}$

(d) $y^3 = a^2x$

5. Each of the following is the equation of a conic section. Draw the graph and find the values of y' and y'' at the given point.

(a) $x^2 + y^2 = 10x$, $(8, 4)$

(b) $x^2 + 9y^2 = 25$, $(-4, 1)$

(c) $x = 3y - \dfrac{1}{2}y^2$, $(4, 2)$

(d) $xy + 2y = 2x - 1$, $(1/2, 0)$

6. Find values of y' and y'' at the point indicated, where $\sqrt{x} + \sqrt{y} = 4$, $(1, 9)$.

7. Find f''' if $f(x) = \dfrac{3}{x - 1}$.

8. Find $\dfrac{d^2y}{dx^2}$, if $y = uv$, where u and v are functions of x.

9. Find $\dfrac{d^ny}{dx^n}$, if $y = \dfrac{1}{x^2}$.

10. Show that if $x^2 - y^2 = a^2$, (a is a constant) then

(a) $\dfrac{dy}{dx} = \dfrac{x}{y}$

(b) $\dfrac{d^2y}{dx^2} = -\dfrac{a^2}{y^3}$

(c) $\dfrac{d^3y}{dx^3} = \dfrac{3a^2x}{y^5}$

11. Show that $y = 2x^3 + 3x^2$ satisfies the equation

$$x^2y'' - 4xy' + 6y = 0$$

Summary and Review   4.5

This chapter describes how to find the derivative of products, quotients, and more complicated functions of x. You have seen that it is possible to differentiate and find dy/dx without first solving for y as a function of x, a process referred to as implicit differentiation. Lastly, the notion of repeated differentiation to obtain higher order derivatives has been introduced.

The differentiation rules are few in number and their application is straightforward. Difficulties arise more out of the algebraic manipulation required in some problems. To master this material, there is no substitute for working a great many problems of all kinds. As you work, keep the following suggestions in mind:

1. Consider rewriting an expression in a simpler way before differentiating.

2. Observe the way in which an expression is organized into factors, quotients, powers, roots, etc. Apply the differentiation rules in the correct sequence.

3. Pay close attention to details. Do not overlook the extra factors that may arise out of the chain rule. Handle parentheses carefully. Watch for errors in sign.

4. Display your calculations in a neat and organized manner. Work down the page placing each step in the algebra on a new line. Avoid trying to include too much in one step.

A disciplined approach to problem solving will reduce errors and yield clearer solutions with less work.

Exercise 4.5

1. Differentiate each of the following functions with respect to x. Simplify as far as possible.

(a) $y = 4x^2(1 - x^3)$

(b) $y = (3x + 4)(x^2 - 5)$

(c) $y = (2x - 1)(8x^2 - 3x + 1)$

(d) $y = (1 - x)(1 + x^2)(5 - x^2)$

(e) $y = (x^2 - 3x - 1)(x^3 + 4x - 3)$

2. Find dy/dx:

(a) $y = (1 - 3x)(x^2 - 2)^3$

(b) $y = (5 - 2x)^2(x - 1)^3$

(c) $y = x(1 - x^2)(2 + x)^3$

(d) $y = (3x^2 - 4x)^2(7 - \sqrt{x})$

(e) $y = x\sqrt{x - 1}$

3. Find $f'(x)$:

(a) $f(x) = x\sqrt[3]{1 - 3x}$

(b) $f(x) = (x - 1)\sqrt{x + 1}$

(c) $f(x) = (x + 1)^2\sqrt{x^2 + 1}$

(d) $f(x) = (1 - 3x)\sqrt{x^2 + 3x - 2}$

(e) $f(x) = \left(1 + \dfrac{2}{x}\right)\left(3 - \dfrac{1}{x}\right)$

(f) $f(x) = (2x)^{1/3}(x - 4)^{-2/3}$

4. Given that p and q are functions of x, find:

(a) $\dfrac{d}{dx}(px)$

(b) $\dfrac{d}{dx}(x^2q)$

(c) $\dfrac{d}{dx}(2q^3)$

(d) $\dfrac{d}{dx}\left(\dfrac{p^2}{q}\right)$

5. Differentiate:

(a) $y = \dfrac{5x}{10 - x}$

(b) $y = \dfrac{x^3 - 3x}{\sqrt{x}}$

(c) $y = \dfrac{3 - x}{6 + x^2}$

(d) $y = \dfrac{2x^3 - 7x^2}{3x^2 + 4}$

(e) $y = \dfrac{c^2 + x^2}{c^2 - x^2}$, c is a constant

(f) $y = 5x^2 + \dfrac{10}{10 - x}$

6. Differentiate:

(a) $f(x) = \dfrac{1}{\sqrt{3 - 4x^2}}$

(b) $f(x) = \dfrac{x}{\sqrt{x^2 - 1}}$

(c) $f(x) = \dfrac{\sqrt{x} - 1}{\sqrt{x} + 1}$

(d) $f(x) = \dfrac{\sqrt{3 - x^2}}{x}$

(e) $f(x) = \dfrac{a^2 + x^2}{\sqrt{a^2 - x^2}}$, a is a constant

(f) $f(x) = \dfrac{a^2b^2c^2}{(x - a)(x - b)(x - c)}$, a, b, c are constants

(g) $f(x) = \dfrac{\sqrt{9 - x^2}}{\sqrt{9 + x^2}}$

(h) $f(x) = \dfrac{\sqrt{x^2 - 4} + x}{\sqrt{x^2 - 4} - x}$

7. Find y' in each of the following cases:

(a) $x^2y + y^2 = 8$

(b) $x^3 - 6xy + 6 = 0$

(c) $x^3 + y^3 - 3axy = 0$, a is a constant

(d) $y^3 = (1 - x^2)^2$

(e) $(x - y)^2 - (x + y)^2 = x^4 + y^4$

(f) $\dfrac{x}{y^2 - 1} - \dfrac{y^2}{4} = 5$

(g) $2\sqrt{x} + 4\sqrt{y} = \sqrt{a}$, a is a constant

(h) $(6x)^{1/3} + (2y)^{1/2} = x$

(i) $\dfrac{y}{x} - \sqrt{x^2 + y^2} = 1$

8. Find the second derivative $\dfrac{d^2y}{dx^2}$:

(a) $y = 2 - x^2 + 3x$

(b) $y = (2 - 3x^2)(5 + 2x)$

(c) $y = \dfrac{3x - 1}{3x + 1}$

(d) $x^2 - 3x + y^2 = 0$

(e) $x^3 + y^3 = 3$

9. Find $\dfrac{d^3x}{dx^3}$ where $y = \dfrac{x}{1 - x}$, and evaluate it at $x = 2$.

10. Find $f^{(n)}(x)$ where $f(x) = 5x^3 - 8x^2 + 6$.

11. Find the equation of the tangent to the graph of each of the following at the given point.

(a) $f(x) = \dfrac{x - 1}{x + 1}$, $(3, 0.5)$

(b) $f(x) = \dfrac{x}{1 + x^2}$, $(0, 0)$

(c) $f(x) = \dfrac{(x^2 + 1)(2x - 3)}{(6x + 1)}$, $(-1, 2)$

(d) $f(x) = \dfrac{5x - 6}{\sqrt{x + 1}}$, $(0, -6)$

(e) $x^2 + 4xy - y^2 = 11$, $(6, -1)$

(f) $x + x^2y^2 - y = 1$, $(1, 1)$

12. If $y = ax + b/x$, show that $x^2y'' + xy' - y = 0$.

13. Show that $y + (2x - 3y)y' = 0$, when $xy^2 - y^3 = C$, where C is a constant.

14. Show that $y = 5x + \sqrt{26}$ satisfies $y = xy' + \sqrt{1 + (y')^2}$.

15. Given that $f(x)g(x) = C$, where C is a constant, express $f'(x)$ in terms of $g(x)$ and $g'(x)$.

16. Find the equation of the tangent to the circle $(x + 1)^2 + (y - 3)^2 = 25$ at $(2, -1)$.

17. The tangent to the graph of $5xy = k$ at $x = a$ is parallel to the secant through the points where $x = b$ and $x = c$. Show that $a^2 = bc$.

18. The total surface area of a right circular cylinder of radius r and height h is $A = 2\pi rh + 2\pi r^2$. Find dh/dr if A has a fixed value.

D. Graphing by Computer

The graph of a function gives an overall picture of how a function behaves. Using the program given in the Figure D.1, a computer can display the graph of a function in a matter of seconds.

Figure D.1

```
10      REM FUNCTION PLOTTER
100     :
110     DEF FNF(X) = X ↑ 2
120     :
130     NC = 40 : REM NUMBER OF COLUMNS
140     NR = 25 : REM NUMBER OF ROWS
200     :
210     (clear screen)
220     INPUT "X RANGES FROM";XA
230     INPUT "            TO";XB
240     INPUT "Y RANGES FROM";YA
250     INPUT "            TO";YB
260     :
270     DX = (XB − XA)/(NC − 1)
280     DY = (YB − YA)/(NR − 1)
300     :
310     (clear screen)
320     FOR C = 1 TO NC
330        X = XA + (C − 1)∗DX
340        Y = FNF(X)
350        R = (YB − Y)/DY + 1
360        IF 1<=R and R<=NR THEN plot (C,R)
370     NEXT C
400     :
999     END
```

There are a few things you will have to do to tailor the program for your computer.

1. In lines 130 and 140, put the number of rows and columns on which points can be plotted. A **low resolution** screen has only 25 rows and 40 columns, and a plotted point is represented by a character such as an asterisk. A graph produced on a low resolution screen will look like that in Figure D.2.

   High resolution screens on the other hand may have as many as 400 rows and 640 columns. In high resolution, a graph will look much smoother, because the points are smaller and can be plotted much closer together. You may need to place an

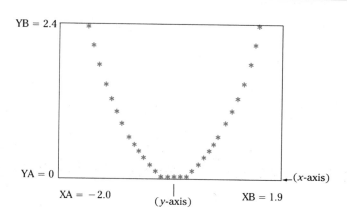

Figure D.2

additional statement at the start of the program to set the resolution high or low.

2. In lines 210 and 310, put in statements to clear the screen.

3. In line 360 replace the statement "plot (C, R)" by the statement(s) your computer uses to place a point at column C, row R. It may be necessary to make the value of R an integer by writing line 350 $R = INT((YB - Y)/DY + 1)$.

The program converts from Cartesian coordinates (x, y) to screen coordinates (C, R), using the fact that the lengths in the plane and on the screen are proportional. In Figure D.3, you can see that

$$\frac{x - x_a}{x_b - x_a} = \frac{C - 1}{NC - 1}$$

and

$$\frac{y_b - y}{y_b - y_a} = \frac{1 - R}{1 - NR}$$

Solving the former for x and the latter for R gives the formulas in lines 330 and 350 respectively.

Before the program is run, you must type on line 110 the equation of the function which is to be plotted. When the program is run, you must declare what portion of the xy-plane you wish to see by inputting the ranges of x and y.

The scales on the x- and y-axes may be different depending on what values are entered for x_a, x_b, y_a, and y_b. If you want the scales

Figure D.3

A portion of the xy-plane

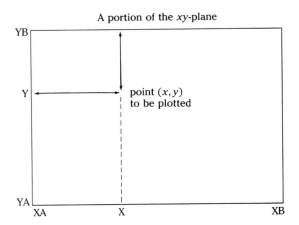

The computer screen

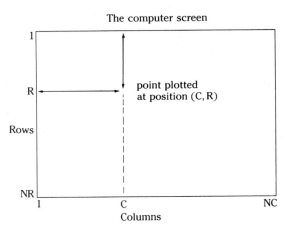

to be the same, that is, DX = DY, then you must calculate y_b from the other values using

$$\frac{x_b - x_a}{NC - 1} = \frac{y_b - y_a}{NR - 1}$$

which gives

$$y_b = y_a + \frac{(NR - 1)}{(NC - 1)} \cdot (x_b - x_a)$$

You may or may not want to put this equation in the program in place of line 250.

This is an unsophisticated program which you can improve in a number of ways, by adding x- and y-axes, for instance. It gives no protection against faulty input. It is up to you to make sense of what you see on the screen.

Exercise D

1. Derive the formulas given in Figure D.1 on lines 330 and 350.

2. If x ranges from -3 to 4.8 and $y_a = -2$,

    (a) find the value of y_b that makes the scales of the x- and y-axes the same on a low resolution screen.

    (b) determine the row and column in which the x- and y-axes are located.

3. Using the program in Figure D.1, display the graph of the function

$$f(x) = \frac{6x}{x^2 + 1}$$

in the intervals

    (a) $x_a = -12, x_b = 11.40, y_a = -7.2, y_b = 7.2$

    (b) $x_a = -9, x_b = 8.55, y_a = -5.4, y_b = 5.4$

    (c) $x_a = -6, x_b = 5.70, y_a = -3.6, y_b = 3.6$

    (d) DX (units per column) and DY (units per row) are the scales on the x- and y-axes, respectively. What are DX and DY for the values given in (a), (b), and (c)?

    (e) Describe what effect decreasing the scale has on the display.

4. Display the graph of the function $f(x) = \sin x$ in the intervals

    (a) $x_a = -6, x_b = 5.70, y_a = -3.6, y_b = 3.6$

    (b) $x_a = -6, x_b = 5.70, y_a = -7.2, y_b = 7.2$

    (c) $x_a = -6, x_b = 5.70, y_a = -1.8, y_b = 1.8$

    (d) What are DX and DY for the values given in (a), (b), and (c)?

    (e) What is the effect on the display when the y-scale is greater than the x-scale? When the y-scale is less than the x-scale?

5. When you display the graph of the function

$$f(x) = \frac{1}{x}$$

in the interval

$$x_a = -6, \ x_b = 5.70, \ y_a = -3.6, \ y_b = 3.6$$

the program may halt with a division by zero error due to the vertical asymptote at $x = 0$. Try it.

You can successfully display graphs of functions that have vertical asymptotes, if you alter the value of x_a slightly. Use for example $x_a = -5.9999$ instead of exactly -6. Try it.

6. Display graphs of the following functions. Record the values of x_a, x_b, y_a, and y_b which produce the best results.

    (a) $f(x) = x^2 - 2x$

    (b) $f(x) = x^3 - 4x$

    (c) $f(x) = \dfrac{2(x^2 - 1)}{x^2 + 1}$

    (d) $f(x) = \dfrac{1}{2 - x}$

    (e) $f(x) = \dfrac{1}{x^2 - 1}$

    (f) $f(x) = \cos x$

Augustin-Louis Cauchy

ALTHOUGH THE CALCULUS was invented in the seventeenth century, a hundred and fifty years passed before mathematicians developed the modern foundation based on the concept of limit. In three textbooks published in the 1820s the French mathematician Augustin-Louis Cauchy presented research that began a new era in the history of analysis.

At the end of the eighteenth century, mathematicians had become increasingly concerned with finding a logically sound foundation for the calculus. At that time the subject was justified either as a set of operations on formulas that yielded acceptable results or as an analysis of the geometry of curves that was intuitively satisfying. In 1784 Berlin Academy of Sciences offered a prize for the best exposition of the role of the infinite in the differential and integral calculus. Although several memoirs were submitted and the prize was awarded, none of the proposals was entirely satisfactory and none achieved widespread acceptance.

Cauchy came to the problem of the foundation of the calculus as a young lecturer at the Ecole polytechnique. He realized that the right approach involved the concepts of limit and continuity, which he formulated clearly for the first time. Beginning with these notions he systematically developed a theory of differentiable and integrable functions of a real variable. In attending to logical rigour and basing the calculus on the continuum of real numbers he began a new period in mathematical analysis. Cauchy is deservedly known as the first "modern" mathematician.

Cauchy and his German contemporary Carl Friedrich Gauss were the two leading mathematicians of the early nineteenth century. Although an innovator in mathematics, Cauchy remained in religion and politics a devout Catholic and absolutist. He refused to pledge allegiance to the Orleanist monarch Louis-Phillipe, installed during the July Revolution of 1830, and followed the deposed Charles X into exile to Prague. He eventually returned to Paris.

Cauchy continued to impress his contemporaries with his odd behaviour. He would sometimes begin research on a subject, forget that he had worked on it years before, and completely duplicate his earlier results. Unlike Gauss, who made public only his deepest and most mature work, Cauchy during his long career published all research in which he became engaged. In addition to his work in the calculus he made fundamental contributions to complex analysis, the theory of elasticity, and algebra. More concepts and theorems have been named for Cauchy than for any other mathematician.

Functions and Their Graphs

5

A function f relates two varying quantities: an **independent variable**, x, and a **dependent variable**, y, which is the value of f at x. The graph of f reveals the main features of this relationship at a glance.

One of the first questions to ask is whether the graph of f has asymptotes. This is answered by examining the limit of f in certain special cases.

Other questions have to do with how the value of f changes when x changes. Does it increase or does it decrease?

Are there points at which the value of f is a maximum or a minimum? How can such points be discovered? Such questions can be answered by examining the slope of the graph of f.

A further question concerns the rate at which changes in the value of f take place. As x changes, does the rate at which the value of f changes increase or decrease? Answers to questions of this sort have to do with the manner in which the graph of the function bends or curves.

Asymptotes: Limits Involving ∞   5.1

The notion of **infinity**, as it is used in mathematics, refers to cases where a quantity increases without limits or bounds. The symbol for infinity, ∞, does not represent a real number. It is a notation used in limit problems to describe unbounded behaviour.

Infinite limits: vertical asymptotes

If the value $f(x)$ of a function increases without bound, as x approaches some real number a, such a case is referred to as an **infinite limit**. As an example of an infinite limit, examine the behaviour of the function $f(x) = 1/x$ near $x = 0$. Table 5.1 shows some values of the

Table 5.1

x	$f(x)$
1	1
0.1	10
0.01	100
0.001	1000
0.0001	10000
. . .	. . .
10^{-20}	10^{20}
. . .	. . .

function as x approaches zero from the right. Clearly, in the limit as x approaches zero, the function increases through positive values beyond all bounds. This behaviour is described by writing

$$\lim_{x \to 0^+} f(x) = \lim_{x \to 0^+} \frac{1}{x}$$

$$= \infty$$

It simply means that there is no limit to the value of the function as $x \to 0$; the limit does not exist. Likewise, as x approaches zero from the left, this function *decreases* through *negative* values beyond all bounds.

$$\lim_{x \to 0} f(x) = \lim_{x \to 0} \frac{1}{x}$$

$$= -\infty$$

Figure 5.1

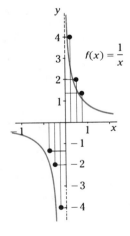

The behaviour of the function $y = 1/x$ near $x = 0$ is illustrated graphically in Figure 5.1. As x approaches zero from either side, the slope of the graph becomes steeper, and the graph approaches the y-axis. The y-axis, $x = 0$, is therefore a **vertical asymptote** for this function.

In general, if the absolute value of a function f increases without bound as x approaches some value a, this behaviour is indicated by writing

$$\boxed{\lim_{x \to a} f(x) = \infty}$$

and the line $x = a$ is a vertical asymptote of the function.

EXAMPLE A

Determine the vertical asymptotes of the function

$$f(x) = \frac{x}{x^2 - 4x}$$

SOLUTION

A rational function may possibly have a vertical asymptote at a value of x, which makes the denominator zero. In this example, the denominator

$$x^2 - 4x = x(x - 4)$$

$$= 0$$

when $x = 0$ or $x = 4$

Since $\lim\limits_{x \to 4} f(x) = \lim\limits_{x \to 4} \dfrac{x}{x(x - 4)}$

$\qquad\qquad = \lim\limits_{x \to 4} \dfrac{1}{x - 4}$

$\qquad\qquad = \infty$

then the line $x = 4$ is a vertical asymptote. On the other hand,

$\lim\limits_{x \to 0} f(x) = \lim\limits_{x \to 0} \dfrac{x}{x(x - 4)}$

$\qquad\qquad = \lim\limits_{x \to 0} \dfrac{1}{x - 4}$

$\qquad\qquad = -\dfrac{1}{4}$

so there is no asymptote at $x = 0$. The function is not defined at $x = 0$ (Figure 5.2).

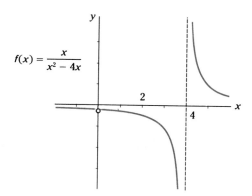

$$f(x) = \dfrac{x}{x^2 - 4x}$$

Figure 5.2

Table 5.2

x	$f(x)$
1	1
10	0.1
100	0.01
1000	0.001
10000	0.0001
. . .	. . .
10^{20}	10^{-20}
. . .	. . .

Limits at infinity: horizontal asymptotes

If the value $f(x)$ of a function approaches a limit, as x increases without bound, the function is said to have a **limit at infinity**. Consider once again the function $f(x) = 1/x$, this time letting x increase beyond all bounds, as in Table 5.2. You can see in Figure 5.3 that as x increases, the graph of f approaches the x-axis from above. Hence the limit of the fraction is zero and the positive x-axis is a **horizontal asymptote**.

Figure 5.3

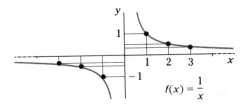

$$f(x) = \frac{1}{x}$$

Expressed in the proper mathematical notation, this limit is written

$$\lim_{x \to +\infty} f(x) = \lim_{x \to +\infty} \frac{1}{x}$$

$$= 0$$

Similarly, as x decreases through negative values without bound,

$$\lim_{x \to -\infty} f(x) = \lim_{x \to -\infty} \frac{1}{x}$$

$$= 0$$

In general, if the value of a function f approaches some constant L as x increases without bound, that is if

$$\boxed{\lim_{x \to \infty} f(x) = L}$$

then the function has a limit at infinity and the line $y = L$ is a horizontal asymptote.

EXAMPLE B

Find the horizontal asymptote of the function

$$f(x) = \frac{4x^2 - 6}{2x^2 + 1}$$

(a) by computation
(b) algebraically.

SOLUTION

(a) A table of values of this function is shown in Table 5.3. Note that as x becomes larger, the numerator and the denominator both become larger. The value of the function, however, seems to approach 2. The same is true if $x \to -\infty$.

Table 5.3

x	numerator	denominator	f(x)
0	-6	1	-6
1	-2	3	-0.666667
2	10	9	1.111111
3	30	19	1.578947
10	394	201	1.960199
100	39994	20001	1.999600
1000	3999994	2000001	1.999996

(b) To find the horizontal asymptote, let $x \to \infty$. In this limit, the function has the indeterminate form ∞/∞. Proceed, therefore, by first dividing numerator and denominator by the highest power of x. This will produce powers of $(1/x)$, which go to zero as $x \to \infty$:

$$\lim_{x \to \infty} f(x) = \lim_{x \to \infty} \frac{(4x^2 - 6)}{(2x^2 + 1)} \cdot \frac{(1/x^2)}{(1/x^2)}$$

$$= \lim_{x \to \infty} \frac{4 - \dfrac{6}{x^2}}{2 + \dfrac{1}{x^2}}$$

$$= \frac{4 - 0}{2 + 0}$$

$$= 2$$

Therefore, the line $y = 2$ is a horizontal asymptote. This function has no vertical asymptotes. Its graph is shown in Figure 5.4. ◻

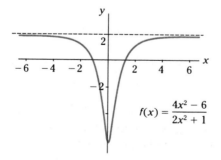

Figure 5.4

$$f(x) = \frac{4x^2 - 6}{2x^2 + 1}$$

EXAMPLE C

Find the asymptotes and graph the function

$$f(x) = \frac{x}{\sqrt{4x - 1}}$$

SOLUTION

The function is not defined for $x < 1/4$. As $x \to 1/4^+$, the denominator approaches zero, but the numerator does not. Therefore,

$$\lim_{x \to 1/4^+} \frac{x}{\sqrt{4x - 1}} = \infty$$

and a vertical asymptote exists at $x = 1/4$.

To find if there is a horizontal asymptote, take the limit as $x \to \infty$:

$$\lim_{x \to \infty} f(x) = \lim_{x \to \infty} \frac{x}{\sqrt{4x - 1}}$$

$$= \lim_{x \to \infty} \frac{x}{\sqrt{x(4 - 1/x)}}$$

$$= \lim_{x \to \infty} \frac{x}{\sqrt{x}\sqrt{4 - 1/x}}$$

$$= \lim_{x \to \infty} \frac{\sqrt{x}}{\sqrt{4 - 1/x}}$$

$$= \infty$$

Figure 5.5

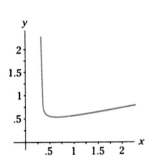

Since the limit as $x \to \infty$ does not exist, there is no horizontal asymptote. The graph is shown in Figure 5.5. For large values of x, the function looks approximately like $y = \sqrt{x}/2$. □

EXAMPLE D

Investigate the limits of the exponential function $y = k^x$, $k > 0$, as $x \to \pm\infty$.

SOLUTION

Graphs of the exponential function for several values of k are shown in Figure 5.6a and b. The graphs illustrate that

$$\lim_{x \to +\infty} k^x = \begin{cases} +\infty, & k > 1 \\ 1, & k = 1 \\ 0, & 0 < k < 1 \end{cases}$$

and

$$\lim_{x \to -\infty} k^x = \begin{cases} 0, & k > 1 \\ 1, & k = 1 \\ +\infty, & 0 < k < 1 \end{cases}$$

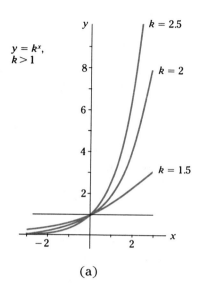

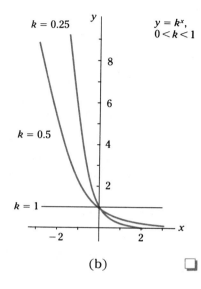

(a) (b)

Figure 5.6

Oblique asymptotes

An **oblique asymptote** is one that is neither horizontal nor vertical. The line $y = mx + b$ is an oblique asymptote to the graph of f, if the difference between the value of the function and the value of $mx + b$ approaches zero as $x \to \infty$:

$$\lim_{x \to \pm\infty} \left(f(x) - (mx + b) \right) = 0$$

Note that a horizontal asymptote is a special case of an oblique asymptote.

EXAMPLE E

Find the asymptotes to the graph of

$$f(x) = \frac{2x^2 + 6x + 5}{x + 2}$$

SOLUTION

By long division, the function f can be written

$$f(x) = 2x + 2 + \frac{1}{x + 2}$$

Therefore,

$$\lim_{x \to \pm\infty} \Big(f(x) - (mx + b) \Big)$$

$$= \lim_{x \to \infty} \Big((2x + 2) + \frac{1}{x + 2} - (mx + b) \Big)$$

It is evident now that the line $y = 2x + 2$ is an oblique asymptote. If $mx + b = 2x + 2$, then this limit reduces to

$$\lim_{x \to \pm\infty} \left(\frac{1}{x + 2} \right) = 0$$

This function also has a vertical asymptote at $x = -2$ because

$$\lim_{x \to -2} \frac{2x^2 + 6x + 5}{x + 2} = \infty$$

Its graph is shown in Figure 5.7.

Figure 5.7

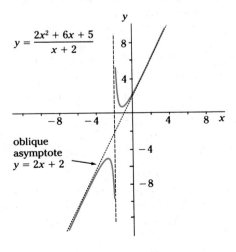

$$y = \frac{2x^2 + 6x + 5}{x + 2}$$

oblique
asymptote
$y = 2x + 2$

Exercise 5.1

1. Evaluate the following limits.

(a) $\lim\limits_{x \to \infty} \dfrac{1 - 2x + 3x^2}{5 + x - 4x^2}$ *horizontal*

(b) $\lim\limits_{x \to 2} \dfrac{4}{(x - 2)^2}$ *verticle*

(c) $\lim\limits_{x \to 3^+} \dfrac{x + 3}{x^2 - x - 6}$ *verticle*

(d) $\lim\limits_{x \to \infty} \dfrac{x^2}{7 - x}$ *horizontal*

(e) $\lim\limits_{x \to \infty} \dfrac{x^5 + 1}{x^4 + x^3 - 1}$ *horizontal*

(f) $\lim\limits_{x \to \infty} \dfrac{3 - x + 7x^2}{1 - 2x^2}$ *horizontal*

2. Evaluate the following limits.

(a) $\lim\limits_{x \to \infty} \dfrac{3x - 1}{2x}$ *horizontal*

(b) $\lim\limits_{x \to \infty} \dfrac{3 + 2x}{x^2 + 7x}$ *horizontal*

(c) $\lim\limits_{x \to 3} \left(\dfrac{8}{x - 3} - \dfrac{8}{x^2 - 9} \right)$ *verticle*

(d) $\lim\limits_{x \to 0} \left(5x - \dfrac{4}{x} \right)$ *verticle*

(e) $\lim\limits_{x \to 0} \dfrac{a}{2} \left(e^{x/2} + e^{-x/2} \right)$ *verticle*

(f) $\lim\limits_{x \to 1} \dfrac{x^2 + x - 2}{x^3 - x^2 - x + 1}$ *verticle*

3. Evaluate the following limits.

(a) $\lim\limits_{x \to 0} \left(\dfrac{x^3 - 2x + 3}{x - 5} - 1 \right)$

(b) $\lim\limits_{x \to 0} \dfrac{1 - \dfrac{1}{x^2}}{1 - \dfrac{1}{x}}$

(c) $\lim\limits_{x \to 1} \left(1 - \dfrac{1}{x} \right) \left(\dfrac{1}{x - 1} \right)$

(d) $\lim\limits_{x \to \infty} \left(\dfrac{x^3}{2x^2 - 1} - \dfrac{x^2}{2x + 1} \right)$

4. Evaluate the following limits.

(a) $\lim\limits_{x \to +\infty} \dfrac{x}{\sqrt{x - 2}}$

(b) $\lim\limits_{x \to 2^+} \dfrac{x}{\sqrt{x - 2}}$

(c) $\lim\limits_{x \to 0} (\sqrt{x^2 + 2} - x)$

(d) $\lim\limits_{x \to \sqrt{2}} (\sqrt{x^2 + 2} - x)$

(e) $\lim\limits_{x \to \infty} (\sqrt{x^2 + 2} - x)$

(f) $\lim\limits_{x \to \infty} \left(4 - 3x - \dfrac{5}{x + 1} \right)$

5. Determine if the following functions have vertical or horizontal asymptotes. Sketch the graph of the function in each case.

(a) $y = x + 1 + \dfrac{3}{x - 4}$

(b) $y = \dfrac{2x - 1}{x + 4}$

(c) $y = \dfrac{3x^2 + 5x - 6}{x + 2}$

(d) $y = \dfrac{x^2 + 4x - 3}{(x - 1)^2}$

(e) $y = \dfrac{x^2 - x - 6}{x^2 - 9}$

(f) $y = \dfrac{3x}{\sqrt{x - 2}}$

(g) $y = \dfrac{4}{x^2 - 1}$

6. Which of the functions in question 5 has an oblique asymptote? State the equation of the asymptote.

7. In certain cases, the limit of a function may give rise to an indeterminate form that cannot be evaluated by the methods described in this text. However, by calculating the value of the function at several values of x, it is generally possible to infer what the limit must be. Evaluate the following limits:

(a) $\lim\limits_{x \to +\infty} (1/x)^x$

(b) $\lim\limits_{x \to 0^+} x^{1/x}$

(c) $\lim\limits_{x \to \infty} x \cdot 2^{-x}$

(d) $\lim\limits_{x \to \infty} (2^{1/x} - 1)$

(e) $\lim\limits_{x \to 0} \left(\dfrac{1}{x} - \dfrac{1}{2x - 1} \right)$

5.2 Qualitative Features of Graphs

The first step in analyzing the graph of a function is to compare the value of the function at one point to its value at another point. Compare, for instance, the values of the function

$$f(x) = \frac{6}{3 - x} \qquad \text{at } x = 1 \text{ and } x = 2:$$

$$f(1) = \frac{6}{3 - 1}, \qquad f(2) = \frac{6}{3 - 2}$$

$$= 3 \qquad\qquad = 6$$

A comparison shows that $f(2) > f(1)$. In other words, the value of the function f has *increased* from 3 to 6.

The range of values $1 \le x \le 2$ is called a closed interval and is denoted by $[1, 2]$. How do the values of f at other points in the interval $[1, 2]$ compare? Check $f(1.5)$ and $f(1.8)$, for example:

$$f(1.5) = \frac{6}{3 - 1.5} \qquad f(1.8) = \frac{6}{3 - 1.8}$$

$$= 4 \qquad\qquad = 5$$

Comparing these values shows that $f(1.8) > f(1.5)$. In a similar manner you can verify that when $x_2 > x_1$, $f(x_2) > f(x_1)$ for any values of x in the interval. This function is said to be *increasing* (its value is increasing) on the interval $[1, 2]$ (see Figure 5.8). In general, as Figure 5.9 shows,

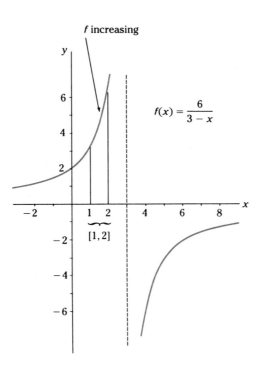

Figure 5.8

A function *f* defined on the interval [*a*, *b*] is
increasing on [*a*, *b*] if $f(x_2) > f(x_1)$ whenever $x_2 > x_1$,
decreasing on [*a*, *b*] if $f(x_2) < f(x_1)$ whenever $x_2 > x_1$,
where x_1 and x_2 are any numbers in [*a*, *b*].
f is *stationary* if it is neither increasing nor decreasing.

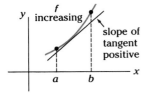
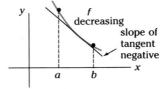

Figure 5.9

The graphs in Figure 5.9 suggest that when a function is increasing,
the slope of the tangent to its graph is positive, and when it is decreasing, the slope is negative. A convenient way, therefore, to discover
whether a function is increasing or decreasing is to examine values
of its first derivative. In the above example,

$$\frac{d}{dx}\frac{6}{3-x} = \frac{d}{dx}6(3-x)^{-1}$$ ⟵———————— use the chain rule

$$= 6(-1)(3-x)^{-2}(-1)$$

$$= \frac{6}{(3-x)^2}$$

The derivative is clearly positive for all values of x in the interval $[1, 2]$. The conclusion is that the function is increasing on that interval. Note that this function is increasing on every interval, except for ones that include the value $x = 3$, since the function is not defined at $x = 3$.

EXAMPLE A

Determine the interval on which the function $f(x) = 2x^2 - 8x$ is increasing.

SOLUTION

$$\frac{d}{dx}f(x) = \frac{d}{dx}(2x^2 - 8x)$$

$$= 4x - 8$$

The function is increasing when $f'(x) > 0$:

Figure 5.10

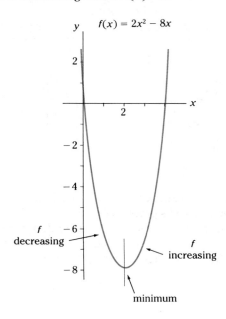

$$4x - 8 > 0$$

$$4x > 8$$

$$x > 2$$

Thus, the function f is increasing when $x > 2$ (see Figure 5.10). You can show in a similar way that $f'(x) < 0$, that is, the function is decreasing when $x < 2$. ❑

Observe in Example A, that at $x = 2$, the value of the function stops decreasing and begins to increase. The value of f at this point is called a **minimum** of the function. At the minimum point, the value of f is *less* than the value at any nearby point.

The value of a function reaches a **maximum** when it stops increasing and begins to decrease, as shown in Figure 5.11. At a maximum, the value of f is *greater* than that at any nearby point. Section 5.3 will describe how to find the values of x at which maximum and minimum values of a function occur.

Figure 5.11

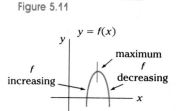

EXAMPLE B

The function $f(x) = 2x^3 - 15x^2 + 24x$ has a maximum at $x = 1$ and a minimum at $x = 4$.

(a) Determine the maximum and minimum values of f.

(b) Sketch the graph of f.

(c) State the intervals on which f is increasing and decreasing.

SOLUTION

(a) $f(1) = 2(1)^3 - 15(1)^2 + 24(1)$

$\quad = 2 - 15 + 24$

$\quad = 11$

$f(4) = 2(4)^3 - 15(4)^2 + 24(4)$

$\quad = 128 - 240 + 96$

$\quad = -16$

Therefore, the value of f at the maximum is 11. At the minimum, the value of f is -16.

(b) Around the maximum point, the graph of the function has a shape like ⌒. Around the minimum point, it has a shape like ⌣. These shapes are often called "hills" and "valleys," respectively.

Figure 5.12

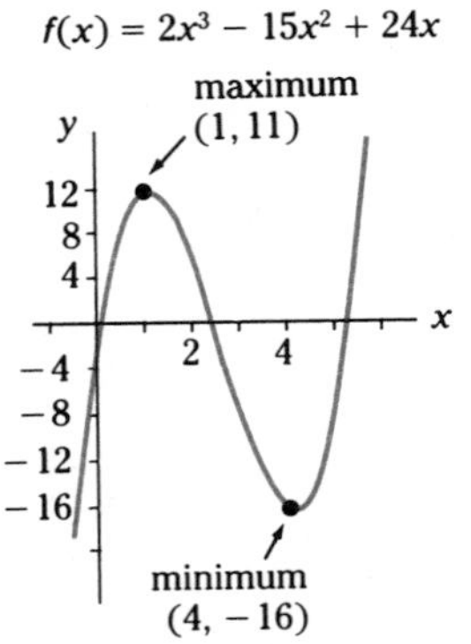

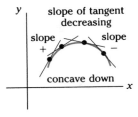

$f(x) = 2x^3 - 15x^2 + 24x$

By smoothly joining a hill, which "peaks" at $(1, 11)$, to a valley, which "bottoms out" at $(4, -16)$, you can get an qualitative idea of what the graph of the function looks like. Such a sketch is shown in Figure 5.12.

(c) The value of the function is

increasing when $x < 1$ (up to the maximum),
decreasing on $[1, 4]$ (from the maximum to the minimum),

and increasing when $x > 4$ (beyond the minimum).   ❑

Figure 5.13

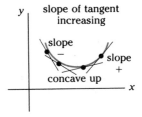

There are several things about the function described in Example B that should be noted. At large values of x, the function has values that are greater than the maximum value of 11 at $x = 1$. For instance $f(6) = 36$. For this reason, the maximum at $(1, 11)$, is referred to as a **relative maximum**. To be a relative maximum, the value of f need only be greater than the value at any *nearby* point. For a similar reason, the minimum at $(4, -16)$ is called a **relative minimum**.

The terms "hill" and "valley" were used in Example B to describe parts of the graph, which were curved in a particular way. Imagine a point moving along a curve over a hill. The slope of the tangent to the curve at any point on the left side of the hill is positive, as shown in Figure 5.13. As the point moves, the slope decreases to zero at the top of the hill. It continues to decrease, becoming negative as the point moves on down the right side of the hill. A part of any curve on which the slope of the tangent is decreasing is said to be **concave down**.

Figure 5.14

If a point moves along a curve through a valley, as shown in Figure 5.14, the slope of the tangent increases from negative, through zero, to positive values. A part of any curve on which the slope of the tangent is increasing is said to be **concave up**.

In sketching the graph of the function of Example B in Figure 5.12, it was necessary to join a hill to a valley. Someplace between the maximum and the minimum there must be a point where the curve changes from being concave down to being concave up. This point is called a **point of inflection**. In Example B, the point of inflection occurs at $x = 2.5$. It is indicated in Figure 5.15. Section 5.4 will describe how to find points of inflection.

EXAMPLE C

The function $f(x) = \dfrac{24}{x^2 + 12}$ has a maximum at $x = 0$ and inflection points at $x = \pm 2$.

(a) Determine the maximum value of f, and the y-coordinates of the inflection points.

(b) Sketch the graph of f.

(c) State the intervals on which f is concave up and concave down.

SOLUTION

(a) $f(0) = \dfrac{24}{0^2 + 12}$

$\qquad = 2$

$f(\pm 2) = \dfrac{24}{(\pm 2)^2 + 12}$

$\qquad = \dfrac{3}{2}$

Therefore, the maximum value of the function is 2. The inflection points both have a y-coordinate of $3/2$.

(b) Around the maximum, the graph is concave down. It is joined at the inflection points on the left and the right to parts of the graph that are concave up, as shown in Figure 5.16. The curve has no minima. It approaches the x-axis asymptotically.

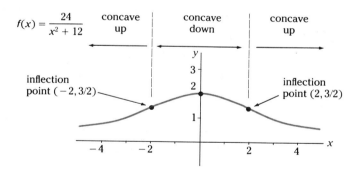

(c) The function is
concave up for $x < -2$ (up to the left inflection point),
concave down in the interval $[-2, 2]$ (around the maximum),
and concave up for $x > 2$ (beyond the right inflection point). ◻

In considering the qualitative features of the graph of a function f, it is important to understand the role played by the derivative f'.

Figure 5.15

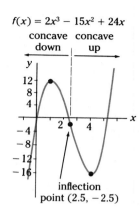

$f(x) = 2x^3 - 15x^2 + 24x$

inflection
point $(2.5, -2.5)$

Figure 5.16

This role is summarized in the following statements:

The value of f is increasing,	when the value of f' is positive.
The value of f is decreasing,	when the value of f' is negative.
The graph of f is concave up,	when the value of f' is increasing.
The graph of f is concave down,	when the value of f' is decreasing.

Exercise 5.2

1. Examine the graphs in Figure 5.17. In each case state whether

    (a) the value of f is increasing or decreasing

    (b) the graph of f is concave up or concave down.

Figure 5.17

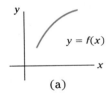

(a)

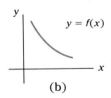

(b)

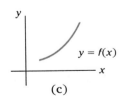

(c)

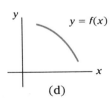

(d)

2. At which point on the graph in Figure 5.18 is the value of

    (a) y negative and y' positive?

    (b) y zero   and y' negative?

    (c) y positive and y' positive?

    (d) y zero   and y' positive?

    (e) y negative and y' zero?

Figure 5.18

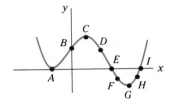

3. An increasing function passes through the origin. Sketch its graph if it is

    (a) concave down for $y < 0$, and concave up for $y > 0$

    (b) concave down for $y > 0$, and concave up for $y < 0$.

4. Repeat question 3 for a decreasing function.

5. Find the derivative of y with respect to x in each of the following cases. Determine the intervals on which the function is increasing and decreasing. Sketch the graphs.

    (a) $y = -5x + 2$

    (b) $y = 2x - x^2$

    (c) $y = \dfrac{1}{x^2}$

    (d) $y = \dfrac{x - 5}{x}$

    (e) $y = \sqrt{x^2 + 4}$

    (f) $y = x^3 + 3x^2 - 24x$

6. The function

$$f(x) = \frac{x^2}{x - 6}$$

has a maximum at $x = 0$ and a minimum at $x = 12$.

(a) Determine the maximum and minimum values of f.

(b) Sketch the graph of f.

(c) State the intervals on which f is increasing and decreasing.

7. The function

$$f(x) = \frac{x^2 - 4}{x^2 + 4}$$

has a minimum at $x = 0$ and inflection points at $x = \pm 2/\sqrt{3}$.

(a) Determine the minimum value of f and the y-coordinates of the inflection points.

(b) Sketch the graph of f.

(c) State the intervals on which the function is concave up and concave down.

8. In each of the following cases, sketch the graph of a function satisfying the given conditions and state the intervals on which the function is increasing, decreasing, concave up, or concave down.

(a) The function $y = f(x)$ has a maximum at $x = 5$, a minimum at $x = -2$, and the point of inflection at $x = 1$.

(b) The function $y = g(x)$ is an odd function having no roots, maxima, or minima. The x-axis and the y-axis are asymptotes.

(c) The function $y = h(x)$ is an even function which is positive everywhere except at $x = 0$, where its value is zero. One maximum is at $x = 2$. Two of its inflection points are at $x = 1$ and $x = 4$.

(d) The function $y = u(x)$ has an x-intercept at $x = 4$, a minimum at $x = -1$, and a point of inflection at $x = -2$. The lines $x = 0$ and $y = -1$ are asymptotes.

9. A function f is concave down for $x < 3$ and concave up for $x > 3$. It has a maximum at $x = 2$.

(a) Must it have a minimum? Sketch graphs showing the possibilities.

(b) Is it possible for it to have no x-intercepts? One x-intercept? Two? Three? More? Sketch graphs showing the possibilities.

10. Make a sketch of an odd function having one maximum and one minimum, but no roots. Must it have an asymptote? Explain.

Extreme Values of Functions   5.3

A maximum or minimum value of a function is referred to as an **extreme value** or simply as an **extremum**. The relative maxima and minima discussed in the previous section are examples of extrema. At these points, the slope of the tangent to the graph of the function is zero. Therefore, to find the relative extrema of a function, you should examine points where $dy/dx = 0$. You can distinguish between a maximum and a minimum by considering how the slope of the tangent changes.

> The graph of f may have an extreme value at $x = a$, if $f'(a) = 0$

EXAMPLE A

(a) Find the coordinates of the point at which the relative minimum of the function $f(x) = x^2 - 2x - 2$ occurs.

(b) Sketch the graphs of f and f'. On the graph of f' identify the point corresponding to the minimum point of f.

SOLUTION

(a) The function f is a quadratic function whose graph opens up. The relative minimum is found where the tangent to the graph has a slope of zero, that is where $f'(x) = 0$.

$$f(x) = x^2 - 2x - 2$$
$$f'(x) = 2x - 2$$
$$0 = 2x - 2$$
$$x = 1$$

The relative minimum occurs at the position $x = 1$. The minimum value of the function is

$$f(1) = (1)^2 - 2(1) - 2$$
$$= -3$$

Therefore, the minimum point is $(1, -3)$. This is the vertex of the parabola.

(b) The graph of f is shown in Figure 5.19 with the graph of f' aligned directly below it. The x-intercept on the graph of f' is the point corresponding to the minimum of f. To the left of the minimum point, f is decreasing and f' is negative. To the right of the minimum, f is increasing and f' is positive. Note furthermore that the value of f' is increasing. This corresponds to the fact that the graph of f is concave up.   ❑

EXAMPLE B

Find the relative extrema and sketch the graph of the function

$$f(x) = \frac{4(x - 1)}{x^2}$$

Figure 5.19

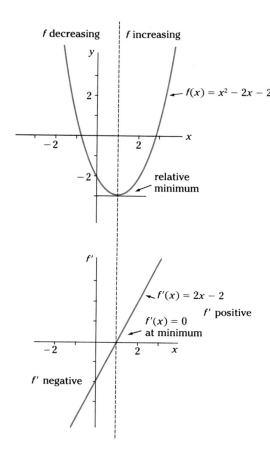

SOLUTION

Look for the relative extrema where $f'(x)$ is zero:

$$f(x) = \frac{4(x - 1)}{x^2}$$

$$f'(x) = 4 \cdot \frac{x^2(1) - (x - 1)(2x)}{x^4} \qquad \longleftarrow \text{using the quotient rule}$$

$$= 4 \cdot \frac{2 - x}{x^3}$$

The first derivative is zero at only one value of x, namely, $x = 2$.

Figure 5.20

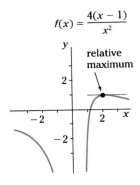

$$f(x) = \frac{4(x-1)}{x^2}$$

Therefore, the function has one extreme value, which is $f(2) = 1$. One way to discover if this is a maximum or a minimum is to examine the first derivative near $x = 2$:

for $0 < x < 2$, f' is positive, so f is increasing

and for $x > 2$, f' is negative, so f is decreasing

Since f is increasing up to $x = 2$ and decreasing thereafter, f has a maximum at $x = 2$ (see Figure 5.20). Another way to distinguish between a relative maximum and a relative minimum involves the second derivative. It will be discussed in Section 5.4.

A point on the graph of a function at which $f'(x) = 0$ is called a **stationary point** of the function. The requirement that $f'(x) = 0$ does not by itself guarantee that the stationary point is a maximum or a minimum. It could also be a point of inflection at which the tangent is horizontal. Thus $f'(x) = 0$ is a *necessary* but not a *sufficient* condition for there to be a maximum or a minimum.

EXAMPLE C

Determine the stationary points of the function $f(x) = x^3 - 3x^2 + 3x$ and sketch its graph.

SOLUTION

Stationary points are the points at which $f'(x) = 0$:

$$f(x) = x^3 - 3x^2 + 3x$$

$$f'(x) = 3x^2 - 6x + 3$$

$$0 = 3(x^2 - 2x + 1)$$

$$0 = 3(x - 1)^2$$

$$x = 1$$

Figure 5.21

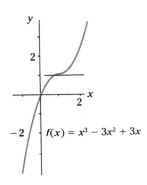

$f(x) = x^3 - 3x^2 + 3x$

This function has only one stationary point. It occurs at the position $x = 1$. Since f' is positive for values of x on either side of the stationary point, the point is neither a maximum nor a minimum. It must be a point of inflection at which the tangent is horizontal. The y-coordinate of the point of inflection is $f(1) = 1$. From this analysis together with the observation that the graph of the function passes through the origin, a sketch of the graph can be made (see Figure 5.21).

EXAMPLE D

Analyze the function $y = \dfrac{x^2}{x - 1}$ and sketch its graph.

SOLUTION

Intercepts: At $x = 0$, $y = 0$
the curve passes through the origin.

Asymptotes: Since $\lim\limits_{x \to 1} \dfrac{x^2}{x - 1} = \infty$,
The line $x = 1$ is a vertical asymptote.
There are no horizontal asymptotes.

Stationary
Points:

$$\dfrac{dy}{dx} = \dfrac{(x - 1)(2x) - x^2(1)}{(x - 1)^2}$$ ◄—— quotient rule

$$= \dfrac{x^2 - 2x}{(x - 1)^2}$$

$$0 = \dfrac{x^2 - 2x}{(x - 1)^2}$$ ◄—— set $y' = 0$

$$0 = x(x - 2)$$

∴ $x = 0$ or 2

This function has two stationary points.

When $x < 0$, y' is positive, so y is increasing.
When $0 < x < 1$, y' is negative, so y is decreasing.
Therefore, there is a relative maximum at $x = 0$.
The value of y at the maximum is 0.

When $1 < x < 2$, y' is negative, so y is decreasing.
When $x > 2$, y' is positive, so y is increasing.
Therefore, there is a relative minimum at $x = 2$.
The value of y at the minimum is 4.

Concavity: This function has no inflection points
(see Section 5.4). Therefore, the graph is
concave down for $x < 1$
(in the interval containing the maximum)
and concave up   for $x > 1$
(in the interval containing the minimum).
The graph of the function is shown in Figure 5.22.

Figure 5.22

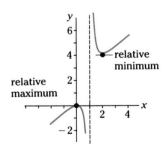

Another type of extreme value is known as an **absolute** or **global maximum** or **minimum**. It is the greatest value or the least value of all the values of a function. The relative minimum of the parabola shown in Figure 5.19, for example, is also an absolute minimum since no points on the graph are lower. The relative maximum shown in Figure 5.22, however, is not an absolute maximum. That function has no absolute maximum; its value increases without bound as $x \to 1$, and as $x \to \infty$.

It is possible for the absolute maximum or minimum of a function to occur at a point which is not a relative maximum or minimum. For instance, an absolute minimum occurs at the left end of the domain in Figure 5.23a, and at the cusp in Figure 5.23b. To discover the absolute maximum or minimum value of a function, you must examine the graph of a function as a whole.

Figure 5.23

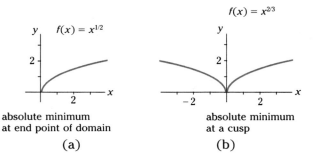

absolute minimum   absolute minimum
at end point of domain   at a cusp
(a)   (b)

EXAMPLE E

Locate the absolute maximum and minimum values of the function $f(x) = x^3 - 7x^2 + 11x, 0 \le x \le 6$.

SOLUTION

First find the stationary points of the function by solving $f'(x) = 0$.

$$f(x) = x^3 - 7x^2 + 11x$$
$$f'(x) = 3x^2 - 14x + 11$$
$$0 = (3x - 11)(x - 1)$$
$$\therefore \qquad x = 1, 11/3$$

This function has two stationary points.

When $x < 1$, y' is positive, so y is increasing.
When $1 < x < 11/3$, y' is negative, so y is decreasing.
When $x > 11/3$, y' is positive, so y is increasing.
Therefore, there is a relative maximum at $x = 1$, and a relative minimum at $x = 11/3$. The value of f at the relative maximum is $f(1) = 5$, and at the relative minimum, $f(11/3) = -121/27$.

The question is now whether these relative extrema are absolute extrema. At the endpoints of the interval, $f(0) = 0$, and $f(6) = 30$. Therefore, the relative maximum at $x = 1$ is *not* the absolute maximum. The absolute maximum occurs at $x = 6$ at the right endpoint of the domain of f. On the other hand, the relative minimum *is* the absolute minimum in the domain of f. The graph of the function is shown in Figure 5.24. ☐

$f(x) = x^3 - 7x^2 + 11x$

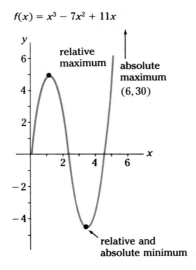

Figure 5.24

Exercise 5.3

1. Find the relative maximum of each of the following functions. Sketch the graph in each case.

(a) $f(x) = 4 - x^2$

(b) $f(x) = 12(x^2 - 4)^{-1}$

(c) $f(x) = \dfrac{4}{x^2 - 4x + 8}$

(d) $f(x) = \dfrac{\sqrt{x - 3}}{x}$

2. Find the relative minimum of each of the following functions. Sketch the graph in each case.

(a) $f(x) = 2x^2 - 10x$

(b) $f(x) = 3x + 48x^{-1/2}$

(c) $f(x) = \dfrac{x^3 + 16}{x}$

(d) $f(x) = (x - 2)\sqrt{x}$

3. Find the vertex of each of the following parabolas. State whether the vertex is a maximum or a minimum.

(a) $y = 18 - 15x - 3x^2$

(b) $y = x^2 + 7x$

(c) $y = (x - 2)(x + 5)$

4. Copy the graph of each of the functions shown in Figure 5.25 and sketch the graph of its derivative. Identify corresponding points on the two graphs.

Figure 5.25

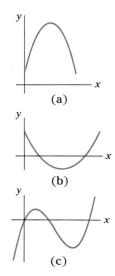

(a)

(b)

(c)

5. Find the relative extremum of each of the following functions. Decide in each case if the extremum is a maximum or a minimum and sketch the graph.

(a) $y = 5 - (x - 4)^2$

(b) $y = (x - 2)^4 + 16$

(c) $y = \dfrac{x - 3}{x^2}$

(d) $y = 8x + \dfrac{2}{\sqrt{1 + x}}$

6. Find the position of the maximum and the minimum values of each of the following functions. Sketch the graph in each case.

(a) $y = x^3 + 24x$

(b) $y = 9x^4 - 2x^2$

(c) $y = x^2 + \dfrac{1}{x^2}$

(d) $y = \dfrac{x^2 - 4}{x^2 - 1}$

(e) $y = \dfrac{x^2 - 3}{x^3}$

(f) $y = \dfrac{x^2}{x - 3}$

7. Sketch the graphs of f and f'. In each case, identify the points of the graph of f' that correspond to the stationary points of f.

(a) $f(x) = (x - 2)^2 + 25$

(b) $f(x) = x^3 - 2x^2 + x + 6$

(c) $f(x) = 3x^4 - 4x^3 - 12x^2$

(d) $f(x) = (x - 2)^3$

8. Determine the stationary points of the following functions. State whether the stationary point is a maximum, a minimum, or a point of inflection.

(a) $y = x^4 + 4x + 5$

(b) $y = x^3 - 3x^2 + 3x + 1$

(c) $y = x^4 - 4x^3$

(d) $y = \sqrt{25 - (x - 3)^2}$

(e) $y = \dfrac{x^2 + 5}{x + 2}$

(f) $y = 2x^5 + 15x^4 + 30x^3$

9. Find the absolute maximum and minimum values of the following functions in the given domain.

(a) $y = -3x + 6, \qquad 1 \le x \le 4$

(b) $y = 5 - (x - 2)^2, \qquad 0 \le x \le 5$

(c) $y = \dfrac{1}{x - 4}, \qquad 0 \le x \le 3$

(d) $y = x^3 - 12x^2 + 45x, \ 2 \le x \le 6$

(e) $y = \sqrt{x(x + 2)}, \qquad 0 \le x \le 3$

Curvature and Points of Inflection 5.4

A point of inflection was described in Section 5.2 as a point at which the curvature of the graph of a function changes from concave up to concave down, or vice versa. How can the coordinates of such points be found?

Consider the curve shown in Figure 5.26 on which point C is an inflection point. Imagine a point moving from left to right along the curve from one side of the inflection point to the other. As the point moves from A to B to C on the part of the curve which is concave up, the slope of the tangent increases. Then, as the point moves on past C to D and E along the part of the curve which is concave down, the slope of the tangent decreases. The inflection point C is the point at which the slope of the tangent is a maximum.

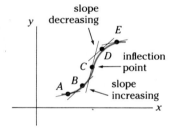

Figure 5.26

If a curve bends in the opposite way as in Figure 5.27, the inflection point C is the point at which the slope of the tangent is a minimum.

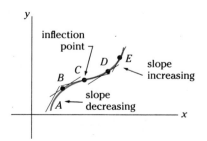

Figure 5.27

Inflection points of a graph, therefore, are located at points where the slope of the tangent has an extreme value. These are points for which

$$\frac{dy'}{dx} = 0, \text{ that is, the } \textit{second} \text{ derivative } \frac{d^2y}{dx^2} = 0$$

> The graph of *f* may have
> an inflection point at $x = a$,
> if $f''(a) = 0$.

EXAMPLE A

Determine the point of inflection of the function

$$y = -x^3 + 6x^2 - 15x \text{ and sketch its graph.}$$

SOLUTION

Look for a point of inflection where $y'' = 0$:

$$y = -x^3 + 6x^2 - 15x$$

$$y' = -3x^2 + 12x - 15$$

$$y'' = -6x + 12$$

$$0 = -6x + 12$$

$$x = 2$$

Figure 5.28

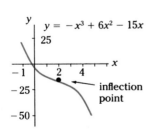

$$y = -x^3 + 6x^2 - 15x$$

inflection point

A point of inflection occurs at $x = 2$. The *y*-coordinate of the point of inflection is

$$y = -(2)^3 + 6(2)^2 - 15(2)$$

$$= -14$$

There is no value of *x*, at which $y' = 0$, so this function has no maximum or minimum. Its graph is shown in Figure 5.28. ❑

When the slope of the tangent to a curve is increasing, that part of the curve is concave up. You can tell when the slope of the tangent is increasing by examining the sign of *second* derivative. The function *f'* is increasing when *its* derivative *f''* is positive. This means that the sign of the second derivative can be used to distinguish between a minimum where the graph of *f* is concave up and a maximum where it is concave down:

> *f* has a minimum at $x = a$, if $f'(a) = 0$ and $f''(a) > 0$
> *f* has a maximum at $x = a$, if $f'(a) = 0$ and $f''(a) < 0$

These are *sufficient* conditions for a maximum or a minimum. They guarantee the existence of a maximum or a minimum when they are satisfied. On the other hand, no conclusions can be drawn when *both* derivatives of *f* are zero. In such cases you must look at how the function behaves in the neighbourhood of the point in question.

EXAMPLE B

Consider the graph of the function $y = f(x)$ shown in Figure 5.29a. Use the maximum, minimum, and inflection points of *f* to sketch graphs of its first and second derivatives.

SOLUTION

Place axes for the graphs of y' and y'' vertically beneath the graph of *f* as in Figures 5.29b and c.

At the maximum, f' is zero and f'' is negative.
At the minimum, f' is zero and f'' is positive.
At the inflection point, f' has an extreme value,
    in this case a minimum, and f'' is zero.

The required graphs of the derivatives can now be sketched from these observations. ❑

EXAMPLE C

Determine the maximum, minimum, and inflection points of the function $y = x^3 - 3x^2 - 9x + 15$ and sketch its graph.

SOLUTION

First, find the first and second derivatives:

$$y = x^2 - 3x^2 - 9x + 15$$

$$y' = 3x^2 - 6x - 9$$

$$y'' = 6x - 6$$

Next, look for relative maxima and minima at points where $y' = 0$:

Figure 5.29

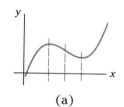

(a)

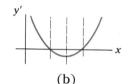

(b)

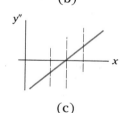

(c)

$$0 = 3x^2 - 6x - 9$$

$$0 = 3(x^2 - 2x - 3)$$

$$0 = 3(x + 1)(x - 3)$$

$$\therefore \quad x = -1 \text{ or } x = 3$$

At $x = -1,\ y'' = 6(-1) - 6$

$$= -12$$

$\therefore \quad y'' < 0$ so there is a maximum at $x = -1$.

The y-coordinate of the maximum is

$$y = (-1)^3 - 3(-1)^2 - 9(-1) + 15$$

$$= 20$$

The maximum point is $(-1, 20)$.

At $x = 3,\quad y'' = 6(3) - 6$

$$= 12$$

$\therefore \quad y'' > 0$ so there is a minimum at $x = 3$.

The minimum point is $(3, -12)$.

Now, look for points of inflection at places where $y'' = 0$:

$$0 = 6x - 6$$

$$x = 1$$

$\therefore$ An inflection point exists at $x = 1$,
The inflection point is $(1, 4)$.

The analysis of the function is now complete. A sketch of its graph is shown in Figure 5.30. ❑

Figure 5.30

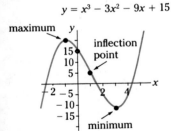

$y = x^3 - 3x^2 - 9x + 15$

Exercise 5.4

1. State the sign of y' and y'' for each of the graphs in Figure 5.31.

Figure 5.31

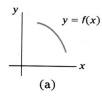

(a)

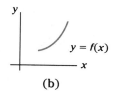

(b)

(c)

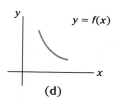

(d)

2. At which point on the graph in Figure 5.32 is the value of

(a) y positive, y' positive, and y'' negative?

(b) y positive, y' negative, and y'' negative?

(c) y negative, y' positive, and y'' positive?

(d) y negative, y' zero,   and y'' positive?

(e) y zero, y' positive, and y'' negative?

Figure 5.32

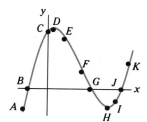

3. Sketch the graph of the function $y = f(x)$ in the neighbourhood of $x = a$ where

(a) $a = 3$, $f(a) = 2$, $f'(a) = 0$, and $f''(a) > 0$

(b) $a = -1$, $f(a) = 0$, $f'(a) > 0$, and $f''(a) > 0$

(c) $a = 2$, $f(a) = -1$, $f'(a) = 0$, and $f''(a) < 0$

(d) $a = 0$, $f(a) = 3$, $f'(a) < 0$, and $f''(a) < 0$

(e) $a = -3$, $f(a) = -1$, $f'(a) > 0$, and $f''(a) = 0$
(two possibilities)

4. Sketch the graphs of y' and y'' for the functions whose graphs are shown in Figure 5.33.

Figure 5.33

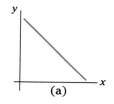

(a)

(b)

(c)

5. Find the points of inflection, if any, of the following curves:

(a) $y = x^3 - 3x + 2$

(b) $y = 3x^3 - x^2 - 12x + 4$

(c) $y = 2x^2 - 5x + 6$

(d) $y = x^4 - 2x^2 + 1$

(e) $y = x^4 + 2x^2 + 1$

(f) $y = \dfrac{1}{x^2 + 1}$

6. Find the stationary points and determine if they are maxima or minima by evaluating the second derivative at those points.

(a) $y = (x - 3)^2 + 5$

(b) $y = \frac{1}{3}x^3 - \frac{1}{2}x^2 + 6x$

(c) $y = 3x^4 - 8x^3 + 6x^2$

7. Find the maximum and minimum points and the points of inflection, and then graph the following functions.

(a) $y = x^3 - 3x^2 + 4x$

(b) $y = x^4 - 4x^3$

(c) $y = \dfrac{x}{x^2 + 2}$

(d) $y = \dfrac{x^2 - 1}{x^3}$

(e) $y = \dfrac{2}{\sqrt{x^2 + 4}}$

8. Show that the x-coordinate of the point of inflection of the curve $y = 2x^3 + x^2 - 5x + 1$ is midway between the x-coordinates of the maximum and minimum points.

9. If the slope of the curve $y = ax^3 + bx^2 + cx$ is 4 at its point of inflection $(1, 5)$, find a, b, and c.

5.5 Summary and Review

The graph of a function $y = f(x)$ gives an overview of the relation between the independent and dependent variables x and y. There are many things to examine for clues as to the nature of the graph.

The graph of f:

—has a y-intercept at $x = 0$

—has an x-intercept at any value of x
for which $f(x) = 0$

—has a vertical asymptote at $x = a$
if $\lim\limits_{x \to a} f(x) = \infty$

—has a horizontal asymptote at $y = L$
if $\lim\limits_{x \to \pm\infty} f(x) = L$

—is symmetric about the y-axis
if it is an even function, that is, $f(-x) = f(x)$

—is symmetric about the origin
if it is an odd function, that is, $f(-x) = -f(x)$

—is increasing in an interval
where $f'(x) > 0$

—is decreasing in an interval
where $f'(x) < 0$

—is concave up in an interval
where $f''(x) > 0$

—is concave down in an interval
where $f''(x) < 0$

—has a relative maximum at $x = a$
if $f'(a) = 0$ and $f''(a) < 0$

—has a relative minimum at $x = a$
if $f'(a) = 0$ and $f''(a) > 0$

—has an inflection point at $x = a$
if $f''(a) = 0$ and $f'(a) \neq 0$

—has either a relative extremum or
an inflection point at $x = a$
if both $f'(a) = 0$ and $f''(a) = 0$.

You can increase your skill in graphing

(a) by being thorough and systematic in looking for intercepts, asymptotes, symmetry, extreme values, and inflection points. Any of these items may be significant in a particular graph.

(b) by being aware of the implications of each thing you find out. If, for instance, you discover that a continuous function has a maximum and a minimum, you can infer that it *must* have an inflection point between them.

Exercise 5.5

1. Evaluate the following limits:

(a) $\displaystyle\lim_{x \to -5} \left(\frac{2x}{x + 5} + \frac{10}{x + 5} \right)$

(b) $\displaystyle\lim_{x \to 1} \frac{x^3 - 1}{x - 1}$

(c) $\displaystyle\lim_{x \to 1} \frac{x - 1}{\sqrt[3]{x - 1}}$

(d) $\displaystyle\lim_{x \to \infty} (\sqrt{x^2 + 9} - x)$

(e) $\displaystyle\lim_{x \to \infty} x(\sqrt{x^2 + 1} - x)$

(f) $\displaystyle\lim_{x \to 0} \frac{\sqrt{4 + x^2} - 2}{x}$

(g) $\displaystyle\lim_{x \to 0} \frac{\sqrt{1 + x} - 1}{x^2}$

2. A function f is defined only for $x > 2$. As $x \to +\infty$ $f(x) \to 2$ and as $x \to 2^+$, $f(x) \to \infty$.

(a) Sketch the graph of a function that satisfies these conditions.

(b) Is it possible for the graph of a function satisfying these conditions to cross the x-axis? Explain.

3. (a) Sketch the graph of a function that satisfies the following two conditions:

i. the x-axis and y-axis are asymptotes.

ii. it has an x-intercept at $x = 1$.

(b) Invent a function, that satisfies these conditions. Express it in the form $f(x) = \cdots$.

4. (a) Sketch the graph of a function that satisfies the following three conditions:

i. The graph passes through the origin.

ii. $\displaystyle\lim_{x \to \pm\infty} f(x) = 0$

iii. there is no value a for which $\displaystyle\lim_{x \to \pm a} f(x) = \infty$

(b) Write down an equation that would define a function satisfying these conditions.

5. Show that the line $y = 2x/3$ is an oblique asymptote of the graph of

$$f(x) = 2\sqrt{\frac{x^2}{9} - 1}$$

6. The function f has a relative maximum at $x = 3$. What can you say about the value of $f(3)$? $f'(3)$? $f''(3)$? Where is f increasing? Decreasing? Concave up? Concave down?

7. Sketch the graph of a function f in the neighbourhood of an inflection point at which the value of f' is

   (a) positive and a maximum

   (b) positive and a minimum

   (c) negative and a maximum

   (d) negative and a minimum.

8. Sketch the graph of a function that satisfies the conditions in each of the following cases.

9. The function $y = f(x)$ is positive for all values of x, and continuous except at $x = -2$, where it has a vertical asymptote. If it has a relative maximum at $x = 2$, what other relative extrema and inflection points *must* it have? Make a sketch.

10. Find the maximum, the minimum, and the point of inflection of the graph of $y = x^5 - 5x^4$. Sketch the graph.

11. Show that the graph of the function

$$f(x) = \frac{1}{1 + 3x^2}$$

has one maximum and two points of inflection. Sketch the graph.

12. Find the stationary points and the points of inflection of each of the following functions. Sketch the graph in each case.

   (a) $y = \dfrac{3x}{x^2 + 1}$

   (b) $y = \dfrac{x^2}{x + 2}$

   (c) $y = \dfrac{x^2 - 1}{x^2 + 1}$

   (d) $y = \dfrac{5x}{5 + x}$

13. Analyze each of the following functions and sketch their graphs.

   (a) $y = \dfrac{1}{\sqrt{1 - x^2}}$

   (b) $y = x - \sqrt{x}$

   (c) $y = \dfrac{4x^2}{\sqrt{x - 1}}$

14. Determine the absolute maximum and minimum values of the following functions.

   (a) $y = (x - 2)(x - 4)(x - 6), \quad 0 \le x \le 4$

   (b) $y = \dfrac{x}{\sqrt{x + 1}}, \quad x \le 8$

E.   Problem Solving

The first five chapters of this textbook have concentrated on the basic concepts of calculus and on the mechanical process of finding the derivative of a function. Now you are ready to use these skills in applying calculus to many types of problems.

Certain general steps should already be a part of your problem-solving routine:

1. Read the question carefully. What is given? What is required?

2. Determine the relationship between the variables. What is the equation?

3. Solve the equation. What method of solution is appropriate?

4. Express the result in a suitable form. Does the answer make sense?

The object in this section is to focus on what is sometimes the most troublesome aspect of solving applications problems: getting the equation. Three things illustrated in the following examples will help you to determine the equation for a problem:

1. the use of a diagram

2. the use of word equations

3. the use of units

EXAMPLE A

Equal rectangular pieces are cut from two adjacent corners of a square piece of material 90 cm on a side in such a way that the remaining T-shaped piece can be folded to form a six-sided rectangular box. Express the volume of the box in terms of one of its edges.

SOLUTION

The dimensions of the box cannot all be independent. If the box is tall, its base must be small. The base could be larger if the height is low. An equation relating the volume of the box to its height, for example, would enable you to determine the shape that would enclose the maximum volume under the given conditions.

A diagram of how the material could be cut and folded is essential. One possibility is shown in Figure E.1, where the folds are indicated by dotted lines, and l, w, and h represent the length, width, and height of the box respectively.

Figure E.1

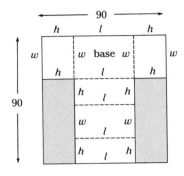

The volume of the box is given by

$$v = l \cdot w \cdot h$$

The folds are made in such a way, that horizontally, across the top of the T,

$$90 = 2h + l$$

and vertically,

$$90 = 2h + 2w$$

Thus, both l and w and ultimately V can be expressed in terms of h:

$$l = 90 - 2h$$

$$w = \frac{90 - 2h}{2}$$

$$V = \frac{90 - 2h}{2} \cdot (90 - 2h) \cdot h$$

$$= \frac{1}{2}(90 - 2h)^2 h$$

Now that you have an equation which gives the volume of the box as a function of its height, you can find the height at which the volume is a maximum by setting the derivative $dV/dh = 0$. ❏

EXAMPLE B

The fuel consumption rate of a tractor-trailer rig, measured in litres per 100 km, is equal to $0.15v$, where v is the speed of the truck. If drivers are paid \$10/h and fuel costs 48¢/L, find an equation that expresses the total cost of a 450 km trip as a function of the speed of the truck.

SOLUTION

Observe that fuel costs are the greatest for fast trips made at high speeds, whereas a driver's wages are the greatest for slow trips made at low speeds. An equation relating the total cost C to the speed v would enable a trucking company to find the most economical speed at which to transport goods.

To account for all the quantities affecting the cost, it helps to state the relationships in words and to take note of the units. Two things contribute to the total cost:

$$\text{total cost} = \text{fuel cost} + \text{driver's wages}$$

The way in which each of these depends on the speed is determined as follows:

$$
\begin{aligned}
\frac{\text{cost}}{\text{of fuel}} &= \frac{\text{price}}{\text{per litre}} \left(\frac{\$}{L}\right) \times \frac{\text{number}}{\text{of litres}} (L) \\[2mm]
&= \frac{\text{price}}{\text{per litre}} \left(\frac{\$}{L}\right) \times \frac{\text{rate of fuel}}{\text{consumption}} \left(\frac{L}{100 \text{ km}}\right) \times \text{distance} (100 \text{ km}) \\[2mm]
&= (0.48) \times (0.15v) \times (4.5) \\[2mm]
&= 0.324v
\end{aligned}
$$

$$
\begin{aligned}
\frac{\text{driver's}}{\text{wages}} &= \text{hourly rate} \left(\frac{\$}{h}\right) \times \text{travel time (h)} \\[2mm]
&= \text{hourly rate} \left(\frac{\$}{h}\right) \times \frac{\text{distance}}{\text{speed}} \left(\frac{\text{km}}{\text{km/h}}\right) \\[2mm]
&= (10.00) \times \frac{450}{v} \\[2mm]
&= \frac{4500}{v}
\end{aligned}
$$

Therefore, the total cost is given by the equation

$$C = 0.324v + \frac{4500}{v}$$

Now that you know how the cost depends on the speed, you can go on to determine the most economical speed, that is, the speed at which the cost is a minimum, by setting the derivative $dC/dv = 0$. ◻

Exercise E

1. A water trough 5 m long and 80 cm wide at its base has a trapezoidal cross section, the sides of which make an angle of 120° with the base.

   (a) Draw a diagram of the trough partly filled with water.

   (b) How does the width of the water surface depend on its depth?

   (c) Find an equation that relates the volume of water in the trough to its depth.

2. A ladder of length l, leaning against a building at a steep angle, begins to slide.

   (a) Draw a diagram of the ladder in reference to an xy-coordinate system. Denote the positions of the top and the bottom of the ladder by suitable variables.

   (b) Find an equation relating the position of the top of the ladder to the position of the bottom of the ladder.

   (c) When the ladder slides, what quantities change? What quantities do not change?

3. A rectangular container made from 3 m² of plywood has a height h, a square base of side x, and no top.

   (a) Draw a diagram of the container.

   (b) How does the height change if the area of the base is increased? Decreased?

   (c) Write an equation for the volume of the container in terms of h and x.

   (d) Write an equation that expresses the 3 m² plywood surface area of the container in terms of h and x.

   (e) Using the formula in (d), express the volume as a function of x only.

   (f) How does the volume change if the area of the base is increased? Decreased?

4. A chain of variety stores has 64 outlets in the city and surrounding area, each averaging $5000 income per week. An efficiency study reveals that the average weekly income per store drops $50 for each new store opened.

   (a) What is the total company income per week at present?

   (b) What would the total company income per week be if one new store is opened?

   (c) Express the total company income per week as a function of the number of new stores opened.

   (d) Why would each store's income decline with the opening of more stores?

Extreme Value Problems

6

You have learned to look for the extreme values of a function at points where the first derivative of the function is zero. This is a straightforward mathematical process. In applications of calculus, there are additional requirements. Often the function in question is not explicitly given; you must first construct it. Furthermore, when calculations have been completed, the result should be interpreted or explained in a meaningful way in the context of the original problem.

Extreme Values of Areas and Volumes

6.1

It is important to understand in a qualitative way how a maximum or minimum value arises in a given situation. Consider the problem of enclosing a rectangular plot of ground by a given length of fencing. What shape should the rectangular plot have in order to maximize the area enclosed? Since the length of the fence around the perimeter is constant, the length and width of the rectangular plot cannot be varied independently. If the width is increased, the length must decrease, and vice versa. Thus, the rectangular plot could be wide and short, or narrow and long as in Figure 6.1. In either case, as the length or the width approaches zero, the area approaches zero.

Figure 6.1

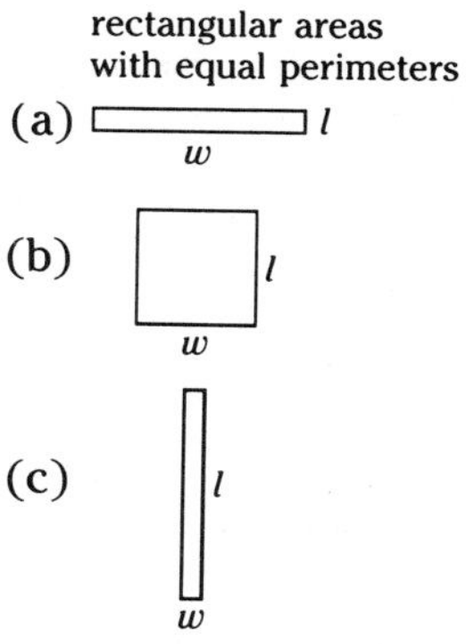

rectangular areas
with equal perimeters

(a)

(b)

(c)

The point is that somewhere between these two limiting cases, there *must* be a certain shape for which the area is a maximum. The problem then is to *maximize* the area subject to the *constraint* that the perimeter is a constant. The equation for the perimeter, $P = 2l + 2w$, is known as an **equation of constraint** in this context.

Figure 6.2

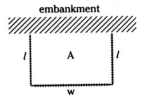

embankment

l A l

w

EXAMPLE A

A rectangular field is to be enclosed using 300 m of fencing along three of its sides, the fourth side being a natural embankment. Determine the dimensions and area of the field having a maximum area.

SOLUTION

Figure 6.2 shows the area to be maximized. It is

$$A = lw$$

The variables l and w are not independent, however, The constraint in this problem is that the fencing surrounding three sides of the field must have a length of 300 m. The equation of constraint in this case is

$$300 = 2l + w$$

The equation of constraint is used to eliminate one of the variables:

$$w = 300 - 2l$$

$$\therefore \quad A = l(300 - 2l)$$

$$= 300l - 2l^2$$

The area A is now expressed as a function of a single independent variable. To maximize the area, differentiate and look for the value of l that makes $dA/dl = 0$:

$$\frac{dA}{dl} = 300 - 4l$$

$$0 = 300 - 4l$$

$$\therefore \quad l = 75$$

$$w = 300 - 2(75)$$

$$= 150$$

$$A = (75)(150)$$

$$= 11\ 250$$

Therefore, a field of length 75 m and width 150 m will have the maximum area, which is 11 250 m². ◻

Usually you know already from the nature of a problem if a result is a maximum or a minimum. To be certain, remember that you can always check the sign of the second derivative.

EXAMPLE B

A manufacturer wishes to produce cylindrical fruit juice cans with a capacity of 250 mL. What dimensions will minimize the amount of material required for a can? (1 mL = 1 cm³)

SOLUTION

The amount of material required is proportional to the total surface area of the container, which depends on its height h and its radius r. The problem is to minimize the surface area $A = 2\pi rh + 2\pi r^2$ subject to the constraint that the volume $V = \pi r^2 h$ is a constant. Note that if the radius of the container is increased, the height must be decreased, and vice versa, in order to keep the volume constant. The container, therefore, could be flat and wide, or tall and narrow as in Figure 6.3. Somewhere between these two extremes is a shape for which the surface area of the cylinder is a minimum.

The equation of constraint is

$$250 = \pi r^2 h$$

Solving for h (which is simpler to eliminate than r) gives

$$h = \frac{250}{\pi r^2}$$

$$\therefore \quad A = 2\pi r\left(\frac{250}{\pi r^2}\right) + 2\pi r^2$$

$$= \frac{500}{r} + 2\pi r^2$$

To minimize the surface area, differentiate and look for the value of r that makes $dA/dr = 0$:

$$\frac{dA}{dr} = -\frac{500}{r^2} + 4\pi r$$

$$0 = -\frac{500}{r^2} + 4\pi r$$

$$r^3 = \frac{500}{4\pi}$$

$$= \frac{125}{\pi}$$

$$\therefore r = \frac{5}{\sqrt[3]{\pi}}$$

$$\doteq 3.414$$

Figure 6.3

cylindrical containers with equal volumes

(a)

(b)

(c)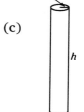

If the radius is 3.414 cm, the surface area is minimized. The height of the can is then

$$h = \frac{250}{\pi r^2}$$

$$= \frac{250}{\pi} \cdot \left(\frac{\sqrt[3]{\pi}}{5}\right)^2$$

$$= \frac{10}{\sqrt[3]{\pi}}$$

$$\doteq 6.828$$

The height of the can is 6.828 cm. Therefore, if the can is made with its height equal to twice its radius, the amount of material required for its manufacture will be minimized. ◻

EXAMPLE C

Determine the dimensions of the cylinder of maximum volume that can be inscribed in a sphere of radius 10 cm.

SOLUTION

The volume to be maximized is

$$V = \pi r^2 h$$

Figure 6.4

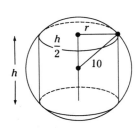

where h is the height and r is the radius of the cylinder. However, h and r cannot vary independently if the cylinder is constrained to fit inside a sphere of radius 10 cm as shown in Figure 6.4. An equation of constraint that relates h and r can be found by noting that $h/2$ and r form the legs of a right triangle, the hypotenuse of which must be the radius of the sphere:

$$\left(\frac{h}{2}\right)^2 + r^2 = 10^2$$

Thus,

$$r^2 = 100 - \frac{h^2}{4}$$

$$V = \pi\left(100 - \frac{h^2}{4}\right)h$$

$$= 100\pi h - \frac{\pi h^3}{4}$$

The volume of the cylinder is maximized by looking for the value of h which makes $dV/dh = 0$:

$$\frac{dV}{dh} = 100\pi - \frac{3}{4}\pi h^2$$

$$0 = 100\pi - \frac{3}{4}\pi h^2$$

$$h^2 = \frac{400}{3}$$

$$h = \frac{20}{\sqrt{3}}$$

$$\doteq 11.55$$

$$r^2 = 100 - \frac{1}{4}\left(\frac{400}{3}\right)$$

$$= \frac{200}{3}$$

$$r = \frac{10\sqrt{2}}{\sqrt{3}}$$

$$\doteq 8.17$$

Thus, the volume of the cylinder will be a maximum when it has a radius of 16.33 cm and a height of 11.55 cm. In other words, the maximum volume occurs when the ratio of height to radius is $\sqrt{2}$.

□

Exercise 6.1

1. Consider the problem of finding the largest rectangular plot of ground that can be enclosed with 300 m of fencing.

   (a) What is the equation of constraint?

   (b) What quantity is to be maximized? Express it as a function of a single variable.

   (c) Find the dimensions of the largest plot of ground.

2. Three hundred metres of fencing is available to enclose a rectangular field and divide it into two smaller rectangular plots. What should the dimensions of the field be for the area to be a maximum?

3. Find the lengths of the sides of an isosceles triangle which has a perimeter of 12 m and a maximum area.

4. A gable window has the form of a rectangle topped by an equilateral triangle, the sides of which are equal to the width of the rectangle. Find the maximum area of the window if the perimeter is 600 cm.

5. A farmer has 54 m of fencing with which to build two animal pens with a common side. One pen is rectangular; the other is square. If the area of the pens is to be maximized, what are their dimensions?

6. Find the dimensions and area of the largest rectangle having its two lower corners on the x-axis and its two upper corners on the parabola $y = 16 - x^2$.

7. Consider the problem of finding the largest open-topped rectangular box that has a square base and a total surface area of 3600 cm^2.

   (a) What is the equation of constraint?

   (b) What quantity is to be maximized? Express it as a function of a single variable.

   (c) Find the dimensions of the largest box.

   (d) What is its volume?

8. A rectangular open-topped box is to be made from a piece of material 18 cm by 48 cm by cutting a square from each corner and turning up the sides. What size squares must be removed to maximize the capacity of the box?

9. A right circular cone is inscribed in a sphere of radius 15 cm. Find the dimensions of the cone that has the maximum volume. Hint: The volume of a cone is $V = \pi r^2 h/3$.

10. A cone-shaped paper drinking cup is to hold 36 cm^3 of water. Find the height and radius of the cup that will require the least amount of paper to make. The lateral surface area of a cone is $A = \pi rs$, where r is the radius and s is the slant height of the cone.

11. A cylinder of radius r is inscribed in a cone of height H and base radius R. Show that the maximum volume of the cylinder is 4/9 the volume of the cone.

12. Find the equation of the tangent to the graph of $y = 10 - x^2, x, y \geq 0$, that forms the triangle of minimum area with the positive x-axis and y-axis.

6.2 Extreme Values of Distances and Times

Certain applications of calculus require that the positions and the paths of moving objects be described. It is helpful to approach all such problems from the same standpoint in a systematic way.

Organize your work around an xy-coordinate system. Relate compass directions to this system by taking the position of an object north or east of the origin to be positive, and south or west of the origin to be negative. Consider an object that is travelling north or east to have a positive velocity, and one that is travelling south or west to have a negative velocity. Remember that when an object is moving at a constant speed,

$$\text{distance travelled} = \text{speed} \times \text{travel time}$$

Making diagrams such as those in Figure 6.5 and labelling them carefully will help you to arrive at the required functional relationships.

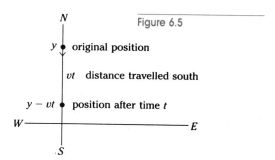

Figure 6.5

original position

vt distance travelled south

$y - vt$ position after time t

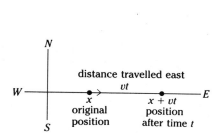

distance travelled east

original position

position after time t

EXAMPLE A

A light airplane flying east at 300 km/h passes over Buttonville airport 10 min before a second airplane flying south at 420 km/h passes over the same point. Assuming that both the airplanes are at the same altitude,

(a) at what time is the distance between them a minimum?

(b) what is the minimum distance?

Figure 6.6

SOLUTION

Choose the origin of the coordinate system to be at the airport, where the flight paths of the two airplanes intersect. Denote the positions of the two airplanes by x and y, respectively. Initially, the first airplane is at the origin: $x = 0$. The second airplane, travelling at a speed of 420 km/h for 10 min covers a distance of

$$y = \frac{420\,\text{km}}{\text{h}} \cdot \frac{1\,\text{h}}{60\,\text{min}} \cdot 10\,\text{min}$$

$$= 70\,\text{km}$$

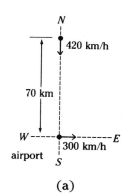

(a)

Therefore, the second airplane must initially be 70 km north of the airport: $y = 70$. This situation is shown in Figure 6.6a.

In t hours, the first airplane travels east a distance of $300t$ km, and the second airplane travels south a distance of $420t$ km. After t hours, each aircraft has changed its position to that shown in Figure 6.6b. The distance l between the two airplanes is given by

$$l^2 = x^2 + y^2$$

where $x = 300t$

and $y = 70 - 420t$

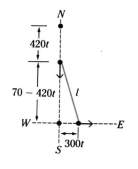

(b)

$$\therefore \quad l^2 = (300t)^2 + (70 - 420t)^2$$

(a) The distance between the airplanes will be a minimum at a time t for which $dl/dt = 0$. In calculating the derivative, it is simpler to use the chain rule, rather than multiply out the quadratic expressions, and to use implicit differentiation, rather than deal with radicals:

$$\frac{d}{dt}l^2 = \frac{d}{dt}(300t)^2 + \frac{d}{dt}(70 - 420t)^2$$

$$2l \cdot \frac{dl}{dt} = 2(300t)(300) + 2(70 - 420t)(-420)$$

After dividing out a common factor of 2,

$$\frac{dl}{dt} = \frac{90\,000t - 29\,400 + 176\,400t}{l}$$

$$0 = \frac{266\,400t - 29\,400}{l}$$

$$\therefore \quad t = \frac{29\,400}{266\,400}$$

$$\doteq 0.11$$

In 0.11 h or 6.62 min after the first airplane passes over the airport, the distance between the two airplanes is a minimum.

(b) When $t = 0.11$

$$x = (300)(0.11)$$

$$= 33$$

$$y = 70 - (420)(0.11)$$

$$= 23.8$$

$$l^2 = (33)^2 + (23.8)^2$$

$$= 1655.4$$

$$l = 40.7$$

The minimum distance between the airplanes is 40.7 km.   ❏

EXAMPLE B

A pair of scuba divers wish to dive on a wreck that lies 0.4 km off the shore from a point 0.8 km from their present position. If they can walk carrying their gear at 5 km/h and swim at 3 km/h, what course should they follow to reach the wreck in the minimum time?

SOLUTION

Choose the origin of a coordinate system to be at the point on shore directly opposite the wreck, as in Figure 6.7. The total time it takes to reach the wreck is

$$\frac{\text{total}}{\text{time}} = \frac{\text{walking}}{\text{time}} + \frac{\text{swimming}}{\text{time}}$$

Since travel time = distance/speed when speed is a constant,

$$\frac{\text{total}}{\text{time}} = \frac{\text{distance walked}}{\text{walking speed}} + \frac{\text{distance swum}}{\text{swimming speed}}$$

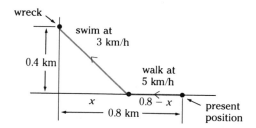

Figure 6.7

Let x represent the position on shore at which the divers begin their swim. Then,

$$T = \frac{0.8 - x}{5} + \frac{\sqrt{x^2 + 0.4^2}}{3}$$

Now look for the point x which minimizes the total time by setting $dT/dx = 0$:

$$\frac{dT}{dx} = -\frac{1}{5} + \frac{1}{3} \cdot \frac{1}{2} \cdot \frac{2x}{\sqrt{x^2 + 0.4^2}}$$

$$0 = -\frac{1}{5} + \frac{x}{3\sqrt{x^2 + 0.4^2}}$$

$$5x = 3\sqrt{x^2 + 0.4^2}$$

$$25x^2 = 9(x^2 + 0.16)$$

$$16x^2 = 1.44$$

$$x^2 = .09$$

$$x = 0.3$$

Therefore, to minimize the travel time, the divers should walk 0.5 km until they are 0.3 km from the point on shore opposite the wreck, and then swim from there directly to the wreck.   ❑

You may discover in some problems that an extreme value of a function cannot be found by setting the first derivative equal to zero. This simply means that there is no relative maximum or minimum. In such cases, you should look for an *absolute* maximum or minimum at the endpoints of the domain of the function.

EXAMPLE C

Find the point on the circle $x^2 + y^2 = 1$ that is closest to the point $(3, 0)$.

SOLUTION

The distance from any point (x, y) to $(3, 0)$ is given by

$$s^2 = (x - 3)^2 + y^2$$

This distance is to be minimized subject to the constraint that the point (x, y) lies on the circle $x^2 + y^2 = 1$, as in Figure 6.8. Using $y^2 = 1 - x^2$ to eliminate y^2 gives

Figure 6.8

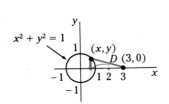

$$s^2 = (x - 3)^2 + 1 - x^2$$

$$= x^2 - 6x + 9 + 1 - x^2$$

$$= 10 - 6x$$

$$s = \sqrt{10 - 6x}$$

Upon differentiating,

$$\frac{ds}{dx} = \frac{1}{2} \cdot \frac{1}{\sqrt{10 - 6x}} \cdot (-6)$$

$$= \frac{-3}{\sqrt{10 - 6x}}$$

There is *no* value of x for which $ds/dx = 0$. By itself, the function $f(x) = \sqrt{10 - 6x}$ appears to have domain $x \leq 10/6$. However, the constraint that (x, y) lies on the circle further restricts the domain of f to $-1 \leq x \leq 1$. At the right end of the domain, at $x = 1$, the distance is 2 units. This corresponds to an absolute minimum of f in the domain. Thus the point $(1, 0)$ is the point on the circle closest to $(3, 0)$. ◻

Exercise 6.2

1. At 1300 hours, a merchant ship sailing south at 18 knots is 40 nautical miles due east of a patrol boat travelling east at 24 knots. When will they be closest to each other?
   (1 knot = 1 nautical mile per hour)

2. A toy tugboat is launched from the side of a pond and travels north at 5 cm/s. At the same moment, a toy launch starts from a point $8\sqrt{2}$ m northeast of the tugboat and travels west at 7 cm/s. How closely do the two boats approach each other?

3. At 11 a.m., a 747 jet is travelling east at 800 km/h. At the same instant, a DC-8 is 45 km east and 90 km north of the 747. It is at the same altitude travelling south at 600 km/h. What is the closest distance of approach of the planes, and at what time does it occur? What are the relative positions of the two planes at this moment?

4. Find the point on the graph of $y = x^2/\sqrt{2}$ that is closest to the point $(30, 0)$. What is the minimum distance?

5. Find the point on the parabola $x = y^2 - 8y + 18$ closest to $(-2, 4)$.

6. Find the point on the hyperbola

$$\frac{x^2}{1^2} - \frac{y^2}{2^2} = 1 \qquad \text{closest to}$$

   (a) $(6, 0)$
   (b) $(5, 0)$
   (c) $(4, 0)$

7. Two isolated farms are situated 12 km apart on a straight country road that runs parallel to the main highway 20 km away. The power company decides to run a wire from the highway to a junction box, and from there, wires of equal length to the two houses. Where should the junction box be placed to minimize the length of wire needed?

8. Given the function $f(x) = \sqrt{9 - x}, 0 \leq x \leq 9$, show that there is no value of x for which $dy/dx = 0$. Determine the maximum and minimum values of the function.

9. A sailor in a boat 8 km off a straight coastline wants to reach a point on shore 10 km from the point directly opposite her present position in the shortest possible time. Toward what point on the shore should she steer and how long does it take her to reach her destination if

   (a) she can row at 4 km/h and run at 6 km/h?
   (b) she can row at 4 km/h and walk at 5 km/h?

6.3 Extreme Values in Economics

A prime objective in any business enterprise is to maximize profits or revenues and minimize costs. This section will describe how to construct functions for these quantities and determine their extreme values.

The total revenue or income received from selling a product depends on the selling price of the product and on the number of units of the product sold:

revenue = (price per unit) × (number of units)

The price of a product may itself depend in some way on the number of units sold. If, for instance, the demand for a product is high, then a large number of units can be sold and the price per unit may go down. Any function which describes how the revenue depends on the number of units sold is called a **revenue function**.

EXAMPLE A

A railway company offers excursions at $120 per person for tour groups of up to 100 persons. If the size of a group is greater than 100, the company will reduce the price of *every* ticket by 50 cents for each person in excess of 100. What size tour group would produce the greatest revenue?

SOLUTION

The total revenue R depends on the price of a ticket and on the number of tickets sold. For instance, if there are 100 persons on the tour, then

$R = \$120$ per person $\times 100$ persons

However, if there are x persons in excess of 100 on the tour, the ticket price is reduced to $(120 - 0.5x)$. The revenue function is therefore

$$R(x) = (120 - 0.5x)(100 + x)$$

The greatest revenue is found at the value of x which makes $dR/dx = 0$:

$$\frac{dR}{dx} = (120 - 0.5x) + (100 + x)(-0.5) \qquad \longleftarrow \text{product rule}$$

$$0 = 120 - 0.5x - 50 - 0.5x$$

$$x = 70$$

Under this pricing scheme, the company would receive the greatest revenue from a tour group of 170 people. For a group of this size, the ticket price would be $85, and the total revenue, $14 450. ❏

A **cost function** in simple cases can be worked out in a manner much like the revenue function:

Total cost = cost per unit × number of units

EXAMPLE B

An open-topped storage box with a square base is to have a capacity of 5 m³. Material for the sides costs $1.60/m², while that for the bottom costs $2.00/m². Find the dimensions that will minimize the cost of the material. What is the minimum cost?

SOLUTION

The total cost is the cost of the material for the sides plus the cost of the material for the bottom. If the box has a height h and a base of side x, then

$$\text{cost of sides} = \text{cost per unit area} \times \text{area of sides}$$

$$= \$1.60 \times 4hx$$

$$\text{cost of bottom} = \text{cost per unit area} \times \text{area of bottom}$$

$$= \$2.00 \times x^2$$

Therefore, the total cost C is

$$C = 1.60\,(4hx) + 2.00(x^2)$$

$$= 6.4hx + 2x^2$$

The cost of the box is to be minimized subject to the constraint that the volume hx^2 is 5 m². The variable h can thus be eliminated using

$$h = \frac{5}{x^2}$$

so that

$$C = 6.4 \cdot \frac{5}{x^2} \cdot x + 2x^2$$

$$= \frac{32}{x} + 2x^2$$

The minimum cost is found by looking for the value of x that makes $dC/dx = 0$:

$$\frac{dC}{dx} = -\frac{32}{x^2} + 4x$$

$$0 = -\frac{32}{x^2} + 4x$$

$$4x^3 = 32$$

$$x = 2$$

Thus, when the base of the storage box is 2 m on a side and the height is $5/2^2$, or 1.25 m, the cost is a minimum.

$$C = 6.4(1.25)(2) + 2(2)^2$$

$$= 24$$

The minimum cost is \$24.   ❏

Generally, the costs incurred in producing x units of a commodity are of two kinds. **Fixed costs** are cost items such as rent, depreciation on machinery, etc., which continue even when nothing is produced. **Variable costs** include the cost of materials and labour, which depend on the actual number of items produced. In general, the **cost function** is therefore

$$C(x) = \text{variable costs} + \text{fixed costs}$$

The cost of producing *one* unit is the **average cost** defined by

$$\text{Average cost} = \frac{\text{total cost}}{\text{number of units}}$$

$$A(x) = \frac{C(x)}{x}$$

In manufacturing, it is of interest to minimize the average cost of production.

EXAMPLE C

The variable cost of producing x refrigerators is $C(x) = 0.2x^2 + 130x$. If the fixed costs are \$50 000, for what number of refrigerators will the average production cost be a minimum? What is the minimum average cost?

SOLUTION

Total cost = variable costs + fixed costs

$$C(x) = (0.2x^2 + 130x) + (50\,000)$$

The average cost is

$$A(x) = \frac{C(x)}{x}$$

$$= 0.2x + 130 + \frac{50\,000}{x}$$

The minimum value of the average cost is found at the value of x which makes $dA/dx = 0$:

$$\frac{dA}{dx} = 0.2 - \frac{50\,000}{x^2}$$

$$0 = 0.2 - \frac{50\,000}{x^2}$$

$$x^2 = \frac{50\,000}{0.2}$$

$$= 250\,000$$

$$x = 500$$

Thus, the average cost of production will be a minimum when 500 refrigeration units are produced. Then,

$$A(500) = 0.2(500) + 130 + \frac{50\,000}{500}$$

$$= 330$$

The minimum value of the average cost to produce one refrigerator is $330. ❏

While it is reasonable to want to minimize the average cost of production, a person in business will more likely want to maximize the profit. The profit function is found by subtracting the costs from the revenue:

$$P(x) = R(x) - C(x)$$

The value of x which minimizes the cost is not necessarily the one which maximizes the profit.

EXAMPLE D

A manufacturer of calculators produces x calculators per day at a daily cost in dollars of $C(x) = 40x - 0.035x^2 + 1250$. If the calculators are sold for $60 - 0.05x$ each, find the value of x that maximizes the daily profit. What is the selling price when the profit is a maximum?

SOLUTION

The total revenue is given by

$$R(x) = \begin{matrix} \text{price per} \\ \text{calculator} \end{matrix} \times \begin{matrix} \text{number of} \\ \text{calculators} \\ \text{sold} \end{matrix}$$

$$= (60 - 0.05x)x$$

The profit is therefore

$$\begin{aligned} P(x) &= R(x) - C(x) \\ &= (60 - 0.05x)x - (40x - 0.035x^2 + 1250) \\ &= -0.015x^2 + 20x - 1250 \end{aligned}$$

To find the maximum profit, set $dP/dX = 0$:

$$\frac{dP}{dx} = -0.03x + 20$$

$$0 = -0.03x + 20$$

$$x = \frac{20}{0.03}$$

$$= 667$$

The profit will be a maximum when 667 calculators per day are made and sold. (Although in calculus, x is a continuous variable, in the present context, only positive integer values of x are meaningful.) The selling price is \$26.65. An increase in the selling price would result in fewer calculators sold and less profit. A decrease in the selling price would result in more calculators sold, but not enough to make up for the loss in profit due to the lower price. ❑

Exercise 6.3

1. A real estate firm owns 250 apartments that can be rented at \$460 per month each. For each \$5 per month increase in rent there are two vacancies created that cannot be filled. What should the monthly rent be to maximize the total revenue? What is the maximum revenue?

2. The Bigger Battery Company sells car batteries at \$60 each. They give a discount of $36n$ cents per battery, where n is the number of batteries purchased. Find the number of batteries the company should persuade Marky's Car Repairs to purchase in order to make the greatest income.

3. The cost in dollars of producing x tonnes of coal per day is

$$C(x) = 35x + 0.02x^2 + 200$$

If the coal can be sold at a price of \$39 per tonne, how many tonnes should be produced each day to maximize the profits?

4. It costs C dollars per kilometre to operate an intercity bus, where $C = 80/v + v/80$ and v is the average speed in km/h. Determine the aver-age speed at which the bus should be operated to minimize the operating costs. What is the minimum operating cost?

5. A craftsperson has the capacity to produce up to 16 pieces of ceramic pottery per week. Experience has shown that x pieces per week can be sold at a price of $(140 - 2x)$ each and the cost of producing x pieces is $(600 + 20x + x^2)$. How many pieces of pottery should be made each week to give the greatest profit?

6. Electric power for a community on one bank of a straight river 200 m wide must come from a power station on the opposite bank of the river 500 m downstream. If it costs twice as much to lay cable underwater as it does on land, what path should be chosen to minimize the cost?

7. A wooden chest is rectangular in shape with its length along the front twice as long as its width. The top, front, and two sides of the chest are made of oak; the back and the bottom are made of pine. The chest is to have a volume of 0.25 m³. Oak costs *three* times as much as pine. Find the dimensions that minimize the cost of the chest.

Summary and Review   6.4

In solving extreme value problems, you must (a) identify the quantity whose extreme value is desired; (b) decide which quantity to use as the independent variable; and (c) determine the relation between the two quantities.

Getting the equation is a large step toward getting a solution. After that, it is a straightforward matter to look for extreme values at points where the first derivative is zero.

To discover what equation to use, it is helpful to

(a) draw a diagram and carefully label it with variables and given values. Use capital letters for constants and lower case letters for varying quantities, for example.

(b) use word equations to describe relationships, before introducing variables.

(c) look for equations of constraint that can be used to eliminate all but a single independent variable.

(d) rely on units to tell you if an equation is properly formed. For example, from the units alone you can see that

cost per hour = cost per km × velocity

because $\dfrac{\$}{h} = \dfrac{\$}{km} \times \dfrac{km}{h}$

Exercise 6.4

1. The sum of the squares of two numbers x and y is to be minimized subject to the constraint that the sum of the numbers is 12.

   (a) What quality is to be minimized?

   (b) What is the equation of constraint?

   (c) Find the numbers.

2. Express 24 as the sum of two positive numbers such that the product of one by the square of the other is a maximum.

3. Find a positive number for which the sum of the number and its reciprocal is as small as possible.

4. Find the number which exceeds its square by the greatest amount.

5. If the sum of two positive numbers is 28, prove that the square of one added to the cube of the other is at least 640.

6. During the first 5 s of its motion, the displacement of an object from a fixed point O is given by the equation $s = 4t(27 - t)^2$, where s is in centimetres and t is in seconds. What is the maximum displacement?

7. A church window consists of a blue, semicircular section of radius r on top of a clear rectangular section of width $2r$. Blue glass lets through half as much light as clear glass. Find the dimensions of the window that admits the most light if the perimeter of the entire window is 5m.

8. A fixed volume of metal is melted down and made into a solid consisting of a circular cylinder of radius r and height h, surmounted by a circular cone, the radius and height of which are both equal to r. Prove that the surface area of the solid is a minimum when $h/r = \sqrt{2}$.

9. Find the dimensions of the open-topped rectangular box that can be made from 3600 cm² of cardboard, if the length of the base is three times as long as its width. What is the maximum volume?

10. Find the maximum possible area of an isosceles triangle, the two equal sides of which are each 6 cm long.

11. Find the point(s) on the hyperbola $x^2 - y^2 = 1$ closest to $(2, 0)$.

12. The cost per hour for fuel to operate a cargo vessel is proportional to the cube of the speed in km/h. In addition, there are fixed costs of $1728/h during each hour of operation. If the cost of diesel fuel at a speed of 8 km/h is $256/h, find the most economical speed for a trip of 500 km.

13. The strength of a rectangular beam is proportional to its width w, and its thickness t, that is, $S = kwt^2$, where k is a constant. Find the dimensions of the strongest beam that can be cut from a cylindrical log of radius 20 cm.

14. A large aquarium tank in the shape of an open-topped rectangular box is to measure 5 m from front to back and have a volume of 300 m³. The base, ends, and back are to be made of slate. The front is to be made of plate glass which costs 4 times as much as slate per square metre. What dimensions will minimize the total cost?

15. When travelling at highway speeds, a truck burns diesel oil at a rate given by $(v + 900/v)/300$ L/km where v is the velocity in km/h. If diesel oil costs $0.40/L, find

(a) the steady speed that will minimize the cost of the fuel on a 500 km trip

(b) the steady speed that will minimize the total cost of a 500 km trip during which the driver is paid $10/h.

16. Murray can hire a dump truck for $40/h (driver included). The truck takes 30 min to deliver a load of topsoil and return. It would take 40 h for one person to load the truck with topsoil. Labourers get $10/h whether they are loading the truck or standing idle while they await its return. How many labourers should be hired to minimize the cost per load. What is the minimum cost?

Rate of Change Applications

7

I n previous chapters, the derivative has been used as a tool to assist in the graphing of relations and in the solution of maximum or minimum problems. The main focus has been the use of the derivative in the calculation of slope. In some applications what is important is the effect of the change in one variable on a second variable. For example, how does a change in price affect sales or how does a change in time affect position? The definition of the derivative involves dividing the change in the dependent variable by the change in the independent variable. Therefore, the derivative can be considered to be a measure of the relationship between the changes in the variables. In this chapter the derivative is interpreted as a rate of change.

Average and Instantaneous Rate of Change

7.1

Consider a function $y = f(x)$ where x changes from x_1 to x_2. The **change in x** is defined as

the change in $x = \Delta x$

$$= x_2 - x_1$$

The **change in y** that corresponds to this change in x from x_1 to x_2 is defined as

the change in $y = \Delta y$

$$= f(x_2) - f(x_1)$$

EXAMPLE A

The cost $C(x)$ in dollars of importing x litres of wine is given by the formula $C(x) = 50\sqrt{x}$. What is the change in cost as the amount imported increases from 2400 L to 2500 L? What is the change in the amount purchased?

SOLUTION

$$\text{the change in cost} = \Delta C$$

$$= C(2500) - C(2400)$$

$$= 50\sqrt{2500} - 50\sqrt{2400}$$

$$\doteq 50.51$$

$$\text{the change in the amount} = \Delta x$$

$$= 2500 - 2400$$

$$= 100$$

The change in cost is $50.51 and the change in the amount purchased is 100 L. ☐

Consider the graph of the cost in dollars $C(t)$ of a barrel of oil against time t in months over a period of one year shown in Figure 7.1.

Figure 7.1

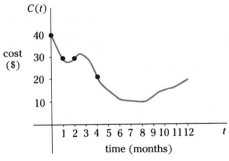

From the graph the following information is obtained:

At time $t = 0$ the cost $C(0) = \$40$.

At time $t = 1$ the cost $C(1) = \$30$.

From time $t = 0$ to time $t = 1$, that is in one month, the cost dropped $10.

At time $t = 2$ the cost $C(2) = \$30$.

At time $t = 4$ the cost $C(4) = \$20$.

From time $t = 2$ to time $t = 4$, that is in two months, the cost dropped $10.

There is a difference in the two situations. A drop of \$10 in one month is more significant than a drop of \$10 in two months. To describe this kind of information, the average rate of change is employed.

For a function $y = f(x)$ the average rate of change of $f(x)$ with respect to x is the ratio of the change in y to the change in x. If x changes from x_1 to x_2 the **average rate of change** is defined as

$$\text{the average rate of change} = \frac{\Delta y}{\Delta x}$$

$$= \frac{f(x_2) - f(x_1)}{x_2 - x_1}$$

The average rate of change can be interpreted as the slope of the secant to the graph of $y = f(x)$ through the points $(x_1, f(x_1))$ and $(x_2, f(x_2))$.

EXAMPLE B

The temperature on a particular day in January is given by $T(h) = -0.1h^2 + 2h - 20$, where $T(h)$ is the temperature in degrees Celsius at time h in hours measured from 12:00 midnight, $0 \le h \le 24$.

(a) What is the average rate of change of temperature with respect to time from 1:00 a.m. to 7:00 a.m.?

(b) Sketch the graph of $T(h)$ and give a geometric interpretation of this average rate of change.

SOLUTION

(a) The average rate of change $= \dfrac{\Delta T}{\Delta h}$

$$= \frac{T(7) - T(1)}{7 - 1}$$

$$= \frac{-10.9 - (-18.1)}{6}$$

$$= 1.2$$

The average rate of change from 1:00 a.m. to 7:00 a.m. is 1.2°C/h.

Figure 7.2

Figure 7.2

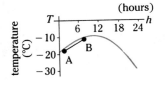

Figure 7.3

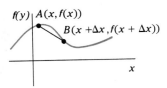

(b) The derivative $\dfrac{dT}{dh} = -0.2h + 2$, which equals zero at $h = 10$, so the maximum point for this parabola occurs where $h = 10$. The graph of $T(h)$ is shown in Figure 7.2.

The geometric interpretation of the average rate of change is that the average rate of change from 1:00 a.m. to 7:00 a.m. is the slope of the secant AB. ❑

For any differentiable function $y = f(x)$ such as the one in Figure 7.3 the average rate of change as x goes from x to $x + \Delta x$ is the slope of the secant AB.

The instantaneous rate of change with respect to x at point A is given by the limit of the average rate of change as B gets closer and closer to A. The **instantaneous rate of change** is defined as

$$\text{the instantaneous rate of change} = \lim_{\Delta x \to 0} \frac{\Delta y}{\Delta x}$$

$$= \lim_{\Delta x \to 0} \frac{f(x + \Delta x) - f(x)}{\Delta x}$$

This is the slope of the tangent to $y = f(x)$ at point A. Since this limit and the slope of the tangent are equivalent to the derivative, the instantaneous rate of change of y with respect to x for any value of x is defined as

$$\text{the instantaneous rate of change} = \frac{dy}{dx}$$

$$= \frac{df(x)}{dx}$$

EXAMPLE C

The volume V of water in a small humidifier at any time t is given by $V(t) = 0.08(4 + t)^{-1}, 0 \le t \le 24$. The volume is in cubic metres (m^3) and the time is in hours (h). Determine the instantaneous rate of change of volume with respect to time at any time t. What is the significance of the instantaneous rate of change?

SOLUTION

$$\text{The instantaneous rate of change} = \frac{dV}{dt}$$

$$= (-1)(0.08)(4 + t)^{-2}$$

The instantaneous rate of change of volume with respect to time is $-0.08(4 + t)^{-2}(\text{m}^3/\text{h})$. Since the instantaneous rate of change of volume with respect to time is negative for all values of t, this means that the volume of water in the humidifier is always decreasing as time elapses. ❑

EXAMPLE D

Find the instantaneous rate of change of the volume of a cube with respect to the length of its side when the side is 15 cm.

SOLUTION

Let the length of the side of the cube be s cm, and let the volume be $V\,\text{cm}^3$, where

$$V = s^3$$

$$\text{The instantaneous rate of change} = \frac{dV}{ds}$$

$$= 3s^2$$

When $s = 15$ $$\frac{dV}{ds} = 3(15)^2$$

$$= 450$$

When the side length is 15 cm, the instantaneous rate of change of volume with respect to side length is 450 cm³/cm. ❑

In some applications, such as the number of people riding a bus as a function of price, or the revenue of a company as a function of the number of items sold, variables are not continuous. In these situations a graph of the data would be a series of separated points rather than a continuous curve. The actual data would not represent a differentiable function. To solve these problems using calculus, an assumption is made that a continuous curve is drawn through the data and this relationship is used in the calculation of the instantaneous rates of change. This assumption is made in the next example where money is treated as a continuous variable.

EXAMPLE E

The revenue $R(x)$ in thousands of dollars of a firm producing diskettes depends on the price x in dollars charged for each box according to the formula $R(x) = 100x - 2x^2$ where $0 \leq x \leq 30$.

(a) What is the rate of change of revenue with respect to price when the price per box is $15 and when the price is $20?

(b) What do the values of the above rates of change mean to the firm?

(c) For what prices do revenues continue to increase?

(d) At what price is the rate of change of revenue $0 per dollar?

(e) What is the significance of the price at which the rate of change of revenue is $0 per dollar?

SOLUTION

(a) The instantaneous rate of change $= \dfrac{dR}{dx}$

$$= 100 - 4x$$

when $x = 15$ $\dfrac{dR}{dx} = 100 - 4(15)$

$$= 40$$

when $x = 20$ $\dfrac{dR}{dx} = 100 - 4(20)$

$$= 20$$

When the price is $15, the rate of change of revenue with respect to price is $40 000 per dollar. When the price is $20 the rate of change of revenue with respect to price is $20 000 per dollar.

(b) Both rates of change are positive so an increase in price in either situation would lead to an increase in revenue. However, when the price is $20, a given price increase would not affect the revenue as much as it would when the price is $15. For instance, a price increase of 10¢ per box would increase revenues by approximately $4000, if the original price were $15, but only by approximately $2000 if the original price were $20.

(c) Revenues continue to increase as long as the instantaneous rate of change of revenue with respect to price is positive.

$$\frac{dR}{dx} > 0$$

$$100 - 4x > 0$$

$$x < 25$$

Revenues continue to increase for as long as the price per box is less than $25.

(d) $$\frac{dR}{dx} = 0$$

$$100 - 4x = 0$$

$$x = 25$$

The rate of change of revenue is $0 per dollar when $25 is charged for a box of diskettes.

(e) The price at which the rate of change of revenue is $0 per dollar is the price at which the firm attains its maximum revenue of $1 250 000.

Exercise 7.1

1. If $f(x) = 2x^3 - 9$ what is the change in $f(x)$ as x changes from 1 to 4? What is the change in x?

2. The formula for the volume of a sphere is $V = (4/3)\pi r^3$. What is the change in volume as the radius changes from 3 cm to 6 cm?

3. The price $P(t)$ of one share in an oil company at time t is given by the formula $P(t) = -t^2 + 14t + 10$ for $0 \leq t \leq 14$. The price is measured in dollars and the time in years. Determine the average rate of change in price from $t = 1$ year to $t = 5$ years.

4. Water is being pumped from a tank so that the volume V in litres remaining after t minutes is determined by $V(t) = 1000(10 - t^2)$ for $0 \leq t \leq \sqrt{10}$. Over the first 2 minutes what is the average rate of change of the volume remaining?

5. Find the rate of change of y with respect to x when $x = 2$ for each of the following:

(a) $y = \dfrac{1}{x(x + 1)}$ (b) $y = \sqrt{3x + 19}$

(c) $y = (x^2 + 1)(3x^3 + 2x^2)$

6. A designer is experimenting with a cylindrical can with a fixed height of 15 cm. Find the rate of change of volume with respect to radius when the radius is 4 cm. The volume of a cylinder is given by $V = \pi r^2 h$.

7. Find the rate of change of the area of a circle with respect to radius when $r = 10$ cm.

8. The value V of a certain mutual fund share over the past 10 years is given by $V(t) = t^3 + 15t^2 + 72t + 241$, $-10 \leq t \leq 0$. The value is measured in dollars and the time t in years. Compare the present rate of change of value with respect to time to the rate of change of value with respect to time 6 months ago.

9. The number N of people riding a bus is dependent upon the price c of tickets according to

the formula $N = 1000 + 40c - 0.5c^2$, where c is measured in cents and $0 \le c \le 100$. Find the rate of change of the number riding with respect to price. What is the rate of change of the number riding with respect to price when the price is 40¢? What is the significance of this rate of change?

10. For what value of t does the rate of change s with respect to t equal 5 if $s(t) = 10t - 5t^2$?

11. The period $T(x)$ of a pendulum depends upon its length x. If the period is measured in seconds and the length in metres $T(x) = 6.2\sqrt{0.1x}$, what is the rate of change of period with respect to length when the length is 250 cm?

12. The profit $P(x)$ of a pen company in dollars is a function of the price x in cents that is charged for their pens. At what price is the rate of change of profit with respect to price equal to 400 dollars/cent? $P(x) = -20x^2 + 3200x - 280\,000$, $0 \le x \le 100$.

13. Find the rate of change of the area A of a square with respect to its side length s.

14. Find the rate of change of the area A of a square with respect to the length of its diagonal d.

15. Find the rate of change of the area A of an equilateral triangle with respect to the change in its side length s.

7.2 Rate of Change of Position per Unit Time—Velocity

Consider an object which moves along a straight line so that its position relative to a fixed point O at any time $t \ge 0$ is given by the function $s(t)$. The convention used is that positions to the right of O (or above O in the case of a vertical line) are positive. Positions to the left of O (or below it in the case of a vertical line) are negative. At times t_1 and t_2 the positions of the object are $s(t_1)$ and $s(t_2)$ respectively. The displacement of the object in this time interval is its change in position from time t_1 to time t_2, that is, the interval of length between the two positions. The **displacement** is defined as

the displacement $= \Delta s$

$$= s(t_2) - s(t_1)$$

In travelling from position $s(t_1)$ to position $s(t_2)$ the object may have moved directly from one position to the other or it may have moved in a variety of directions before ending up at position $s(t_2)$. Therefore, *the distance travelled may not be the same as the displacement.* In the special case when the object is initially located at the reference point O, the displacement from the reference point at any time t is $s(t) - s(0)$ which equals $s(t)$.

EXAMPLE A

The position in centimetres, relative to a fixed point O, at any time

t seconds of an object moving along a horizontal straight line is given by $s(t) = t^2 - 4t + 8, t \geq 0$.

(a) Determine the initial position of the object.

(b) Determine the displacement from time $t = 1$ s to $t = 3$ s.

(c) Is the distance travelled from time $t = 1$ s to $t = 3$ s the same as the displacement?

SOLUTION

(a) the initial position $= s(0)$

$$= 8$$

The initial position is 8 cm to the right of 0.

(b) the displacement $= s(3) - s(1)$

$$= 5 - 5$$

$$= 0$$

The displacement in the interval from $t = 1$ s to $t = 3$ s is 0 cm.

(c) The distance travelled is not the same as the displacement. At time $t = 2$, the position $s(2) = 4$. Therefore, the object moved during the time interval, so its distance travelled is not 0. $\square$

For the position function $s(t)$ the rate of change of position with respect to time is called **velocity**. The **average velocity** v_{avg} from time t_1 to time t_2 is defined as

$$v_{\text{avg}} = \frac{\text{change in position}}{\text{change in time}}$$

$$= \frac{\Delta s}{\Delta t}$$

$$= \frac{s(t_2) - s(t_1)}{t_2 - t_1}$$

The **instantaneous velocity** $v(t)$ at time t is defined as

$$v(t) = \frac{ds(t)}{dt}$$

A positive velocity indicates motion to the right or upward and a negative velocity indicates motion to the left or downward relative to the fixed point O.

___

EXAMPLE B

The position in centimetres, relative to the origin, at any time t in seconds of an object moving along the x-axis is given by

$$s(t) = -2t^2 + 8t + 1 \quad t \geq 0$$

(a) What is the average velocity from time $t = 2$ s to time $t = 5$ s?

(b) What is the position of the object at time $t = 4$ s?

(c) What is the instantaneous velocity at time $t = 4$ s?

(d) Is the object moving away from or toward the origin at time $t = 4$ s?

SOLUTION

(a) The average velocity from time $t = 2$ s to $t = 5$ s is given by

$$v_{avg} = \frac{s(5) - s(2)}{5 - 2}$$

$$= \frac{-9 - 9}{3}$$

$$= -6$$

The average velocity during this three-second time interval is -6 cm/s.

(b) At time $t = 4$ s the position is given by

$$s(4) = -2(4)^2 + 8(4) + 1$$

$$= 1$$

At time $t = 4$ s the object is 1 cm to the right of the origin.

(c) The instantaneous velocity at time $t = 4$ s is given by $v(4)$.

$$v(t) = \frac{ds(t)}{dt}$$

$$= -4t + 8$$

$$v(4) = -16 + 8$$

$$= -8$$

At the instant when $t = 4$ s, the velocity is -8 cm/s.

(d) At time $t = 4$ s the object is moving to the left because its velocity is negative. Since at this moment the object is located to the right of the origin, it is moving toward the origin. ◻

EXAMPLE C

The graph shown in Figure 7.4 is a position-time graph of the motion of an object along a straight line. Use this graph to determine the average velocity from time $t = 1$ s to time $t = 5$ s. Use the graph to draw a velocity-time graph.

SOLUTION

At time $t = 1$ s the position is -2 and at time $t = 5$ s the position is 14.

$$v_{avg} = \frac{14 - (-2)}{5 - 1}$$

$$= 4$$

The average velocity from time $t = 1$ s to $t = 5$ s is 4 cm/s.

The instantaneous velocity $v(t)$ is given by the derivative of the position function $s(t)$. The value of $v(t)$ at any time t is equivalent to the slope of the tangent to $s(t)$. To determine the shape of $v(t)$ a number of tangents must be drawn to $s(t)$, and their slopes calculated. The slopes of the tangents are shown in the Table 7.1.

The graphing process can be simplified by drawing each graph with the same time scale and locating the position-time graph directly above the velocity-time graph as shown in Figure 7.5. ◻

EXAMPLE D

The function $s(t) = t^2 - 10t + 16$ describes the position relative to a fixed point O of a particle moving horizontally along a straight line for $t \geq 0$. When is the particle moving toward the point O? Describe its motion in detail.

SOLUTION

For the particle to move toward O two quantities need to be considered, its position and its velocity. If the particle is located to the right of O, then it must move left to move toward O. In other words, if its position is positive, its velocity must be negative. Conversely, if the particle is located to the left of O it must move right. That is, if it has a negative position it must have a positive velocity. Therefore, when a particle moves toward O the signs of position $s(t)$ and velocity $v(t)$

Figure 7.4

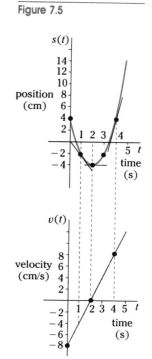

Table 7.1

time t	1	2	4
slope $v(t)$	-8	0	8

Figure 7.5

are opposite. In other words, for the particle to be moving toward O the product of position and velocity is negative. That is,

$$s(t)v(t) < 0$$

$$s(t) \cdot \frac{ds(t)}{dt} < 0$$

In this example,

$$(t^2 - 10t + 16)(2t - 10) < 0$$

$$(t - 2)(t - 8)(t - 5) < 0$$

To determine the values of t for which this expression is negative, consider the intervals created by the values of t that make the expression equal to zero. The expression equals zero for $t = 2, 5,$ or 8. Since $t \geq 0$ these intervals are $0 \leq t < 2, 2 < t < 5, 5 < t < 8,$ and $t > 8$. Choose a value from each interval and substitute it into the expression $(t - 2)(t - 8)(t - 5)$ and determine if it satisfies the inequality or not. When $t = 9$, for example, the expression has a positive value, 28. Therefore points in the interval $t > 8$ do not satisfy the inequality, and this interval is not included in the answer, However, $t = 6$ in the interval $5 < t < 8$ produces a value of -8 which satisfies the inequality so this interval is included in the answer. By continuing this process you can determine that the particle moves toward the point O when $0 \leq t < 2$ or $5 < t < 8$.

An alternate method of solving this problem is to diagram the signs of the factors (see A. Preparation for Calculus, Example B, pages 19–20). The product $(t - 2)(t - 8)(t - 5)$ will be negative when exactly one or exactly three of its factors are negative. From the diagram in Figure 7.6, you can see that this will be the case when $0 \leq t < 2$ or $5 < t < 8$.

Figure 7.6

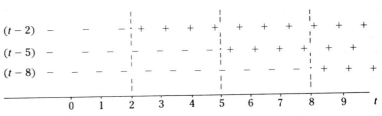

To describe the motion of the particle more fully, calculate its position at various special points along its path. Initially it is at the position $s(0) = 16$. It is at the origin when $s(t) = 0$:

$$t^2 - 10t + 16 = 0$$

$$(t - 2)(t - 8) = 0$$

$$t = 2 \text{ or } 8$$

It comes to rest when $v(t) = 0$:

$$2t - 10 = 0$$

$$t = 5$$

This motion is shown schematically in Figure 7.7. At $t = 0$ it starts (A), passes the origin (B) going left at $t = 2$, stops and turns around (C) at $t = 5$, passes the origin again (D) at $t = 8$, and continues to the right (E). ☐

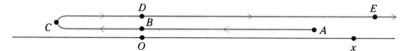

Figure 7.7

Exercise 7.2

1. An object moves on a straight line relative to a fixed point O so that its position $s(t)$ in cm at time t in s is $s(t) = 3t^2 - 9t + 5$. What is its initial position? What is its displacement from time $t = 1$ s to $t = 3$ s?

2. For each of the following position-time graphs sketch the corresponding velocity-time graphs.

Figure 7.8

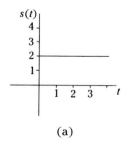

(a)

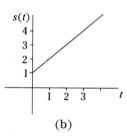

(b)

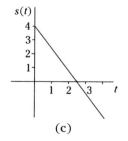

(c)

3. The position relative to the origin of an object moving along the x-axis at any time $t \geq 0$ seconds is given by $s(t) = t^2 - 3t - 5$. Position is measured in centimetres and t in seconds. Sketch the graph of this function for $0 \leq t \leq 6$. What is the average velocity over this period? What is the instantaneous velocity at $t = 1.5$ s? What is the relationship between the velocity at $t = 1.5$ s and the position-time graph?

4. For each of the following position functions determine the instantaneous velocity at any time $t \geq 0$.

   (a) $s(t) = \sqrt[3]{t^2 + 2}$ (b) $s(t) = \dfrac{2}{\sqrt{8t + 1}}$

   (c) $s(t) = \dfrac{4t}{1 + t^2}$

5. For each of the following position functions determine the instantaneous velocity at time $t = 3$.

   (a) $s(t) = 3t - 2t^3$ (b) $s(t) = t(2t^2 - t)^{-1}$

   (c) $s(t) = \sqrt{2t^2 + 7}$

6. When will the velocity of a car travelling along a straight road be 100 km/h if its position in metres at time t seconds is $s(t) = 5t + 2t^2$?

7. What will be the velocity of a toboggan that slides down a hill 18 m long when it reaches the bottom? The toboggan's position in metres at time t seconds is $s(t) = 0.5t^2$.

8. A bicycle rider travelling along a straight road applies the brakes, and the rider's position in metres at any time t seconds is given by $s(t) = 3t - 0.75t^2$. What is the rider's initial velocity? How long does it take the rider to stop?

9. Two objects starting from the same point at the same time and moving in the same straight line are located at time t seconds at positions $a(t)$ and $b(t)$ measured in metres from their starting place. If $a(t) = 2t^2 - 3t$ and $b(t) = 3t - t^2$, determine when the objects will have the same velocity. What are their velocities when they have the same position?

10. Two objects starting from the same point at the same time and moving in the same straight line are located at time t seconds at positions $a(t)$ and $b(t)$ measured in metres from their starting place. If $a(t) = -t^3 - 4t^2 + 5t$ and $b(t) = -2t^3 + 6t^2 - 4t$ what are their speeds when they last coincide?

11. The function $s(t) = 3t^2 - 12t + 8$ describes the position relative to a fixed point O of a particle moving along a straight line. Fill in the following table of position and instantaneous velocity for the times shown. State those times when the particle is moving toward the point O. What are the conditions on $s(t)$ and $v(t)$ for motion to be toward the point O?

Table 7.2

t	0	1	2	3	4
$s(t)$					
$v(t)$					

12. The position of a ball, relative to the ground, which is thrown vertically upward is given by $s(t) = 6 + 11.9t - 4.9t^2, t \geq 0$. The time t is in seconds and the position in metres. From what height is the ball thrown? What is the initial velocity? When does the ball begin to come down? How high does the ball go? When does the ball hit the ground and what is its velocity at this time?

Rate of Change of Velocity per Unit Time—Acceleration

7.3

In most physical situations, such as the motion of a car or the launching of a satellite, the velocity changes with time. The rate of change of velocity is called the **acceleration**.

The **average acceleration** a_{avg} from time t_1 to time t_2 for an object moving in a straight line with velocity $v(t)$ is defined as

$$a_{avg} = \frac{\text{change in velocity}}{\text{change in time}}$$

$$= \frac{\Delta v}{\Delta t}$$

$$= \frac{v(t_2) - v(t_1)}{t_2 - t_1}$$

The **instantaneous acceleration** $a(t)$ at any time t for an object moving in a straight line with velocity $v(t)$ is defined as

$$a(t) = \frac{dv(t)}{dt}$$

Velocity $v(t)$ is the derivative of position $s(t)$ and instantaneous acceleration $a(t)$ is the derivative of velocity. Instantaneous acceleration can therefore be thought of as the second derivative of $s(t)$:

$$a(t) = \frac{dv(t)}{dt}$$

$$= \frac{d}{dt}\left(\frac{ds(t)}{dt}\right)$$

$$= \frac{d^2 s(t)}{dt^2}$$

Acceleration to the right or upward is positive and acceleration toward the left or downward is negative.

EXAMPLE A

The position in centimetres, relative to a fixed point O, at any time t seconds of an object moving along a straight line is given by $s(t) = t^3 - 8t^2 + 9$.

(a) What is the average acceleration from $t = 0$ s to $t = 5$ s?

(b) What is the instantaneous acceleration at $t = 2.5$ s?

(c) What is the graphical significance of the answers to the previous parts of this example?

SOLUTION

(a) To determine the acceleration, the velocity function must be determined.

$$v(t) = \frac{ds(t)}{dt}$$

$$v(t) = 3t^2 - 16t$$

The average acceleration from $t = 0$ s to $t = 5$ s is

$$a_{avg} = \frac{\Delta v}{\Delta t}$$

$$= \frac{v(5) - v(0)}{5 - 0}$$

$$= \frac{-5 - 0}{5}$$

$$= -1$$

The average acceleration from $t = 0$ s to $t = 5$ s is -1 cm/s^2.

(b) The instantaneous acceleration at $t = 2.5$ s is $a(2.5)$.

$$a(t) = \frac{dv(t)}{dt}$$

$$= 6t - 16$$

$$a(2.5) = 6(2.5) - 16$$

$$= -1$$

The instantaneous acceleration at $t = 2.5$ s is -1 cm/s^2.

(c) The average acceleration from $t = 0$ s to $t = 5$ s equals the instantaneous acceleration at $t = 2.5$ s. This means that the slope of the secant to the velocity-time graph from $t = 0$ to $t = 5$ equals the slope of the tangent to the velocity-time graph at $t = 2.5$ (see Figure 7.9). The secant and the tangent are parallel. ☐

Figure 7.9

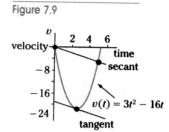

EXAMPLE B

If $s(t) = (8t + 5)^3$ where $s(t)$ is in metres and t is in hours find the acceleration at any time t.

SOLUTION

The acceleration at any time t is given by the second derivative of the position $s(t)$.

$$\frac{ds(t)}{dt} = 24(8t + 5)^2$$

$$\frac{d^2s(t)}{dt^2} = 384(8t + 5)$$

$$a(t) = 3072t + 1920$$

The acceleration at any time t is $(3072t + 1920)$ m/h^2. ☐

EXAMPLE C

An object is moving along a straight line. Its position s relative to a fixed point O at any time t is given by $s(t) = t^3 - 6t^2 - 15t$, where s is in metres and t is in seconds. Analyze the motion when $t = 1$.

SOLUTION

$$s(t) = t^3 - 6t^2 - 15t$$

$$v(t) = 3t^2 - 12t - 15$$

$$a(t) = 6t - 12$$

At $t = 1$ the position is $s(1) = -20$
        the velocity is $v(1) = -24$
        the acceleration is $a(1) = -6$

The object's position is -20 m, its velocity -24 m/s, and its acceleration -6 m/s^2 at $t = 1$ s. The object is 20 m to the left of the origin, and since the velocity is negative it is travelling away from the origin.

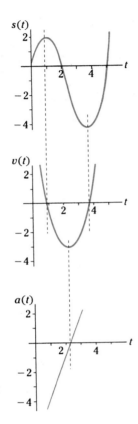

The acceleration is in the same direction as the velocity and so the object is speeding up. ☐

EXAMPLE D

Given the graph of $s(t)$ shown in Figure 7.10 sketch the graphs of $v(t)$ and $a(t)$.

Figure 7.10

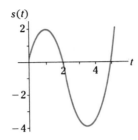

SOLUTION

Particular points on these graphs can be obtained by noting that the velocity is 0 when the position-time graph has a local maximum or minimum point and the acceleration is 0 when the velocity-time graph has a local maximum or minimum point. Drawing tangents to the position-time curve and estimating their slope provide data for the velocity-time graph. Drawing tangents to the velocity-time graph and estimating their slope provide data for the acceleration-time graph. The graphing process can be simplified by drawing each graph with the same time scale and positioning the graphs as shown in Figure 7.11. ☐

Exercise 7.3

1. For each of the following velocity-time graphs sketch the corresponding acceleration-time graphs.

Figure 7.12

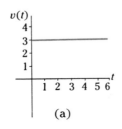

(a)

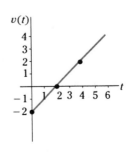

(b)

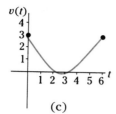

(c)

2. For each of the following position-time graphs sketch the corresponding velocity-time graphs and acceleration-time graphs.

Figure 7.13

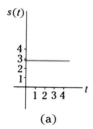

(a)

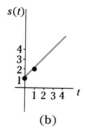

(b)

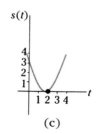

(c)

3. For each of the following what is the average acceleration from time $t = 1$ to $t = 4$?

(a) $v(t) = 100 - 9.8t$ (b) $v(t) = 25 - 3t^2$

(c) $v(t) = 3\sqrt{t}$

4. A motorcycle travelling against an increasing wind has its position in kilometres at any time t hours given by $s(t) = 80t - 2t^3, 0 \le t \le 5$. What is its average acceleration from time $t = 1$ h to $t = 3$ h?

5. Find the acceleration for each of the following at the instant specified. Time is measured in seconds and position in metres.

(a) $s(t) = \dfrac{3}{1 + t^2}$ at $t = 5$

(b) $s(t) = \sqrt{9t - 11}$ at $t = 3$

(c) $s(t) = (4t^2 + 5)^3$ at $t = 1$

6. Express the following as an equation or in-equation using $s(t)$, ds/dt, and/or d^2s/dt^2:

(a) The particle's acceleration is 8.

(b) The particle is 5 units to the left of the reference point.

(c) The velocity is 9.

(d) The particle is travelling toward the reference point.

(e) The particle is slowing down.

7. State the sign of d^2s/dt^2 and the sign of ds/dt if an object is located 3 units to the left of a reference point O, is travelling toward O, and is slowing down.

8. Determine whether a particle travelling on a straight line according to $s(t) = 9t - 8t^3$ is speeding up or slowing down at $t = 4$.

9. Determine the velocity of a particle when its acceleration is zero if its position in metres at time t seconds is given by $s(t) = 2t^3 - 48t^2$.

10. The position of a braking car at any time t is $s(t) = 28t - 0.2t^3$. The position is measured in metres and the time in seconds. Determine the acceleration when $t = 1$ s. How long does it take the car to stop?

11. A snowmobiler travelling down a narrow bush road comes over a hill and sees another machine stalled 15 m directly ahead. The brakes are applied immediately and the snowmobiler's position is determined by $s(t) = 12t - t^3$ $t \ge 0$ where t is in seconds and $s(t)$ in metres. Will a crash occur?

12. A bullet is fired into a bale of paper. The position of the bullet $s(t)$ in centimetres at time t in seconds is given by $s(t) = 12t - 6t^2 + t^3$ $0 \le t \le 2$. Determine the acceleration when the position is 8 cm.

13. Two objects starting from the same point at the same time and moving in the same straight line are located at time t seconds at positions $x(t)$ and $y(t)$ measured in metres from their starting place, where $x(t) = t^3 - 6t^2 + 19$ and $y(t) = -1.5t^3 + 9t^2 + 5$. Find their velocities when they have the same acceleration.

14. Determine when a particle whose position in metres at time t seconds is given by $s(t) = 2t^3 - 15t^2 + 36t + 5$ is speeding up.

7.4　Summary and Review

For a function $y = f(x)$ the rate of change of y with respect to x is the ratio of the change in y to the corresponding change in x. This definition has lead to two distinct types of rates, an average rate of change and an instantaneous rate of change. Their formulas and their geometric interpretation are summarized in Table 7.3.

Table 7.3

For $y = f(x)$

average rate of change	$\dfrac{\Delta y}{\Delta x}$ or $\dfrac{\Delta f}{\Delta x}$	slope of a secant to $y = f(x)$
instantaneous rate of change	$\dfrac{dy}{dx}$ or $\dfrac{df}{dx}$	slope of a tangent to $y = f(x)$

When the independent variable is time and the dependent variable is position, $y = s(t)$ can be used to describe motion in a straight line. The displacement is defined as the change in position. The rate of change of position with respect to time is the velocity. The rate of change of velocity with respect to time is the acceleration. Formulas

Table 7.4

For $y = s(t)$

average velocity	$v_{avg} = \dfrac{\Delta s}{\Delta t}$	slope of a secant to the position-time graph
instantaneous velocity	$v(t) = \dfrac{ds}{dt}$	slope of a tangent to the position-time graph
average acceleration	$a_{avg} = \dfrac{\Delta v}{\Delta t}$	slope of a secant to the velocity-time graph
instantaneous acceleration	$a(t) = \dfrac{dv}{dt} = \dfrac{d^2s}{dt^2}$	slope of a tangent to the velocity-time graph

for average velocity, instantaneous velocity, average acceleration, and instantaneous acceleration are given in Table 7.4 along with their geometric interpretation.

In motion problems it is important to be consistent about direction and sign. The convention usually used is that right and up are positive and left and down are negative. For an object to speed up, its instantaneous velocity and acceleration must be in the same direction; that is, the product of velocity and acceleration is positive. When the instantaneous velocity and acceleration are in opposite directions; that is, when the product of velocity and acceleration is negative, the object is slowing down.

To master the material in this chapter, it is helpful to:

1. Read a problem carefully. Determine if it is about an average or instantaneous rate of change. Questions about average velocity, for instance, can be recognized by the fact that the problem includes a time interval. Questions that require the calculation of an instantaneous velocity, on the other hand, usually refer to only one specific time.

2. Pay attention to the algebraic signs of quantities. Signs can be interpreted as a direction of motion (left or right, up or down), or as a direction of change (increasing or decreasing).

3. Express answers with units. If the position s of an object, for example, is given in centimetres, the first derivative ds/dt may have units of cm/s, and the second derivative d^2s/dt^2, units of cm/s^2.

Exercise 7.4

1. If $f(x) = 3x^2 - 9$ what is the change in x as x changes from 2 to 5? What is the corresponding change in $f(x)$?

2. The formula for the surface area of a sphere is $s = 4\pi r^2$. What is the change in surface area as the radius changes from 1 cm to 3 cm?

3. The height of a kite is a function of the length of the string that has been let out. Assuming the string is at an angle of 60° to the ground $H(s) = 0.866s$. The string length s and height $H(s)$ are in metres. What is the change in $H(s)$ as s changes from 50 m to 100 m? What is the average rate of change of height with respect to string length as s changes from 50 m to 100 m.

4. If $f(x) = 3x^2 + 2$ what is the average rate of

change of $f(x)$ with respect to x as x changes from 1 to 5?

5. For each of the following find the instantaneous rate of change of y with respect to x.

   (a) $y = 4x^2 + 3\sqrt{x}$ (b) $y = \sqrt[3]{2x^2 + 5}$

   (c) $y = (3x^2 + 4)^2$ (d) $y = \dfrac{9x + 4}{3x + 5}$

   (e) $y = (2x^2 - 3)^{4.3}$

6. For each of the following find the instantaneous rate of change of y with respect to x when $x = 2$.

   (a) $y = 3x^2 - 3\sqrt{2x}$ (b) $y = \sqrt{x^2 + 5}$

   (c) $y = (2 - 3x^2)^3$

7. For each of the following position-time graphs sketch the corresponding velocity-time graphs.

Figure 7.14

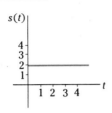

(a)

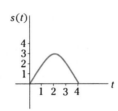

(b)

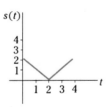

(c)

8. For each of the following velocity-time graphs sketch the corresponding acceleration-time graphs.

Figure 7.15

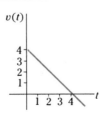

(a)

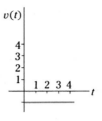

(b)

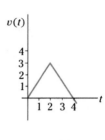

(c)

9. For each of the following position-time graphs sketch the corresponding velocity-time graphs and acceleration-time graphs.

Figure 7.16

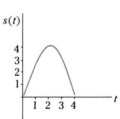

(a)

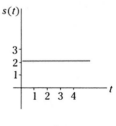

(b)

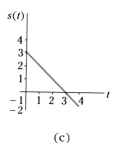

(c)

10. For each of the following what is the average velocity from time $t = 0$ s to $t = 4$ s. The position $s(t)$ is measured in metres and time t in seconds.

(a) $s(t) = 5 + \sqrt{t}$

(b) $s(t) = -3 + 8t - t^2$

(c) $s(t) = (2t + 5)^2$

11. For each of the following what is the average acceleration from time $t = 1$ s to $t = 3$ s. Assume position in metres and time in seconds.

(a) $v(t) = 80 + 12t$ (b) $v(t) = 9 - 3t^2$

(c) $s(t) = 4t - 2t^3$

12. For each of the following position functions determine the instantaneous velocity at any time $t \geq 0$.

(a) $s(t) = \dfrac{2t - 6}{9 - 8t}$ (b) $s(t) = (t^2 - 4)^3$

(c) $s(t) = \sqrt{3t + 4}$

13. For each of the following position functions determine the instantaneous velocity at time $t = 2$.

(a) $s(t) = 4t + 3t^2$ (b) $s(t) = (t + t^2)^{-1}$

(c) $s(t) = \sqrt{3t^2 + 4}$

14. For each of the following position functions determine the instantaneous velocity and acceleration at time $t = 1$. Position is measured in metres and time in seconds.

(a) $s(t) = 3t^3 - 4t^2 + 9t$

(b) $s(t) = t(t + 1)^{-1}$

(c) $s(t) = 3\sqrt{t}$

15. Express each of the following as an equation or inequation using $s(t)$, ds/dt, and/or d^2s/dt^2.

(a) The particle's velocity is 8.

(b) The particle is at the reference point.

(c) The acceleration is toward the left.

(d) The particle is travelling away from the reference point.

(e) The particle is speeding up.

(f) The acceleration is 4.

16. The number of bacteria $N(t)$ in a culture after t hours is approximated by the formula $N(t) = 10^5(1 + 2t^2)$. What is the rate of change of the number of bacteria after 2 h?

17. The yield of barley from 1 ha of land depends on the amount of seed planted. The yield $Y(s)$ in kilograms is related to the amount of seed used in kilograms by the formula $Y(s) = -0.075s^2 + 30s + 3700$ for $50 \leq s \leq 200$. What is the rate of change of yield with respect to the amount of seed used when 100 kg of seed is used? Would increasing the amount of seed planted from 100 kg improve the yield?

18. If the temperature of a confined gas is kept constant, its pressure P and volume V are related by $PV = k$ where k is a constant. Show that the rate of change of pressure with respect to volume is $-P/V$.

19. When x dollars per day are spent on advertising, the daily profit of a retail store is found to depend on x according to the formula $P(x) = 250\,000 + 3600x - 20x^2, 0 \leq x \leq 180$. If the current advertising budget is \$100 per day, could the store increase its profit by spending more money on advertising?

20. The demand for a product $D(p)$ is dependent on its price p in dollars according to $D(p) = 100 - 2p, 0 < p \leq 50$. The revenue is found by multiplying the demand by the price. What is the rate of change of revenue with respect to price? Can revenue be increased by increasing the price from \$10?

21. The cost of water $C(w)$ in dollars is dependent on the consumption according to the formula $C(w) = 15 + 0.1\sqrt{w}$ where w is the number of litres of water used. What is the rate of change of cost with respect to consumption when the cost is $20?

22. A border of width x cm is put around a rectangle 100 cm by 50 cm. Write an expression $A(x)$ for the area of the border and determine the rate of change of $A(x)$ when $x = 5$ cm.

23. The intensity of illumination I on a surface is inversely proportional to the square of the distance r from the surface to the source of light. If the intensity is 1000 cd when r is 1 m, what is the rate of change of intensity when $r = 2$ m?

24. The value of a painting $V(t)$ at time t is given by $V(t) = 10\,000 - 500t^2 + 200t^3$, $0 \leq t \leq 10$. $V(t)$ is measured in dollars and t is measured in years from the time of the initial purchase. What is the initial purchase price of the painting? What is the rate of change of the value when $t = 1$? What is the average rate of change over the first 3 years?

25. The population of a small northern community is given by $P(t) = 1000(1 + 0.01t)^{-1}$ where t is measured in years. What is the rate of change of the population when $t = 5$ years? What is the meaning of this rate of change?

26. When will the velocity of a car travelling along a straight road be 90 km/h if its position in metres at time t seconds is $s(t) = 4t + 3t^2$?

27. The position of an object falling from rest is given by $s(t) = -4.9t^2$. The position is measured in metres and the time in seconds. How far has the object fallen when its velocity is -30 m/s?

28. What will be the velocity at the bottom of a 200 m long hill of a skier who skis straight down? The skier's position in metres at time t seconds is $s(t) = 0.4t^2$.

29. Two objects starting from the same point at the same time and moving in the same straight line are located at time t seconds at positions $x(t)$ and $y(t)$ measured in metres from their starting place. If $x(t) = 3t^2 - 8t$ and $y(t) = -2t - t^2$, determine when the objects will have the same velocity. What are their velocities when they have the same position?

30. The function $s(t) = -t^2 + 10t - 16, t \geq 0$ describes the position relative to a fixed point 0 of a particle moving along a straight line. When does the particle move toward 0?

31. A windsurfer, out in winds that are gradually increasing, decides to sail from the shore directly across a small lake. If the position in metres at any time t in seconds is given by $s(t) = 0.02t^3 + 0.1t^2 + 3t, 0 \leq t \leq 15$, what is his acceleration at $t = 10$ s?

32. Determine when a particle whose position in metres at time t seconds is given by $s(t) = 2t^3 - 21t^2 + 60t + 6, t \geq 0$ is speeding up.

33. If the position of an object is given by $s(t) = 4\sqrt{9 + t}$ determine if the velocity increases or decreases with time.

34. Two objects starting from the same point at the same time and moving in the same straight line are located at time t seconds at positions $x(t)$ and $y(t)$ measured in metres from their starting place. If $x(t) = 2t^2 - t$ and $y(t) = -3t - t^2$, how fast is the distance between them changing when $t = 2$ s?

35. Determine when the average acceleration in the first three seconds equals the instantaneous acceleration if $s(t) = t^3 - t^2 - t$ for $t \geq 0$.

Related Rates

8

If a quantity x changes with time, then in general, any quantity y that depends on x also changes with time. When y is related to x, the *rate* at which y changes is related to the *rate* at which x changes. According to the chain rule,

if $y = f(x)$

then $\dfrac{dy}{dt} = \dfrac{dy}{dx} \cdot \dfrac{dx}{dt}$

This chapter is concerned with questions about how the rate of change dy/dt of one quantity is related to the rate of change dx/dt of another.

Related Rates Involving Areas and Volumes

8.1

The area of a region depends on its dimensions. The area A of a circle, for instance, is related to its radius r through $A = \pi r^2$. If the radius changes, then the area changes accordingly. Knowing how A is related to r makes it possible to find out how the rate of change of area dA/dt is related to the rate of change of radius dr/dt.

You may not know in detail *how* the dimensions of a region vary with time, but simply *that* they vary with time. This means that in the case of the circle, for instance, you should treat both the radius r and the area A as *implicit* functions of t.

EXAMPLE A

When a small pebble is dropped in a pool of still water, it produces a circular wave that travels outward at a constant speed of 15 cm/s. At what rate is the area inside the wave increasing

Figure 8.1

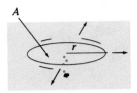

A

(a) when the radius is 5 cm?

(b) when the area is 400π cm²?

(c) when 4 s have elapsed?

SOLUTION

The circular area A inside the wave and the radius r of the wave both change with time. At any time t, however, A and r are related by the formula $A = \pi r^2$ (see Figure 8.1). Differentiating with respect to t gives a relation between the rate of change of area and the rate of change of radius.

$$A = \pi r^2$$

$$\frac{dA}{dt} = \frac{d}{dt}\pi r^2 \qquad\longleftarrow\qquad \text{use implicit differentiation and the chain rule}$$

$$= 2\pi r \frac{dr}{dt}$$

The rate of change dr/dt is the speed of the wave, given as 15 cm/s. To find the value of dA/dt, it is necessary to know the value of r.

(a) When $r = 5$ cm,

$$\frac{dA}{dt} = 2\pi(5)(15)$$

$$= 150\pi$$

Therefore the area is increasing at a rate of 150π cm²/s, when the radius is 5 cm.

(b) When the area is 400π cm², then

$$400\pi = \pi r^2$$

$$r = 20$$

$$\therefore \quad \frac{dA}{dt} = 2\pi(20)(15)$$

$$= 600\pi$$

Therefore the area is increasing at a rate of 600π cm²/s, when the area is 400π cm².

(c) Since the speed of the wave is constant, the radial distance the wave has travelled is found by multiplying speed by time. The speed of the wave is 15 cm/s so that when 4 s have elapsed, the radius is

$$r = (15)(4)$$

$$= 60$$

$$\therefore \quad \frac{dA}{dt} = 2\pi(60)(15)$$

$$= 1800\pi$$

Therefore, 4 s after the pebble is dropped, the area is increasing at a rate of 1800π cm²/s. ⌴

In some cases when three variables in a relation vary with time, you may use an equation of constraint to eliminate one of them, as in the following example.

EXAMPLE B

Sand pouring from a conveyor belt forms a conical pile, the radius of which is 3/4 of the height. If the sand is piling up at a constant rate of 1/2 m³/min, at what rate is the height of the pile growing 3 min after the pouring starts?

SOLUTION

The volume V, the height h, and the radius r of the sand pile all change with time. Since the pile retains its conical shape as it grows, these three quantities are related by the formula $V = \pi r^2 h/3$ (see Figure 8.2).

The rate of change of volume dV/dt is given. The rate of change of height dh/dt is required. To determine the relation between these two rates of change, it is necessary before differentiating to eliminate the variable r using the constraint that $r = (3/4)h$.

$$V = \frac{1}{3}\pi r^2 h$$

$$= \frac{1}{3}\pi \left(\frac{3}{4}h\right)^2 h$$

$$= \frac{3}{16}\pi h^3$$

Figure 8.2

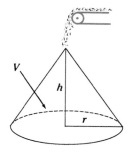

The relation between the rates of change of volume and height is, therefore,

$$\frac{dV}{dt} = \frac{9}{16}\pi h^2 \frac{dh}{dt}$$

To determine dh/dt, it is necessary to find the height after 3 min. Since the volume increases at a constant rate, the volume at any time is found by multiplying the rate $1/2$ m³/min by the time:

$$V = \left(\frac{1}{2}\right)(3)$$

$$= \frac{3}{2}$$

The volume after 3 min is $3/2$ m³. The height at that moment is therefore

$$\frac{3}{2} = \frac{3}{16}\pi h^3$$

$$h^3 = \frac{8}{\pi}$$

$$h = \frac{2}{\sqrt[3]{\pi}}$$

$$\doteq 1.37$$

Thus, after 3 min, the height is 1.37 m. The rate of change of height is found from

$$\frac{dV}{dt} = \frac{9}{16}\pi h^2 \frac{dh}{dt}$$

$$\frac{1}{2} = \frac{9}{16}\cdot\pi\left(\frac{2}{\sqrt[3]{\pi}}\right)^2 \cdot\frac{dh}{dt}$$

$$\frac{dh}{dt} = \frac{2}{9\sqrt[3]{\pi}}$$

$$\doteq 0.152$$

The rate of change of height is 0.152 m/min.

The derivative dV/dt can also be interpreted as the rate at which a volume fluid flows into or out of a container.

EXAMPLE C

Water is flowing out of a cylindrical storage tank at a rate of $2 \text{ m}^3/\text{min}$. If the tank has a radius of 8 m, how fast is the water level falling?

SOLUTION

The volume V and the height h of the water in the tank are both changing with time as the tank empties. Due to the geometry of the tank, however, V and h are always related by the formula $V = \pi R^2 h$, where R is the constant radius of the tank (see Figure 8.3). Differentiating with respect to t gives a relation between dV/dt, the rate of outflow, and dh/dt, the rate at which the water level falls:

Figure 8.3

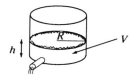

$$V = \pi R^2 h$$

$$\frac{dV}{dt} = \frac{d}{dt} \pi R^2 h \qquad \longleftarrow \qquad \text{use implicit differentiation}$$

$$= \pi R^2 \frac{dh}{dt} \qquad \longleftarrow \qquad R \text{ is a constant}$$

The rate at which the water level falls can now be calculated. Note that the rate of outflow dV/dt is negative because the volume of water in the tank is decreasing with time.

$$-2 = \pi (8)^2 \cdot \frac{dh}{dt}$$

$$\frac{dh}{dt} = \frac{-2}{64\pi}$$

$$\doteq -0.0099$$

Therefore, the water level in the tank is falling at a rate of approximately 1 cm/min. ❏

In Example C, the rate of change of volume was directly proportional to the rate of change of height because the area of the water surface πR^2 was constant. The same proportionality can be used in the case of any body of fluid during a short period of time when changes in height are small and changes in the area of the fluid surface are negligible. In the limit, the instantaneous rates of change are related by

$$\left(\begin{array}{c}\text{rate of inflow}\\ \text{or outflow}\end{array}\right) = \text{area} \cdot \left(\begin{array}{c}\text{rate of change}\\ \text{of height}\end{array}\right)$$

$$\frac{dV}{dt} = A \cdot \frac{dh}{dt}$$

EXAMPLE D

During the spring run-off, the flow of water into the reservoir of a small town averages out to be $3360 \, \text{m}^3/\text{d}$. If the water consumption rate is $2730 \, \text{m}^3/\text{d}$, at what rate does the water level in the reservoir rise? By how much will the water level rise in one week? The area of the water surface in the reservoir is $18\,000 \, \text{m}^2$.

SOLUTION

The rate of change of volume is proportional to the rate of change of height:

$$\frac{dV}{dt} = A \cdot \frac{dh}{dt}$$

The rate of change of volume is the difference between the inflow and the outflow rates:

$$\frac{dV}{dt} = 3360 - 2730$$

$$= 630$$

Therefore,

$$630 = 18\,000 \cdot \frac{dh}{dt}$$

$$\frac{dh}{dt} = \frac{630}{18\,000}$$

$$= 0.035$$

The water level increases at a rate of $0.035 \, \text{m/d}$ or $3.5 \, \text{cm/d}$. At this rate, the water level would rise $24.5 \, \text{cm}$ in one week.   ❑

Exercise 8.1

1. In Example A, explain why the rate of change of the area inside the circular wave is greater for larger radii.

2. The side of a square is increasing at a rate of 5 cm/s. At what rate is the area changing, when the side is 10 cm long?

3. The edge of a cube is expanding at a rate of 3 cm/s.

   (a) How fast is the volume changing when the edge is 1 cm? 10 cm?

   (b) At what rate is the surface area changing when the edge is 1 cm? 10 cm?

4. An oil tanker ruptures and begins to leak oil in a circular pattern, the radius of which is changing at a rate of 3 m/s. How fast is the area of the spill changing when the radius of the spill has reached 30 m?

5. The area of an equilateral triangle is decreasing at a rate of 5 cm²/s. Find the rate at which the length of a side is changing when the area is 250 cm².

6. The local newspaper purchases its newsprint in large rolls 1.25 m in diameter. If the cross-sectional area of a roll decreases at a constant rate of 0.5 m² per hour when the presses are rolling, how fast is the radius of the roll decreasing when the radius is 0.75 m?

7. The length of a rectangle is changing at a rate of 2 cm/s, while the width is changing in such a way that the area remains constant at 20 cm². What is the rate of change of the perimeter when the length is 5 cm?

8. The length of a rectangle is increasing at a rate of 4 cm/s and the width is increasing at a rate of 5 cm/s. At what rate is the area increasing when the length is 10 cm and the width is 12 cm?

9. A spherical snowball increases its volume at a rate of 0.75 m³/min as it rolls. At what rate is its radius increasing when the diameter is 1.25 m?

10. A spherical balloon is being filled with helium at a rate of 8 cm³/s. At what rate is its radius increasing

   (a) when the radius is 12 cm?

   (b) when the volume is 1500 cm³?

   (c) when it has been filling for 50 s?

11. A cylindrical tank with height 3 m and diameter 1 m is being filled with gasoline at a rate of 5 L/min. At what rate is the fluid level in the tank rising? (1 L = 1000 cm³)

12. Gas is escaping from a spherical weather balloon at a rate of 50 cm³/min. How fast is the surface area S shrinking when the radius is 15 m? ($S = 4\pi r^2$)

13. Grain is emptying from the bottom of a funnel-shaped hopper at a rate of 1.2 m³/min. If the diameter of the top of the hopper is 5 m and the sides make an angle of 30° with the vertical, determine the rate at which the level of grain in the hopper is changing, when it is half full.

14. Water is being pumped into a trough that is 4.5 m long and has a cross section in the shape of an equilateral triangle 1.5 m on a side. If the rate of inflow is 2 m³/min, how fast is the water level rising when the water is 0.5 m deep?

15. The amount of cleaning fluid in a partially filled, spherical storage tank is given by $V = \pi R h^2 - \pi h^3/3$, where R is the radius of the tank, and h is the depth of the fluid in the centre of the tank. If cleaning fluid is pumped into a tank of radius 12 m at a rate of 8 m³/min, how fast is the fluid level rising

   (a) when the depth is 3 m?

   (b) when the tank is 3/4 full?

16. The tide pushes sea water into a narrow inlet at the rate of 7200 m³/min. If the area of the inlet is 1.5 km², at what rate does the water level rise?

17. Gasoline is pumped from the tank of a tanker truck at a rate of 20 L/s. If the tank is a cylinder 2.5 m in diameter and 15 m long, at what rate is the level of gasoline falling when the gasoline in the tank is 0.5 m deep? (1 L = 1000 cm³)

8.2 Related Rates Involving Lengths

The crucial step in a related rates problem is to find a relationship between the variables in the problem. Once that is done, differentiating implicitly with respect to t gives the relationship between the rates of change of the variables.

Note in the following examples that derivatives like dx/dt are interpreted in two different ways: as the rate at which a length or distance changes with time, and as the speed of a moving object.

EXAMPLE A

A ladder 6 m long leaning against a wall begins to slide. How fast is the top of the ladder falling, when the bottom of the ladder is 4 m from the wall and sliding at a speed of 1 m/s?

SOLUTION

Figure 8.4

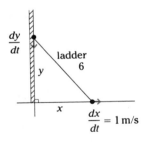

Figure 8.4 is a diagram of the ladder at some instant. The bottom of the ladder is x m from the base of the wall. Its top touches the wall at a point y m from the base. The speed dx/dt at which the bottom of the ladder slides is given. To find the speed dy/dt at which the top of the ladder falls, it is necessary first to have a relation between x and y. According to the Pythagorean theorem:

$$x^2 + y^2 = 6^2$$

Differentiating implicitly with respect to t,

$$\frac{d}{dt}x^2 + \frac{d}{dt}y^2 = \frac{d}{dt}6^2$$

$$2x\frac{dx}{dt} + 2y\frac{dy}{dt} = 0$$

When $x = 4$, the value of y is found from

$$4^2 + y^2 = 6^2$$

$$\therefore \qquad y = \sqrt{20}$$

It follows that

$$2(4)(1) + 2(\sqrt{20})\frac{dy}{dt} = 0$$

$$\frac{dy}{dt} = \frac{-4}{\sqrt{20}}$$

$$\doteq -0.89$$

The negative sign means that the value of y is decreasing with time; the top of the ladder is falling at a speed of 0.89 m/s. ∎

EXAMPLE B

Two small private planes take off from the same airport at the same time. One travels north at 200 km/h and the other, west at 150 km/h. If the planes fly at the same altitude, how fast are they separating after 2 h?

Figure 8.5

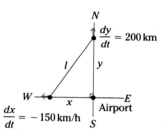

SOLUTION

The positions x and y of the two planes at some time t after they have taken off are shown in Figure 8.5. The distance l between them is given by the Pythagorean theorem

$$l^2 = x^2 + y^2$$

Differentiating implicitly gives the rate of separation dl/dt in terms of the positions and speeds of the two planes:

$$\frac{d}{dt}l^2 = \frac{d}{dt}x^2 + \frac{d}{dt}y^2$$

$$2l \cdot \frac{dl}{dt} = 2x \cdot \frac{dx}{dt} + 2y \cdot \frac{dy}{dt}$$

Since the speeds of the two planes are constant, their positions after 2 h are given by speed × time:

$$x = (-150)(2)$$

$$= -300$$

and

$$y = (200)(2)$$

$$= 400$$

The distance between them at that time is

$$l^2 = (-300)^2 + (400)^2$$

$$l = 500$$

Note that both the position and speed of the westbound plane are negative since the plane is to the left of the origin travelling left. Everything that is needed to find the rate of separation is now known:

$$2(500)\frac{dl}{dt} = 2(-300)(-150) + 2(400)(200)$$

$$\frac{dl}{dt} = \frac{90\,000 + 160\,000}{1000}$$

$$= 250$$

After 2 h, the planes are separating at a rate of 250 km/h.   ❏

EXAMPLE C

A spotlight at a school dance is fastened to a wall 8 m above the dance floor. A girl 1.75 m tall moves away from the wall at a speed of 0.75 m/s.

(a) At what rate is the length of her shadow increasing?

(b) At what speed is the tip of her shadow moving?

SOLUTION

The situation at an arbitrary time t is shown in Figure 8.6, where x is the position of the girl relative to the wall, and l is the length of her shadow on the floor. The position s of the tip of her shadow is given by

Figure 8.6

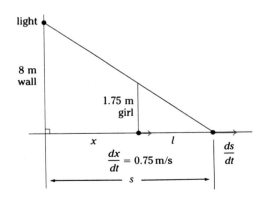

$$s = x + l$$

from which it follows that

$$\frac{ds}{dt} = \frac{dx}{dt} + \frac{dl}{dt}$$

The speed dx/dt at which the girl moves is 0.75 m/s. If you can determine the rate dl/dt at which her shadow is increasing in length, then you can find the speed ds/dt at which the tip of her shadow moves across the floor.

(a) From similar triangles,

$$\frac{8}{x + l} = \frac{1.75}{l}$$

$$8l = 1.75(x + l)$$

$$l = \frac{1.75}{6.25}x$$

$$= 0.28x$$

Therefore,

$$\frac{dl}{dt} = 0.28\frac{dx}{dt}$$

$$= 0.28(0.75)$$

$$= 0.21$$

The shadow of the girl is increasing in length at a rate of 0.21 m/s.

(b) The speed at which the tip of the shadow is moving is thus

$$\frac{ds}{dt} = 0.75 + 0.21$$

$$= 0.96\,\text{m/s}$$

Observe that the tip of the girl's shadow is moving faster than the girl herself.   ❏

Exercise 8.2

1. Two small stunt planes take off from an airport at the same time. One flies north at a speed of 300 km/h. The other flies east at a speed of 225 km/h. How fast are the planes separating after 4 min?

2. In question 1, the plane travelling north leaves 2 min before the plane travelling east. How fast are the planes separating 4 min after the second plane leaves? Would you expect the answer to be the same if the plane travelling east left first? Explain.

3. A passenger car approaches a railway crossing from the east at 40 km/h, while a locomotive approaches the crossing from the north at 50 km/h. How fast is the distance between them changing at the moment when the car is 30 m and the train is 40 m from the intersection?

4. Vince is at the top of a 5 m ladder, which is leaning against a vertical wall. If the bottom of the ladder slides away from the wall at a rate of 0.25 m/s, with what speed does he fall when

   (a) the bottom of the ladder is 3 m from the wall?

   (b) the top of the ladder is 1 m from the ground?

5. A rocket rising vertically is being tracked by a radar command post on the ground 10 km from the launch pad. How fast is the rocket rising when it is 8 km high and its distance from the radar station is increasing at a rate of 1000 km/h?

6. At 2:00 p.m., an oil tanker is 25 nautical miles due south of a navy frigate. If the tanker is travelling west at a speed of 16 knots and the frigate is moving south at a speed of 20 knots, find the rate at which the distance between the ships is changing at 2:30 p.m. (1 knot = 1 nautical mile per hour)

7. A baseball diamond has the shape of a square 27.4 m on a side. Mary starts running from first to second base at a rate of 2 m/s at the same time that Steve starts running from second to third base at a rate of 1.5 m/s. How fast is the distance between them changing 5 s after they leave the bases?

8. A kite flying 120 m off the ground is blown horizontally by the wind at a speed of 2 km/h. At what rate is the string unwinding at the moment when 400 m of string have unwound?

9. A boat is pulled toward a dock by means of a rope that is attached to the bow of a boat and runs through a pulley on the dock 4 m above the boat. If the rope is pulled through the pulley at a rate of 6 m/min, at what rate does the boat approach the dock at the moment when the rope between the boat and the pulley is 10 m in length?

10. A police helicopter is flying north at 60 km/h at a constant altitude of 1 km. On the highway below, a car is travelling east at 45 km/h. When the chopper passes over the highway, the car is 2 km west of the point directly below it. At this moment, how fast is the distance between the car and the chopper changing? Is this distance increasing or decreasing? Explain.

Summary and Review   8.3

The initial and most important step in solving a related rates problem is to determine a relationship between the variables which are changing with time. Differentiating this relationship implicitly with respect to t then gives the relation between the rates.

To master the material in this chapter, it is helpful to

(a) draw appropriate diagrams, on which you indicate what the variables stand for;

(b) be familiar with the formulas for the areas and volumes of geometrical figures, the Pythagorean theorem, and the properties of similar triangles, and recognize which one is the correct one to use;

(c) use the sign convention in which positions and velocities to the right or up are positive and to the left or down are negative.

Exercise 8.3

1. An air traffic controller spots two planes at the same altitude flying on perpendicular paths converging on a point. One plane is 150 km from the point and is moving at 450 km/h. The other plane is 200 km from the point and has a speed of 600 km/h.

  (a) At what rate is the distance between the planes decreasing?

  (b) How much time does the controller have to get one of the planes on a different flight path?

2. The height of a conical icicle formed by the dripping of water is twelve times the radius of its base. If the volume is increasing at a rate of 1 cm³/h, at what rate is the height increasing when $h = 8$ cm?

3. A kite is being carried aloft by the wind in such a way that the string makes an angle of 30° with the vertical. How fast is the height of the kite changing if the string is unwinding at a rate of 0.75 m/s?

4. The combined electrical resistance R resulting from resistances R_1 and R_2 connected in parallel is

$$\frac{1}{R} = \frac{1}{R_1} + \frac{1}{R_2}$$

R_1 and R_2 are increasing at rates of 1 and 1.5 ohms/s respectively. At what rate is R increasing when $R_1 = 50$ ohms and $R_2 = 75$ ohms?

5. Two cars approach an intersection at equal speeds, one from the north and one from the east. When one car is 200 m north and the other is 150 m east of the intersection, the distance between them is decreasing at a rate of 20 m/s. At what speed are the cars travelling?

6. A man 1.8 m tall approaches a lamp-post at 1.5 m/s. If the streetlight is hanging 4.5 m above the ground, at what speed is the end of his shadow moving when he is 3 m from the post? At what rate is the length of his shadow changing at that moment?

7. A helicopter takes off from a landing pad at 10:44 a.m. and travels eastward at a speed of 270 km/h. A second helicopter travelling south at 180 km/h is planning to land at 10:56 a.m.

(a) How far apart are the helicopters at 10:48 a.m.?

(b) Find the rate at which the distance between them is changing at 10:48 a.m. Are they approaching or receding from each other at this moment? Explain.

8. Let p be the distance between a convex lens and an object and q be the distance between the lens and the image. p and q are related to the focal length f of the lens by the formula

$$\frac{1}{p} + \frac{1}{q} = \frac{1}{f}$$

Find the rate at which the position of the image is changing for a lens of focal length 5 cm when the object is 8 cm from the lens and moving away from it at a rate of 4.9 cm/s.

9. When a gas is compressed adiabatically (with no gain or loss of heat), the volume and the pressure are related by the formula $PV^{1.4} = k$, where k is a constant. If at a given instant, a gas has a volume of 56 L and a pressure of 36 kPa that is increasing at a rate of 0.3 kPa/min, at what rate is the volume changing?

10. A meteorite enters the earth's atmosphere and burns up at a rate that, at each instant, is proportional to its surface area. Assuming that the meteorite is always spherical, show that the radius decreases at a constant rate.

F. The Trigonometric Functions

These pages contain a review of the trigonometric functions. You will find here a convenient collection of definitions, formulas, and identities that you may want to refer to from time to time. If this material is already familiar to you from previous courses, you may turn directly to Chapter 9.

Radian Measure

One **radian** is the measure of a sector angle that subtends an arc equal in length to the radius. Since two arcs on the same circle are proportional to their sector angles,

Figure F.1

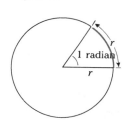

$$\frac{1 \text{ radian}}{360°} = \frac{r}{2\pi r}$$

$$\therefore \quad 1 \text{ radian} = \frac{360°}{2\pi}$$

$$\doteq 57.3°$$

$$2\pi \text{ radians} = 360°$$

$$\pi \text{ radians} = 180°$$

$$\pi/2 \text{ radians} = 90°$$

$$\pi/3 \text{ radians} = 60°$$

$$\pi/4 \text{ radians} = 45°$$

$$\pi/6 \text{ radians} = 30°$$

In calculus, all angles are understood to be measured in radians (the unit "radian" is usually not written). When angles are expressed in radians, formulas involving angles have their simplest form. For instance, for a sector of a circle,

Figure F.2

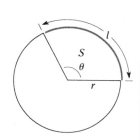

$$\frac{\text{sector angle}}{1 \text{ revolution}} = \frac{\text{arc length}}{\text{circumference}} = \frac{\text{sector area}}{\text{area of circle}}$$

$$\frac{\theta}{2\pi} = \frac{l}{2\pi r} = \frac{S}{\pi r^2}$$

$$\therefore \quad \text{arc length} \quad l = r\theta$$

$$\text{and} \quad \text{sector area} \quad S = \frac{1}{2}r^2\theta$$

Figure F.3

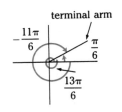

Rotation Angles

An angle whose initial arm lies along the positive x-axis and whose vertex is at the origin is said to be in **standard position**. By convention, an angle is positive if the position of the terminal arm is determined by a counterclockwise rotation from the initial arm, and negative if the rotation is clockwise. Angles in standard position are **coterminal** if their terminal arms coincide as in Figure F.3. In general, θ and $\theta \pm 2n\pi$, $n \in I$, are coterminal angles.

Definitions of the Trigonometric Functions

Let θ be an angle in standard position and $P(x, y)$ be a point on its terminal arm, a distance r from the origin O, where $r = \sqrt{x^2 + y^2}$, as in Figure F.4. Then the trigonometric functions of θ are defined in terms of the following ratios:

Figure F.4

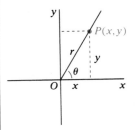

sine:	$\sin \theta = \dfrac{y}{r}$	cosecant:	$\csc \theta = \dfrac{r}{y}$
cosine:	$\cos \theta = \dfrac{x}{r}$	secant:	$\sec \theta = \dfrac{r}{x}$
tangent:	$\tan \theta = \dfrac{y}{x}$	cotangent:	$\cot \theta = \dfrac{x}{y}$

If $P(x, y)$ lies in the first quadrant, then θ is an acute angle of a right triangle, in which case these definitions are equivalent to

Figure F.5

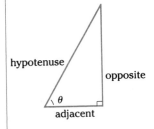

$$\sin \theta = \frac{\text{opposite}}{\text{hypotenuse}} \qquad \csc \theta = \frac{\text{hypotenuse}}{\text{opposite}}$$

$$\cos \theta = \frac{\text{adjacent}}{\text{hypotenuse}} \qquad \sec \theta = \frac{\text{hypotenuse}}{\text{adjacent}}$$

$$\tan \theta = \frac{\text{opposite}}{\text{adjacent}} \qquad \cot \theta = \frac{\text{adjacent}}{\text{opposite}}$$

Signs and Values of the Functions

The value of a trigonometric function is positive or negative depending on the position of the terminal arm. For instance, if the terminal arm of θ lies in the second quadrant, then x is negative and y is positive (r is always positive). It follows that the values of the sine and cosecant functions are positive, and the values of the other functions are negative. The CAST rule, Figure F.6, is a convenient memory aid that tells which functions have positive values in which quadrants.

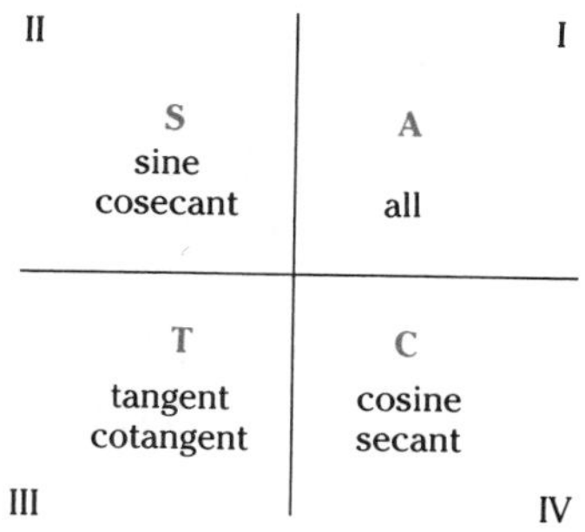

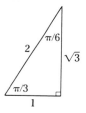

Figure F.6

Exact values of the trigonometric functions for angles of $\pi/6$, $\pi/4$, and $\pi/3$ and their multiples can be found with the help of special triangles.

Figure F.7

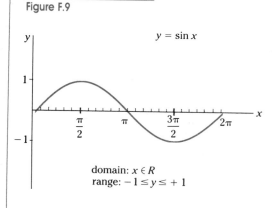

For instance, according to the diagram in Figure F.8, $\sin 7\pi/6 = -1/2$, that is, $\sin 7\pi/6$ is equal to $\sin \pi/6$ except for the algebraic sign, which is negative because $7\pi/6$ lies in the third quadrant.

Figure F.8

Graphs of the Trigonometric Functions

The cosine and the secant functions are even functions. The others are odd functions. Their graphs for $0 \le \theta \le 2\pi$ are shown below:

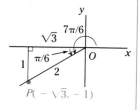

Figure F.9

Figure F.10

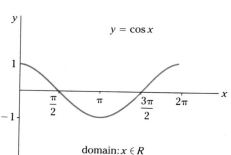

$y = \sin x$

domain: $x \in R$
range: $-1 \le y \le +1$

$y = \cos x$

domain: $x \in R$
range: $-1 \le y \le +1$

Figure F.11

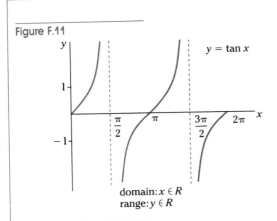

$y = \tan x$

domain: $x \in R$
range: $y \in R$

Figure F.12

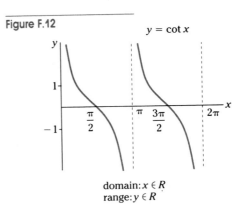

$y = \cot x$

domain: $x \in R$
range: $y \in R$

Figure F.13

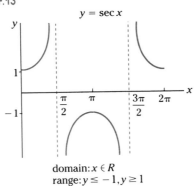

$y = \sec x$

domain: $x \in R$
range: $y \le -1, y \ge 1$

Figure F.14

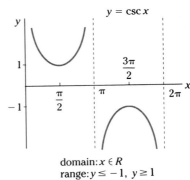

$y = \csc x$

domain: $x \in R$
range: $y \le -1, \ y \ge 1$

The sine and cosine functions are called **sinusoidal functions**. When a sinusoidal function is written, for example, in the form

$$y = A \sin k(x - \phi)$$

then the **amplitude**   is $|A|$
the **period**   is $2\pi/k$
and the **phase shift** is ϕ, relative to $\sin(kx)$.

Figure F.15

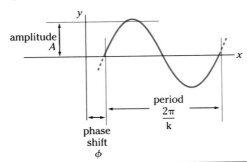

Trigonometric Identities

The following basic trigonometric identities can be derived from the definitions of the trigonometric functions.

Reciprocal identities:

$$\csc \theta = \frac{1}{\sin \theta} \qquad \sec \theta = \frac{1}{\cos \theta} \qquad \cot \theta = \frac{1}{\tan \theta}$$

Quotient identities:

$$\tan \theta = \frac{\sin \theta}{\cos \theta} \qquad \cot \theta = \frac{\cos \theta}{\sin \theta}$$

Pythagorean identities:

$$\cos^2 \theta + \sin^2 \theta = 1 \qquad 1 + \tan^2 \theta = \sec^2 \theta \qquad \cot^2 \theta + 1 = \csc^2 \theta$$

Two further identities that will prove useful are the Complementary Angle Identities:

$$\sin \theta = \cos\left(\frac{\pi}{2} - \theta\right) \qquad \cos \theta = \sin\left(\frac{\pi}{2} - \theta\right)$$

These can be proved by reflecting a point P in the line $y = x$, as in Figure F.16, and then comparing the coordinates of P with the coordinates of its image P'.

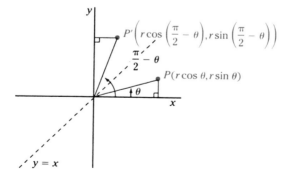

Figure F.16

Trigonometric Equations

An equation containing trigonometric functions is solved by first using identities to transform the equation into one or more equations containing a single trigonometric function. For example,

$$2 \cos^2 \theta + \sin \theta = 2$$

$$2(1 - \sin^2 \theta) + \sin \theta = 2$$

$$2 - 2 \sin^2 \theta + \sin \theta = 2$$

$$2 \sin^2 \theta - \sin \theta = 0$$

$$\sin \theta (2 \sin \theta - 1) = 0$$

$$\therefore \quad \sin \theta = 0 \quad \text{or} \quad 2 \sin \theta - 1 = 0$$

Then by using the properties of the trigonometric functions, graphs, special triangles, or a calculator, as the situation requires, find the values of θ. In the above example

$$\theta = 0, \pi, 2\pi \quad \text{and} \quad \theta = \pi/6, 5\pi/6$$

Two formulas of special value in the solution of triangles are

The sine law: $\quad \dfrac{\sin A}{a} = \dfrac{\sin B}{b} = \dfrac{\sin C}{c}$

The cosine law: $\quad c^2 = a^2 + b^2 - 2ab \cos C$

where $\angle A$ is opposite side a, etc.

Exercise F

1. Convert to radians: $120°, 135°, 225°, 240°, 330°,$ $450°$. Express the answers in terms of π.

2. Convert to degrees: $3\pi/2,\ 4\pi/3,\ 5\pi/6,\ 6\pi/5,$ $17\pi/12,\ 23\pi/10$.

3. Sketch the following angles in standard position: $150°,\ -300°,\ 5\pi/3,\ -5\pi/4$

4. Give the radian measures of a positive and a negative angle that are coterminal with $7\pi/6$.

5. State the measure of angles θ and ϕ shown in Figures F.17 and F.18 in both degrees and radians.

Figure F.17

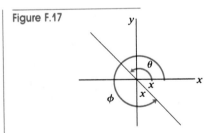

Figure F.18

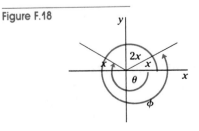

6. A gear 6 cm in diameter revolves around the outside of a fixed gear 10 cm in diameter. Through how many radians does the smaller gear revolve (about its own centre) in going once around the fixed gear?

7. The area of a sector of a circle is $180\,\text{cm}^2$, and the length of its arc is 15 cm. Find the radius of the circle, and the measure of the sector angle.

8. Find the value:

(a) $\cos\dfrac{5\pi}{6}$　(b) $\csc\dfrac{3\pi}{2}$　(c) $\tan\dfrac{7\pi}{4}$

(d) $\sin\dfrac{11\pi}{6}$　(e) $\cot\dfrac{4\pi}{3}$　(f) $\sec\dfrac{7\pi}{6}$

9. Find the values of the other five trigonometric functions in each of the following:

(a) $\sin\theta = \dfrac{12}{13}$,　$0 < \theta < \dfrac{\pi}{2}$

(b) $\cos\theta = \dfrac{-2\sqrt{6}}{7}$,　$\dfrac{\pi}{2} < \theta < \pi$

(c) $\tan\theta = \dfrac{24}{7}$,　$\pi < \theta < \dfrac{3\pi}{2}$

10. State the amplitude, period, and phase shift, and sketch the graphs:

(a) $y = 5\cos(2x + \pi/2)$

(b) $y = 3\sin(3x - 2\pi/3)$

(c) $y = 2\sin(x/3 + \pi/6)$

11. Prove the identities:

(a) $\sin\theta\cot\theta\sec\theta = 1$

(b) $(1 - \cos^2\theta)(1 + \tan^2\theta) = \tan^2\theta$

(c) $\sin^4\theta - 2\sin^2\theta = \cos^4\theta - 1$

(d) $(\sec\theta + \csc\theta)(\sin\theta + \cos\theta) = \sec\theta\csc\theta + 2$

(e) $\tan\theta + \cot\theta = \sec\theta\csc\theta$

(f) $\dfrac{1}{1 - \sin\theta} + \dfrac{1}{1 + \sin\theta} = 2\sec^2\theta$

12. Solve for θ, $0 \le \theta \le 2\pi$:

(a) $\sin^2\theta - \sin\theta = 0$

(b) $2\cos\theta\csc\theta - \sqrt{3}\csc\theta = 0$

(c) $\sin\theta + \cos\theta = 0$

(d) $2\sin^2\theta - 5\sin\theta - 3 = 0$

Compound Angle Formulas

9

In the preceding chapters, you have learned how to find the derivatives of algebraic functions and studied applications to extreme value and rate problems. The object of Chapters 9 to 11 is to do the same for the trigonometric functions. Definitions and a review of basics are to be found in F. The Trigonometric Functions, pages 223–229.

Before getting into the derivatives of the trigonometric functions, it is necessary to become acquainted with the compound angle formulas. These are a collection of trigonometric identities involving sums of angles, multiple angles, and so on. They will be discussed in this chapter with a particular emphasis on their use in calculus.

Addition Formulas for Sin θ and Cos θ

9.1

Consider the problem of expressing a function such as $\sin(\theta + \phi)$, as functions of the angles θ and ϕ separately. To begin with, observe carefully that the sine of a sum is *not* equal to the sum of the sines. For instance,

$$\sin(30° + 45°) = \sin 75°$$

$$\doteq 0.9659$$

whereas

$$\sin 30° + \sin 45° \doteq 0.5 + 0.7071$$

$$= 1.2071$$

The following identities are known as the addition formulas for cosine:

$$\cos(\theta + \phi) = \cos\theta\cos\phi - \sin\theta\sin\phi$$

$$\cos(\theta - \phi) = \cos\theta\cos\phi + \sin\theta\sin\phi$$

The addition formulas for sine are

$$\sin(\theta + \phi) = \sin\theta\cos\phi + \cos\theta\sin\phi$$

$$\sin(\theta - \phi) = \sin\theta\cos\phi - \cos\theta\sin\phi$$

Of the four, the formula for $\cos(\theta + \phi)$ is the most straightforward to derive. It in turn can be used to establish the other three. All other compound angle identities are derived from these four.

To prove the addition formula for $\cos(\theta + \phi)$, start by taking three points A, B, and C on a unit circle as shown in Figure 9.1a, where $\angle BOA = \theta$ and $\angle BOC = -\phi$. The proof depends on the fact that the distance between points A and C stays the same when the figure is rotated.

Figure 9.1

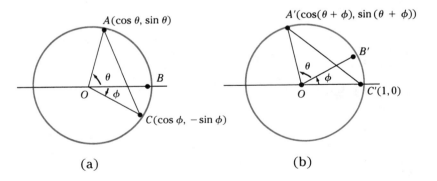

(a)   (b)

The coordinates of A can be found by noting that

since $\dfrac{y}{r} = \sin\theta$, $\dfrac{x}{r} = \cos\theta$,   and $r = 1$

then $y = \sin\theta$, $x = \cos\theta$

The coordinates of C can be found in a similar manner. Therefore, points A and C are

$$A(\cos\theta, \sin\theta) \qquad C(\cos\phi, -\sin\phi)$$

The square of the length AC is

$$
\begin{aligned}
|AC|^2 &= (\cos\theta - \cos\phi)^2 + (\sin\theta + \sin\phi)^2 \\
&= \cos^2\theta - 2\cos\theta\cos\phi + \cos^2\phi \\
&\quad + \sin^2\theta + 2\sin\theta\sin\phi + \sin^2\phi \\
&= 2 - 2\cos\theta\cos\phi + 2\sin\theta\sin\phi
\end{aligned}
$$

Now rotate the figure counterclockwise through an angle ϕ so that $A \rightarrow A', B \rightarrow B'$, and $C \rightarrow C'$ as in Figure 9.1b. $\angle C'OA' = (\theta + \phi)$ and the points A' and C' are

$$A'(\cos(\theta + \phi), \sin(\theta + \phi)) \qquad C'(1, 0)$$

The square of the length $A'C'$ is

$$
\begin{aligned}
|A'C'|^2 &= (\cos(\theta + \phi) - 1)^2 + (\sin(\theta + \phi) - 0)^2 \\
&= \cos^2(\theta + \phi) - 2\cos(\theta + \phi) + 1 + \sin^2(\theta + \phi) \\
&= 2 - 2\cos(\theta + \phi)
\end{aligned}
$$

Since the lengths AC and $A'C'$ are equal, it follows that

$$2 - 2\cos(\theta + \phi) = 2 - 2\cos\theta\cos\phi + 2\sin\theta\sin\phi$$

$$\therefore \qquad \cos(\theta + \phi) = \cos\theta\cos\phi - \sin\theta\sin\phi$$

The formula for $\cos(\theta - \phi)$ results from replacing ϕ by $-\phi$ in the above identity:

$$
\begin{aligned}
\cos(\theta - \phi) &= \cos(\theta + (-\phi)) \\
&= \cos\theta\cos(-\phi) - \sin\theta\sin(-\phi) \\
&= \cos\theta\cos\phi + \sin\theta\sin\phi
\end{aligned}
$$

The formula for $\sin(\theta + \phi)$ can be found with the help of the complementary angle identities (see F. The Trigonometric Functions, pages 223–229). Since

$$\cos(\pi/2 - x) = \sin x \quad \text{and} \quad \sin(\pi/2 - x) = \cos x$$

then

$$\begin{aligned}
\sin(\theta + \phi) &= \cos[\pi/2 - (\theta + \phi)] \\
&= \cos[(\pi/2 - \theta) - \phi] \\
&= \cos(\pi/2 - \theta)\cos\phi + \sin(\pi/2 - \theta)\sin\phi \\
&= \sin\theta\cos\phi + \cos\theta\sin\phi
\end{aligned}$$

$$\therefore \quad \sin(\theta + \phi) = \sin\theta\cos\phi + \cos\theta\sin\phi$$

Replacing ϕ by $-\phi$ produces the remaining identity:

$$\begin{aligned}
\sin(\theta - \phi) &= \sin(\theta + (-\phi)) \\
&= \sin\theta\cos(-\phi) + \cos\theta\sin(-\phi) \\
&= \sin\theta\cos\phi - \cos\theta\sin\phi
\end{aligned}$$

EXAMPLE A

Using the fact that $75° = 30° + 45°$, find the exact value of $\sin 75°$.

SOLUTION

$$\begin{aligned}
\sin 75° &= \sin(30° + 45°) \\
&= \sin 30°\cos 45° + \cos 30°\sin 45° \\
&= \frac{1}{2}\cdot\frac{1}{\sqrt{2}} + \frac{\sqrt{3}}{2}\cdot\frac{1}{\sqrt{2}} \\
&= \frac{1 + \sqrt{3}}{2\sqrt{2}} \\
&= \frac{\sqrt{2} + \sqrt{6}}{4}
\end{aligned}$$

☐

EXAMPLE B

Prove the identity:

$$\sin(\alpha + \beta)\sin(\alpha - \beta) = \sin^2\alpha - \sin^2\beta$$

SOLUTION

$$\begin{aligned}
&\sin(\alpha + \beta)\sin(\alpha - \beta) \\
&= (\sin\alpha\cos\beta + \cos\alpha\sin\beta)(\sin\alpha\cos\beta - \cos\alpha\sin\beta)
\end{aligned}$$

$$= \sin^2 \alpha \cos^2 \beta - \cos^2 \alpha \sin^2 \beta$$

$$= \sin^2 \alpha (1 - \sin^2 \beta) - (1 - \sin^2 \alpha) \sin^2 \beta$$

$$= \sin^2 \alpha - \sin^2 \alpha \sin^2 \beta - \sin^2 \beta + \sin^2 \alpha \sin^2 \beta$$

$$= \sin^2 \alpha - \sin^2 \beta \qquad \square$$

The function $y = \sin(x - \phi)$ is a sine function with a phase shift ϕ. Its graph is that of a sine function that has been translated parallel to the x-axis a distance ϕ to the right (or to the left, if ϕ is negative). As the following example shows, it is possible to express a function of this sort as the sum of two sinusoidal functions.

EXAMPLE C

Write the function $y = \sin(x + \pi/3)$ as a sum of a sine and a cosine function. Compare their graphs.

SOLUTION

$$y = \sin(x + \pi/3)$$

$$= \sin x \cos(\pi/3) + \cos x \sin(\pi/3)$$

$$= \frac{1}{2} \sin x + \frac{\sqrt{3}}{2} \cos x$$

In Figure 9.2, you can see graphically that the function $y = \sin(x + \pi/3)$, which is a sine function shifted left by $\pi/3$, is the sum of the other two functions.

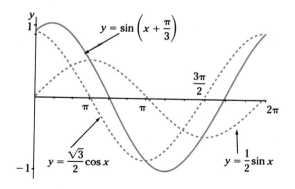

Figure 9.2

EXAMPLE D

Express the function $y = 3 \sin x + 4 \cos x$ as a single trigonometric function with a phase shift.

SOLUTION

Start with a general function having amplitude A and phase shift ϕ, for instance,

$$y = A \sin(x - \phi)$$
$$= A \sin x \cos \phi - A \cos x \sin \phi$$
$$= (A \cos \phi) \sin x + (-A \sin \phi) \cos x$$

This will be equal to $y = 3 \sin x + 4 \cos x$ if

$$A \cos \phi = 3 \quad \text{and} \quad -A \sin \phi = 4$$

To find the amplitude A, observe that

$$(A \cos \phi)^2 + (-A \sin \phi)^2 = (3)^2 + (4)^2$$
$$A^2(\cos^2 \phi + \sin^2 \phi) = 9 + 16$$
$$A^2 = 25$$
$$A = \pm 5$$

Using $+5$ for A, it follows that

$$\sin \phi = \frac{-4}{5} \quad \text{and} \quad \cos \phi = \frac{3}{5}$$

which means that ϕ must be an angle in the fourth quadrant, $\phi \doteq -53°$. These relationships are illustrated in Figure 9.3. Thus,

$$y = 5 \sin(x - \phi) \text{ with } \phi \doteq -53°$$

Note that we could equally well have started with $y = A \cos(x - \phi)$ producing the equivalent result

$$y = 5 \cos(x - \phi) \text{ with } \phi \doteq 37°$$

In other words, the original function could be viewed as a sine function shifted left, Figure 9.4a, or a cosine function shifted right, Figure 9.4b.   ❑

Figure 9.3

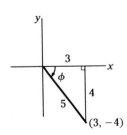

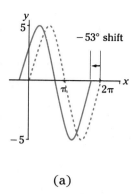

$-53°$ shift

(a)

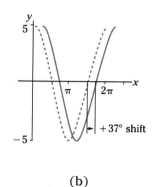

$+37°$ shift

Figure 9.4

(b)

In this example, the two parts of the original function, $3 \sin x$ and $4 \cos x$, had the same period and differ in phase by $90°$. The ability to combine sinusoidal functions having arbitrary periods and phases is of considerable importance in applied mathematics. Such problems will be discussed in Section 9.5, once the required identities have been established.

EXAMPLE E

Solve for x:

$$24 \cos x - 10 \sin x = 13, 0 < x < 2\pi$$

SOLUTION

Following the method outlined in Example D, set

$$24 \cos x - 10 \sin x = A \sin(x - \phi)$$

$$= A \sin x \cos \phi - A \cos x \sin \phi$$

$$= (-A \sin \phi) \cos x + (A \cos \phi) \sin x$$

Therefore, $-A \sin \phi = 24, A \cos \phi = -10$, and the amplitude is

$$A = \sqrt{(24)^2 + (-10)^2} = 26$$

The phase shift is a third quadrant angle found from either

$$\sin \phi = \frac{-24}{26} \quad \text{or} \quad \cos \phi = \frac{-10}{26}$$

giving, $\phi \doteq 4.3176$ radians or $247°$. The given equation thus reduces to

$$26 \sin(x - \phi) = 13$$

Consequently,

$$\sin(x - \phi) = 0.5 \quad \text{and} \quad x - \phi = \pi/6 \quad \text{or} \quad 5\pi/6$$

$$\therefore \quad x = \pi/6 + \phi \doteq 0.5236 + 4.3176 = 4.8412 \text{ radians or } 277°$$

and $x = 5\pi/6 + \phi \doteq 2.6180 + 4.3176 = 6.9356$ radians or $397°$

Since $397°$ is outside the interval $0 < x < 2\pi$, the correct answer is the coterminal angle obtained by subtracting 2π:

$$x \doteq 0.6524 \text{ radians or } 37°$$

Exercise 9.1

1. Show that:

(a) $\cos\left(\dfrac{\pi}{2} - \theta\right) = \sin \theta$

(b) $\sin(\pi - \theta) = \sin \theta$

(c) $\sin\left(\theta + \dfrac{3\pi}{2}\right) = -\cos \theta$

(d) $\cos(\theta - \pi) = -\cos \theta$

2. Verify that:

(a) $\sqrt{2} \sin\left(x + \dfrac{\pi}{4}\right) = \cos x + \sin x$

(b) $\sqrt{2} \cos\left(x + \dfrac{\pi}{4}\right) = \cos x - \sin x$

3. Find $\sin x$ if $x = y + \pi/3$, and y is an acute angle such that $\sin y = 5/13$.

4. Find the exact value of each of the following:

(a) $\sin 105°$

(b) $\sin 15°$

(c) $\cos 165°$

(d) $\sin\left(\dfrac{11\pi}{12}\right)$

(e) $\cos\left(\dfrac{7\pi}{12}\right)$

5. Establish the identities:

(a) $\cos(x + y)\cos(x - y) = \cos^2 x - \sin^2 y$

(b) $\cos(x + y) \sin(x - y) = \sin x \cos x - \sin y \cos y$

6. (a) Prove that:

$$\cot \theta - \cot \phi = \frac{-\sin(\theta - \phi)}{\sin \theta \sin \phi}$$

(b) Work out similar identities for $\tan \theta - \tan \phi$ and $\tan \theta - \cot \phi$.

7. (a) Show that:

$$\sin(A + B + C) = \sin A \cos B \cos C$$
$$+ \cos A \sin B \cos C$$
$$+ \cos A \cos B \sin C$$
$$- \sin A \sin B \sin C$$

Hint: write $\sin(A + B + C)$ as $\sin[(A + B) + C]$ and use an addition formula.

(b) Derive a formula for $\cos(A + B + C)$.

(c) Discuss the patterns you find in the formulas of parts (a) and (b).

8. If $\angle C$ in $\triangle ABC$ is a right angle, then show that:

$$\sin(A - B) = \frac{a^2 - b^2}{a^2 + b^2}$$

and

$$\cos(A - B) = \frac{2ab}{c^2}$$

9. Show that in any triangle:

$$\sin C = \sin A \cos B + \cos A \sin B$$

10. Express each of the following functions in the form $y = A \sin(x - \phi)$. Find the amplitude and phase shift of each, and sketch their graphs.

(a) $y = \sqrt{3} \cos x - \sqrt{2} \sin x$

(b) $y = 4 \sin x - \cos x$

(c) $y = 9 \cos x + 3\sqrt{11} \sin x$

11. Write the function $y = 3 \sin x - 2 \cos x$ in the form $y = A \sin(x - \phi)$ and also in the form $y = A \cos(x - \phi)$. In each case, find the phase shift and sketch the graph.

12. State the equation of the function whose graph is shown in Figure 9.5. Show that this function is the sum of two sinusoidal functions, $f(x) = A \sin x$ and $g(x) = B \cos x$. Find A and B and sketch the graphs of f and g.

Figure 9.5

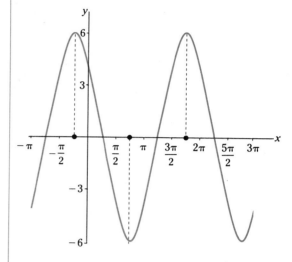

13. Solve for x:

(a) $\sqrt{6} \cos x - \sqrt{6} \sin x = 3$

(b) $\sqrt{3} \sin x + 3 \cos x = -\sqrt{6}$

(c) $\sin x + 2 \cos x = 1/2$

14. Solve for θ:

$$\cos(40° + \theta) = 3 \sin(50° + \theta)$$

Hint: Write $40° + \theta = 90° - (50° - \theta)$

9.2 An Addition Formula for Tan θ

A formula for $\tan(\theta + \phi)$ is readily obtained from the addition formulas for sine and cosine.

$$\tan(\theta + \phi) = \frac{\sin(\theta + \phi)}{\cos(\theta + \phi)}$$

$$= \frac{\sin\theta\cos\phi + \cos\theta\sin\phi}{\cos\theta\cos\phi - \sin\theta\sin\phi}$$

Divide the numerator and the denominator by $\cos\theta\cos\phi$:

$$\tan(\theta + \phi) = \frac{\dfrac{\sin\theta\cos\phi}{\cos\theta\cos\phi} + \dfrac{\cos\theta\sin\phi}{\cos\theta\cos\phi}}{\dfrac{\cos\theta\cos\phi}{\cos\theta\cos\phi} - \dfrac{\sin\theta\sin\phi}{\cos\theta\cos\phi}}$$

$$= \frac{\dfrac{\sin\theta}{\cos\theta} + \dfrac{\sin\phi}{\cos\phi}}{1 - \dfrac{\sin\theta}{\cos\theta} \cdot \dfrac{\sin\phi}{\cos\phi}}$$

$$\therefore \quad \tan(\theta + \phi) = \frac{tan\,\theta + \tan\phi}{1 - \tan\theta\tan\phi}$$

Replacing ϕ by $-\phi$ results in

$$\tan(\theta - \phi) = \frac{\tan\theta - \tan\phi}{1 + \tan\theta\tan\phi}$$

Therefore, the addition formulas for tangent are

$$\tan(\theta + \phi) = \frac{\tan\theta + \tan\phi}{1 - \tan\theta\tan\phi}$$

$$\tan(\theta - \phi) = \frac{\tan\theta - \tan\phi}{1 + \tan\theta\tan\phi}$$

EXAMPLE A

Find the exact value of $\tan 105°$.

SOLUTION

$$\tan 105° = \tan(45° + 60°)$$

$$= \frac{\tan 45° + \tan 60°}{1 - \tan 45° \tan 60°}$$

$$= \frac{1 + \sqrt{3}}{1 - 1 \cdot \sqrt{3}}$$

$$= \frac{(1 + \sqrt{3})}{(1 - \sqrt{3})} \cdot \frac{(1 + \sqrt{3})}{(1 + \sqrt{3})} \qquad \longleftarrow \qquad \text{rationalize to simplify}$$

$$= \frac{1 + 2\sqrt{3} + 3}{-2}$$

$$= -2 - \sqrt{3} \qquad \qquad \square$$

The addition formula for tangent can be applied to the problem of finding the angle between two intersecting lines. Define the angle of inclination θ of a line to be the positive angle from the x-axis to the line, as shown in Figure 9.6. It is evident from the figure that

Figure 9.6

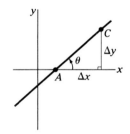

$$\tan \theta = \frac{\Delta y}{\Delta x} = \text{slope } m$$

Thus, the tangent of the angle of inclination is equal to the slope of the line.

Suppose that the angles of inclination of lines l_1 and l_2 are, respectively, θ_1 and θ_2, as shown in Figure 9.7. Since θ_1 is an exterior angle of $\triangle ABC$, then $\theta_1 = \theta_2 + \alpha$. Therefore, the angle α between the lines is given by

$$\alpha = \theta_1 - \theta_2$$

Figure 9.7

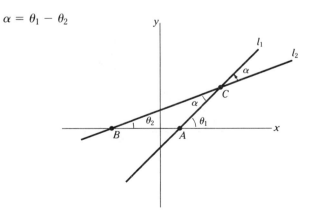

It follows that

$$\tan \alpha = \tan(\theta_1 - \theta_2)$$

$$= \frac{\tan \theta_1 - \tan \theta_2}{1 + \tan \theta_1 \tan \theta_2}$$

$$\therefore \quad \tan \alpha = \frac{m_1 - m_2}{1 + m_1 m_2}$$

Observe that if $m_1 < m_2$ as in Figure 9.8, then $\tan \alpha$ is negative and the angle between the lines is obtuse.

Figure 9.8

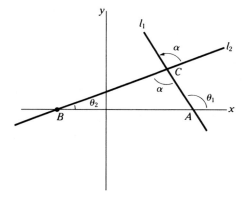

A special case of this formula occurs when the denominator $1 + m_1 m_2 = 0$. Then $\tan \alpha = \infty$ and $\alpha = \pi/2$ or 90°. Therefore, the two lines are perpendicular.

Another special case occurs if one of the lines, say l_1, is parallel to the y-axis, that is, m_1, is undefined (see Figure 9.9). As $m_1 \to \infty$, what is the limit of $\tan \alpha$?

Figure 9.9

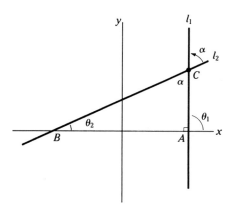

$$\lim_{m_1 \to \infty} \tan \alpha = \lim_{m_1 \to \infty} \frac{m_1 - m_2}{1 + m_1 m_2} \cdot \frac{(1/m_1)}{(1/m_1)}$$

$$= \lim_{m_1 \to \infty} \frac{1 - \dfrac{m_2}{m_1}}{\dfrac{1}{m_1} + m_2}$$

$$= \frac{1}{m_2}$$

Therefore, in the limit,

$$\tan \alpha = \frac{1}{\tan \theta_2}$$

$$= \cot \theta_2$$

$\therefore$ α is the complement of θ_2

EXAMPLE B

Find the angle of intersection of the lines
$3x + 4y = 8$ and $x - 2y = 8$.

SOLUTION

The slopes of the lines (see Figure 9.10) are, respectively,

$$m_1 = -\frac{3}{4}, \quad m_2 = \frac{1}{2}$$

$$\therefore \quad \tan \alpha = \frac{(-3/4) - (1/2)}{1 + (-3/4) \cdot (1/2)}$$

$$= \frac{-5/4}{5/8}$$

$$= -2$$

Figure 9.10

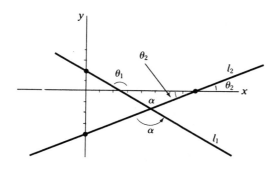

$$\therefore \quad \alpha \doteq -1.1071 \text{ radians or } -63°.$$

This is the measure of the *acute* angle between the lines. The negative sign, represents a *clockwise* rotation from l_2 to l_1. ❑

EXAMPLE C

Find the slope of the line that intersects the tangent to the curve

$$y = x^2 - 6x + 10$$

at the point $(4, 2)$ at an angle of $\pi/6$.

SOLUTION

There are actually two possible lines, as you can see in **Figures 9.11a and b**. Let l_1 be the tangent to the curve at C. Its slope is

Figure 9.11

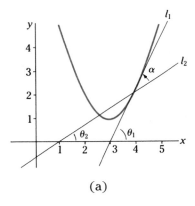

(a)

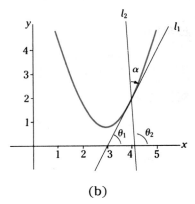

(b)

$$y' = 2x - 6$$

$$\therefore \quad m_1 = 2(4) - 6$$

$$= 2$$

The slope of m_2 line l_2 in Figure 9.11a is found as follows:

$$\tan \frac{\pi}{6} = \frac{2 - m_2}{1 + 2m_2}$$

$$\frac{1}{\sqrt{3}} = \frac{2 - m_2}{1 + 2m_2}$$

$$1 + 2m_2 = 2\sqrt{3} - m_2\sqrt{3}$$

$$(2 + \sqrt{3})m_2 = 2\sqrt{3} - 1$$

$$m_2 = \frac{2\sqrt{3} - 1}{2 + \sqrt{3}}$$

On rationalizing, this becomes

$$m_2 = -8 + 5\sqrt{3}$$

$$\tan\theta_2 = -8 + 5\sqrt{3}$$

making the angle of inclination, $\theta_2 \doteq 0.5835$ radians or 33°.
   The slope of the other possible line, line l_2 in Figure 9.10b, can be found in a similar fashion. Since this line lies on the opposite side of the tangent, $\alpha = -\pi/6$. The calculation yields $m_2 = -8 - 5\sqrt{3}$, and $\theta_2 \doteq 1.6307$ radians or 93°.

Exercise 9.2

1. Evaluate:

   (a) $\tan(60° - 45°)$

   (b) $\tan\left(\dfrac{\pi}{4} + \dfrac{\pi}{6}\right)$

2. Find the exact value of the following, leaving the answer in radical form:

   (a) $\tan 165°$

   (b) $\tan\dfrac{13\pi}{12}$

3. Show that:

   (a) $\tan\left(\dfrac{\pi}{4} - \theta\right) = \dfrac{1 - \tan\theta}{1 + \tan\theta}$

   (b) $\tan(\pi - \theta) = -\tan\theta$

   (c) $\tan\left(\dfrac{\pi}{2} + \theta\right) = -\cot\theta$

4. If $\tan y = 2$ and $x + y = 3\pi/4$, what is $\tan x$?

5. Find B, given that $A - B = \pi/4$ and $\tan A = 1/3$.

6. Express $\tan\theta$ in terms of $\tan(\theta + \phi)$ and $\tan\phi$.

7. Obtain a formula for $\cot(\theta + \phi)$

   (a) in terms of $\tan\theta$ and $\tan\phi$

   (b) in terms of $\cot\theta$ and $\cot\phi$.

8. Find a formula for $\tan(A + B + C)$.

9. Prove that in any triangle:

$$\tan A + \tan B + \tan C = \tan A \tan B \tan C$$

10. Solve for x, $(0 \le x \le \pi/2)$:

$$\tan\left(x - \dfrac{2\pi}{3}\right) = 14\tan x.$$

11. Show that if $m_1 < m_2$, the tangent of the acute angle between two intersecting lines with slopes m_1 and m_2 is

$$\tan\alpha = \left|\dfrac{m_1 - m_2}{1 + m_1m_2}\right|$$

12. Find the equation of the line with positive slope that intersects the line $y = -2x + 6$ at $x = 2$ at an angle of $\pi/4$ radians.

13. In Figure 9.11b of Example C, show that the slope of l_2 is $-8 - 5\sqrt{3}$.

14. Find the acute angle between the tangents to the curves $y = -2x^2 + 5$ and $y = x^2 + 2$ at the point of intersection in the first quadrant.

15. Show that the parabolas

$$y = \frac{1}{2a}(x^2 - a^2), a > 0$$

and

$$y = -\frac{1}{2b}(x^2 - b^2), b > 0$$

always intersect at right angles, regardless of the values of a and b.

9.3 Double-Angle Formulas

Formulas for trigonometric functions of 2θ are easily obtained from the addition formulas, by setting $\phi = \theta$. Thus

$$\sin(\theta + \theta) = \sin\theta\cos\theta + \cos\theta\sin\theta$$

$$\sin 2\theta = 2\sin\theta\cos\theta$$

There are three different formulas for $\cos 2\theta$. One arises from the addition formula for cosine:

$$\cos(\theta + \theta) = \cos\theta\cos\theta - \sin\theta\sin\theta$$

$$\cos 2\theta = \cos^2\theta - \sin^2\theta$$

Use of the Pythagorean identity, $\sin^2\theta + \cos^2\theta = 1$, produces the other two:

$$\cos 2\theta = 2\cos^2\theta - 1$$

$$\cos 2\theta = 1 - 2\sin^2\theta$$

Similarly, the addition formula for tangent gives:

$$\tan(\theta + \theta) = \frac{\tan\theta + \tan\theta}{1 - \tan\theta\tan\theta}$$

$$\tan 2\theta = \frac{2\tan\theta}{1 - \tan^2\theta}$$

EXAMPLE A

Given $\sin\theta = 3/5, \pi/2 < \theta < \pi$. Find $\sin 2\theta$ and $\cos 2\theta$.

SOLUTION

If $\sin \theta = 3/5$, then $\cos \theta = -4/5$, since θ is in the second quadrant (Figure 9.12).

$$\sin 2\theta = 2 \sin \theta \cos \theta$$

$$= 2 \cdot \left(\frac{3}{5}\right) \cdot \left(-\frac{4}{5}\right) = -\frac{24}{25}$$

$$\cos 2\theta = 1 - 2 \sin^2 \theta$$

$$= 1 - 2 \cdot \left(\frac{3}{5}\right)^2 = \frac{7}{25}$$

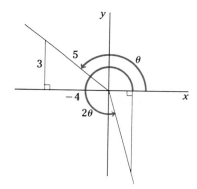

Figure 9.12

Note carefully that $2 \sin \theta \neq \sin 2\theta$ and $2 \cos \theta \neq \cos 2\theta$. Recall that multiplying the function by 2 doubles the amplitude, whereas multiplying the argument by 2 cuts the period in half, so these expressions cannot possibly be equal for all values of θ.

EXAMPLE B

Solve for θ, $0 \leq \theta \leq 2\pi$:

$$-5 \sin \theta = \cos 2\theta + 2$$

SOLUTION

Replace $\cos 2\theta$ using the identity $\cos 2\theta = 1 - 2 \sin^2 \theta$. This is the best choice, because it leads to a quadratic equation in $\sin \theta$.

$$-5 \sin \theta = \cos 2\theta + 2$$

$$-5 \sin \theta = (1 - 2 \sin^2 \theta) + 2$$

$$-5 \sin \theta = 3 - 2 \sin^2 \theta$$

$$\therefore \quad 2\sin^2\theta - 5\sin\theta - 3 = 0$$

$$(2\sin\theta + 1)(\sin\theta - 3) = 0$$

$$2\sin\theta + 1 = 0, \quad \text{or} \quad \sin\theta - 3 = 0$$

$$\sin\theta = -\frac{1}{2} \qquad\qquad \sin\theta = 3 \quad\longleftarrow \quad \begin{array}{l}\text{no solution—}\\ \text{the range of}\\ \text{the sine function}\\ \text{is } |\sin\theta| \leq 1\end{array}$$

$$\theta = \frac{7\pi}{6}, \frac{11\pi}{6}$$

Exercise 9.3

1. If $\theta = \pi/3$, what are $\sin 2\theta$, $\cos 2\theta$, and $\tan 2\theta$?

2. $\angle A$ is an acute angle such that $\sin A = 1/3$. Determine the exact value of the following:

(a) $\sin 2A$

(b) $\cos 2A$

(c) $\sin 4A$

(d) $\cos 4A$

3. Find the exact values of the trigonometric functions of 2θ if $\sin\theta = 4/5$ and θ is

(a) a first quadrant angle

(b) a second quadrant angle.

4. If $\tan\theta = \sqrt{2}, 0 < \theta < \pi/2$, find the exact values of $\tan 2\theta$ and $\tan 4\theta$.

5. Find $\tan\alpha$ if $\alpha + 2\beta = \pi/4$ and $\tan\beta = 1/3$.

6. Prove the identities:

(a) $\dfrac{2\tan\theta}{1 + \tan^2\theta} = \sin 2\theta$

(b) $\dfrac{1 - \tan^2\theta}{1 + \tan^2\theta} = \cos 2\theta$

(c) $\cot\theta + \tan\theta = 2\csc 2\theta$

(d) $\cot\theta - \tan\theta = 2\cot 2\theta$

7. Show that:

(a) $\sin 3\theta = 3\sin\theta - 4\sin^3\theta$

(b) $\cos 3\theta = 4\cos^3\theta - 3\cos\theta$

(c) $\tan 3\theta = \dfrac{3\tan\theta - \tan^3\theta}{1 - 3\tan^2\theta}$

8. In certain calculus problems it is helpful to express powers of sines or cosines in terms of functions of multiple angles. Show that:

(a) $\cos^2\theta = \dfrac{1}{2} + \dfrac{1}{2}\cos 2\theta$

(b) $\cos^4\theta = \dfrac{3}{8} + \dfrac{1}{2}\cos 2\theta + \dfrac{1}{8}\cos 4\theta$

(c) Find similar expressions for $\sin^2\theta$ and $\sin^4\theta$.

9. $\triangle ABC$ is isosceles with $\angle B = \angle C$. Show that $2\cot A = \tan B - \cot B$

10. Solve for the following equations, $0 \leq \theta \leq 2\pi$.

(a) $\sin\theta\cos\theta = 0.25$

(b) $\sin^2\theta - \cos^2\theta = \dfrac{1}{2}$

(c) $2\sin^2\theta - 3\cos 2\theta = 3$

(d) $3\cos 2\theta + \cos\theta + 1 = 0$

(e) $3 + \sin\theta = 5\cos 2\theta$

11. In the double angle formulas for cosine, replace θ by $\theta/2$. Then show that:

(a) $\cos\dfrac{\theta}{2} = \pm\sqrt{\dfrac{1 + \cos\theta}{2}}$

(b) $\sin \dfrac{\theta}{2} = \pm\sqrt{\dfrac{1 - \cos \theta}{2}}$

12. Given $\cos \theta = -119/169, \pi < \theta < 3\pi/2$, find $\sin(\theta/2), \cos(\theta/2)$, and $\tan(\theta/2)$.

13. Find an exact value for $\cos 15°$, expressed in radicals. Hint: $\cos 15° = \cos(30°/2)$.

14. Prove:

(a) $\tan \dfrac{\theta}{2} = \pm\sqrt{\dfrac{1 - \cos \theta}{1 + \cos \theta}}$

(b) $\tan \dfrac{\theta}{2} = \dfrac{1 - \cos \theta}{\sin \theta}$

(c) $\tan \dfrac{\theta}{2} = \dfrac{\sin \theta}{1 + \cos \theta}$

15. If $z = \tan(\theta/2)$, show that:

(a) $\cos \theta = \dfrac{1 - z^2}{1 + z^2}$

(b) $\sin \theta = \dfrac{2z}{1 + z^2}$

Sums and Products of Sines and Cosines

9.4

In advanced work, it is often useful to be able to convert a product of sines and/or cosines into a sum of functions, or conversely, a sum into a product. Formulas to accomplish this are readily derived from the addition formulas.

Start, for instance, with the addition formulas for cosine:

$$\cos(\theta + \phi) = \cos \theta \cos \phi - \sin \theta \sin \phi$$

$$\cos(\theta - \phi) = \cos \theta \cos \phi + \sin \theta \sin \phi$$

Adding these gives

$$\cos(\theta + \phi) + \cos(\theta - \phi) = 2 \cos \theta \cos \phi$$

Dividing by 2 and rearranging,

$$\cos \theta \cos \phi = \frac{1}{2}[\cos(\theta + \phi) + \cos(\theta - \phi)]$$

Similarly, you can show that

$$\sin \theta \sin \phi = -\frac{1}{2}[\cos(\theta + \phi) - \cos(\theta - \phi)]$$

These formulas make it possible to transform a product of functions into a sum.

You can obtain formulas for changing a sum of functions into a product by replacing θ and ϕ in the above formulas by α and β using

$$\theta + \phi = \alpha \quad \text{and} \quad \theta - \phi = \beta$$

or

$$\theta = \frac{\alpha + \beta}{2} \quad \text{and} \quad \phi = \frac{\alpha - \beta}{2}$$

These substitutions result in

$$\cos \alpha + \cos \beta = 2 \cos \frac{\alpha + \beta}{2} \cos \frac{\alpha - \beta}{2}$$

$$\cos \alpha - \cos \beta = -2 \sin \frac{\alpha + \beta}{2} \sin \frac{\alpha - \beta}{2}$$

In a similar manner, starting with the addition formulas for sine,

$$\sin(\theta + \phi) = \sin \theta \cos \phi + \cos \theta \sin \phi$$
$$\sin(\theta - \phi) = \sin \theta \cos \phi - \cos \theta \sin \phi$$

one obtains the product formulas

$$\sin \theta \cos \phi = \frac{1}{2}[\sin(\theta + \phi) + \sin(\theta - \phi)]$$

$$\cos \theta \sin \phi = \frac{1}{2}[\sin(\theta + \phi) - \sin(\theta - \phi)]$$

and the sum formulas

$$\sin \alpha + \sin \beta = 2 \sin \frac{\alpha + \beta}{2} \cos \frac{\alpha - \beta}{2}$$
$$\sin \alpha - \sin \beta = 2 \cos \frac{\alpha + \beta}{2} \sin \frac{\alpha - \beta}{2}$$

EXAMPLE A

Find $\cos 15°$ by applying a sum formula to $\cos 60° + \cos 30°$

SOLUTION

$$\cos 60° + \cos 30° = 2 \cos \frac{60° + 30°}{2} \cos \frac{60° - 30°}{2}$$

$$= 2 \cos 45° \cos 15°$$

$$\frac{1}{2} + \frac{\sqrt{3}}{2} = 2 \cdot \frac{1}{\sqrt{2}} \cdot \cos 15°$$

$$\cos 15° = \frac{\sqrt{6} + \sqrt{2}}{4}$$ ❏

EXAMPLE B

Convert the product $\cos 6\theta \sin 2\theta$ into a sum of functions.

SOLUTION

$$\cos 6\theta \sin 2\theta = \frac{1}{2}[\sin(6\theta + 2\theta) - \sin(6\theta - 2\theta)]$$

$$= \frac{1}{2}\sin 8\theta - \frac{1}{2}\sin 4\theta$$ ❏

This ability to convert products of functions into sums will be useful later in working out certain problems in integration.

EXAMPLE C

Solve for θ: $\sin\theta - \cos\theta = \dfrac{1}{\sqrt{2}}, 0 \le \theta \le 2\pi$

SOLUTION

None of the sum formulas involve the difference of a sine and a cosine. One must first use a complementary angle identity:

$$\sin\theta - \cos\theta = \frac{1}{\sqrt{2}}$$

$$\sin\theta - \sin\left(\frac{\pi}{2} - \theta\right) = \frac{1}{\sqrt{2}}$$

$$2\cos\frac{\pi}{4}\sin\left(\theta - \frac{\pi}{4}\right) = \frac{1}{\sqrt{2}}$$

$$\frac{2}{\sqrt{2}}\sin\left(\theta - \frac{\pi}{4}\right) = \frac{1}{\sqrt{2}}$$

$$\sin\left(\theta - \frac{\pi}{4}\right) = \frac{1}{2}$$

$$\theta - \frac{\pi}{4} = \frac{\pi}{6} \quad \text{and} \quad \theta - \frac{\pi}{4} = \frac{5\pi}{6}$$

$$\theta = \frac{5\pi}{12} \quad \text{and} \quad \theta = \frac{13\pi}{12} \qquad \square$$

EXAMPLE D

Describe the result of adding two sinusoidal functions that have the same amplitude and period but which differ in phase by an amount ϕ, e.g.,

$$f(x) = \sin x + \sin(x - \phi)$$

SOLUTION

It would be straightforward (although tedious) to construct a table of values for a given value of ϕ and plot the function $y = f(x)$ in the usual way. The essential features of the curve, however, can be discovered by converting the sum into a product:

$$f(x) = \sin x + \sin(x - \phi)$$

$$= 2 \sin \frac{2x - \phi}{2} \cos \frac{\phi}{2}$$

$$= \left(2 \cos \frac{\phi}{2}\right) \sin\left(x - \frac{\phi}{2}\right)$$

Thus the sum of two sinusoidal functions is another sinusoidal function, whose amplitude depends on the phase difference. This is an example of a physical phenomenon known as the interference of two waves. If the phase difference is zero, the two waves are exactly in phase and the amplitude is twice the original amplitude. As the phase difference increases, the waves interfere to an ever greater extent, and the amplitude drops (Figure 9.13). Ultimately, when the phase difference becomes π, the amplitude is zero and the two waves exactly cancel each other out. $\square$

EXAMPLE E

Describe the result of adding two sinusoidal functions which have the same amplitude but different periods, e.g.,

$$f(x) = \sin 11x + \sin 9x$$

Figure 9.13

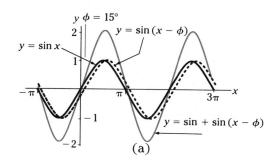

(a)

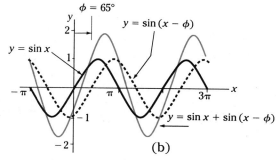

(b)

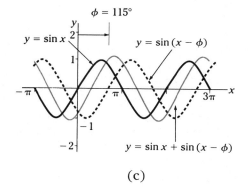

(c)

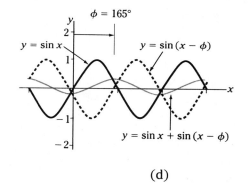

(d)

SOLUTION

Convert this sum into a product:

$$f(x) = \sin 11x + \sin 9x$$

$$= 2 \sin \frac{11x + 9x}{2} \cos \frac{11x - 9x}{2}$$

$$= (2 \cos x) \sin 10x$$

This is another example of the interference of two waves. Here, the period of $\sin 10x$ is $2\pi/10$, a comparatively short period. The period of $\cos x$ at 2π is much longer. Thus, the graph of f consists of the graph of $\sin 10x$, with an amplitude that varies from 0 to 2, as the factor $(2 \cos x)$ slowly runs through its full cycle of values. The graph of this function is shown in Figure 9.14. This type of behaviour is known as the phenomenon of *beats*. It occurs whenever two periodic functions having approximately the same frequency are superimposed. If, for instance, two musical instruments are slightly out of tune, beats add a slow, pulsating quality to the sound. ❏

Figure 9.14

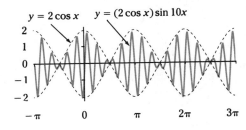

Exercise 9.4

1. Express as a sum:

    (a) $\sin 3\theta \sin 6\theta$

    (b) $\sin 4\theta \cos 2\theta$

2. Express as a product:

    (a) $\sin 8\theta - \sin 2\theta$

    (b) $\cos 5\theta + \cos 3\theta$

3. Show that:

    (a) $\sin 85° - \sin 25° = \cos 55°$

    (b) $\cos 36° \cos 54° = \dfrac{1}{2}\cos 18°$

    (c) $\cos\left(\dfrac{\pi}{8}\right) \sin\left(\dfrac{7\pi}{8}\right) = \dfrac{1}{2\sqrt{2}}$

    (d) $\cos\left(\dfrac{5\pi}{12}\right) - \cos\left(\dfrac{11\pi}{12}\right) = \dfrac{\sqrt{6}}{2}$

    (e) $\sin 43° + \cos 73° = \cos 13°$

    (f) $\cos\left(\dfrac{4\pi}{5}\right) + \sin\left(\dfrac{\pi}{5}\right) = \sqrt{2}\cos\left(\dfrac{11\pi}{20}\right)$

4. (a) Prove:

$$(\sin\theta + \sin\phi)^2 + (\cos\theta + \cos\phi)^2$$
$$= 4\cos^2\left(\dfrac{\theta - \phi}{2}\right)$$

    (b) Discover a similar identity for:

$$4\sin^2\left(\dfrac{\theta - \phi}{2}\right)$$

5. (a) By applying a sum formula to $\sin(\pi/2) + \sin A$, show that $1 + \sin A$ can be written as a perfect square:

$$1 + \sin A = \left(\cos\dfrac{A}{2} + \sin\dfrac{A}{2}\right)^2$$

    (b) Prove the identity:

$$\sqrt{1 + \sin A} - \sqrt{1 - \sin A} = 2\sin\dfrac{A}{2}$$

6. If A is *not* a multiple of $2\pi/5$, prove that

$$\dfrac{\sin 2A + \sin 3A}{\cos 2A - \cos 3A} = \cot\dfrac{A}{2}$$

7. Solve for x:

$$\sin x + \cos x = \sqrt{\dfrac{3}{2}}$$

8. Find the amplitude, period, and phase and sketch the graph of the following:

    (a) $y = \cos 3\theta - \sin 3\theta$

    (b) $y = \cos(5\theta) + \cos(5\theta + 3\pi/4)$

9. Two sine waves having the same amplitude and period are out of phase by 60°. Find the amplitude, period, and phase, and sketch a graph of their sum.

10. Two sine waves with the same amplitude have periods of $2\pi/4$ and $2\pi/5$, respectively. Sketch a graph of their sum.

Summary and Review   9.5

This chapter has introduced a number of trigonometric identities and some applications. The principal ones are

Addition formulas:

$$\cos(\theta \pm \phi) = \cos\theta\cos\phi \mp \sin\theta\sin\phi$$

$$\sin(\theta \pm \phi) = \sin\theta\cos\phi \pm \cos\theta\sin\phi$$

$$\tan(\theta \pm \phi) = \frac{\tan\theta \pm \tan\phi}{1 \mp \tan\theta\tan\phi}$$

Double angle formulas:

$$\sin 2\theta = 2\sin\theta\cos\theta$$

$$\cos 2\theta = \cos^2\theta - \sin^2\theta$$

$$= 2\cos^2\theta - 1$$

$$= 1 - 2\sin^2\theta$$

$$\tan 2\theta = \frac{2\tan\theta}{1 - \tan^2\theta}$$

Product Formulas:

$$\cos\theta\cos\phi = \frac{1}{2}[\cos(\theta + \phi) + \cos(\theta - \phi)]$$

$$\sin\theta\sin\phi = -\frac{1}{2}[\cos(\theta + \phi) - \cos(\theta - \phi)]$$

$$\sin\theta\cos\phi = \frac{1}{2}[\sin(\theta + \phi) + \sin(\theta - \phi)]$$

Sum Formulas:

$$\cos\theta + \cos\phi = 2\cos\frac{\theta+\phi}{2}\cos\frac{\theta-\phi}{2}$$

$$\cos\theta - \cos\phi = -2\sin\frac{\theta+\phi}{2}\sin\frac{\theta-\phi}{2}$$

$$\sin\theta + \sin\phi = 2\sin\frac{\theta+\phi}{2}\cos\frac{\theta-\phi}{2}$$

$$\sin\theta - \sin\phi = 2\cos\frac{\theta+\phi}{2}\sin\frac{\theta-\phi}{2}$$

In order to master the material in this chapter, you must first be thoroughly acquainted with the basic trigonometric identities mentioned in F. The Trigonometric Functions, pages 223–229. In addition, you should know or be able to find the exact values of the trigonometric functions of multiples of 30°, 45°, 60°, and 90°.

We suggest that you memorize the addition formulas and the double angle formulas. The remaining identities can be derived on the spot whenever they are needed. The importance of these compound angle identities is that they reveal further relationships among the trigonometric functions, which are useful in calculus and in areas of applied mathematics.

When working with identities, try to anticipate the consequences of a step before you actually carry it out. For example, will the use of a particular identity allow you to cancel or combine terms? Will it enable you to express a function in a form that is easier to interpret and graph? Will it bring you closer to your objective, even if only in a small way?

Working and trying to work many different problems is the only sure way to gain familiarity with these identities and to become skilled in their use.

Exercise 9.5

1. If $\sin\theta = 5/8$ and $\sin\phi = 4/5$ find $\cos(\theta+\phi) - \cos(\theta-\phi)$.

2. Simplify.

(a) $\sin(A+\pi/6) - \sin(A-\pi/6)$

(b) $\cos(A+\pi/3) + \cos(A-\pi/3)$

(c) Construct an expression similar to those in (a) and (b) which simplifies to $\sin A$.

3. If $\cos A = 3/5, 0 < A < \pi/2$, find $\sin(A+\pi/6)$.

4. If $\cos\theta = 0.5$, $3\pi/2 < \theta < 2\pi$, and $\sin\phi = -0.4$, $\pi < \phi < 3\pi/2$. find $\sin(\theta-\phi)$ and $\cos(\theta+\phi)$.

5. If $A > 0, B > 0$, and $A+B < \pi/2$, show that $\sin(A+B) < \sin A + \sin B$.

6. Expand $\sin(A+B+C)$ and $\cos(A+B+C)$. Notice the patterns in the resulting expressions.

7. If $\sin A = 3/5$, $\sin B = 12/13$, and $\sin C = 7/25$ and A, B, C are acute angles, find the value of

(a) $\sin(A + B + C)$

(b) $\cos(A + B + C)$

8. Use an appropriate identity to find the exact value of

(a) $\tan 195°$

(b) $\sin(7\pi/12)$

(c) $\sin(3\pi/8)$

(d) $\cos(9\pi/8)$

(e) $\sin(5\pi/8) + \sin(\pi/8)$

9. Simplify:

(a) $\sin 75° - \sin 15°$

(b) $\sin 75° + \sin 15°$

(c) $\cos 105° - \cos 15°$

(d) $\cos 105° + \cos 15°$

10. Express in the form of a product:

(a) $\sin 2A + \sin 4A$

(b) $\cos 2A + \cos 4A$

(c) $\sin A - \sin 2A$

(d) $\cos A - \sin 2A$

(e) $\sin(A + B) - \sin A$

(f) $\cos(A + B) - \sin A$

11. Prove that:

$$\frac{\cos 3A + \cos A}{\sin 3A - \sin A} = \frac{\cos A}{\sin A}$$

12. Solve:

$$\sin 5\theta + \sin 3\theta = \sqrt{3}\cos\theta, 0 < \theta < 2\pi$$

13. Express as a product of sines and/or cosines:

(a) $\cos\dfrac{\theta + \phi}{2} + \cos\dfrac{\theta - \phi}{2}$

(b) $\sin^2\theta - \sin^2\phi$

(c) $\tan^2\theta - \tan^2\phi$

(d) $\sin\theta\cos\theta + \sin\phi\cos\phi$

(e) $\sqrt{1 + \cos\theta} + \sqrt{1 - \cos\theta}$

(f) $\sin\theta + \sin\phi + \sin(\theta + \phi)$

14. Prove:

(a) $\dfrac{\sin A + \sin B}{\cos A - \cos B} = \cot\dfrac{B - A}{2}$

(b) $\dfrac{\cot A - \tan B}{\cot A + \cot B} = \cot(A + B)\tan B$

(c) $\dfrac{1}{1 + \tan A \tan 2A} = \cos 2A$

15. Show that:

$$\sin 80° - \sin 20° + \sin 40° - \sin 100°$$
$$= \sqrt{3}\sin 10°$$

16. Prove that:

(a) $\sin 3\theta = 3\sin\theta - 4\sin^3\theta$

(b) Find a similar formula for $\cos 3\theta$.

17. If $\theta = \pi/5$, then $\sin 5\theta = 0$. Expand $\sin 5\theta$ in powers of $\sin\theta$. Factor the resulting expression, and solve for $\sin\theta$ to find the exact value of $\sin\pi/5$ in radical form.

Leonhard Euler

I N THE EIGHTEENTH CENTURY an enormous effort was devoted to extending and applying the new calculus of Leibniz and Newton. The Swiss mathematician Leonhard Euler stood at the centre of this activity, engaging in and organizing research at the academies of science in Berlin and St. Petersburg. Unsurpassed in the history of mathematics in formal calculation and transformations, Euler (1707-1783) dominated European mathematics during his long career.

From its beginning the calculus had been closely connected to problems in the geometry of curves. In a series of textbooks published in the 1740s, 1750s, and 1760s Euler systematically accomplished the separation of the calculus from geometry. In 1748 he introduced and made central the concept of a function as an analytical expression composed from variables and constants using algebraic and transcendental operations. Much of the standard notation of mathematics was invented by Euler, including the symbol e for the base of the natural logarithms, the use of i for the square root of minus one and the notation $f(x)$ to denote a function of x.

Euler's contributions to analysis are illustrated by his development of the calculus of trigonometric functions in the 1740s. Trigonometric tables had existed since antiquity, and the relations between sines and cosines were commonly used in mathematical astronomy. In the early eighteenth century, mathematicians had derived in their study of periodic mechanical phenomena the differential equation

$$dy/dx = 1/\sqrt{1 - x^2}$$

and were able to interpret its solution geometrically in terms of lines and angles in the circle. Euler was the first to introduce the sine and cosine functions as quantities involving letters and numbers, the relation of which to other such quantities could be studied independently of any relation to a geometrical diagram. In his *Introductio in analysin infinitorum* (1748) he presented the familiar angle-sum formulas and derived the standard infinite expansions for the sine and cosine functions:

$$\sin x = x - \frac{x^3}{3!} + \frac{x^5}{5!} - \cdots,$$

$$\cos x = 1 - \frac{x^2}{2!} + \frac{x^4}{4!} - \cdots$$

The son of a Lutheran minister, Euler was pious and devoted to his large family during his career. In 1765 he suffered a brief illness which left him almost completely blind. During the last eighteen years of his life he remained extremely productive in mathematics, dictating his work to his daughter. In addition to mathematical research he wrote popular accounts of science, including his famous *Letters to a German Princess* (1768), a book that originated in lessons given to a relative of the Prussian king. Euler was the most prolific mathematician of all time; his collected works run to seventy-five volumes and his mathematical research during his lifetime averaged 800 pages a year.

Derivatives of Trigonometric Functions

10

A t the start of this chapter, the limit formula for find- ing the derivative of a function will be applied to the trigonometric functions, in par- ticular to the sine function. You will see that to obtain the derivatives of the trigonometric functions, it will be necessary to work out certain special trigonometric limits. Once that has been accomplished, we will turn to problems and applications involving the trig- onometric functions.

Trigonometric Limits

10.1

The purpose of this section is to determine the derivative of the sine function from first principles. To accomplish this, it will be necessary to work out certain limits involving trigonometric functions.

Recall that the definition of the derivative is

$$\frac{d}{dx}f(x) = \lim_{h \to 0} \frac{f(x+h) - f(x)}{h} \qquad \longleftarrow \quad \text{here we have set } \Delta x = h \text{ for convenience}$$

If $\qquad f(x) = \sin(x)$ then

$$\frac{d}{dx}\sin(x) = \lim_{h \to 0} \frac{\sin(x+h) - \sin(x)}{h} \qquad \longleftarrow \quad \text{addition formula}$$

$$= \lim_{h \to 0} \frac{\sin x \cos h + \cos x \sin h - \sin x}{h} \qquad \longleftarrow \quad \text{regrouping terms}$$

$$= \lim_{h \to 0} \left(\sin x \cdot \frac{\cos h - 1}{h} + \cos x \cdot \frac{\sin h}{h} \right)$$

Since $\sin x$ and $\cos x$ do not involve h, they remain constant as $h \to 0$. Consequently

$$\frac{d}{dx}\sin(x) = \sin x \cdot \lim_{h \to 0} \frac{\cos h - 1}{h} + \cos x \cdot \lim_{h \to 0} \frac{\sin h}{h}$$

provided the limits of the following two trigonometric expressions exist:

$$\lim_{h \to 0} \frac{\cos h - 1}{h} \quad \text{and} \quad \lim_{h \to 0} \frac{\sin h}{h}$$

It is not immediately clear what these limits are, since by inspection, both lead to the indeterminate form 0/0.

In what follows, the limit of $(\sin h)/h$ will be dealt with first. Once this limit has been found, it can be used to work out the limit of $(\cos h - 1)/h$. It will then be possible to complete the job of finding the derivative of the sine function, and from that result, the derivatives of the remaining trigonometric functions can easily be determined.

According to the values shown in Table 10.1, the function $y = (\sin h)/h$ seems to approach the value 1, as h approaches 0. To see that the limit is indeed equal to 1, consider the diagram shown in Figure 10.1, where h is represented by the angle θ, and $0 < \theta < \pi/2$. Compare the area of sector AOB to the areas of the right triangles $\triangle AOP$ and $\triangle QOB$:

Table 10.1

h	$\dfrac{\sin h}{h}$
1.0	0.841 471
0.1	0.998 334
0.01	0.999 983
0.001	0.999 999
0	?

Figure 10.1

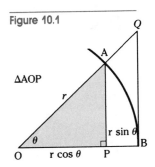

$\triangle AOP$

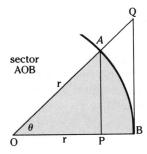

sector AOB

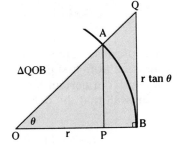

$\triangle QOB$

$$\begin{array}{ll} \text{Area of} \\ \triangle AOP \end{array} = \frac{1}{2} \cdot r \sin \theta \cdot r \cos \theta = \frac{1}{2}r^2 \sin \theta \cos \theta$$

$$\begin{array}{ll} \text{Area of} \\ \text{Sector } AOB \end{array} = \frac{\theta}{2\pi} \cdot \pi r^2 = \frac{1}{2}r^2 \theta$$

$$\begin{array}{ll} \text{Area of} \\ \triangle QOB \end{array} = \frac{1}{2} \cdot r \tan \theta \cdot r = \frac{1}{2}r^2 \tan \theta$$

From Figure 10.1, it can be seen that, in magnitude, the area of the sector lies between the areas of the two triangles.

$$\underset{\triangle AOP}{\text{Area of}} < \underset{\text{Sector } AOB}{\text{Area of}} < \underset{\triangle QOB}{\text{Area of}}$$

$$\frac{1}{2}r^2 \sin\theta\cos\theta < \quad \frac{1}{2}r^2\theta \quad < \frac{1}{2}r^2\tan\theta$$

Dividing out $r^2/2$ and then dividing by $\sin\theta$ leads to

$$\sin\theta\cos\theta < \quad \theta \quad < \frac{\sin\theta}{\cos\theta}$$

$$\cos\theta < \frac{\theta}{\sin\theta} < \frac{1}{\cos\theta}$$

On taking the limit now as $\theta \to 0$, observe an unusual situation: since the limit of $\cos\theta$ is 1 as $\theta \to 0$, then

Figure 10.2

$$1 < \lim_{\theta\to 0}\frac{\theta}{\sin\theta} < 1$$

In other words, as $\theta \to 0$, the limit of $\theta/(\sin\theta)$ is squeezed between two other limits, both of which become 1 (see Figure 10.2). There is no escape. The limit of $\theta/(\sin\theta)$ must also be 1.

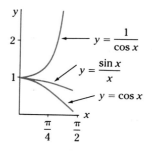

$$\lim_{\theta\to 0}\frac{\theta}{\sin\theta} = 1$$

It follows that

$$\lim_{\theta\to 0}\frac{\sin\theta}{\theta} = \lim_{\theta\to 0}\frac{1}{(\theta/\sin\theta)}$$

$$= 1$$

Thus, one of the trigonometric limits needed for determining the derivative of the sine function has now been found.

The other limit, $\lim_{\theta\to 0}\dfrac{\cos\theta - 1}{\theta}$, can be found by relating it to the limit of $(\sin\theta)/\theta$. To start, multiply the numerator and the denominator by the factor $(\cos\theta + 1)$:

$$\lim_{\theta \to 0} \frac{\cos \theta - 1}{\theta} = \lim_{\theta \to 0} \frac{(\cos \theta - 1)}{\theta} \cdot \frac{(\cos \theta + 1)}{(\cos \theta + 1)} \quad \longleftarrow \quad \begin{array}{l} \text{note that} \\ (\cos \theta + 1) \neq 0 \\ \text{as } \theta \to 0 \end{array}$$

$$= \lim_{\theta \to 0} \frac{(\cos^2 \theta - 1)}{\theta(\cos \theta + 1)}$$

$$= \lim_{\theta \to 0} \frac{-\sin^2 \theta}{\theta(\cos \theta + 1)}$$

$$= \lim_{\theta \to 0} \frac{\sin \theta}{\theta} \cdot \frac{-\sin \theta}{\cos \theta + 1}$$

$$= 1 \cdot \frac{0}{2}$$

$$= 0$$

Thus,

$$\lim_{\theta \to 0} \frac{\cos \theta - 1}{\theta} = 0$$

With these results, it is now possible to complete the process of finding the derivative of the sine function. Recall that the derivation had been interrupted at this point:

$$\frac{d}{dx} \sin(x) = \sin x \cdot \lim_{h \to 0} \frac{\cos h - 1}{h} + \cos x \cdot \lim_{h \to 0} \frac{\sin h}{h}$$

Now the limits are known, therefore,

$$\frac{d}{dx} \sin(x) = \sin x \cdot (0) + \cos x \cdot (1)$$

$$\boxed{\frac{d}{dx} \sin(x) = \cos x}$$

Derivatives of the trigonometric functions will be further explored in the next section. This section concludes with some additional examples of trigonometric limits. The procedure is to relate them to the two basic limits derived above.

EXAMPLE A

Find the limit:

$$\lim_{\theta \to 0} \frac{\sin 2\theta}{\theta}$$

SOLUTION

Using an identity,

$$\lim_{\theta \to 0} \frac{\sin 2\theta}{\theta} = \lim_{\theta \to 0} \frac{2 \sin \theta \cos \theta}{\theta}$$

$$= \lim_{\theta \to 0} 2 \cdot \frac{\sin \theta}{\theta} \cdot \cos \theta$$

$$= 2 \cdot 1 \cdot 1$$

$$= 2$$

On the other hand, the use of an identity is not necessary, if you put the expression into the correct *form*. On multiplying and dividing by 2, for instance, and observing that if $\theta \to 0$ then certainly $2\theta \to 0$, then

$$\lim_{\theta \to 0} \frac{\sin 2\theta}{\theta} = \lim_{\theta \to 0} \frac{2 \sin 2\theta}{2\theta}$$

$$= 2 \cdot \lim_{2\theta \to 0} \frac{\sin(2\theta)}{(2\theta)} \quad \text{which has the form } \lim_{h \to 0} \frac{\sin h}{h}$$

$$= 2 \cdot 1$$

$\square$

EXAMPLE B

Find the limit:

$$\lim_{\theta \to 0} \frac{\tan 3\theta}{2\theta}$$

SOLUTION

$$\lim_{\theta \to 0} \frac{\tan 3\theta}{2\theta} = \lim_{\theta \to 0} \frac{1}{2\theta} \cdot \frac{\sin 3\theta}{\cos 3\theta}$$

$$= \lim_{\theta \to 0} \frac{1}{2} \cdot \frac{\sin 3\theta}{\theta} \cdot \frac{1}{\cos 3\theta}$$

$$= \lim_{3\theta \to 0} \frac{3}{2} \cdot \frac{\sin 3\theta}{3\theta} \cdot \frac{1}{\cos 3\theta}$$

$$= \frac{3}{2} \cdot 1 \cdot \frac{1}{1}$$

$$= \frac{3}{2}$$

$\square$

EXAMPLE C

Find the limit:

$$\lim_{\theta \to \pi} \frac{1 + \cos \theta}{2 \sin \theta}$$

SOLUTION

This problem asks for the limit as $\theta \to \pi$, *not* zero. It is necessary to use a substitution to express this limit in a more suitable form. Let $h = \theta - \pi$. Then as $\theta \to \pi$, $h \to 0$. Thus,

$$\lim_{\theta \to \pi} \frac{1 + \cos \theta}{2 \sin \theta} = \lim_{h \to 0} \frac{1 + \cos(h + \pi)}{2 \sin(h + \pi)}$$

$$= \lim_{h \to 0} \frac{1 - \cos h}{-2 \sin h}$$

$$= \lim_{h \to 0} \frac{1}{2} \cdot \frac{\cos h - 1}{h} \cdot \frac{h}{\sin h}$$

$$= \frac{1}{2} \cdot 0 \cdot 1$$

$$= 0$$

Exercise 10.1

1. Determine the limits:

(a) $\lim_{\theta \to 0} \dfrac{\sin \theta}{2\theta}$

(b) $\lim_{\theta \to 0} \dfrac{\sin^2 \theta}{\theta}$

(c) $\lim_{\theta \to 0} \dfrac{\sin \theta}{\theta^2}$

(d) $\lim_{\theta \to 0} \dfrac{\sin^2 \theta}{\theta^2}$

(e) $\lim_{\theta \to 0} \dfrac{1 - \cos \theta}{\sin \theta}$

(f) $\lim_{\theta \to 0} \dfrac{\theta}{1 - \cos \theta}$

(g) $\lim_{\theta \to \pi} \dfrac{1 - \sin \theta}{1 - \cos \theta}$

(h) $\lim_{\theta \to \pi/2} \dfrac{\cos \theta - \sin \theta}{\cos 2\theta}$

2. Determine the limits with the help of a trigonometric identity:

(a) $\lim_{\theta \to 0} \dfrac{\cos^2 \theta - 1}{2\theta}$

(b) $\lim_{\theta \to 0} \dfrac{1 - \cos \theta}{\theta^2}$

(c) $\lim_{\theta \to 0} \dfrac{\tan \theta}{2\theta}$

(d) $\lim_{\theta \to 0} \theta^2(1 + \cot^2 \theta)$

3. Determine the limits by rationalizing the radical:

(a) $\lim\limits_{x \to 0} \dfrac{x}{\sqrt{1 - \cos x}}$

(b) $\lim\limits_{x \to 0} \dfrac{\sqrt{1 - \cos x}}{\sin x}$

4. Determine the limits with the help of a suitable substitution:

(a) $\lim\limits_{\theta \to \pi} \dfrac{\sin \theta}{\pi - \theta}$

(b) $\lim\limits_{\theta \to \pi} \dfrac{\sin 3\theta}{\sin 2\theta}$

(c) $\lim\limits_{\theta \to \pi/2} \dfrac{\cot \theta}{\theta - \pi/2}$

5. (a) Let θ be a sector angle in a unit circle as shown in Figure 10.3. Express the length of the chord BC and the arc BC in terms of θ.

Figure 10.3

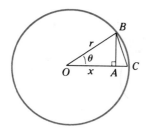

(b) Using the fact that $0 < \text{chord } BC < \text{arc } BC$

show that $0 < \dfrac{1 - \cos \theta}{\theta} < \dfrac{\theta}{2}$

and therefore that $\lim\limits_{\theta \to 0} \dfrac{1 - \cos \theta}{\theta} = 0$

6. Find the derivative of $\cos x$ from first principles.

7. Find the derivative of $\sin 2x$ from first principles.

8. Find the derivative of $\tan x$ from first principles.

9. Determine the limits:

(a) $\lim\limits_{x \to 0} \dfrac{\tan kx}{x}$

(b) $\lim\limits_{x \to 0} \dfrac{\sin ax}{\sin bx}$

(c) $\lim\limits_{x \to 0} \dfrac{\tan x - \sin x}{x^3}$

(d) $\lim\limits_{x \to \pi/2} \left(x - \dfrac{\pi}{2} \right) \tan x$

(e) $\lim\limits_{x \to \pi} \dfrac{\sin x}{1 - (x/\pi)^2}$

(f) $\lim\limits_{x \to 0} \dfrac{\cos(a + x) - \cos(a - x)}{x}$

(g) $\lim\limits_{x \to 0} \dfrac{\sqrt{2} - \sqrt{1 + \cos x}}{\sin^2 x}$

(h) $\lim\limits_{x \to 0} \dfrac{\sqrt{1 + \sin x} - \sqrt{1 - \sin x}}{\tan x}$

(i) $\lim\limits_{x \to \pi/4} \dfrac{\tan x - \cot x}{x - \pi/4}$

(j) $\lim\limits_{x \to a} \sin \dfrac{x - a}{2} \tan \dfrac{\pi x}{2a}$

10.2 Derivatives of the Trigonometric Functions

We have established from first principles that the derivative of $\sin x$ is $\cos x$:

$$\frac{d}{dx}\sin x = \cos x$$

It follows that if v is any differentiable function of x, then the function $y = \sin v$ is the composition of two functions, and according to the chain rule

$$\frac{dy}{dx} = \frac{dy}{dv} \cdot \frac{dv}{dx}$$

or

$$\frac{d}{dx}\sin v = \cos v \cdot \frac{dv}{dx}$$

Although, the derivatives of the other trigonometric functions could also be found from first principles, they are more easily found by other means. For instance, the derivative of the cosine function can be found with the help of the complementary angle identities

$$\cos x = \sin(\pi/2 - x) \quad \text{and} \quad \sin x = \cos(\pi/2 - x)$$

and the use of the chain rule as follows:

$$\frac{d}{dx}\cos x = \frac{d}{dx}\sin(\pi/2 - x)$$

$$= \cos(\pi/2 - x) \cdot \frac{d}{dx}(\pi/2 - x)$$

$$= \cos(\pi/2 - x) \cdot (-1)$$

$$= -\sin x$$

The derivative of the tangent function can now be found by writing

$$\tan x = \frac{\sin x}{\cos x}$$

and using the quotient rule. Formulas for differentiating all of the trigonometric functions are listed below. The proofs are left to the exercises.

$$\frac{d}{dx}\sin v = \cos v \cdot \frac{dv}{dx}$$

$$\frac{d}{dx}\cos v = -\sin v \cdot \frac{dv}{dx}$$

$$\frac{d}{dx}\tan v = \sec^2 v \cdot \frac{dv}{dx}$$

$$\frac{d}{dx}\cot v = -\csc^2 v \cdot \frac{dv}{dx}$$

$$\frac{d}{dx}\sec v = \tan v \sec v \cdot \frac{dv}{dx}$$

$$\frac{d}{dx}\csc v = -\cot v \csc v \cdot \frac{dv}{dx}$$

When finding the derivative of a function which involves a trigonometric function, you should be prepared to use the chain rule, the product rule, and the quotient rule, in addition to the preceding formulas. The examples below will illustrate some of the details.

EXAMPLE A

Differentiate:

$$y = \sin x \cos x$$

SOLUTION

Differentiation of this function can be done with the product rule:

$$\frac{dy}{dx} = \frac{d}{dx}(\sin x \cos x)$$

$$= \sin x \frac{d}{dx}\cos x + \cos x \frac{d}{dx}\sin x$$

$$= \sin x(-\sin x) + \cos x(\cos x)$$

$$= \cos^2 x - \sin^2 x$$

EXAMPLE B
Differentiate:

$$y = x^2 \cot x$$

SOLUTION
Use the product rule:

$$\frac{dy}{dx} = \frac{d}{dx}(x^2 \cot x)$$

$$= x^2 \frac{d}{dx} \cot x + \cot x \frac{d}{dx} x^2$$

$$= x^2(-\csc^2 x) + \cot x(2x)$$

$$= 2x \cot x - x^2 \csc^2 x$$

EXAMPLE C
Differentiate:

$$y = \frac{1 + \tan x}{\sec x}$$

SOLUTION
Use the quotient rule:

$$\frac{dy}{dx} = \frac{d}{dx}\left(\frac{1 + \tan x}{\sec x}\right)$$

$$= \frac{(\sec x)\dfrac{d}{dx}(1 + \tan x) - (1 + \tan x)\dfrac{d}{dx}(\sec x)}{\sec^2 x}$$

$$= \frac{(\sec x)(\sec^2 x) - (1 + \tan x)(\tan x \sec x)}{\sec^2 x} \quad \longleftarrow \text{ divide out a factor } \sec x$$

$$= \frac{\sec^2 x - (1 + \tan x)(\tan x)}{\sec x}$$

$$= \frac{\sec^2 x - \tan x - \tan^2 x}{\sec x}$$

$$= \frac{1 - \tan x}{\sec x}$$

EXAMPLE D

Differentiate:

$$y = \tan 4x$$

SOLUTION

Differentiation of this function calls for the use of the chain rule. You should look upon this function as $y = \tan v$, where $v = 4x$. Differentiate first the *tangent function*:

$$\frac{dy}{dx} = \frac{d}{dx}\tan 4x$$

$$= \sec^2 4x \frac{d}{dx}(4x)$$

$$= 4\sec^2 4x$$

EXAMPLE E

Contrast the derivatives of:

(a) $y = \sin x^2$

$\quad = \sin(x^2)$

(b) $y = \sin^2 x$

$\quad = (\sin x)^2$

SOLUTION

In differentiating a composite function, you should work from the "outside" to the "inside." In (a), x^2 is "inside" the sine function, whereas in (b), the situation is reversed; $\sin x$ is "inside" the square function. The derivatives are found as follows:

(a) $\dfrac{dy}{dx} = \dfrac{d}{dx}\sin x^2$

$\quad = \cos x^2 \dfrac{d}{dx}x^2$

$\quad = 2x\cos x^2$

(b) $\dfrac{dy}{dx} = \dfrac{d}{dx}\sin^2 x$

$\quad = 2(\sin x)\dfrac{d}{dx}\sin x$

$\quad = 2\sin x \cos x$

EXAMPLE F

Differentiate:

$$y = \sqrt{\sec 3x}$$

SOLUTION

This function is the composition of three functions: the square root, the secant, and multiplication by 3. Apply the chain rule one step at a time:

$$\frac{d}{dx} = \sqrt{\sec 3x}$$

$$= \frac{1}{2} \frac{1}{\sqrt{\sec 3x}} \frac{d}{dx} \sec 3x$$

$$= \frac{1}{2} \frac{1}{\sqrt{\sec 3x}} (\tan 3x \sec 3x) \frac{d}{dx} 3x$$

$$= \frac{1}{2} \frac{1}{\sqrt{\sec 3x}} (\tan 3x \sec 3x)(3)$$

$$= \frac{3}{2} \tan 3x \sqrt{\sec 3x}$$

Exercise 10.2

1. Find $f'(x)$ and simplify:

(a) $f(x) = \sin 2x$

(b) $f(x) = 2 \cos x - 3 \sin x$

(c) $f(x) = \tan^2 x$

(d) $f(x) = \cot(2x + 1)$

(e) $f(x) = \csc^2 x - \cot^2 x$

(f) $f(x) = \dfrac{1}{\cos 3x}$

2. Find $f'(x)$ and simplify:

(a) $f(x) = (\sec 3x)^2$

(b) $f(x) = \sin \sqrt{x}$

(c) $f(x) = \sec x \tan x$

(d) $f(x) = \cos^2 2x \cot^2 x$

(e) $f(x) = \dfrac{\sec x}{1 + \tan x}$

(f) $f(x) = \dfrac{1 - \cos x}{\sin x}$

3. Once the derivative of the sine function is known, show that the derivative of the cosine function can be found by differentiating implicitly

$$\cos^2 x = 1 - \sin^2 x$$

4. Work out the derivative of the tangent function starting from the identity

$$\tan x = \frac{\sin x}{\cos x}$$

5. Assuming that the derivatives of $\sin x$, $\cos x$, and $\tan x$ are known, determine the derivatives of $\cot x$, $\sec x$, and $\csc x$ by differentiating the reciprocal identities:

(a) $\cot x = \dfrac{1}{\tan x}$

(b) $\sec x = \dfrac{1}{\cos x}$

(c) $\csc x = \dfrac{1}{\sin x}$

6. Assuming that the derivatives of $\tan x$, and $\cot x$ are known, determine the derivatives of $\sec x$, and $\csc x$ by differentiating the Pythagorean identities:

(a) $\sec^2 x = \tan^2 x + 1$

(b) $\csc^2 x = \cot^2 x + 1$

7. Simplify first, then differentiate the following:

(a) $y = (\sin x + \cos x)^2$

(b) $y = \dfrac{2\cos x}{\sin 2x}$

(c) $y = \dfrac{1}{\sec x - \tan x}$

(d) $y = \dfrac{\cos x + \sin x}{\cos x - \sin x}$

8. Find the slope of the tangent line at the point indicated:

(a) $y = (\sin x - \cos x)^2$, $x = 5\pi/4$

(b) $y = \sqrt{\cos x}$ $x = \pi/3$

(c) $y = \dfrac{\sin^2 x}{(1 - \cos x)^2}$, $x = 3\pi/4$

9. Find $f'(x)$:

(a) $f(x) = x \sin x$

(b) $f(x) = x^2 \cos 4x$

(c) $f(x) = x^4 \tan x$

(d) $f(x) = x \sin x + \cos x$

(e) $f(x) = \dfrac{\sin 3x}{x}$

(f) $f(x) = \dfrac{\sin x}{x \cos x}$

10. Find dy/dx using implicit differentiation:

(a) $x \sin y = 1$

(b) $\sin x = \tan y$

(c) $(\sin x + \cos y)^2 = 1$

(d) $x + y = \cot(x - y)$

11. Find the equation of the tangent to each of the following functions at the point indicated.

(a) $y = 2 + \cos x$, $x = 5\pi/6$

(b) $y = \sin(x/2)$, $x = \pi/3$

(c) $y = 3 \csc x$, $x = \pi/4$

12. Differentiate and simplify:

(a) $y = 7 \sin(3\theta + 5)$

(b) $y = \dfrac{1}{3} \sin^3 4\theta$

(c) $y = \sec^2 x + \csc^2 x$

(d) $y = \dfrac{1}{4} \tan^4 \theta$

(e) $y = \cos \theta - \dfrac{1}{3} \cos^2 \theta$

(f) $y = (1 + \sin^2 x)^4$

(g) $y = \dfrac{\sin x}{1 + \cos x}$

(h) $y = x \sec^2 x - \tan x$

(i) $y = \dfrac{x}{1 - \cos x}$

(j) $y = \dfrac{\tan \theta}{\theta}$

(k) $y = \dfrac{\sin \theta}{\theta} + \dfrac{\theta}{\sin \theta}$

(l) $y = x \cos \dfrac{1}{x}$

(m) $y = \dfrac{\theta \sin \theta}{1 + \tan \theta}$

(n) $y = \sqrt{\tan(x/2)}$

(o) $y = \sqrt{3 + 2\tan\theta}$

(p) $y = \sin\sqrt{1 + x^2}$

(q) $y = \cot\sqrt[3]{1 + \theta^2}$

10.3 Applications

Now that formulas for the derivatives of the trigonometric functions have been established, it is possible to consider applications, in which trigonometric functions play a central role. The methods and tools of differential calculus will be used in much the same way as they were for problems involving algebraic functions to find maxima and minima, to graph curves, to calculate rate of change, etc. The following examples illustrate the uses of these functions in these types of problems.

EXAMPLE A

Determine the position of the maxima, minima, and inflection points, if any, and graph the function:

$$y = 2x - \tan x \quad \text{for} \quad 0 \le x \le 2\pi$$

SOLUTION

If

$$y = 2x - \tan x$$

then

$$y' = 2 - \sec^2 x$$

and

$$y'' = -(2\sec x)(\sec x \tan x)$$

$$= -2\sec^2 x \tan x$$

The first derivative is 0 when $\sec x = \pm\sqrt{2}$. This occurs when $x = \pi/4$, $3\pi/4$, $5\pi/4$, or $7\pi/4$. At these values, the second derivative is alternately negative, then positive. Consequently, at the four values of x there are, respectively, a maximum, a minimum, a maximum, and a minimum. Inflection points are located at the points where the second derivative is zero, namely, at $x = 0$, π, and 2π. The function has vertical asymptotes at $x = \pi/2$, and $3\pi/2$. With this information, the graph of the function can now be drawn (Figure 10.4).

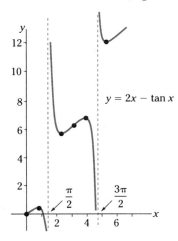

$y = 2x - \tan x$

Figure 10.4

EXAMPLE B

An isosceles triangle has two equal sides of length 15 cm. What is its maximum possible area?

SOLUTION

Drawing several isosceles triangles, as in Figure 10.5, reveals two limiting cases: the area approaches zero as the apex angle approaches 0° or 180°. Somewhere between these two extremes, there must be an angle for which the area is a maximum.

Figure 10.5

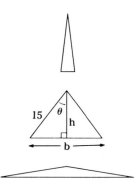

Since

$$h = 15 \cos \theta$$

and

$$\frac{1}{2}b = 15 \sin \theta$$

then

$$A = \frac{1}{2}bh$$

$$= (15 \sin \theta)(15 \cos \theta)$$

$$= \frac{225}{2} \sin 2\theta$$

Therefore

$$\frac{dA}{d\theta} = \frac{225}{2}(\cos 2\theta)(2)$$

Thus, the derivative is zero when $2\theta = \pi/2$, that is, the apex angle is a right angle. The maximum area is $225/2$ cm^2. ◻

EXAMPLE C

A ladder is to be moved horizontally around a corner from a hallway 3 m wide into a corridor 2 m wide. Determine the length of the longest ladder that can negotiate the turn. Neglect the thickness of the ladder.

SOLUTION

Figure 10.6

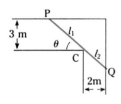

The line PQ in Figure 10.6 represents the ladder at an angle θ partway through the turn. When θ approaches 0° or 90°, the length of PQ increases beyond all bounds. At some angle θ between these extremes, the length of PQ will be a minimum—the required length of the longest ladder.

Point C divides PQ into two segments l_1 and l_2, where $l_1 = 3 \csc \theta$ and $l_2 = 2 \sec \theta$. Expressed as a function of θ, the length l of PQ is, therefore,

$$l = l_1 + l_2$$

$$= 3 \csc \theta + 2 \sec \theta$$

To find the minimum length of PQ, set the derivative of l equal to zero:

$$\frac{dl}{d\theta} = 3(-\csc \theta \cot \theta) + 2(\sec \theta \tan \theta)$$

$$= 0$$

$$\therefore \quad \frac{\sec \theta \tan \theta}{\csc \theta \cot \theta} = \frac{3}{2}$$

$$\tan^3 \theta = \frac{3}{2}$$

$$\tan \theta = \sqrt[3]{1.5}$$

$$\doteq 1.145$$

$$\therefore \quad \theta \doteq 0.853 \,\text{radians} \quad \text{or} \quad 48.86°$$

The maximum length of the ladder is thus

$$l \doteq 3(1.328) + 2(1.520)$$

$$= 7.024 \text{ m.}$$

◻

Exercise 10.3

1. Find the relative maxima and minima, and points of inflection, and sketch the graphs of the following functions. $(0 < x < 2\pi)$

   (a) $y = -\sin 2x$

   (b) $y = x + \sin x$

   (c) $y = \sin x + \cos 2x$

2. The motion of a particle is described by the equation

   $$s = 10\cos(5t - \pi/4)$$

   What are the maximum values of the displacement, the velocity, and the acceleration? At what times t do these maxima occur? What are the minimum values of these quantities?

3. A ladder of length 8 m leaning against a wall starts to slide. If its upper end slides down the wall at a rate of 0.25 m/s, at what rate is the angle between the ladder and the ground changing when the foot of the ladder is 5 m from the wall?

4. A radar antenna on a ship that is 4 km from a straight shoreline revolves at 32 rpm. How fast is the radar beam moving along the shore line when the beam makes an angle of 30° with the shore?

5. A steel girder 9 m long is to be moved horizontally around a corner from one corridor 2.5 m wide into a second corridor at right angles to the first. How narrow can the second corridor be and still permit the girder to go around

   the corner? Neglect the horizontal width of the girder.

6. A right circular cylinder is to be inscribed in a sphere of radius R. Find the ratio of the altitude to the radius of the base of the cylinder having the largest lateral surface area.

7. The range of a javelin thrown with velocity v at an angle θ with the horizontal is given by

   $$R = \frac{v^2}{g}\sin 2\theta$$

   where g is the acceleration due to gravity.

   (a) Calculate the range of a throw for $\theta = 30°$ and $\theta = 60°$, if $v = 24.5$ m/s and $g = 9.8$ m/s². Sketch the flight path of the javelin in these two cases.

   (b) At what angle will the range be a maximum?

8. Find the maximum value of $y = 2\tan x - \tan^2 x$.

9. Show that the maximum value of $y = a\sin x + b\cos x$ is $\sqrt{a^2 + b^2}$, and the minimum value is $-\sqrt{a^2 + b^2}$.

10. A wall is 1.8 m high and 1.2 m from a building. Find the length of the shortest ladder that will touch the building, the top of the wall, and the ground beyond the wall.

11. A conical container is to be made from a circular piece of tin of radius 24 cm, by cutting out a sector with central angle θ and then soldering the cut edges of the remaining piece together. What is the maximum possible volume of the resulting container?

10.4 Summary and Review

Two fundamental trigonometric limits are

$$\lim_{h \to 0} \frac{\sin h}{h} = 1 \quad \text{and} \quad \lim_{h \to 0} \frac{\cos h - 1}{h} = 0$$

The basic formulas for differentiating the trigonometric functions are

$$\frac{d}{dx} \sin v = \cos v \frac{dv}{dx}$$

$$\frac{d}{dx} \cos v = -\sin v \frac{dv}{dx}$$

$$\frac{d}{dx} \tan v = \sec^2 v \frac{dv}{dx}$$

$$\frac{d}{dx} \cot v = -\csc^2 v \frac{dv}{dx}$$

$$\frac{d}{dx} \sec v = \tan v \sec v \frac{dv}{dx}$$

$$\frac{d}{dx} \csc v = -\cot v \csc v \frac{dv}{dx}$$

To gain facility with the use of trigonometric functions in solving problems, you must commit these rules to memory. To accomplish this, it helps to take note of similarities and patterns, for instance, that the derivatives of all the cofunctions are negative. Furthermore, in applying these rules, keep in mind that the chain rule must be used to differentiate the argument v of a trigonometric function in cases where the argument is not just x but some function of x.

Exercise 10.4

1. Find the relative maxima and minima, and points of inflection, and sketch the graphs of the following functions. $(0 < x < 2\pi)$

   (a) $y = \sin^2 x - \dfrac{x}{2}$

   (b) $y = \cos x + \cos 3x$

2. If the position function of a moving point on a line satisfies the differential equation

   $$\frac{d^2s}{dt^2} = -k^2s, \quad (k, \text{a constant})$$

then the point is said to undergo simple harmonic motion. Show that

$$s(t) = A \sin kt + B \cos kt$$

satisfies the differential equation for any constants A and B.

3. The cross section of an eavestrough has the shape of a trapezoid, in which the two sides are inclined at equal angles to the horizontal bottom. If the sides and the bottom are each 4 cm in width, determine the width across the top of the trough which maximizes the cross-sectional area.

4. A right circular cone is to be inscribed in a sphere of radius R. Find the ratio of the altitude to the radius of the base of the cone having the largest volume.

5. Prove that if a and b are positive, the minimum value of $y = a^2 \csc^2 \theta + b^2 \sec^2 \theta$ is $(a + b)^2$.

6. Show that the function $y = 2 \sin x + \tan x$ has a maximum in the interval $\pi/2 \le x \le \pi$.

7. A paper drinking cup is to be made in the shape of a right circular cone. What vertex angle will result in a cone requiring the least material to hold a given volume?

8. The illumination from a point source of light varies directly as the cosine of the angle of incidence (measured from the perpendicular),

Figure 10.7

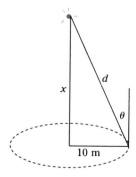

and inversely as the square of the distance from the source. How high above a circle of radius 10 m should a street light be placed so that the illumination of the circumference of the circle will be a maximum? (See Figure 10.7.)

9. A strip of metal 25 cm wide is to be bent into a gutter. If the cross section of the gutter is a circular arc, what radius will give the maximum capacity?

10. The corner of a rectangular piece of paper is folded over so that the corner C touches the opposite edge at C'. Find the value of x which makes the length of the fold a minimum. (See Figure 10.8.)

Figure 10.8

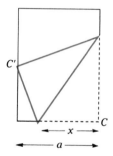

11. An isosceles triangle has area A, perimeter P, and apex angle 2θ.

(a) Show that the height can be expressed in terms of A and θ by

$$h = \sqrt{A \cot \theta}$$

(b) Prove that

$$P^2 = 4A \frac{(\tan \theta + \sec \theta)^2}{\tan \theta}$$

(c) Show that for a given area, the perimeter is a minimum when $\theta = 30°$.

Inverse Trigonometric Functions

I n the work you have done so far with the trigonometric functions, it has been assumed that you can evaluate the inverse of a trigonometric function. You have used an inverse trigonometric function, for example, to find the measure of an unknown angle, when, say, its sine is given. In this chapter, the inverse trigonometric functions are explicitly defined. Their properties are briefly investigated, then their derivatives are introduced along with some typical applications. Later, in Chapter 14, you will see that certain integrals are best expressed in terms of the inverse sine or inverse tangent functions.

Properties of Inverse Trigonometric Functions

Inverse Functions

When a function is expressed in the form $y = f(x)$, the notation emphasizes the fact that there is a relationship between the two variables x and y. We say that y depends on x, that is, y is the dependent, and x, the independent variable. Sometimes, however, it is desirable to reverse the roles of the dependent and the independent variables. Thus, given the function

$$y = 2x + 1$$

it might be more appropriate for some purposes to solve for x and express the relationship between the variables in the form

$$x = \frac{y - 1}{2}$$

If we now rename the variables, calling the independent variable x and the dependent variable y, as is customary, the result is a new relation

$$y = \frac{x - 1}{2}$$

called the **inverse** of the original function. It is denoted by the symbol f^{-1}. Thus, if

$$f(x) = 2x + 1$$

then the inverse of f is

$$f^{-1}(x) = \frac{x - 1}{2}$$

The "f" in the name of this new relation reminds us of its intimate connection with the original function. The "$^{-1}$" is a part of the *name* of the new relation. It stands for "the inverse of" and is *not* to be interpreted as an exponent. Note: To indicate the negative first power of f, write either $[f(x)]^{-1}$, or simply $1/f(x)$.

Exchanging the variables x and y to find the inverse of a function has an important consequence. The graph of the inverse relation f^{-1} is the reflection of the original graph of f in the line $y = x$, as Figure 11.1 indicates. Furthermore, if the graph of f^{-1} is itself reflected again, the result is the original graph of f. This is a graphical representation of the fact that the composition of f and f^{-1} is the identity function:

$$f^{-1}(f(x)) = f(f^{-1}(x)) = x$$

Figure 11.1

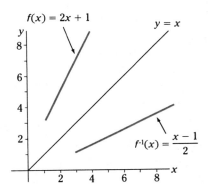

Inverse Trigonometric Functions

In accord with the notation for inverse relations described above, the inverse of a trigonometric function will be denoted by attaching "-1" to the name of the function, e.g., $\sin^{-1}$, $\tan^{-1}$, etc. Thus, the statement

$$y = \sin x \qquad\qquad (y \text{ is the sine of } x)$$

can be written in the equivalent form

$$x = \sin^{-1} y \qquad\qquad (x \text{ is the angle whose sine is } y)$$

It follows that

$$y = \sin(\sin^{-1} y) \qquad\qquad (y \text{ is the sine of the angle whose sine is } y)$$

and

$$x = \sin^{-1}(\sin x) \qquad\qquad (x \text{ is the angle whose sine is } \sin x)$$

For example,

if

$$\sin 30° = \frac{1}{2}$$

then

$$30° = \sin^{-1}\left(\frac{1}{2}\right)$$

Also

$$\sin\left(\sin^{-1}\left(\frac{1}{2}\right)\right) = \frac{1}{2}$$

and

$$\sin^{-1}(\sin 30°) = 30°$$

You should be aware, that the inverse trigonometric functions are sometimes denoted by "$\arcsin x$," "$\arctan x$," etc. We shall not use that notation in this text.

The graph of the inverse of the sine function is the reflection of the graph of $y = \sin x$ in the line $y = x$. It is shown in Figure 11.2. Since no angle can have a sine which is more than 1 or less than -1, the domain of the inverse is $-1 \le x \le 1$. In other words, the inverse $y = \sin^{-1} x$ has a value only when x is between -1 and 1.

Figure 11.2

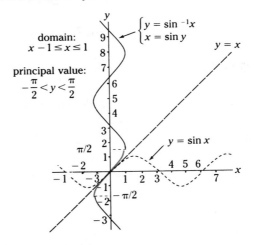

Figure 11.3

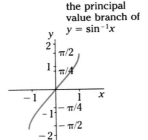

the principal value branch of $y = \sin^{-1} x$

You can see from the graph that the inverse of $y = \sin x$ is not a function. There are infinitely many values of y for each value of x in the domain. These are angles that are coterminal with either y or $\pi - y$, all of which have the same sine. In order to deal with an inverse that is itself a function, we define the **inverse sine function** to be the portion of the graph of $y = \sin^{-1} x$ between $y = -\pi/2$ to $y = \pi/2$. This is called the **principal value** branch. It is indicated in Figure 11.2 by a heavy coloured line, and is shown separately in Figure 11.3. Hereafter, the value of $\sin^{-1} x$ will always be the value of y on the principal value branch, unless a problem specifically demands otherwise.

EXAMPLE A

Given $\sin^{-1} x = \pi/4$, express x as a trigonometric function and determine its value.

SOLUTION

Since

$$\sin^{-1} x = \frac{\pi}{4}$$

then

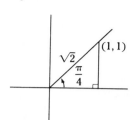

Figure 11.4

$$x = \sin \frac{\pi}{4}$$

$$= \frac{1}{\sqrt{2}}$$

EXAMPLE B

Given $\sin \theta = -\sqrt{3}/2$, express θ as an inverse trigonometric function and determine its value.

SOLUTION

Since

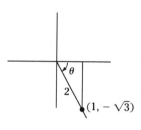

Figure 11.5

$$\sin \theta = -\frac{\sqrt{3}}{2}$$

then

$$\theta = \sin^{-1}\left(-\frac{\sqrt{3}}{2}\right)$$

$$= -\frac{\pi}{3} \quad \text{or} \quad -60°$$

Although $\sin(4\pi/3)$ also equals $-\sqrt{3}/2$, the angle $4\pi/3$ in the third quadrant is *not* in the range of principal values.

EXAMPLE C

Graph the function $f(x) = \sin^{-1}(x - 2)$.

SOLUTION

The graph of this function is the graph of $f(x) = \sin^{-1}(x)$ translated two units to the right (Figure 11.6).

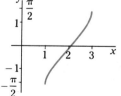

Figure 11.6

$$y = \sin^{-1}(x - 2)$$

The inverse cosine function is defined much like the inverse sine function. If $y = \cos x$ then $x = \cos^{-1} y$. On exchanging the variable names, the result is

$$y = \cos^{-1} x$$

The graph of the inverse of the cosine function is produced by reflecting the graph of $y = \cos x$ in the line $y = x$. It is shown in Figure 11.7. Like all inverse trigonometric functions, it is infinitely many valued. The **inverse cosine function** is therefore defined to be the part of the graph lying between $y = 0$ and $y = \pi$. This is the principal value branch of $y = \cos^{-1} x$.

Figure 11.7

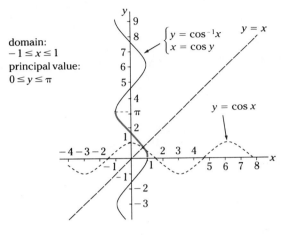

domain:
$-1 \leq x \leq 1$
principal value:
$0 \leq y \leq \pi$

The inverse tangent function is

$$y = \tan^{-1} x$$

Its graph is shown in Figure 11.8. Its principal value branch is taken to be the part of the graph lying between $y = -\pi/2$ and $y = \pi/2$. Note the horizontal asymptote, which corresponds to the limit

$$\lim_{x \to \infty} \tan^{-1} x = \frac{\pi}{2}$$

Figure 11.8

domain: $x \in R$
principal value: $\dfrac{-\pi}{2} < y < \dfrac{\pi}{2}$

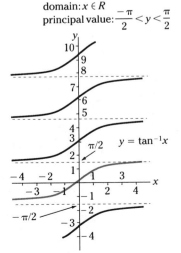

EXAMPLE D

Find the exact value of

$$\cos\left(\sin^{-1}\frac{1}{3}\right) + \sin\left(\cos^{-1}\frac{1}{3}\right)$$

Figure 11.9

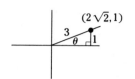

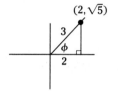

SOLUTION

Let

$$\theta = \sin^{-1}\frac{1}{3}, \qquad\qquad \phi = \cos^{-1}\frac{2}{3}$$

then

$$\sin\theta = \frac{1}{3}, \qquad\qquad \cos\phi = \frac{2}{3}$$

and

$$\cos\theta = \frac{2\sqrt{2}}{3}, \qquad\qquad \sin\phi = \frac{\sqrt{5}}{3} \text{ (see Figure 11.9)}$$

$$\therefore \quad \cos\left(\sin^{-1}\frac{1}{3}\right) + \sin\left(\cos^{-1}\frac{2}{3}\right)$$

$$= \cos\theta + \sin\phi$$

$$= \frac{2\sqrt{2}}{3} + \frac{\sqrt{5}}{3}$$

$$= \frac{2\sqrt{2} + \sqrt{5}}{3}$$

❏

EXAMPLE E

Prove the identity

$$2\tan^{-1}x = \tan^{-1}\left(\frac{2x}{1-x^2}\right)$$

SOLUTION

Let

$$\alpha = \tan^{-1}x$$

then

$$\tan\alpha = x$$

Therefore,

Left side $= 2\tan^{-1}x$

$\qquad = 2\alpha$

$\qquad = \tan^{-1}(\tan 2\alpha)$ ⟵ use the double angle formula for $\tan 2\alpha$

$\qquad = \tan^{-1}\left(\dfrac{2\tan\alpha}{1 - \tan^2\alpha}\right)$ ⟵ substitute $x = \tan\alpha$

$\qquad = \tan^{-1}\left(\dfrac{2x}{1 - x^2}\right)$

$\qquad = $ Right Side

Thus, the identity has been proved.   □

Exercise 11.1

1. Rewrite each of the following using an inverse trigonometric function:

   (a) $\sin\dfrac{\pi}{6} = \dfrac{1}{2}$

   (b) $\tan\left(-\dfrac{\pi}{4}\right) = -1$

   (c) $\cos\dfrac{\pi}{2} = 0$

2. Rewrite each of the following using a trigonometric function:

   (a) $\tan^{-1}\sqrt{3} = \dfrac{\pi}{3}$

   (b) $\cos^{-1}\left(-\dfrac{\sqrt{3}}{2}\right) = -\dfrac{5\pi}{6}$

   (c) $\sin^{-1}\left(-\dfrac{1}{2}\right) = -\dfrac{\pi}{6}$

3. Express each of the following in radian measure:

   (a) $\sin^{-1}1$

   (b) $\cos^{-1}\left(-\dfrac{1}{2}\right)$

   (c) $\tan^{-1}\left(\dfrac{1}{\sqrt{3}}\right)$

   (d) $\tan^{-1}\left(\sin\dfrac{\pi}{2}\right)$

   (e) $\sin^{-1}\left(\cos\dfrac{3\pi}{2}\right)$

4. Find the exact value of each of the following:

   (a) $\tan(\sin^{-1}(0))$

   (b) $\sin\left(\cos^{-1}\left(\dfrac{4}{5}\right)\right)$

   (c) $\cot\left(\sin^{-1}\left(-\dfrac{1}{4}\right)\right)$

5. Solve for x:

   (a) $\sin^{-1}\dfrac{x}{4} = \dfrac{\pi}{6}$

   (b) $2\cos^{-1}x = \dfrac{\pi}{2}$

   (c) $\tan^{-1}\dfrac{1}{x} = -\dfrac{\pi}{3}$

6. (a) Determine the value of the quantity

$$\sin^{-1}x + \cos^{-1}x$$

for values of x equal to $0.25, 0.50$, and 0.75.

(b) Prove that

$$\sin^{-1}x + \cos^{-1}x = \frac{\pi}{2}$$

(c) Interpret this formula geometrically.

7. Prove each of the following:

(a) $\cos^{-1}\left(\dfrac{3}{\sqrt{10}}\right) + \cos^{-1}\left(\dfrac{2}{\sqrt{5}}\right) = \dfrac{\pi}{4}$

(b) $\sin^{-1}\left(\dfrac{3}{5}\right) = \sin^{-1}\left(\dfrac{77}{85}\right) - \sin^{-1}\left(\dfrac{8}{17}\right)$

(c) $2\tan^{-1}\left(\dfrac{2}{3}\right) = \tan^{-1}\left(\dfrac{12}{5}\right)$

(d) $\tan^{-1}1 + \tan^{-1}2 + \tan^{-1}3 = \pi$

8. Sketch the principal value branch of the following graphs:

(a) $y = \cos^{-1}(2x)$

(b) $y = \tan^{-1}(x) + \pi/2$

(c) $y = 3\sin^{-1}(x)$

(d) $y = \cos^{-1}(2x - 2)$

9. (a) Prove that both $\sin^{-1}x$ and $\tan^{-1}x$ are odd functions, that is,

$$\sin^{-1}(-x) = -\sin^{-1}x$$

and

$$\tan^{-1}(-x) = -\tan^{-1}x$$

(b) Is $\cos^{-1}(-x) = \cos^{-1}(x)$?

10. A camera is positioned d metres from the base of a rocket launching pad. If a rocket of length l is launched vertically, show that when the base of the rocket is x metres above the ground, the angle θ subtended at the lens by the rocket is

$$\theta = \tan^{-1}\left(\frac{l + x}{d}\right) - \tan^{-1}\left(\frac{x}{d}\right)$$

11. Only one of the inverse trigonometric functions is present in the computer programming language BASIC. It is the inverse tangent function, (written ATN(X)). Show that the inverse sine and cosine functions can be calculated from $\tan^{-1}x$ using the formulas

$$\sin^{-1}(x) = \tan^{-1}\left(\frac{x}{\sqrt{1 - x^2}}\right)$$

$$\cos^{-1}(x) = -\tan^{-1}\left(\frac{x}{\sqrt{1 - x^2}}\right) + \pi/2$$

Derivatives of Inverse Trigonometric Functions   11.2

To obtain the derivative of the function

$$y = \sin^{-1}x$$

write it first in the equivalent form

$$x = \sin y$$

Now differentiate using implicit differentiation and the chain rule:

$$\frac{d}{dx}x = \frac{d}{dx}\sin y$$

$$1 = \cos y \cdot \frac{dy}{dx}$$

$$\frac{dy}{dx} = \frac{1}{\cos y}$$

However,

$$\frac{1}{\cos y} = \frac{1}{\sqrt{1 - \sin^2 y}}$$

$$= \frac{1}{\sqrt{1 - x^2}}$$

Therefore,

$$\frac{d}{dx}\sin^{-1}x = \frac{1}{\sqrt{1 - x^2}}$$

Choosing the positive sign for the radical, as we have done here, means that $\cos y$ is positive and y lies in the interval $-\pi/2 \le y \le \pi/2$. This is one reason for selecting this interval in the first place to be the principal value branch of the inverse sine function.

The derivative of the inverse cosine function is almost identical, differing only by a negative sign:

$$\frac{d}{dx}\cos^{-1}x = -\frac{1}{\sqrt{1 - x^2}}$$

The derivative of the inverse tangent function is found by differentiating the tangent function using implicit differentiation:

If
$$y = \tan^{-1}x$$

then
$$x = \tan y$$

and

$$\frac{d}{dx}x = \frac{d}{dx}\tan y$$

$$1 = \sec^2 y \cdot \frac{dy}{dx}$$

$$\frac{dy}{dx} = \frac{1}{\sec^2 y}$$

Since

$$\frac{1}{\sec^2 y} = \frac{1}{1 + \tan^2 y}$$

$$= \frac{1}{1 + x^2}$$

then

$$\frac{d}{dx}\tan^{-1}x = \frac{1}{1 + x^2}$$

When the argument of the inverse function is not x itself but some function v of x, the chain rule is required. General formulas for differentiating the three principal inverse trigonometric functions are therefore as follows:

$$\frac{d}{dx}\sin^{-1}v = \frac{1}{\sqrt{1 - v^2}}\frac{dv}{dx}$$

$$\frac{d}{dx}\cos^{-1}v = -\frac{1}{\sqrt{1 - v^2}}\frac{dv}{dx}$$

$$\frac{d}{dx}\tan^{-1}v = \frac{1}{1 + v^2}\frac{dv}{dx}$$

EXAMPLE A

Find $\dfrac{dy}{dx}$ and simplify:

$$y = \sin^{-1}(3x^2)$$

SOLUTION

Note carefully, that in the rule for differentiating the inverse sine function, it is not just x, but the argument $3x^2$, which appears in the radical. Therefore, using the chain rule

$$\frac{dy}{dx} = \frac{d}{dx}\sin^{-1}(3x^2)$$

$$= \frac{1}{\sqrt{1 - (3x^2)^2}} \cdot \frac{d}{dx}(3x^2)$$

$$= \frac{6x}{\sqrt{1 - 9x^4}}$$

EXAMPLE B

Find $\dfrac{dy}{dx}$ and simplify:

$$y = \tan^{-1}\left(\frac{x + 1}{x - 1}\right)$$

SOLUTION

In this function, the argument v is the quantity $\left(\dfrac{x + 1}{x - 1}\right)$.

Therefore,

$$\frac{dy}{dx} = \frac{d}{dx}\tan^{-1}\left(\frac{x + 1}{x - 1}\right)$$

$$= \frac{1}{1 + \left(\dfrac{x + 1}{x - 1}\right)^2} \cdot \frac{d}{dx}\left(\frac{x + 1}{x - 1}\right) \qquad \text{use the quotient rule}$$

$$= \frac{1}{\dfrac{(x - 1)^2 + (x + 1)^2}{(x - 1)^2}} \cdot \frac{(x - 1) - (x + 1)}{(x - 1)^2}$$

$$= \frac{1}{x^2 - 2x + 1 + x^2 + 2x + 1} \cdot (-2)$$

$$= \frac{-2}{2x^2 + 2}$$

$$= \frac{-1}{x^2 + 1}$$

EXAMPLE C

Determine the relative maximum and minimum values of the function $y = \sin^{-1} x^2$ and sketch its graph.

SOLUTION

If

$$y = \sin^{-1} x^2$$

then

$$y' = \frac{d}{dx} \sin^{-1} x^2$$

$$= \frac{1}{\sqrt{1 - (x^2)^2}} \cdot \frac{d}{dx} x^2$$

$$= \frac{2x}{\sqrt{1 - x^4}}$$

and

$$y'' = \frac{d}{dx} \left(\frac{2x}{\sqrt{1 - x^4}} \right)$$

$$= \frac{\sqrt{1 - x^4} \cdot 2 - 2x \cdot \dfrac{1}{2} \cdot \dfrac{(-4x^3)}{\sqrt{1 - x^4}}}{1 - x^4}$$

$$= \frac{2(1 - x^4) + 4x^4}{(1 - x^4)^{3/2}}$$

$$= \frac{2 + 2x^4}{(1 - x^4)^{3/2}}$$

The first derivative is zero when $x = 0$. At this value of x, $y = 0$ and $y'' > 0$. Therefore, there is a minimum at the origin. There are no points of inflection, since y'' is never zero. Note that the slope of the function becomes infinite as $x \to \pm 1$. The graph is shown in Figure 11.10. ❑

Figure 11.10

$y = \sin^{-1} x^2$

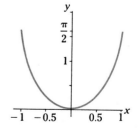

EXAMPLE D

A tracking station is picking up an aircraft flying toward it at an altitude of 6 km and a speed of 600 km/h. How fast is the angle of elevation changing at the moment when the angle of elevation is 30°?

SOLUTION

The diagram in Figure 11.11 shows that θ, the angle of elevation, is the angle whose tangent is $6/x$, where x is the distance to the tracking station:

Figure 11.11

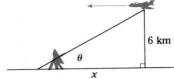

$$\theta = \tan^{-1}\left(\frac{6}{x}\right)$$

Therefore,

$$\frac{d}{dt}\theta = \frac{d}{dt}\tan^{-1}\left(\frac{6}{x}\right)$$

$$= \frac{1}{1 + \left(\dfrac{6}{x}\right)^2} \cdot \frac{d}{dt}\left(\frac{6}{x}\right)$$

$$= \frac{1}{1 + \left(\dfrac{6}{x}\right)^2} \cdot \left(\frac{-6}{x^2}\right) \cdot \frac{dx}{dt}$$

$$= \frac{-6}{x^2 + 36} \cdot \frac{dx}{dt}$$

The negative sign in this expression means that the distance to the station and the angle of elevation change in an opposite sense; when one increases, the other decreases. Since the plane is moving toward the station, the value of x is decreasing, and its rate of change is negative:

$$\frac{dx}{dt} = -600$$

When $\theta = 30°$, the value of x can be found from $\tan^{-1}(6/x) = 30°$, which gives $x = 6\sqrt{3}$. The rate of change of the angle of elevation is therefore

$$\frac{d\theta}{dt} = \frac{-6}{(6\sqrt{3})^2 + 36} \cdot (-600)$$

$$= 25$$

Thus, the angle of elevation is increasing at a rate of 25 radians/h or approximately 0.4°/s, at the moment when it is 30°. ❑

Exercise 11.2

1. If $y = \sin^{-1} v$, where v is a function of x, show by differentiating $v = \sin y$, that

$$\frac{d}{dx} \sin^{-1} v = \frac{1}{\sqrt{1 - v^2}} \cdot \frac{dv}{dx}$$

2. Find $\dfrac{dy}{dx}$:

(a) $y = \tan^{-1}(4x)$

(b) $y = \cos^{-1}\left(\dfrac{x}{2}\right)$

(c) $y = 2\sin^{-1}(1 - x)$

(d) $y = \cos^{-1}(x^2)$

3. Find $\dfrac{dy}{dx}$ and simplify:

(a) $y = \tan^{-1} \sqrt{x^2 - 1}$

(b) $y = x\sin^{-1}(2x)$

(c) $y = x^2 \cos^{-1} x$

(d) $y = \sqrt{\tan^{-1}(2x)}$

4. Find $\dfrac{dy}{dx}$ and simplify:

(a) $y = \sin^{-1}\left(\dfrac{a}{x}\right)$

(b) $y = \cos^{-1}\left(\dfrac{a}{\sqrt{a^2 + x^2}}\right)$

(c) $y = \sqrt{a^2 - x^2} + a\sin^{-1}\left(\dfrac{x}{a}\right),\ a > 0$

(d) $y = a\cos^{-1}\left(1 - \dfrac{x}{a}\right) - \sqrt{2ax - x^2},\ a > 0$

5. Find the second derivative of each of the following:

(a) $y = x\tan^{-1} x$

(b) $y = x\sin^{-1} x$

6. Find the derivative:

(a) $y = \tan^{-1}(\sin x)$

(b) $y = \cos^{-1}(\tan x)$

7. Find the equation of the line tangent to the given curve at the given point:

(a) $y = \tan^{-1} x,\qquad x = -1$

(b) $y = x + \cos^{-1} x,\quad x = \sqrt{3}/2$

(c) $y = (\sin^{-1} x)^2,\qquad x = -1/2$

8. Find the relative maxima and minima and sketch the graphs of the following functions:

(a) $f(x) = \cos^{-1} x^2$

(b) $f(x) = \tan^{-1} x^2$

(c) $f(x) = \tan^{-1} x - x$

(d) $f(x) = 4\sin^{-1}\dfrac{x}{2} + x\sqrt{4 - x^2}$

9. A lighthouse beacon revolves at a rate of 3 rpm. If the lighthouse is located 2 km off a straight shoreline, how fast is the beam moving along the coast, at a point 5 km down the coast?

10. Show that $\sin^{-1} x$ increases at the same rate as x when $x = 0$, and more rapidly than x at every other point. What does this imply about the slope of the graph of $y = \sin^{-1} x$?

11. Sketch the curves $y = \tan^{-1} x$ and $y = \tan^{-1} 2x$. Find their angle of intersection.

12. As a man walks from the centre O of a circular racetrack toward point P at its edge, a light at L casts his shadow on the racetrack fence (see Figure 11.12). If he walks at a rate of $1.5 \, \text{m/s}$, at what rate is his shadow moving along the fence when he is halfway from O to P?

Figure 11.12

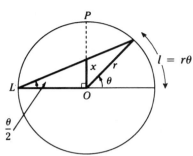

13. The lower edge of a painting 3 m in height is 2 m above an observer's eye level. Assuming that the best view of the painting is obtained when the angle subtended at the eye by the painting is a maximum, how far from the wall should the observer stand?

14. A rocket carrying instruments for studying the upper atmosphere is launched vertically with its height h in metres t seconds after launch given by $h = 600t - t^2$. At what angular rate should the telemetry antenna, located 2000 m from the launch point, be turning to track the rocket 15 s after launch?

11.3 Summary and Review

A function f is changed to its inverse f^{-1} by interchanging the dependent and independent variables. This means that the domain of f is the range of f^{-1}, and vice versa. It also means that the graph of f^{-1} is the reflection of the graph of f in the line $y = x$. Three inverse trigonometric functions, their derivatives, and the graphs of their principal value branches are given in Figure 11.13.

Figure 11.13

$$y = \sin^{-1} x$$
$$y' = \frac{1}{\sqrt{1 - x^2}}$$

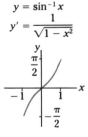

$$y = \tan^{-1} x$$
$$y' = \frac{1}{1 + x^2}$$

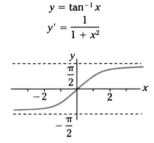

$$y = \cos^{-1} x$$
$$y' = -\frac{1}{\sqrt{1 - x^2}}$$

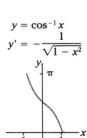

To master this material, it would be helpful to

1. memorize the differentiation formulas
2. keep in mind how the graphs of the inverse trigonometric functions arise
3. begin thinking of an inverse trigonometric function as a legitimate function in its own right, e.g., as the angle whose sine is x, etc.

Exercise 11.3

1. Determine the exact value:

(a) $\sin^{-1}(1/\sqrt{2})$

(b) $\tan^{-1}(-\sqrt{3})$

(c) $\sin^{-1}\left(\sin\left(\dfrac{5\pi}{4}\right)\right)$

(d) $\tan\left(\sin^{-1}\left(\dfrac{\sqrt{3}}{2}\right)\right)$

(e) $\sin\left(\tan^{-1}\left(\dfrac{\sqrt{3}}{2}\right)\right)$

(f) $\sin\left(\cos^{-1}\left(\dfrac{1}{2}\right)\right)$

2. Prove that:

(a) $\cos(\sin^{-1}x) = \sqrt{1-x^2}$

(b) $\cos^{-1}(-x) = \pi - \cos^{-1}x$

3. Differentiate:

(a) $y = \cos^{-1}(2x+1)$

(b) $y = \dfrac{\tan^{-1}x}{x}$

(c) $y = (\sin^{-1}x)^2$

(d) $y = \cos^{-1}x + \dfrac{x}{\sqrt{1-x^2}}$

(e) $y = \sin^{-1}\left(\dfrac{1-x}{1+x}\right)$

4. Show that the function

$$f(x) = \tan^{-1}\left(\frac{x}{9}\right) - \tan^{-1}\left(\frac{x}{16}\right), \ x > 0$$

has a maximum value of $\tan^{-1}(7/24)$ at $x = 12$.

5. For each of the following, locate any relative maxima or minima and points of inflection. Sketch the graph.

(a) $f(x) = \sqrt{x} - \tan^{-1}\sqrt{x}$

(b) $f(x) = \dfrac{x}{\sqrt{a^2 - x^2}} - \sin^{-1}\left(\dfrac{x}{a}\right)$

(c) $f(x) = (\tan^{-1}x)^2$

(Hint: The equation $\tan^{-1}x = 1/(2x)$, which determines the inflection points, can only be solved numerically.

6. Find point A on the x-axis for which the measure of $\angle BAC$ is a maximum, given that B and C are the points $(0, 3)$ and $(0, 4)$, respectively.

7. A ladder 10 m long is leaning against a vertical wall. If the top of the ladder slides down the wall at a constant rate of 1/4 m/s, what is the rate of change of the angle between the ladder and the ground, when the top of the ladder is 5 m above the ground?

8. A kite is 90 m above the ground with 150 m of string out. It is moving horizontally, directly away from the person who holds the string at a speed of 10 m/s. At what rate is the inclination of the string to the horizontal changing?

G.   Properties of Logarithms

Properties of logarithms arise directly out of the fact that the logarithm function is the inverse of the exponential function $f(x) = a^x$, $a > 0$. The inverse function can be written in either of two equivalent forms:

$$x = a^y \qquad \text{exponential form}$$

or

$$y = \log_a x \qquad \text{logarithmic form}$$

Read: "y is the log of x to base a," which means that y is the power to which the base must be raised to obtain x. It follows that

$$\log_a a^x = x \quad \text{and} \quad a^{\log_a x} = x$$

There are two special cases to take note of:

$$\text{since } a^0 = 1 \quad \text{then} \quad \log_a 1 = 0$$

and $\quad$ since $a^1 = a \quad$ then $\quad \log_a a = 1$

Further properties are expressed by the following rules:
Logarithm of a product:

$$\log_a(xy) = \log_a x + \log_a y$$

Logarithm of a quotient:

$$\log_a\left(\frac{x}{y}\right) = \log_a x - \log_a y$$

Logarithm of a power:

$$\log_a x^p = p \log_a x$$

Change-of-base formulas:

$$\log_b a = \frac{1}{\log_a b}$$

$$\log_b x = \frac{\log_a x}{\log_a b}$$

The following examples show how such formulas can be derived.

EXAMPLE A

Show that $\log_a(xy) = \log_a x + \log_a y$.

SOLUTION

Let $\quad p = \log_a x \quad$ and $\quad q = \log_a y$

Then $\quad x = a^p \quad\quad$ and $\quad y = a^q$

$$xy = a^p \cdot a^q$$

$$= a^{p+q}$$

Converting this equation into logarithmic form gives

$$\log_a(xy) = p + q$$

$$= \log_a x + \log_a y$$

EXAMPLE B

Prove that $\quad \log_b a = \dfrac{1}{\log_a b}$

SOLUTION

Let $\quad p = \log_a b$

Then $\quad a^p = b$

Take the logarithm to base b of both sides:

$$\log_b a^p = \log_b b$$

$$p \log_b a = 1$$

$$\log_b a = \frac{1}{p}$$

$$= \frac{1}{\log_a b}$$

When the base of a logarithmic function is not specified, base 10 is implied. Logarithms with base 10 are called **common logarithms**. Thus the symbol log x means $\log_{10} x$. Before calculators and computers came into widespread use, tables of common logarithms were used extensively for making precise calculations.

EXAMPLE C

Solve for x:

$$6^{4x} = 120$$

SOLUTION

Take the common logarithm of both sides of this equation:

$$\log 6^{4x} = \log 120$$

$$(4x)(\log 6) = \log 120$$

$$x = \frac{\log 120}{4 \log 6}$$

$$\doteq \frac{2.0791812}{(4)(0.7781512)} \qquad \longleftarrow \text{using a calculator}$$

$$\doteq 0.6679874 \qquad \qquad \qquad \qquad \Box$$

Exercise G

1. Express in exponential form:

(a) $\log_4 16 = 2$

(b) $\log_{16} 2 = \dfrac{1}{4}$

(c) $\log_2 \dfrac{1}{16} = -4$

2. Express in logarithmic form:

(a) $2^5 = 32$

(b) $36^{1/2} = 6$

(c) $3^{-4} = \dfrac{1}{81}$

3. Evaluate:

(a) $\log_3 81$

(b) $\log_2 \dfrac{1}{64}$

(c) $\log_3 \sqrt{27}$

(d) $\log_{1/2} 4$

(e) $\log_{10} 5 - \log_{10} 50$

(f) $\log_6 18 + \log_6 12$

4. Convert into exponential form and solve for x:

(a) $\log_2 5x = 7$

(b) $\log_{10} x^3 = 18$

(c) $2 \log_4(8x) = 5$

(d) $\log_5(3x^2) - \log_5(15x) = 3$

(e) $\log_x 64 = \dfrac{3}{2}$

5. Which is larger?

(a) $2 \log 5$ or $3 \log 3$

(b) $\dfrac{1}{2} \log_2 16$ or $\dfrac{1}{3} \log_3 27$

6. Prove the following properties of the logarithm function:

(a) $\log_a\left(\dfrac{x}{y}\right) = \log_a x - \log_a y$

(b) $\log_a x^p = p \log_a x$

(c) $\log_b x = \dfrac{\log_a x}{\log_a b}$

7. (a) Construct a table of values for the function $f(x) = \log_2 x$ in which both x and y are whole numbers.

(b) Find $f(1/2)$, $f(1/4)$, and $f(1/8)$.

(c) Sketch the graph of f.

8. Solve for x:

(a) $10^{5x} = 25$

(b) $2^x = 1000$

(c) $4^{2x} = 200$

9. Magnitudes of earthquakes are measured by using the Richter scale. On this scale, the magnitude R of an earthquake is given by

$$R = \log_{10}\left(\frac{I}{I_0}\right)$$

where I is the intensity of the earthquake being measured, and I_0 is a fixed standard intensity used for comparison.

(a) Show that if an earthquake measures $R = 4$ on the Richter scale, then its intensity is 10 000 times the standard.

(b) How many times more intense is an earthquake measuring $R = 6$, than one measuring $R = 4$?

(c) The Mexico City earthquake of 1985 measured 8.1 on the Richter scale. Express its intensity in terms of the standard intensity.

10. The relative intensity β of two sounds is defined by the equation

$$\beta = 10 \log_{10}\left(\frac{I}{I_0}\right)$$

where I and I_0 are the intensities of the two sounds and β is measured in decibels (db).

(a) If one sound is 100 times as loud as another, $(I = 100 I_0)$, show that their relative intensity is 20 db.

(b) How much louder is the sound of ordinary conversation than a whisper, if their relative intensity is 45 db?

Logarithmic and Exponential Functions

12

The purpose of this chapter is to define two new functions—the **natural logarithm function**, and the **natural exponential function**—and to investigate their properties and uses. These functions have important applications in physics, engineering, biology, and economics.

The word "natural" in the names of these functions refers to the number 2.71828 .... The origin of this number, denoted by the symbol "*e*," and the reasons for its choice as the base of a logarithm or an exponent will become clear in what follows.

Basic properties of logarithms are discussed in the pages preceding this chapter. You may wish to review that material at this time.

Exponential and Logarithmic Functions with Base *e*

12.1

An **exponential function** is a function of the form $y = a^x$, $a > 0$, where a is a constant known as the **base** of the function. It is a continuous function that is defined for all real values of x. The graph of a typical exponential function is shown in Figure 12.1. Its y-intercept is 1 and the negative x-axis is an asymptote.

The **logarithm function** is the inverse of the exponential function:

If $f(x) = a^x$ then $f^{-1}(x) = \log_a x$, $a > 0$

The graph of the logarithm function is produced by reflecting the

Figure 12.1

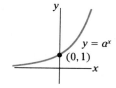

Figure 12.2

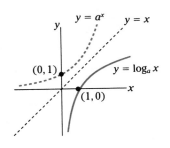

graph of the exponential function in the line $y = x$ as shown in Figure 12.2. It is defined only for positive values of x. It has an x-intercept of 1 and the negative y-axis is an asymptote. Like the exponential function, it is a continuous function that is everywhere increasing.

Two functions that are of particular importance are

$$y = e^x \quad \text{and} \quad y = \log_e x$$

where the base e is the irrational number 2.71828 .... Sometimes, you will see the exponential function with base e written with the symbol "exp." This is often done when the argument is a complicated function of x that is awkward to write as an exponent. For example

$$y = \exp\left(-\frac{x^2}{2}\right) \quad \text{means} \quad y = e^{-x^2/2}$$

A logarithm with base e is known as a **natural logarithm** and is designated by the symbol "ln." For example,

$$y = \ln x \quad \text{means} \quad y = \log_e x$$

The number e is an irrational number like π, that is, a non-terminating, non-repeating decimal. It may seem strange at first to use such a number for the base of a logarithmic or exponential function. However, e is a number with deep theoretical significance, arising in all sorts of seemingly unrelated mathematical situations. As you will see shortly, its use in calculus considerably simplifies many formulas and expressions.

EXAMPLE A

Evaluate $y = e^x$ at $x = 2$, -2, and 1/2.

SOLUTION

A hand calculator gives the following answers:

$$e^2 \doteq (2.71828\ldots)^2$$

$$= 7.389056\ldots$$

$$e^{-2} \doteq \frac{1}{(2.71828\ldots)^2}$$

$$\doteq 0.1353352\ldots$$

$$e^{1/2} \doteq \sqrt{2.71828\ldots}$$

$$\doteq 1.6487212\ldots$$

❏

EXAMPLE B

If $f(x) = \ln x$, determine $f(2)$, $f(1/2)$, and $f(10)$.

SOLUTION

Again relying on a hand calculator for answers:

$$\ln(2) \doteq 0.6931471$$

$$\ln(1/2) \doteq -0.6931471$$

$$\ln(10) \doteq 2.3025851$$

These results of course mean that

$$e^{0.6931471} \doteq 2$$

$$e^{-0.6931471} \doteq 1/2$$

$$e^{2.3025851} \doteq 10$$

❏

The Value of e

The number e is defined as the limit

$$\lim_{u \to 0} (1 + u)^{1/u}$$

It is not at all clear by inspection what the value of this limit could be: the exponent $1/u$ increases without bound, while at the same time, the base $(1 + u)$ approaches 1. If the base were exactly 1, then

$$\lim_{u \to 0} (1)^{1/u} = 1$$

However, if the base were to differ from 1 by any constant amount ϵ, however small, then

$$\lim_{u \to 0} (1 + \epsilon)^{1/u} = \infty$$

You may think of the value of e as the result of a tug-of-war between these two opposing tendencies. Although the base $(1 + u)$ is not exactly 1, it approaches 1 in the limit.

A direct calculation of the limit using a sequence of values of u that approaches zero gives the results shown in Table 12.1. As u approaches 0, there does seem to be a limit in the neighbourhood of 2.71828.

Table 12.1

u	$(1 + u)$	$1/u$	$(1 + u)^{1/u}$
1.	2.	1.	2.00000
0.1	1.1	10.	2.59374
0.01	1.01	100.	2.70481
0.001	1.001	1000.	2.71692
0.0001	1.0001	10000.	2.71815
0.00001	1.00001	100000.	2.71827
0.000001	1.000001	1000000.	2.71828

Another method of determining the value of e relies on the binomial expansion. Recall that (see A. Preparation for Calculus, pages 16–29)

$$(1 + x)^n = 1 + nx + \frac{n(n-1)}{2!}x^2 + \frac{n(n-1)(n-2)}{3!}x^3 + \cdots$$

This expansion has $(n + 1)$ terms if n is an integer, but it is an infinite series if n is not an integer. Applying this expansion to the present problem with x replaced by u and the exponent n replaced by $1/u$ gives

$$(1 + u)^{1/u} = 1 + \left(\frac{1}{u}\right)u + \frac{\left(\frac{1}{u}\right)\left(\frac{1}{u} - 1\right)}{2!}u^2$$

$$+ \frac{\left(\frac{1}{u}\right)\left(\frac{1}{u} - 1\right)\left(\frac{1}{u} - 2\right)}{3!}u^3$$

$$+ \frac{\left(\frac{1}{u}\right)\left(\frac{1}{u} - 1\right)\left(\frac{1}{u} - 2\right)\left(\frac{1}{u} - 3\right)}{4!}u^4 + \cdots$$

or, on simplifying the individual terms,

$$(1 + u)^{1/u} = 1 + 1 + \frac{(1 - u)}{2!} + \frac{(1 - u)(1 - 2u)}{3!}$$

$$+ \frac{(1 - u)(1 - 2u)(1 - 3u)}{4!} + \cdots$$

Now on taking the limit as $u \to 0$,

$$\lim_{u \to 0} (1 + u)^{1/u} = 1 + 1 + \frac{1}{2!} + \frac{1}{3!} + \frac{1}{4!} + \cdots$$

This infinite series provides a way to calculate the value of e to any desired number of decimal places. The sum of the first few terms of this series is shown in Table 12.2. The exact value of e to 25 significant digits is

$$e = 2.7182\ 81828\ 45904\ 52353\ 60287 \ldots$$

Table 12.2

n	$n!$	$\dfrac{1}{n!}$	cumulative sum
0	1	1	1
1	1	1	2
2	2	0.5	2.5
3	6	0.1666667	2.6666667
4	24	0.0416667	2.7083334
5	120	0.0083333	2.7166667
6	720	0.0013889	2.7180556
7	$\cdots$	$\cdots$	$\cdots$

Exercise 12.1

1. If $f(x) = e^x$ determine the values of

(a) $f(3)$

(b) $f(10)$

(c) $f(0.75)$

(d) $f(-3)$

2. (a) Graph the function $f(x) = e^x$.

(b) State the domain and the range.

(c) Describe how the function $y = e^x$ behaves in the limit as $x \to +\infty$ and as $x \to -\infty$.

(d) Describe the slope and the curvature of the function in qualitative terms.

3. If $f(x) = \ln x$, determine the values of

(a) $f(3)$

(b) $f(1/3)$

(c) $f(10)$

(d) $f(100)$

4. (a) Graph the function $f(x) = \ln x$.

(b) State the domain and the range.

(c) Describe how the function $y = \ln x$ behaves in the limit as $x \to \infty$ and as $x \to 0$.

(d) Describe the slope and the curvature of the function in qualitative terms.

5. Determine the next three entries in Table 12.2. How many terms in the sum must be included for the value of e to be accurate to seven significant digits?

6. Estimate how many terms in the series for e must be included in order to compute e correctly to 25 significant digits.

7. Explain why the following are true:

(a) $\exp\left(\frac{1}{3}\ln 64\right) = 4$

(b) $\exp(2\ln x) = x^2$

(c) $\exp\left(-\frac{1}{2}\ln x\right) = \frac{1}{\sqrt{x}}$

(d) $2\ln(e^{x/2}) = x$

(e) $\ln(e^{x-1}) + 1 = x$

8. Compare the graphs of the following functions. (In each case, draw both on the same axes.)

(a) $f(x) = e^x$, $g(x) = 2^x$

(b) $f(x) = e^x$, $g(x) = -e^{-x}$

(c) $f(x) = \ln x$, $g(x) = \log_3 x$

(d) $f(x) = 2^x$, $g(x) = \log_2 x$

9. By translating and/or reflecting the graph of $y = e^x$, sketch the graphs of

(a) $y = e^{x+2}$

(b) $y = e^{-x}$

(c) $y = 2 - e^{-x}$

(d) $y = e^{1-x}$

10. Sketch the graphs of

(a) $y = \ln(x - 2)$

(b) $y = -\ln x$

(c) $y = 3 + \ln x$

(d) $y = \ln(-x)$

11. (a) Show that the graph of $y = \ln(kx)$, $k > 0$ is the graph of $y = \ln x$ translated in the y direction by an amount $\ln k$.

(b) How is the graph of $y = e^{a+x}$ related to the graph of $y = e^x$?

12. Show that the function $y = C2^{ax}$ is equivalent to $y = Ce^{kx}$ where C is a constant and $k = a \ln 2$.

12.2 Derivatives of Logarithmic Functions

The derivative of the function $y = \log_a x$ can be worked out from first principles. Recall that the derivative of a function is by definition

$$\frac{d}{dx} f(x) = \lim_{h \to 0} \frac{f(x + h) - f(x)}{h}$$

where, for convenience, Δx has been replaced by h. If $f(x) = \log_a x$, then

$$\frac{d}{dx} \log_a x = \lim_{h \to 0} \frac{\log_a(x + h) - \log_a x}{h}$$

To find the limit, it is necessary first to rewrite the expression on the right in a different form using the laws of logarithms:

$$\frac{d}{dx}\log_a x = \lim_{h \to 0}\frac{1}{h}\cdot(\log_a(x + h) - \log_a x)$$

$$= \lim_{h \to 0}\frac{1}{h}\cdot\log_a\left(\frac{x + h}{x}\right)$$

$$= \lim_{h \to 0}\frac{1}{h}\cdot\log_a\left(1 + \frac{h}{x}\right)$$

Next, multiply and divide h by x in front of the logarithm:

$$\frac{d}{dx}\log_a x = \lim_{h \to 0}\frac{1}{x}\cdot\frac{1}{(h/x)}\cdot\log_a\left(1 + \frac{h}{x}\right)$$

$$= \lim_{h \to 0}\frac{1}{x}\cdot\log_a\left(1 + \frac{h}{x}\right)^{\frac{1}{(h/x)}}$$

Since $\log_a x$ is a continuous function, the log and the lim operations in the previous equation can be interchanged:

$$\frac{d}{dx}\log_a x = \frac{1}{x}\cdot\log_a\left[\lim_{h \to 0}\left(1 + \frac{h}{x}\right)^{\frac{1}{(h/x)}}\right]$$

The limit now can be recognized as e, since it has the form $(1 + u)^{1/u}$ where $u \to 0$. Consequently,

$$\frac{d}{dx}\log_a x = \frac{1}{x}\log_a e$$

The factor $\log_a e$ is a constant whose numerical value depends on what base is used for the logarithm function. You can now see that choosing e for the base makes the factor $\log_a e = \log_e e = 1$. Thus, the derivative of $\ln x$ is simply

$$\frac{d}{dx}\ln x = \frac{1}{x}$$

In the event that the argument of a logarithm function is not merely x but some function v of x, then according to the chain rule,

$$\frac{d}{dx}\ln v = \frac{1}{v}\frac{dv}{dx}$$

EXAMPLE A

Find the derivative of $y = \ln x^3$ with respect to x.

SOLUTION

Differentiate using the chain rule:

$$\frac{dy}{dx} = \frac{d}{dx} \ln x^3$$

$$= \frac{1}{x^3} \cdot \frac{d}{dx} x^3$$

$$= \frac{1}{x^3} (3x^2)$$

$$= \frac{3}{x}$$

Alternatively, you can use the rule for the logarithm of a power to simplify the function before differentiating:

$$y = \ln x^3$$

$$= 3 \ln x$$

Therefore,

$$\frac{dy}{dx} = 3\left(\frac{1}{x}\right)$$

$$= \frac{3}{x}$$

❏

EXAMPLE B

Find $\frac{dy}{dx}$:

$$y = \ln[(x + 5)(3x - 2)]$$

SOLUTION

Using the chain rule and the product rule:

$$\frac{dy}{dx} = \frac{d}{dx} \ln[(x + 5)(3x - 2)]$$

$$= \frac{1}{(x + 5)(3x - 2)} \cdot \frac{d}{dx}\left[(x + 5)(3x - 2)\right]$$

$$= \frac{1}{(x + 5)(3x - 2)}\left[(x + 5) \cdot 3 + (3x - 2) \cdot 1\right]$$

$$= \frac{6x + 13}{(x + 5)(3x - 2)}$$

If, however, before differentiating, you rewrite the function using the rule for the logarithm of a product, then

$$y = \ln(x + 5)(3x - 2)$$

$$= \ln(x + 5) + \ln(3x - 2)$$

Therefore,

$$\frac{dy}{dx} = \frac{d}{dx}[\ln(x + 5) + \ln(3x - 2)]$$

$$= \frac{1}{x + 5} + \frac{3}{3x - 2}$$

$$= \frac{6x + 13}{(x + 5)(3x - 2)}$$ ❏

EXAMPLE C

Differentiate $y = x^2 \log x$ (a base 10 logarithm) with respect to x.

SOLUTION

Start with the product rule:

$$\frac{dy}{dx} = \frac{d}{dx}x^2 \log x$$

$$= x^2\frac{d}{dx} \cdot \log x + \log x \cdot \frac{d}{dx}x^2$$

$$= x^2 \cdot \frac{1}{x} \cdot \log e + (\log x) \cdot 2x \qquad \longleftarrow \text{remember that a factor } \log_a e \text{ appears when the base is not } e$$

$$= x \log e + 2x \log x, \text{ where } \log e = 0.4342944 \ldots$$ ❏

If a function to be differentiated is the product and/or quotient of several factors, the work is considerably simplified by taking the natural logarithm of the function *before* differentiating. This method is referred to as **logarithmic differentiation**.

EXAMPLE D

Differentiate $y = \dfrac{x\sqrt{4x + 5}}{(3x + 1)^2}$

SOLUTION

Take the natural logarithm of each side of the equation:

$$\ln y = \ln x + \frac{1}{2}\ln(4x + 5) - 2\ln(3x + 1)$$

Now use implicit differentiation:

$$\frac{1}{y} \cdot \frac{dy}{dx} = \frac{1}{x} + \frac{1}{2}\left(\frac{1}{4x + 5}\right) \cdot 4 - 2\left(\frac{1}{3x + 1}\right) \cdot 3$$

$$\frac{dy}{dx} = y\left[\frac{1}{x} + \frac{2}{4x + 5} - \frac{6}{3x + 1}\right]$$

As is usual when implicit differentiation is done, the result is expressed in terms of both x and y. ❏

Exercise 12.2

1. Differentiate y with respect to x:

(a) $y = 6\ln x^3$

(b) $y = \ln(x^3 - 6)$

(c) $y = \ln\left(\dfrac{x^2}{6}\right)$

(d) $y = \dfrac{1}{2}\ln(5x^2 - 10x)$

2. Find $\dfrac{dy}{dx}$:

(a) $y = x\ln x - x + 8$

(b) $y = \ln\sqrt{4 - x^2}$

(c) $y = 4\ln^2(3x)$

(d) $y = \ln(x\sqrt{2 + 3x})$

(e) $y = \ln\left(\dfrac{x^2}{x^2 + 1}\right)$

3. Determine the derivative of the function $y = \ln(5x)$ from first principles.

4. Determine the derivative of the function $y = \ln(x^5)$ from first principles.

5. Differentiate y with respect to x and simplify:

(a) $y = \ln(x + \sqrt{x^2 + a^2})$

(b) $y = \ln\dfrac{x + a}{x - a}$

6. Differentiate y with respect to x and simplify:

(a) $y = \dfrac{1}{2}\sec^2 x + \ln\cos x$

(b) $y = \ln \csc^2 x$

(c) $y = \ln \tan^2(2x)$

(d) $y = \ln(\sec x + \tan x)$

(e) $y = \ln \sqrt{\dfrac{1 + \sin x}{1 - \sin x}}$

7. In each of the following, find d^2y/dx^2:

(a) $y = x^2 \ln x$

(b) $y = x(\ln x)^2$

(c) $y = \dfrac{\ln x^2}{x}$

8. Find $\dfrac{dy}{dx}$ using logarithmic differentiation:

(a) $y = x^2 \sqrt{x^2 - 4}$

(b) $y = x\sqrt{3x + 1}\sqrt{x - 5}$

(c) $y = (2x)^{1/2}\sqrt[3]{9x + 4}$

(d) $y = \dfrac{x^3\sqrt{3x - 7}}{\sqrt{6x + 2}}$

(e) $y = \sqrt{\dfrac{(x + 3)(x - 5)}{(x^2 - 4)(2x + 2)}}$

9. Differentiate the function

$$f(x) = \dfrac{x^2 - a^2}{x^2 + a^2}$$

using the quotient rule and using logarithmic differentiation. Show that the results are equivalent.

10. Determine an equation of the line tangent to each of the following functions at the specified point:

(a) $y = \ln x$, $(1, 0)$

(b) $y = \ln x^3$, $(1, 0)$

(c) $y = \ln(\cos x)$, $(\pi/4, -\ln \sqrt{2})$

(d) $y = \sin(\ln x)$ $(1, 0)$

11. Find the angle at which the curves $y = \ln x$ and $y = \ln x^3$ intersect.

Derivatives of Exponential Functions

12.3

To determine the derivative of the exponential function from first principles is not a straightforward task. You will find in the exercises some suggestions on how that could be done. It is simpler instead to start with the inverse of the exponential function, that is, the logarithm function, the derivative of which is known. The first step is to write the exponential function

$$y = a^x, \; a > 0$$

in the equivalent logarithmic form

$$x = \log_a y$$

Now, differentiate both sides of this equation implicitly with respect to x using the chain rule:

$$\frac{d}{dx}x = \frac{d}{dx}\log_a y$$

$$1 = \frac{1}{y}(\log_a e)\frac{dy}{dx}$$

Solving for the derivative gives

$$\frac{dy}{dx} = \frac{y}{\log_a e}$$

Since $y = a^x$ and $\dfrac{1}{\log_a e} = \log_e a$, the result is

$$\boxed{\frac{dy}{dx} = a^x \ln a}$$

If the base of the exponential function is e, then the numerical factor $\ln e = 1$, and the derivative is simply

$$\boxed{\frac{d}{dx}e^x = e^x}$$

The exponential function $y = e^x$ is the only function that possesses this remarkable property, that the function and its derivative are the same. If the argument of the exponential function is not merely x but some function v of x, then the chain rule gives

$$\boxed{\frac{d}{dx}e^v = e^v \frac{dv}{dx}}$$

EXAMPLE A

Find $\dfrac{dy}{dx}$ where $y = 5e^{x^2}$.

SOLUTION

Think of x^2 as the function v:

$$\frac{dy}{dx} = \frac{d}{dx}5e^{x^2} \qquad\longleftarrow\qquad \text{use the chain rule}$$

$$= 5\,e^{x^2}\frac{d}{dx}x^2$$

$$= 5\,e^{x^2}(2x)$$

$$= 10x\,e^{x^2} \qquad\qquad\qquad \square$$

EXAMPLE B

Differentiate $y = x^3\,10^x$.

SOLUTION

Use the product rule:

$$\frac{dy}{dx} = \frac{d}{dx}\left(x^3 \cdot 10^x\right)$$

$$= x^3 \cdot \frac{d}{dx}10^x + 10^x \cdot \frac{d}{dx}x^3$$

$$= x^3 \cdot 10^x \cdot (\ln 10) + 10^x \cdot 3x^2 \qquad\longleftarrow\qquad \text{a factor (ln 10) appears when the base is not } e$$

$$= (\ln 10)x^3 10^x + 3x^2 10^x, \qquad \text{where } \ln 10 = 2.302585\ldots \quad \square$$

It is incorrect to apply the rule for differentiating exponential functions to functions of the form $y = u^v$, where the base u and the exponent v both vary with x. That rule is applicable only when the base is a constant. When neither the base nor the exponent is a constant, it is necessary to resort to logarithmic differentiation, as the following example shows.

EXAMPLE C

Differentiate $y = (5x + 3)^{2x}$.

SOLUTION

Take the natural logarithm of both sides:

$$\ln y = \ln(5x + 3)^{2x}$$

$$\ln y = 2x \cdot \ln(5x + 3)$$

This transforms the expression on the right into a product of functions. Now use implicit differentiation and the product rule:

$$\frac{d}{dx}\ln y = \frac{d}{dx}\left[2x \cdot \ln(5x + 3)\right]$$

$$\frac{1}{y} \cdot \frac{dy}{dx} = 2x \cdot \frac{d}{dx}\ln(5x + 3) + \ln(5x + 3) \cdot \frac{d}{dx}2x$$

$$\frac{1}{y} \cdot \frac{dy}{dx} = 2x \cdot \left(\frac{1}{5x + 3}\right) \cdot 5 + [\ln(5x + 3)] \cdot 2$$

$$\frac{dy}{dx} = y\left[\frac{10x}{5x + 3} + 2\ln(5x + 3)\right]$$

❏

Exercise 12.3

1. Find the derivative of the following functions with respect to x:

   (a) $y = 5e^{-4x}$

   (b) $y = \frac{1}{2}e^{x^2}$

   (c) $y = x^4 e^x$

   (d) $y = \frac{e^{-x}}{x}$

   (e) $y = \sqrt{1 + e^x}$

   (f) $y = x + e^{\sqrt{x}}$

2. Find $\frac{dy}{dx}$:

   (a) $y = 3^{3x}$

   (b) $y = 10^{-x+2}$

   (c) $y = 2^{-x} + x^{-2}$

   (d) $y = 2^x \cdot x^2$

3. Determine an equation of the line tangent to the curve at the specified point:

   (a) $y = e^{2x}$ $(1, e^2)$

   (b) $y = \frac{e^{x^2}}{x}$ $(1, e)$

   (c) $y = 2^{x^2}$ $(1, 2)$

   (d) $y = e^{\sin x}$ $(\pi/6, \sqrt{e})$

4. In each of the following find d^2y/dx^2:

   (a) $y = x^2 e^{-x}$

   (b) $y = 4xe^{x^2}$

(c) $y = e^{-x} \sin x$

(d) $y = \ln(xe^x)$

5. Show that if $y = e^x \cos 2x$, then

$$\frac{d^2y}{dx^2} - 2\frac{dy}{dx} + 5y = 0$$

6. For what values of a does $x = Ce^{at}$ satisfy the differential equation

$$\frac{d^2x}{dt^2} + \frac{dx}{dt} = 6x$$

7. Differentiate the following implicit functions and find y':

   (a) $e^x + e^y = e^{x+y}$

   (b) $ye^{2x} + xe^{2y} = 1$

8. Find the derivative of y with respect to x:

   (a) $y = x^{\ln x}$

   (b) $y = x^{\sqrt{x}}$

   (c) $y = (6e^x)^{3x}$

   (d) $y = (\sin x)^x$

9. Using logarithmic differentiation, show that the derivative of u^v is given by the formula

$$\frac{d}{dx}u^v = vu^{v-1} \cdot \frac{du}{dx} + u^v \ln u \cdot \frac{dv}{dx}$$

Note that this amounts to differentiating the function as if v were a constant and then as if u were a constant and adding the results.

10. (a) Starting with the binomial expansion and following steps similar to those in Section 12.1, pages 304–305, show that

$$(1 + u)^{x/u} = 1 + x + \frac{x(x - u)}{2!}$$

$$+ \frac{x(x - u)(x - 2u)}{3!}$$

$$+ \frac{x(x - u)(x - 2u)(x - 3u)}{4!}$$

$$+ \cdots$$

(b) By taking the limit as $u \to 0$, show that the exponential function $y = e^x$ can be represented by the infinite series

$$e^x = \lim_{u \to 0}\left[(1 + u)^{1/u} \right]^x$$

$$= 1 + x + \frac{x^2}{2!} + \frac{x^3}{3!} + \frac{x^4}{4!} + \cdots$$

(c) Using the series representation of the exponential function, work out the derivative of e^x from first principles.

Graphical Analysis   12.4

The object of this section is to apply the tools of differential calculus to exponential, logarithmic, and related functions to determine maximum and minimum values, points of inflection, and other properties of their graphs. Of particular interest will be to see how the functions $y = e^x$ and $y = \ln x$ compare to functions of the form $y = x^n$ in the limit as $x \to \pm \infty$.

EXAMPLE A

Analyze the function $f(x) = \ln x$, $x > 0$.

SOLUTION

If

$$y = \ln x$$

then

$$y' = \frac{1}{x}$$

and

$$y'' = -\frac{1}{x^2}$$

On inspection, these equations reveal that

(a) the first derivative is never zero. Therefore, $\ln x$ has no relative maxima or minima. y' is always positive, so the value of $\ln x$ is always increasing.

(b) the second derivative is always negative. Therefore, ln x is always concave down.

By examining values of the function, you can convince yourself that

$$\lim_{x \to \infty} \ln x = \infty$$

so there is no horizontal asymptote, and that

$$\lim_{x \to 0} \ln x = -\infty$$

which means that the negative y-axis is a vertical asymptote. These considerations give the familiar graph of the natural logarithm function, a sketch of which is shown in Table 12.3 along with a summary of these results. ❑

Table 12.3
Properties of $f(x) = \ln x$

value	$y = \ln x$	positive, if $x > 1$ zero, if $x = 1$ negative, if $0 < x < 1$
slope	$y' = \dfrac{1}{x}$	positive
concavity	$y'' = -\dfrac{1}{x^2}$	down
graph		
limit	$\displaystyle\lim_{x \to \infty} \ln x$	$+\infty$
	$\displaystyle\lim_{x \to 0^+} \ln x$	$-\infty$
asymptote		$-y$-axis

EXAMPLE B

Analyze the exponential function $f(x) = a^x$, $a > 0$.

SOLUTION

If

$$y = a^x$$

then

$$y' = a^x(\ln a)$$

and

$$y'' = a^x(\ln a)^2$$

You can conclude from these equations that
1. the values of the function are always positive
2. the first derivative is never zero (provided $a \neq 1$).

Therefore, f has no relative maxima or minima.

3. the second derivative is always positive (provided $a \neq 1$). Therefore, f is concave up everywhere. There are no points of inflection.

Beyond this, there are two cases to consider depending on whether a is greater than or less than 1:

1. When $a > 1$, y' is positive and f is increasing everywhere. Furthermore,

$$\lim_{x \to +\infty} a^x = +\infty$$

The value of f is increasing at an increasing rate. Thus it exceeds all bounds as x becomes very large.

On the other hand:

2. When $0 < a < 1$, y' is negative (the sign comes from the factor $\ln a$) and f is decreasing everywhere. Also,

$$\lim_{x \to +\infty} a^x = 0$$

Consider that even though f is a decreasing function, its value never becomes negative (its graph never crosses the x-axis). In addition, f never reaches a minimum, even though it is concave up. Consequently, f must have a horizontal asymptote. By trying a few specific values of x, you can convince yourself that the limit is 0, and that the horizontal asymptote must be the x-axis.

Note that when $0 < a < 1$, you can write a $= 1/b$, where $b > 1$, which means that

$$y = a^x$$

$$= \left(\frac{1}{b}\right)^x$$

$$= b^{-x}$$

Thus, an exponential function with a base less than 1 is equivalent to an exponential function with a negative exponent. These results are summarized in Table 12.4. ❏

Table 12.4
Properties of $f(x) = a^x$

		$a > 1$	$0 < a < 1$
value	$y = a^x$	positive	positive
slope	$y' = a^x(\ln a)$	positive	negative
concavity	$y'' = a^x(\ln a)^2$	up	up
graph			
limit	$\lim\limits_{x \to +\infty} a^x$	$+\infty$	0
	$\lim\limits_{x \to -\infty} a^x$	0	$+\infty$
asymptote		$-x$-axis	$+x$-axis

It is important to appreciate the difference between the exponential functions like $y = e^x$ and functions that consist of powers of x, e.g., $y = x^2$. For positive values of x, their graphs look superficially alike: both are increasing and concave up. But what, for instance, is the limit

$$\lim_{x \to +\infty} \frac{x^2}{e^x}$$

Is it $+\infty$ because as $x \to +\infty$, the factor x^2 in the numerator increases without bound? Or is it zero because, as shown in Figure 12.3, $1/e^x$ approaches zero in the limit? In other words, which function is the "stronger" for large x? This question is investigated in the next example, in which the quotient of x^2 and e^x is examined.

Figure 12.3

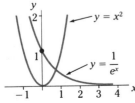

EXAMPLE C

Analyze the function $f(x) = x^2/e^x$, and find the limit of f as $x \to +\infty$.

SOLUTION

Zeros of the function occur when $y = 0$:

$$\frac{x^2}{e^x} = 0$$

$$x = 0$$

Therefore, the function has a zero at the origin. Everywhere else, the value of the function is positive. (This of course means that $(0, 0)$ is a minimum point.)

The first derivative is found using the quotient rule:

$$y' = \frac{d}{dx}\left(\frac{x^2}{e^x}\right)$$

$$= \frac{e^x(2x) - x^2(e^x)}{(e^x)^2}$$

$$= \frac{2x - x^2}{e^x}$$

Maximum or minimum values of the function may occur where the first derivative is zero:

$$\frac{2x - x^2}{e^x} = 0$$

$$x(2 - x) = 0$$

$$\therefore \quad x = 0 \quad \text{or} \quad x = 2$$

To distinguish between a maximum and a minimum, the second derivative is required:

$$y'' = \frac{d}{dx}\left(\frac{2x - x^2}{e^x}\right)$$

$$= \frac{e^x(2 - 2x) - (2x - x^2)(e^x)}{(e^x)^2}$$

$$= \frac{2 - 4x + x^2}{e^x}$$

At $x = 0$, $y'' > 0$, therefore, $(0, 0)$ is a minimum as expected. At $x = 2$, $y'' < 0$, therefore, there is a maximum at $x = 2$.

Inflection points may be found where $y'' = 0$:

$$\frac{(x^2 - 4x + 2)}{e^x} = 0$$

$$\therefore \quad x = \frac{4 \pm \sqrt{4^2 - 4 \cdot 1 \cdot 2}}{2 \cdot 1} \qquad \longleftarrow \text{quadratic formula}$$

$$x = 2 \pm \sqrt{2}$$

The function has inflection points at $x = 2 - \sqrt{2}$ and $x = 2 + \sqrt{2}$, one on each side of the maximum. Thus the function is concave down in the interval $2 - \sqrt{2} < x < 2 + \sqrt{2}$, and concave up outside the interval. Its graph is shown in Figure 12.4.

Figure 12.4

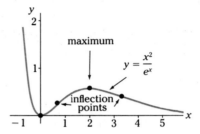

Observe that when $x > 2$, the value of f decreases with x, because the slope is negative. However, the value never reaches another minimum nor becomes negative, so there must be a horizontal asymptote. Thus, despite the presence of the factor x^2, the function f appears to approach zero in the limit as $x \to +\infty$:

$$\lim_{x \to +\infty} \frac{x^2}{e^x} = 0$$

Table 12.5 Values of $f(x) = x^2/e^x$

x	y	
2	0.541	max.
$2 + \sqrt{2}$	0.384	inf.
5	0.168	
10	0.0045	
15	0.000069	
20	0.0000008	

To prove that the limit is indeed zero requires methods that are beyond the scope of this text. However, examining numerical values such as those given in Table 12.5, should be convincing. It is true in fact that

$$\lim_{x \to +\infty} \frac{x^n}{e^x} = 0$$

for any value of n greater than zero, as the exercises will show by the same reasoning. The point is that the exponential function $y = e^x$ dominates any power of x, when x becomes large. ◻

EXAMPLE D

Analyze the function $f(x) = \dfrac{\ln x}{x}$.

Find $\lim\limits_{x \to +\infty} \dfrac{\ln x}{x}$ and $\lim\limits_{x \to 0+} \dfrac{\ln x}{x}$.

SOLUTION

Zeros of the function occur when $y = 0$:

$$\frac{\ln x}{x} = 0$$

$$\therefore \quad x = 1$$

The first derivative is found using the quotient rule:

$$y' = \frac{d}{dx}\left(\frac{\ln x}{x}\right)$$

$$= \frac{x \cdot \dfrac{d}{dx}(\ln x) - (\ln x) \cdot \dfrac{d}{dx}x}{x^2}$$

$$= \frac{1 - \ln x}{x^2}$$

A maximum or minimum may occur when $y' = 0$:

$$\frac{1 - \ln x}{x^2} = 0$$

$$\therefore \qquad \ln x = 1$$

$$x = e, \qquad e = 2.71828\ldots$$

$$\therefore \quad f(e) = \frac{\ln e}{e}$$

$$= \frac{1}{e}$$

To decide whether the point $(e, 1/e)$ is a maximum or a minimum, the second derivative is needed. Using the quotient rule again,

$$y'' = \frac{d}{dx}\left(\frac{1 - \ln x}{x^2}\right)$$

$$= \frac{x^2 \cdot \dfrac{d}{dx}(1 - \ln x) - (1 - \ln x) \cdot \dfrac{d}{dx}x^2}{x^4}$$

$$x^2 \cdot \left(-\frac{1}{x}\right) - (1 - \ln x) \cdot 2x$$
$$= \frac{}{x^4}$$

$$= \frac{2\ln x - 3}{x^3}$$

At $x = e$,

$$y'' = \frac{2\ln e - 3}{e^3} < 0$$

Therefore, the point $(e, 1/e)$ is a maximum point.
Inflection points may occur where $y'' = 0$:

$$\frac{2\ln x - 3}{x^3} = 0$$

$$2\ln x = 3$$

$$\ln x = \frac{3}{2}$$

$$x = e^{3/2} \,(\doteq 4.48)$$

$$y = \frac{\ln e^{3/2}}{e^{3/2}}$$

$$= \frac{3}{2} e^{-3/2} \,(\doteq 0.33)$$

Figure 12.5

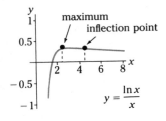

$$y = \frac{\ln x}{x}$$

The graph of f is shown in Figure 12.5. For values of x to the right of the inflection point, $y'' > 0$, so the graph is concave up in this region. The function is decreasing, but there is no minimum and its value never becomes negative. So, just as in Example C, the positive x-axis is a horizontal asymptote and

$$\lim_{x \to +\infty} \frac{\ln x}{x} = 0$$

Thus, even though $\ln x$ increases without bound as $x \to +\infty$, the factor $(1/x)$ dominates the behaviour of f for large values of x.

To the left of the inflection point, the graph of the function is concave down. As $x \to 0^+$, $f(x)$ decreases through negative values without bound. Thus,

$$\lim_{x\to 0^+} \frac{\ln x}{x} = -\infty$$

◻

Exercise 12.4

1. State the limits of the following functions:

(a) $\lim\limits_{x\to\infty} e^{-x}$

(b) $\lim\limits_{x\to\infty} (2^x - 2^{-x})$

(c) $\lim\limits_{x\to 0} \ln x^2$

(d) $\lim\limits_{x\to 0^+} \dfrac{x}{\ln x}$

2. Sketch the graphs of the functions:

(a) $y = e^x$

(b) $y = -e^x$

(c) $y = e^{-x}$

(d) $y = 1 - e^{-x}$

3. Sketch the graphs of the functions:

(a) $y = \ln x$

(b) $y = -\ln x$

(c) $y = \ln\left(\dfrac{1}{x}\right)$

(d) $y = \ln(1 - x)$

4. Determine the coordinates of the point(s) at which the tangents to the following functions have a slope equal to 1.

(a) $y = x^3$

(b) $y = e^{3x}$

(c) $y = \ln(3x)$

5. Are the following functions increasing or decreasing? Are their graphs concave up or concave down?

(a) $f(x) = 2x + e^x$

(b) $f(x) = e^{-2x} - x$

(c) $y = \ln(2x) + x$

(d) $y = \dfrac{1}{2x} - \ln x$

6. Find the relative maxima and minima, inflection points, and asymptotes, if any, and sketch the graphs of the following functions.

(a) $y = x - e^x$

(b) $y = x - \ln x$

(c) $y = xe^{-x/2}$

(d) $y = e^x/x^2$

(e) $y = x \ln x$

(f) $y = \dfrac{\ln x}{x^2}$

(g) $y = (\ln x)^2$

(h) $y = (x - 1)^2 e^x$

7. (a) Show that the function $y = x^n/e^x$ has a maximum at $x = n$ and inflection points at $x = n \pm \sqrt{n}$.

(b) What is the value of $\lim\limits_{x\to\infty} \dfrac{x^n}{e^x}$?

8. (a) Show that the function $y = (\ln x)/x^n$ has a maximum at

$$x = \exp\left(\frac{1}{n}\right)$$

and an inflection point at

$$x = \exp\left(\frac{2n + 1}{n(n + 1)}\right)$$

(b) Show that for any value of $n > 0$, the inflection point lies to the right of the maximum. What does this suggest about the value of

$$\lim_{x\to\infty} \frac{\ln x}{x^n}$$

12.5 Applications

Exponential functions are intimately related to growth and decay phenomena. Consider for instance a colony of simple organisms that reproduce by a process of cell division. In such a system, the more individuals that are present, the more there are to divide and produce new individuals. In other words, the rate of growth of the colony is proportional to its size.

The growth function, $y = f(t)$, which describes how the number of individuals in the colony changes with time, can be derived from this basic premise. As we will show below, it is an exponential function with base e. Applications of this type have led to the use of the term "natural" in reference to logarithmic and exponential functions with base e.

Suppose that the size of the colony of organisms changes from y_0 to $y_0 + \Delta y$, as the time increases by an amount Δt (Figure 12.6). Then $\Delta y/\Delta t$ is the average rate of growth of the colony during the interval of time Δt. Assuming that this quantity is proportional to the number y_0 of individuals present at the start of the interval, then

$$\frac{\Delta y}{\Delta t} = ky_0$$

or

$$\frac{y_1 - y_0}{\Delta t} = ky_0$$

where k is the constant of proportionality and y_1 is the number of individuals present at the end of the interval. Solving for y_1 gives

$$y_1 = y_0(1 + k\Delta t)$$

Figure 12.6

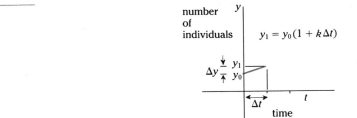

Similarly, in the next interval of time,

$$\frac{\Delta y}{\Delta t} = ky_1$$

$$\frac{y_2 - y_1}{\Delta t} = ky_1$$

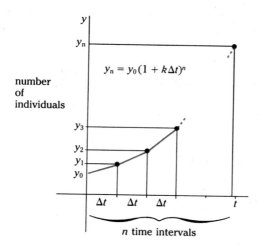

Figure 12.7

Therefore,

$$y_2 = y_1(1 + k\Delta t)$$

$$= [y_0(1 + k\Delta t)](1 + k\Delta t)$$

$$= y_0(1 + k\Delta t)^2$$

With each subsequent interval, the exponent of $(1 + k\Delta t)$ increases by 1. After n intervals (see Figure 12.7), the number of individuals present is

$$y_n = y_0(1 + k\Delta t)^n$$

$$= y_0(1 + k\Delta t)^{t/\Delta t}$$

where $t = n(\Delta t)$, the total elapsed time. Upon multiplying and dividing the exponent by k, this expression can be written

$$y_n = y_0\left((1 + k\Delta t)^{\frac{1}{k\Delta t}}\right)^{kt}$$

Now take the limit as $\Delta t \to 0$. Keep in mind that the period of growth from 0 to t consists of n intervals of length Δt. As Δt decreases in size, the number of intervals must increase correspondingly.

$$y = y_0 \lim_{\Delta t \to 0} \left((1 + k\Delta t)^{\frac{1}{k\Delta t}} \right)^{kt}$$

The quantity in parentheses has the form $(1 + u)^{1/u}$, whose limiting value, as you have seen, is the constant e. Therefore, the function which describes the growth of the colony is the natural exponential function

$$y = y_0 e^{kt}$$

The rate at which the colony grows is given by the derivative dy/dt:

$$\frac{dy}{dt} = \frac{d}{dt} \left(y_0 e^{kt} \right)$$

$$= y_0 e^{kt} \cdot \frac{d}{dx} kt$$

$$= k y_0 e^{kt}$$

or

$$\frac{dy}{dt} = ky$$

This **differential equation** expresses the **law of natural growth**, that the rate of growth at any point in time is proportional to the number of individuals present at that moment. The exponential function

$$y = y_0 \, e^{kt}$$

satisfies this requirement. It is a **solution** of the differential equation.

EXAMPLE A

The number of bacteria in a certain culture increases with time according to the equation

$$y = 1000e^{3t}$$

Determine:

(a) the number present initially and after 2 hours.

(b) the initial rate of growth and the rate of growth after 2 hours.

SOLUTION

(a) Initially, when $t = 0$, $y = 1000$.
Later, when $t = 2$, $y = 1000e^6$.
Thus, there are 1000 bacteria present initially. After 2 hours, there are approximately 403 000 present.

(b) The rate of growth is

$$\frac{dy}{dx} = \frac{d}{dt} 1000e^{3t}$$

$$= 3000e^{3t}$$

$$y'(0) = 3000$$

$$y'(2) = 3000e^6$$

Initially, at $t = 0$, the culture grows at a rate of 3000 bacteria per hour. After 2 hours, the growth rate has increased to approximately 1.2 million bacteria per hour. Keep in mind that these numbers represent instantaneous rates of growth. As time passes, not only does the number of individuals increase, but the rate of increase changes.

Although the number of individuals must be a whole number, and can only change in discrete steps, observe that this number is represented by a continuous variable y in this example. This is a common approximation in mathematical models of this type. ❑

EXAMPLE B

If the population of the town of Newport grows at a rate of 5% per year, how long will it take to double in size?

SOLUTION

The population growth is governed by

$$\frac{dy}{dt} = ky \quad \text{and} \quad y = y_0 e^{kt}$$

A growth of 5% in one year means that when $t = 1$, $y = 1.05y_0$:

$$1.05 y_0 = y_0 e^{k(1)}$$

$$1.05 = e^k$$

$$k = \ln(1.05)$$

$$\doteq 0.0296$$

If in t years, the population has doubled, then $y = 2y_0$, and

$$2y_0 \doteq y_0 e^{(0.0296)t}$$

$$2 \doteq e^{(0.0296)t}$$

$$\ln(2) \doteq (0.0296)t$$

$$t \doteq \frac{\ln(2)}{0.0296}$$

$$\doteq 23.4$$

Therefore, the population doubles in 23.4 years.   □

EXAMPLE C

The **half-life** of a radioactive substance is the time required for half of it to decay. It is known from experiment that the half-life of Strontium-90 is 29 years. How long would it take for 90% of a given amount of Sr^{90} to decay?

SOLUTION

The rate of decay of a radioactive substance is proportional to the amount of the substance present. Therefore,

$$\frac{dy}{dt} = ky \quad \text{and} \quad y = y_0 e^{kt}$$

In 29 years, $y = \frac{1}{2}y_0$, so

$$\frac{1}{2}y_0 = y_0 e^{k(29)}$$

$$0.5 = e^{29k}$$

$$\ln(0.5) = 29k$$

$$k = \frac{\ln(0.5)}{29}$$

$$\doteq -0.0239$$

(In general, the value of k is negative in problems of decay.) When 90% of the Sr^{90} has decayed, then 10% is left. Therefore, in t years,

$$0.1y_0 = y_0 e^{(-0.0239)t}$$

$$\ln(0.1) = -0.0239t$$

$$t = \frac{\ln(0.1)}{-0.0239}$$

$$\doteq 96.3$$

Therefore, it takes approximately 96.3 years for 90% of an amount of Sr^{90} to decay. ☐

Note that in Examples B and C, it is the fractional or percent change which is the relevant quantity. The doubling time and the half-life are not dependent on the actual initial amounts.

There are many other applications of exponential and logarithmic functions. In business and commerce, they may be suitable for describing how costs or profits depend on production. In science and engineering, they are required for describing how certain physical quantities change with time. Further examples will be found in the exercises.

Exercise 12.5

1. Show that the exponential function

$$y = y_0 2^{kt}$$

does *not* satisfy the differential equation for natural growth

$$\frac{dy}{dt} = ky$$

2. The rate of natural growth of a city is proportional to the population. If the population increases from 40 000 to 60 000 in 40 years, when will the population be 80 000?

3. Twenty percent of a radioactive substance disappears in 20 seconds. Find the half-life of the substance.

4. If the half-life of radium is 1690 years, what percent of the amount present now will have decayed (a) in 100 years, and (b) in 1000 years?

5. The amount of a radioactive substance diminishes with time according to the natural law of radioactive decay:

$$\frac{dy}{dx} = -ky$$

(a) Show that the amount y remaining after t years is

$$y = y_0 e^{-kt}$$

where y_0 is the amount present initially.

(b) Show that its half-life is given by

$$t = \frac{\ln 2}{k}$$

6. Lasertronics, which manufactures compact disk systems, finds that its monthly revenue R is

given by the equation

$$R(x) = 100[200x - 50x \ln(x)]$$

where x is the selling price of its product, in dollars. The equation is considered valid for values of x between 15 and 25. At what price should the company sell its product in order to maximize its monthly revenue? What is the maximum monthly revenue?

7. The Sundew Company estimates that the cost of producing x thousand cans of orange juice per month is given by the equation

$$C(x) = 1000x - 1000 \ln(x - 19) - 18\,000$$

This equation is considered valid for values of x between 19.5 and 21. Find

(a) the number of thousands of cans of juice that the company should produce each month to minimize its monthly costs

(b) the minimum monthly production cost

(c) the minimum production cost per can of juice.

8. A biologist finds that when a bacteria culture is treated with an experimental drug, the number of viable bacteria remaining after t hours is given by

$$f(t) = N\left[\frac{\exp(t) + \exp(3 - t)}{1 + e^3} \right]$$

where N is the number originally present. After applying the drug to a colony of 80 000 bacteria, how long does it take for the number of viable bacteria to reach a minimum? When the population is minimum, how many viable bacteria are present in the culture?

9. The Ministry of Health has determined that t days after detection of a communicable disease, the percent P of a city's population that will be infected is given by the equation

$$P(t) = 10t \exp(-t/10)$$

for $1 \le t \le 15$. How many days after detection will the maximum percent of the population be infected? What is the maximum percent of the population that will be infected at any given time?

10. Newton's law of cooling states that the rate at which the temperature of a body changes is proportional to the difference between its temperature, T, and that of the surrounding medium, T_0:

$$\frac{dT}{dt} = k(T - T_0)$$

(a) Show that the function $T = Ce^{kt} + T_0$ satisfies this equation.

(b) If a body in air at a temperature of $0°$ cools from $200°$ to $100°$ in 4 minutes, how many more minutes will it take the body to cool to $50°$?

11. If P dollars are invested at an annual interest rate of $r \times 100\%$, and the interest is credited to the account at the end of one year, the value of the investment at that time would be $A = P(1 + r)$. If the accumulated interest is credited to the account more frequently, then it is said to be *compounded semi-annually*, or *monthly*, or *daily*, as the case may be. The more frequently it is compounded, the better it is for the investor, since more of the interest is itself earning interest.

(a) Show that if interest is compounded n times a year at equally spaced intervals, then the value of the investment A after t years is

$$A = P\left(1 + \frac{r}{n}\right)^{nt}$$

(b) One can imagine interest to be compounded each hour, each minute, and so on, with n increasing as the compounding interval decreases. In the limit, it is possible to conceive of interest being *compounded continuously*, that is, at each instant of time. Show in this case that the value of the investment in (a) is given by

$$A = \lim_{n \to +\infty} P\left(1 + \frac{r}{n}\right)^{nt}$$

$$= Pe^{rt}$$

12. (a) If $1000 is invested at 9% per year compounded continuously (Problem 11), what will the investment be worth after 5 years?

(b) How much should be invested now at 9% per year compounded continuously, in order to have an investment worth $10 000 in 10 years?

13. When a diver swims beneath the surface of the water, the increase in pressure causes some of the nitrogen in the air breathed to dissolve in the body tissues. The amount of nitrogen present in the tissues at any time is expressed in terms of its *partial pressure P*—the pressure it would exert, if it alone were present. The rate at which P changes when a diver dives is proportional to the difference between the partial pressure of nitrogen in the air breathed at a given depth, P_a, and the instantaneous value of P according to the differential equation:

$$\frac{dP}{dt} = k(P_a - P)$$

(a) Show that a solution of this equation is

$$P = P_0 + (P_a - P_0)(1 - e^{-kt})$$

where P_0 is the initial value of P, say, at the surface at the start of the dive.

(b) A diver swims from the surface, at which $P_0 = 0.8$ atm to a depth at which $P_a = 2.4$ atm (about 20 m). Given that $k = 0.0433$, make a graph of P (atm) as a function of t (min). How many minutes does it take for the partial pressure of nitrogen in the tissues to increase to 90% of the value of P_a at that depth?

Summary and Review   12.6

The rules for differentiating the logarithm and exponential functions are

$$\frac{d}{dx}\log_a v = \frac{1}{v}\log_a e \cdot \frac{dv}{dx}$$

$$\frac{d}{dx}a^v = a^v \ln a \cdot \frac{dv}{dx}$$

In the event that the function v is just x itself, then the factor dv/dx is 1. Moreover, if the base is e, the logarithmic factor disappears. The rules then reduce to

$$\frac{d}{dx}\ln x = \frac{1}{x}$$

$$\frac{d}{dx}e^x = e^x$$

The limit properties of these functions are the following:

$$\lim_{x \to +\infty} a^x = \begin{cases} +\infty, & a > 1 \\ 1, & a = 1 \\ 0, & 0 < a < 1 \end{cases}$$

$$\lim_{x \to +\infty} \ln x = \infty$$

$$\lim_{x \to 0+} \ln x = -\infty$$

The following limits, stated without proof, show how the behaviour of the logarithmic and exponential functions compare to that of powers of x.

$$\lim_{x \to +\infty} \frac{x^n}{e^x} = 0 \qquad\qquad e^x \text{ is "stronger"}$$

$$\lim_{x \to +\infty} \frac{\ln x}{x^n} = 0 \qquad\qquad x^n \text{ is "stronger"}$$

$$\lim_{x \to 0+} \frac{\ln x}{x^n} = -\infty \qquad\qquad \ln x \text{ is "stronger"}$$

Exercise 12.6

1. Find $\dfrac{dy}{dx}$:

(a) $y = e^{x \ln x}$

(b) $y = \dfrac{e^x}{\ln x}$

(c) $y = \ln(4 - x)$

(d) $y = \ln(\cos x)$

(e) $y = \ln(1 - xe^{-x})$

(f) $y = e^{\sqrt{1 + x^2}}$

(g) $y = e^{1/x^2}$

2. Find the equation of the line tangent to the graph of $y = 2 \ln x + 2x$ at the point for which $x = 2$. Sketch the graph of the function and the tangent line.

3. At what value of m is the line $y = mx$ through the origin tangent to the curve $y = e^x$? What are the coordinates of the point of tangency?

4. Graph the function in each of the following sets on the same axes and compare their graphs:

(a) (i) $y = e^x$ (ii) $y = 2e^x$ (iii) $y = e^{2x}$

(b) (i) $y = \ln x$ (ii) $y = 2\ln x$ (iii) $y = \ln 2x$

5. Find the inflection points and the relative maxima and minima and sketch the graphs of the following functions:

(a) $y = e^{x^2} + \dfrac{x}{2} - 1$

(b) $y = xe^x$

(c) $y = \dfrac{e^x}{x}$

(d) $y = xe^{x^2}$

(e) $y = x^2 \exp\left(-\dfrac{x^2}{4}\right)$

(f) $y = e^{1/x}$
Hint: check left and right limits as $x \to 0$.

6. Find the inflection points and the relative maxima and minima and sketch the graphs of the following functions:

(a) $y = x(\ln x)^2$

(b) $y = \dfrac{1}{4}x^2(2 \ln x - 3)$

(c) $y = \ln\left(\dfrac{x^2}{x-1}\right)$

(d) $y = \ln\left(\dfrac{x}{1+x^2}\right)$

7. Compare the graphs of

$y = \ln(4-x^2)$ and $y = \ln(4x - x^2)$

8. Sketch the graph of

(a) $y = e^{-x/4}\sin x$, $x > 0$

9. The population of a developing country has doubled in the past 30 years. Its present population is 15 million. If it continues to grow at the same rate, in how many years will the population reach 20 million?

10. Terrific Toys Company, finds that its monthly profit P in dollars is approximated by the function

$P(x) = 10\,000(x-4)^2\exp(x-4) + 1$

where x is the amount in hundreds of dollars spent on advertising each month. The equation is valid for values of x between 1 and 4. How much should the company spend on advertising each month to maximize its monthly profit? What is the maximum monthly profit?

11. A lake is treated by bactericidal agents to reduce pollution due to coliform bacteria. Officials estimate that t days after treatment, the number N of viable bacteria per millilitre will be given approximately by the equation

$N(t) = 24t - 120\ln\left(\dfrac{t}{5}\right) - 70$

How many days after treatment will the minimum number of viable bacteria per millilitre be present? What is this minimum number?

12. Show that $y = e^x$ is a solution of the differential equation $y'' - 2y' + y = 0$. Show that $y = xe^x$ is also a solution. Show that $y = c_1e^x + c_2xe^x$, where c_1 and c_2 are arbitrary constants, is also a solution. (Differential equations of this sort occur in the theory of damped vibration, that is, vibrations with continuously decreasing amplitude.)

13. A fundamental function of mathematical statistics called the normal distribution has the following form:

$f(x) = \dfrac{1}{\sqrt{2\pi}\sigma}\exp\left[-\dfrac{1}{2}\left(\dfrac{x-\bar{x}}{\sigma}\right)^2\right]$

where $\bar{x}$ and σ are constants known, respectively, as the mean and the standard deviation.

(a) Locate the inflection points and the relative maxima and minima.

(b) Find the limit as $x \to \pm\infty$.

(c) Sketch the graph of f.

Differentials

13

I n Chapter 3, the expression dy/dx was introduced as a single symbol for the derivative of a function. Up to now, the dy and the dx in this expression did not have any separate meanings. The quantities dy and dx are known as **differentials**. In what follows, you will see that it is possible to regard a derivative as the ratio of dy to dx. This point of view leads to a method for approximating the value of a function, to an investigation of how small changes in variables are related, and to an analysis of measurement errors.

The Differential of a Function   13.1

The symbol dx is called the **differential of x**. It is a quantity which, like Δx, represents an arbitrary change in the variable x. As a general rule, dx and Δx are taken to be equal in value. The **differential dy of a function $y = f(x)$** is defined as the product of the derivative of f and dx:

$$dy = f'(x) \cdot dx$$

The graph of f with a tangent of x is shown in Figure 13.1. In this

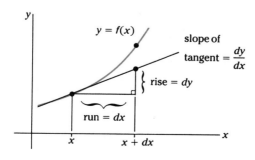

Figure 13.1

diagram, dy and dx can be interpreted, respectively, as the rise and the run of the tangent line at x, whose slope is the derivative dy/dx.

It is important to understand the distinction between the increment Δy and the differential dy. A comparison of the two is shown in Figure 13.2. For a given change Δx in x, the differential dy is the change in the y-coordinate of the tangent line, whereas the increment Δy is the change in the value of the function from $f(x)$ to $f(x + \Delta x)$. In other words,

when the independent variable changes
    from x to $x + \Delta x$
        or $x + dx$

the value of the function changes
    from y to $y + \Delta y$ where $\Delta y = f(x + \Delta x) - f(x)$

and the y-coordinate of a point on the tangent changes
    from y to $y + dy$ where $dy = f'(x)\,dx$

Figure 13.2

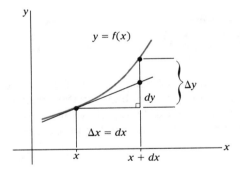

EXAMPLE A

For a given change $\Delta x(= dx)$ in x, find dy and Δy for the function

$$f(x) = 3x^2 - 8x + 2$$

SOLUTION

The differential dy is found by multiplying the derivative by dx:

$$f'(x) = 6x - 8$$

$$dy = f'(x)\,dx$$

$$= (6x - 8)\,dx$$

The increment Δy, on the other hand, is

$$\Delta y = f(x + \Delta x) - f(x)$$

$$= (3(x + \Delta x)^2 - 8(x + \Delta x) + 2) - (3x^2 - 8x + 2)$$

$$= 3x^2 + 6x(\Delta x) + 3(\Delta x)^2 - 8x - 8(\Delta x) + 2 - 3x^2 + 8x - 2$$

$$= 6x(\Delta x) - 8(\Delta x) + 3(\Delta x)^2$$

$$= (6x - 8)\Delta x + 3(\Delta x)^2 \qquad \square$$

Observe that the difference between Δy and dy is a term of second order in Δx.

EXAMPLE B

Given $f(x) = 3\sqrt{x}$.

(a) Find the values of Δy and dy, when $x = 2$, and $\Delta x = dx = 3$.

(b) Make a sketch of the function $f(x) = 3\sqrt{x}$ and of the tangent to f at $x = 2$, showing Δy and dy in the diagram.

SOLUTION

$$\Delta y = f(x + \Delta x) - f(x)$$

$$= 3\sqrt{x + \Delta x} - 3\sqrt{x}$$

$$= 3\sqrt{2 + 3} - 3\sqrt{2}$$

$$= 3\sqrt{5} - 3\sqrt{2}$$

$$\doteq 2.466$$

$$dy = f'(x)\,dx$$

$$= \frac{d}{dx}\left(3\sqrt{x}\right) \cdot dx$$

$$= \frac{3}{2\sqrt{x}}dx$$

$$= \frac{3}{2\sqrt{2} \cdot 3}$$

$$= \frac{9}{2\sqrt{2}}$$

$$\doteq 3.182$$

The graph of f, the tangent at $x = 2$, and lengths corresponding to Δy and dy are shown in Figure 13.3.

Figure 13.3

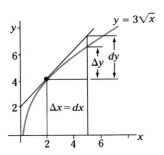

One application of differentials is in approximating the values of a function. The difference between the value of a function at x and its value nearby at $x + \Delta x$ is

$$\Delta y = f(x + \Delta x) - f(x)$$

This relationship can be rewritten

$$f(x + \Delta x) = f(x) + \Delta y$$

When Δx is small, however, as the diagram in Figure 13.4 suggests, the increment Δy is nearly equal to the differential dy. Indeed, their difference is zero in the limit as $\Delta x \to 0$. Thus, near x, $f(x + \Delta x)$ can be approximated by using dy instead of Δy:

$$f(x + \Delta x) \doteq f(x) + dy$$

$$f(x + \Delta x) \doteq f(x) + f'(x)\,dx$$

Figure 13.4

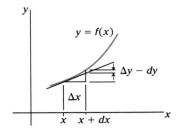

Observe that in the neighbourhood of x, the graph of the function is almost the same as the graph of the tangent at x. The expression $f(x) + f'(x)\,dx$ on the right above is the y-coordinate of the point of the tangent line at $x + \Delta x$. In effect, this approximation is using the y-coordinates of points on the tangent line to approximate the values of the function.

The differential dy of the function $y = f(x)$ thus represents the *approximate* change in the value of f due to a small change dx in x. The smaller the value of dx, the better the approximation is.

EXAMPLE C

Given $f(x) = x^2 - 6x + 11$. Find an approximate value for $f(4.25)$ using differentials. Compare to the exact value of $f(4.25)$.

SOLUTION

Since

$$f(x) = x^2 - 6x + 11$$

Then

$$f'(x) = 2x - 6$$

and

$$f(x + \Delta x) \doteq f(x) + f'(x)\,dx$$

$$= (x^2 - 6x + 11) + (2x - 6)\,dx$$

Since 4.25 is near 4, approximate $f(4.25)$ by taking $x = 4$, and $dx = \Delta x = 0.25$

$$f(4.25) = (4^2 - 6 \cdot 4 + 11) + (2 \cdot 4 - 6)(0.25)$$

$$= 3.5$$

The exact value is

$$f(4.25) = (4.25)^2 - 6(4.25) + 11$$
$$= 3.5625$$

Therefore, the approximate value is 0.0625 less than the exact value, which is a percentage error of about 2%. ❏

EXAMPLE D

Using differentials, approximate the value of $\sin 44°$.

SOLUTION

$$\sin(x + \Delta x) \doteq \sin x + \left(\frac{d}{dx}\sin x\right)dx$$

$$= \sin x + (\cos x)\,dx$$

Since $45° = \pi/4$ radians and the value of $\sin \pi/4$ is exactly equal to $\sqrt{2}/2$, choose

$$x = \frac{\pi}{4},$$

Then

$$44° = 45° - 1°$$

$$= \frac{\pi}{4} - \frac{\pi}{180}$$

$$\therefore \quad dx = -\frac{\pi}{180}(= \Delta x)$$

Thus

$$\sin 44° = \sin(45° - 1°)$$

$$= \sin\left(\frac{\pi}{4} - \frac{\pi}{180}\right)$$

$$\doteq \sin\frac{\pi}{4} + \left(\cos\frac{\pi}{4}\right)\left(-\frac{\pi}{180}\right)$$

$$= \frac{\sqrt{2}}{2} + \frac{\sqrt{2}}{2}\left(-\frac{\pi}{180}\right)$$

$$\doteq 0.69477$$

Compare this result to $\sin 44° = 0.69466$ found with a calculator. ☐

Exercise 13.1

1. Find the differential dy and the increment Δy:

(a) $y = x + \sqrt{x - 5}$

(b) $y = x^3 - 2$

(c) $y = \dfrac{3}{x + 2}$

(d) $y = e^{2x}$

2. $A(x)$ is the area of a square of side x.

(a) Find an expression for each of the quantities: $dA, (A + dA), A(x + \Delta x)$, and ΔA, and $(\Delta A - dA)$.

(b) Draw a square of side $(x + \Delta x)$. Shade in the area corresponding to each of the quantities in (a).

(c) Graph the function $A(x) = x^2$. Draw the tangent at some point $(x, A(x))$. On this graph identify each of the quantities in (a).

3. Find the differential dy:

   (a) $f(x) = (x^2 - a^2)^3$

   (b) $f(x) = 4\sin^2 2x$

   (c) $f(x) = \dfrac{2 - x}{3 - 4x}$

   (d) $f(x) = x\ln(3x^2)$

4. Find an expression which can be used to approximate the value of $f(x + \Delta x)$ in each of the following cases:

   (a) $f(x) = 4 - 2x^3$

   (b) $f(x) = \dfrac{x}{\sqrt{x^2 - 1}}$

   (c) $f(x) = \cos 2x$

   (d) $f(x) = 5xe^{-x}$

5. Using differentials, approximate the value of

   (a) $\sqrt{99}$

   (b) $(3.1)^5$

   (c) $\sqrt[4]{83}$

   (d) $\dfrac{1}{\sqrt[3]{120}}$

   (e) $\cos 59°$

   (f) $\tan 63°$

6. If $y = x^6$, determine the approximate change in y if x changes from 3 to 3.002.

7. Given that $f(x) = x^{2/5}$, estimate the change in f if x is decreased from 1 to 9/10.

8. Sketch the curve $y = x^2/4$. Compute both dy and Δy corresponding to $x = 3$ and $\Delta x = 0.25$. Show both in the diagram.

9. Show that if h is small, $\sqrt[n]{1 + h}$ is approximately equal to $1 + (h/n)$.

Applications   13.2

The relation $dy = f'(x)dx$ between the differentials dx and dy, where $y = f(x)$, shows how *changes* in y are related to *changes* in x. It is to be understood that the changes in question are relatively small: the smaller the change, the better the value of dy approximates the value of Δy. To judge the relative size of a change, it may be expressed as a percentage of the original value of the variable.

EXAMPLE A

What volume of metal is required to manufacture a hollow cylindrical shell of length 18 m, radius 4 m, and wall thickness 0.02 m?

SOLUTION

The volume of metal is equal to the amount by which the volume of a cylinder changes when its radius increases from 4 m to 4.02 m. Using dV as an approximation to this,

$$V = \pi r^2 l$$

$$dV = \frac{dV}{dr} \cdot dr$$

$$= (2\pi r l)(dr)$$

Taking $r = 4$ and $dr = 0.02$,

$$dV = 2\pi(4)(18)(0.02)$$

$$\doteq 9.05$$

Therefore, 9.05 m³ of metal are required.   ❏

EXAMPLE B

The total cost in dollars of producing x barrels of a pesticide is given by

$$C(x) = 0.002x^3 - 0.5x^2 + 50x$$

If production were to be increased from 425 to 450 barrels per week:
(a) What increase in production costs should be expected?
(b) What would be the percentage increases in production and in costs?

SOLUTION

(a) The change in the cost is the differential dC. It is related to dx, the increase in production by

$$dC = C'(x)\,dx$$

$$= (0.006x^2 - x + 50)\,dx$$

Taking $x = 425$ and $dx = 25$,

$$dC = \left(0.006(425)^2 - 425 + 50\right)(25)$$

$$= \$17\,719$$

Thus, it will cost an additional \$17 719 per week, approximately, to increase production by 25 barrels per week.

(b) The total cost of producing 425 barrels per week is

$$C(425) = 0.002(425)^3 - 0.5(425)^2 + 50(425)$$

$$= \$84\,469$$

Relative changes in production and costs are

$$\frac{dx}{x} = \frac{25}{425}$$

$$= 0.059 \text{ or } 5.9\%$$

$$\frac{dC}{C} = \frac{17\,719}{84\,469}$$

$$= 0.21 \text{ or } 21\%$$

Thus, a 5.9% increase in production would increase costs by 21%. ◻

Error Analysis

It is generally not possible to measure a quantity exactly, due to limitations in measuring instruments and many other factors. Consequently, measurements are usually expressed in the form $x \pm \Delta x$, where x is the measured value and Δx is the degree of uncertainty or error in the measurement. This notation implies that the true value lies somewhere in the interval between $x - \Delta x$ and $x + \Delta x$.

A measured value x may be used to calculate the value of a related quantity y. It is of interest to determine the error Δy in the calculated value that results from the measurement error Δx.

EXAMPLE C

The diameter of a cylindrical pipe is found to be 4.8 cm with an error of ± 0.05 cm. What is the error in the computed cross-sectional area?

SOLUTION

The area is calculated from

$$A = \frac{\pi D^2}{4}$$

With $D = 4.8$,

$$A = \frac{\pi(4.8)^2}{4}$$

$$\doteq 18.1$$

The error ΔA is the amount by which the area of the circular cross section changes as the diameter varies between 4.8 and 4.85. Approximating this by dA gives

$$dA = \frac{\pi D}{2} dD$$

With $D = 4.8$ and $dD = 0.05$, the error in the computed area is

$$dA = \frac{\pi(4.8)}{2}(0.05)$$

$$\doteq 0.4$$

The value of the cross-sectional area is therefore expressed as

$$A = 18.1 \pm 0.4\,\text{cm}^2 \qquad\qquad \square$$

The **relative error** of a quantity Q is defined as the ratio of the error ΔQ to the value of Q:

$$\frac{\Delta Q}{Q}$$

When the relative error is expressed as a percentage, then it is called the **percentage error**.

EXAMPLE D

Determine the percentage error in the cross-sectional area of the pipe of Example C.

SOLUTION

The relative error in the cross-sectional area is

$$\frac{dA}{A} \doteq \frac{0.4}{18.1}$$

$$= 0.02$$

The percentage error is therefore 2%.

Observe in this example, that it is possible to find the relative error directly by taking the natural logarithm of each side before differentiating:

$$A = \frac{\pi D^2}{4}$$

$$\ln A = \ln \frac{\pi}{4} + 2 \ln D$$

$$\frac{1}{A} \frac{dA}{dD} = 2 \frac{1}{D}$$

$$\frac{dA}{A} = 2 \cdot \frac{dD}{D}$$

This equation states that the relative error in the area is twice the relative error in the diameter. With $D = 4.8$ and $dD = 0.05$,

$$\frac{dA}{A} = 2 \cdot \frac{0.05}{4.8}$$

$$= .02 \text{ or } 2\%$$

Exercise 13.2

1. The radius of a circle is 32 cm. By approximately how much does the area increase if the radius increases by 1/16 cm?

2. Compute approximately the area of a walk 1.5 m wide around a city square that is 120 m on a side. Make a sketch showing the square and the walk, and indicate the part of the walk that is neglected in the approximation.

3. A wheat farmer whose current production is 5000 bushels per year can produce x bushels of wheat at a cost in dollars given by

$$C(x) = \frac{x^2}{10\,000} + 500$$

What is the increase in cost if the production is increased to 5500 bushels per year? What is the percentage increase?

4. How many litres of paint are required to cover the sides of a cylindrical tank of radius 15 m and height 10 m if the paint is applied with a thickness of 0.02 cm? $(1 \text{ L} = 1000 \text{ cm}^3.)$ Hint: Consider the layer of paint to be a change in volume due to a slight increase in radius.

5. (a) Derive an approximate formula for the volume in a thin spherical shell of radius r and thickness dr.

   (b) Determine the volume of air in the earth's atmosphere, assuming that the atmosphere has a uniform density and extends to a height of 40 km above the surface. The radius of the earth is 6400 km.

6. The diameter of a sphere is to be measured and its volume computed. The diameter can be measured to within ± 0.002 cm, and the allowable error in the volume is 0.5 cm³. For what size spheres is this process satisfactory?

7. Show that when θ is near 60°, the error in $\tan \theta$, due to a small error in θ, is approximately three times as large as it is when θ is near 30°.

8. For what range of values of x can $\sqrt[3]{x + 1}$ be replaced by $\sqrt[3]{x}$ with an error of not more than ± 0.01?

9. The height and diameter of a right circular cone are known to be equal. If the volume is to be determined with an error no greater than 2%,

what is the greatest percentage error that can be tolerated in determining the height?

10. Show that a 25% reduction in the radius of a circular airway corresponds approximately to a 50% reduction in the cross-sectional area of the airway and consequently to a 50% reduction in airflow.

13.3 Summary and Review

The differential dx represents a change in the variable x. The differential dy is related to dx by

$$dy = f'(x) \cdot dx$$

In contrast to this, the increment Δy is given by

$$\Delta y = f(x + \Delta x) - f(x)$$

When Δx is small, dy is a good approximation to Δy. Under these circumstances, the graph of $y = f(x)$ and the graph of the tangent to f at x differ only slightly.

To work out practical problems involving differentials, it is helpful to picture how small changes in one variable can cause corresponding small changes in another.

Exercise 13.3

1. Find dy:

(a) $f(x) = x(1 - 2x)^3$

(b) $f(x) = 3x \cos^2 x$

(c) $f(x) = \dfrac{2x^2}{\sqrt{1 + x}}$

2. Find and compare the values of dy and Δy:

(a) $f(x) = 5x - 15$, $x = 6$, $dx = 0.02$

(b) $f(x) = 2 - \dfrac{x^2}{12}$, $x = -4$, $dx = 0.1$

(c) $f(x) = \dfrac{1}{\sqrt{3x + 1}}$, $x = 8$, $dx = 0.25$

3. A circular area of radius 2.5 km is to be covered by a search and rescue squad. By approximately how much does the area increase if the radius increases by 0.5 km?

4. The volume of a sphere is to be determined with an error no greater than 1%. What is the greatest percentage error that can be tolerated in determining the diameter?

5. A cubical container must hold 1000 cm³ with an error of no more than ±0.01 cm³. What is the maximum tolerance permitted in the linear dimensions?

6. A ladder 5 m in length leans against a wall with its top resting h metres up the wall. Estimate the change in h if the angle between the ladder and the ground changes from 60° to 58°.

7. The pendulum in a grandfather clock swings with a period T (seconds) given by

$$T = 2\pi \sqrt{\frac{l}{g}}$$

where l is the length (metres) and $g = 9.8 \text{ m/s}^2$. By what percent should the length be changed to correct for a loss of 2 min/d?

8. When a metal wire of length l is heated, the coefficient of linear expansion, $\Delta l/l$, is a constant for the metal. Likewise, the volumetric coefficient of expansion, $\Delta V/V$ is constant when a cubical block of the metal is heated. What is the relationship between these two quantities?

9. (a) Show that the area of a right triangle with hypotenuse h and one acute angle θ is given by

$$A = \frac{1}{4}h^2 \sin 2\theta$$

(b) If $h = 12 \text{ cm}$, what is the error in calculating A arising from a measurement of $\theta = 30° \pm 0.5$?

10. Show that the increase in area of a sector of a circle of radius r and central angle θ is $dA = (r\theta) \cdot dr$ if r increases by a small amount dr, θ remaining constant. Draw the figure and show that this added area is approximately a rectangle with dimensions $r\theta$ and dr.

Abraham Robinson

ONE OF THE MOST remarkable developments of recent mathematics has been the rehabilitation of the infinitesimal in the calculus. Abraham Robinson, a mathematical logician who worked in the 1950s and the 1960s at the University of Toronto and Yale University, devised an approach to the calculus known as non-standard analysis that permits calculation with infinitesimals, elements that are non-zero but smaller than any finite quantity.

The idea of the differential dx as an infinitely small increment in the variable x was basic to Leibnizian analysis in the eighteenth century. In later mathematics the continuing heuristic and pedagogical value of infinitesimals was recognized. Nevertheless, it was believed that the use of infinitesimals was not quite rigorous, and that in any sound presentation of the calculus they would need to be eliminated. This belief was apparently confirmed in the nineteenth century with the emergence and acceptance of Cauchy's classical foundation based on the concept of limits.

Robinson expressed the differential and integral calculus as a consistent formal system in which infinitesimal elements of various orders appear. A formal system consists of a clearly specified mathematical language with a definite syntax and rules of inference. Robinson developed the resulting language and model into a comprehensive presentation of the calculus known as non-standard analysis. At the time of his death in 1974 many mathematicians were engaged in the research and development of this new branch of mathematics.

Abraham Robinson was a German Jew who left Germany in the 1930s to escape Fascist persecution. After years in Palestine and England he arrived in 1951 at the University of Toronto where he stayed until 1957 working on the theory of models. The department of mathematics at Toronto was the most prominent one in Canada and included such renowned figures as the applied mathematician J.L. Synge and the geometer H.M.S. Coxeter. The quiet atmosphere of the city in those days is described in the memoirs of Leopold Infeld, another distinguished visitor to the University: "It must be good to die in Toronto. The transition between life and death would be continuous, painless and scarcely noticeable in this silent town."

Integration

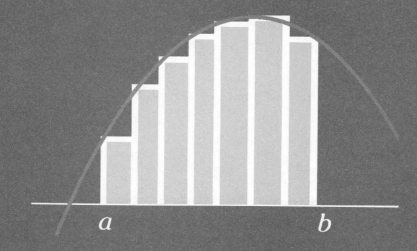

Integration— the Inverse of Differentiation

14

The central problem in the preceding chapters has been to find the derivative of a given function. In many important applications of calculus, one is concerned with the opposite problem: to find the original function when its derivative is given. Take, for example, the law of exponential growth. According to this law, the rate of growth of a population is proportional to the number y of individuals present. In mathematical language, this law is expressed as a **differential equation**:

$$\frac{dy}{dt} = ky$$

To determine how the size of the population changes with time, that is, to find an expression for y as a function of t, it is necessary to "undo" the differentiation by means of an inverse operation called **integration**.

This chapter is concerned primarily with the mechanical aspects of integration. Starting from the familiar rules of differentiation, a number of basic formulas for integration will be established. Applications and further methods of integration will be treated in subsequent chapters.

The Indefinite Integral 14.1

The problem of integration is to find the function whose derivative, call it $f(x)$ here, is given. The operation of integration is denoted by the symbol

$$\int \qquad dx$$

The function to be integrated is placed between the **integral sign** $\int$ and the **differential** dx. The original problem

$$\frac{d}{dx}\underline{\qquad ? \qquad} = f(x)$$

is thus restated as

$$\int f(x)\, dx = \underline{\qquad ? \qquad}$$

Here, the function we are seeking is called an **antiderivative** or an **integral** of $f(x)$. The function f itself is called the **integrand**.

You already know the integrals of many simple functions. They are found by inverting the usual rules for differentiation. For example,

Since $\dfrac{d}{dx}x^3 = 3x^2$
Read: the derivative
of x^3 is $3x^2$

then $\int 3x^2\, dx = x^3$
Read: an integral
of $3x^2$ is x^3

Likewise,

Since $\dfrac{d}{dx}\sin x = \cos x$
Read: the derivative
of $\sin x$ is $\cos x$

then $\int \cos x\, dx = \sin x$
Read: an integral
of $\cos x$ is $\sin x$

You can learn to recognize many integrals just by thinking of the corresponding rule for differentiation. Indeed, as you will see, most integrals are worked out by rearranging them to correspond to certain standard formulas.

Inversion of the rules for differentiation is not quite the entire story, however. The original function could contain a constant term that disappears upon differentiation. This means that any function such as $F(x) = x^3 + 5$ or $F(x) = x^3 - 3$, for instance, could be an integral of $3x^2$:

$$\int 3x^2\, dx = x^3 + 5$$

or $\int 3x^2\, dx = x^3 - 3$

or $\int 3x^2\,dx = x^3 \pm some\ other\ constant$

The functions $F(x) = x^3 + 5$ and $F(x) = x^3 - 3$ are members of a family of functions produced by applying a vertical translation to the function $F(x) = x^3$. As shown in Figure 14.1, these and any other functions so generated all have the same derivative. This is why each of them is an integral of $3x^2$. Therefore, the result of an integration is commonly written as a function plus an arbitrary constant C as follows:

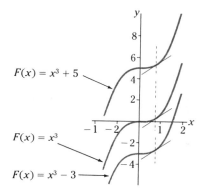

Figure 14.1

$F(x) = x^3 + 5$

$F(x) = x^3$

$F(x) = x^3 - 3$

$\int 3x^2\,dx = x^3 + C$ because $\dfrac{d}{dx}\left(x^3 + C\right) = 3x^2$

and in general

$\int f(x)\,dx = F(x) + C$ because $\dfrac{d}{dx}\left(F(x) + C\right) = f(x)$

Such an integral is called an **indefinite integral**. In order to evaluate C and thus single out a *particular* member of the family of functions, additional information is needed. As you will see in Chapter 15, this additional information is expressed as an **initial condition** or **boundary condition**, depending on the problem at hand.

EXAMPLE A

$\int 5x^4\,dx$

SOLUTION

Think of the rule for differentiating a power of x and recall that the differentiation reduces the exponent by 1. Consider, therefore, the derivative of x^5. Since

$$\frac{d}{dx}x^5 = 5x^4$$

then

$$\int 5x^4\,dx = x^5 + C$$ ❑

EXAMPLE B

$$\int x^{-3}\,dx$$

SOLUTION

Since differentiating a power of x reduces the exponent by 1, integrating a power of x must increase it by 1, from -3 to -2 in this case. Consider, therefore, the derivative of x^{-2}:

$$\frac{d}{dx}x^{-2} = (-2)x^{-3}$$

The integrand in this example does not contain a factor (-2), however, so there must be a factor $(-1/2)$, which cancels the factor (-2) that comes from the exponent. Therefore, from the derivative

$$\frac{d}{dx}\left(-\frac{x^{-2}}{2}\right) = x^{-3}$$

it follows that

$$\int x^{-3}\,dx = -\frac{x^{-2}}{2} + C$$ ❑

EXAMPLE C

$$\int \sec^2 x\,dx$$

SOLUTION

Think of a function whose derivative is $\sec^2 x$.

$$\frac{d}{dx}\underline{\quad\quad ?\quad\quad} = \sec^2 x$$

It is $\tan x$ which gives this result. Therefore,

$$\int \sec^2 x \, dx = \tan x + C$$ ☐

EXAMPLE D

Find a function F that satisfies the following conditions:

$$F'(x) = 3x^2 - 5x \qquad \text{and} \qquad F(2) = 4$$

SOLUTION

Since

$$\frac{d}{dx}F(x) = 3x^2 - 5x$$

$$F(x) = \int (3x^2 - 5x) \, dx$$

$$= \left(x^3 - \frac{5}{2}x^2\right) + C \quad \longleftarrow \quad \text{check by differentiating}$$

At $x = 2$,

$$F(2) = 2^3 - \frac{5}{2}(2)^2 + C$$

$$= 4$$

$$\therefore \quad C = 6$$

$$\therefore \quad F(x) = x^3 - \frac{5}{2}x^2 + 6$$ ☐

Exercise 14.1

1. What type of function is each of the following?
   (a) the derivative of a quadratic function
   (b) the integral of a quadratic function
   (c) the derivative of a linear function
   (d) the integral of a linear function

2. State two different integrals of the function $f(x) = -2x^3$. How are they related?

3. What functions have the following derivatives?
   (a) $F'(x) = 4x^3$

   (b) $F'(x) = e^x$

   (c) $F'(x) = \cos x$

   (d) $F'(x) = 1/x$

4. Calculate the slopes of the following functions at $x = 3$. Show graphically why they are all equal.

   (a) $f(x) = x^2 - 2x$

   (b) $g(x) = (x - 1)^2$

   (c) $h(x) = (x - 4)(x + 2)$

5. Find the function which passes through the point $(3, -4)$, and whose derivative is the same as the derivative of $F(x) = x^2 - 2x$.

6. Determine the following integrals. Verify by differentiating the result.

(a) $\int 3x^2\,dx$

(b) $\int \sin x\,dx$

(c) $\int \sec x \tan x\,dx$

(d) $\int (2x + 3)\,dx$

(e) $\int (4 - x^4)\,dx$

(f) $\int (2x^3 - 4x^2)\,dx$

7. Find the functions which have the following derivatives and which satisfy the additional conditions.

(a) $F'(x) = 5x^4$ $\qquad F(0) = 3$

(b) $F'(x) = \sin x$ $\qquad F(\pi/2) = 1$

(c) $G'(x) = x - 4$ $\qquad G(2) = -3$

(d) $G'(x) = e^x$ $\qquad G(0) = 2$

14.2 Basic Integration Formulas

The first four integration formulas that we shall introduce are those used for integrating powers of x and polynomial functions. To begin with, recall the sum rule for derivatives:

$$\frac{d}{dx}\Big(f(x) + g(x)\Big) = \frac{d}{dx}f(x) + \frac{d}{dx}g(x)$$

The corresponding integration formula is

1. $\int \Big(f(x) + g(x)\Big)\,dx = \int f(x)\,dx + \int g(x)\,dx$

According to this formula, the integral of a sum of functions is equal to the sum of the integrals of the functions.

The rule for the derivative of a function multiplied by a constant is

$$\frac{d}{dx}\Big[a \cdot f(x)\Big] = a \cdot \frac{d}{dx}f(x)$$

The corresponding integration formula is

2. $\int a f(x)\,dx = a \int f(x)\,dx$

which means that the integral of a constant times a function is equal to the constant times the integral of the function. In other words, a constant factor can be removed from an integral just as it can be removed from a derivative. In the exercises for this section, you can convince yourself of the validity of Formulas 1 and 2.

There are two formulas for integrating a power of x. Which one is used depends on whether or not the exponent is -1:

$$3. \quad \int x^n dx = \frac{x^{n+1}}{n+1} + C \qquad (n \neq -1)$$

$$4. \quad \int \frac{dx}{x} = \ln x + C \qquad (x > 0)$$

$$\text{(If } x < 0 \text{ write } \ln |x| + C)$$

It is essential that these formulas be well understood. Study the following examples carefully.

EXAMPLE A

$$\int x^6 \, dx$$

SOLUTION

According to Formula 3, integrating a power of x increases the exponent by 1: In this case, n is 6 and $n + 1$ is 7:

$$\int x^6 \, dx = \frac{x^7}{7} + C$$

EXAMPLE B

$$\int (5x^2 + 4x) \, dx$$

SOLUTION

Formulas 1, 2, and 3 together make it possible to integrate any polynomial function. Using each formula in turn,

$$\int (5x^2 + 4x) \, dx = \int 5x^2 \, dx + \int 4x \, dx \qquad \longleftarrow \quad \text{Formula 1}$$

$$= 5 \int x^2 \, dx + 4 \int x \, dx \qquad \longleftarrow \quad \text{Formula 2}$$

$$= 5 \cdot \frac{x^3}{3} + 4 \cdot \frac{x^2}{2} + C \qquad \longleftarrow \quad \text{Formula 3}$$

$$= \frac{5}{3} x^3 + 2x^2 + C \qquad \longleftarrow \quad \substack{\text{check by} \\ \text{differentiating}}$$

It is important to understand that Formulas 3 and 4 are patterns for integration that do not depend on the variable used. Thus,

$$\int v^n \, dv = \frac{v^{n+1}}{n+1} + C$$

where v could be x itself or any function of x. This means that it is possible to integrate a function to a constant power if its differential $dv = (dv/dx)dx$ is also present in the integrand. Examine the following integrals carefully:

$$\int x^2 \cdot dx = \frac{x^3}{3} + C \qquad\qquad\qquad \text{Here, } v = x, \qquad dv = dx$$

$$\int (2x+1)^2 \cdot 2 \, dx = \frac{(2x+1)^3}{3} + C \quad \text{Here, } v = 2x+1, \quad dv = 2 \, dx$$

$$\int \sin^2 x \cdot \cos x \, dx = \frac{\sin^3 x}{3} + C \qquad \text{Here, } v = \sin x, \qquad dv = \cos x \, dx$$

$$\int (e^x)^2 \cdot e^x \, dx = \frac{(e^x)^3}{3} + C \qquad\qquad \text{Here, } v = e^x, \qquad dv = e^x \, dx$$

$$\int (\ln x)^2 \cdot \frac{1}{x} dx = \frac{(\ln x)^3}{3} + C \qquad\quad \text{Here, } v = \ln x, \qquad dv = \frac{1}{x} \, dx$$

All of these integrals consist of some function $v(x)$ squared, multiplied by the differential of v. Although the integrals look quite different, they all have the form $\int v^2 \, dv$, and therefore all are integrated in exactly the same way.

EXAMPLE C

$$\int (2x^2 + 3x + 4)^3 (4x + 3) \, dx$$

SOLUTION

The key is to recognize that this integral is in the form $\int v^n \, dv$. Taking

$$v = (2x^2 + 3x + 4)$$

then,

$$dv = (4x + 3) \, dx$$

and the integral has exactly the form

$$\int v^3 \, dv$$

Consequently,

$$\int (2x^2 + 3x + 4)^3 (4x + 3) \, dx = \frac{(2x^2 + 3x + 4)^4}{4} + C \qquad \square$$

EXAMPLE D

$$\int (4x^2 - 1)^2 \, dx$$

SOLUTION

It is of great importance to realize that Formula 3 does *not* say that

$$\int (4x^2 - 1)^2 \, dx = \frac{(4x^2 - 1)^3}{3} + C \qquad \textit{incorrect}$$

If v is chosen to be $4x^2 - 1$, then $dv = 8x \, dx$. The integral does not contain $8x \, dx$, however. It is not in the form $\int v^n \, dv$, and therefore, Formula 3 does not apply. The problem is solved by expanding the integrand:

$$\int (4x^2 - 1)^2 \, dx = \int (16x^4 - 8x^2 + 1) \, dx$$

$$= \frac{16x^5}{5} - \frac{8x^3}{3} + x + C \qquad \square$$

EXAMPLE E

$$\int x^3 \sqrt{3x^4 + 4} \, dx$$

SOLUTION

First rewrite the integral:

$$\int (3x^4 + 4)^{1/2} x^3 \, dx$$

Taking $v = (3x^4 + 4)$, then $dv = 12x^3 \, dx$

It is clear that the integral lacks only the constant 12 to be in the form $\int v^n \, dv$. Constants such as this are readily supplied. Multiply the integrand by 12/12. This does not change its value:

$$\int (3x^4 + 4)^{1/2} \frac{12x^3}{12} dx$$

Removing the factor $(1/12)$ now leaves the integral exactly in the form $\int v^n \, dv$. The full solution follows these steps:

$$\int x^3 \sqrt{3x^4 + 4} \, dx = \int (3x^4 + 4)^{1/2} x^3 \, dx$$

$$= \frac{1}{12} \int (3x^4 + 4)^{1/2} 12x^3 \, dx$$

$$= \frac{1}{12} \frac{(3x^4 + 4)^{3/2}}{3/2} + C$$

$$= \frac{1}{18} (3x^4 + 4)^{3/2} + C$$

❏

Exercise 14.2

1. Determine the following integrals using the fact that each can be expressed in the form $\int v^n \, dv$.

(a) $\int 3x^5 \, dx$

(b) $\int (4x)^3 \, dx$

(c) $\int (x + 3)^4 \, dx$

(d) $\int (3 - 2x)^2 \, dx$

(e) $\int \frac{dx}{4x^3}$

(f) $\int \sqrt{3x} \, dx$

(g) $\int \frac{4}{\sqrt{x^3}} \, dx$

(h) $\int 3x^{7.3} \, dx$

2. Determine the following integrals:

(a) $\int (4x^4 - 3x^2 - 2) \, dx$

(b) $\int \frac{x^3 + 2x^2}{\sqrt{x}} \, dx$

(c) $\int (2 - 3x)\sqrt{x} \, dx$

(d) $\int (\sqrt{3} + \sqrt{x})^2 \, dx$

3. Determine the following integrals:

(a) $\int (x^2 + 3x - 4)^2 \, dx$

(b) $\int (x^2 + 3x - 4)^2 (2x + 3) \, dx$

(c) $\int \frac{(4x + 6)}{(x^2 + 3x - 4)^3} \, dx$

(d) $\int \frac{4}{(x^2 - 8x + 16)} \, dx$

(e) $\int \sqrt[3]{2x - 5} \, dx$

(f) $\int x \sqrt{x^2 - 9} \, dx$

4. Determine the following integrals:

(a) $\displaystyle\int \sin^5 \theta \cos \theta \, d\theta$

(b) $\displaystyle\int \sin \theta \sec^5 \theta \, d\theta$ Hint: Use a trigonometric identity.

(c) $\displaystyle\int (\sin \theta + \cos \theta)^2 \, d\theta$

(d) $\displaystyle\int \tan^4 \theta \sec^2 \theta \, d\theta$

(e) $\displaystyle\int \frac{(\ln x)^3}{x} \, dx$

(f) $\displaystyle\int (2 - e^x)^4 e^x \, dx$

(g) $\displaystyle\int (e^{2x} + 3e^x) \, dx$ Hint: Factor.

(h) $\displaystyle\int \frac{e^x}{\sqrt{e^x + 5}} \, dx$

5. Why is Formula 3 invalid when $n = -1$? How is the following integral to be done:

$$\int x^{-1} \, dx$$

6. Integrate the following by reducing each to the form $\int dv/v$.

(a) $\displaystyle\int \frac{2 \, dx}{2x + 3}$

(b) $\displaystyle\int \frac{dx}{2 - 3x}$

(c) $\displaystyle\int \frac{x \, dx}{2x^2 + 1}$

(d) $\displaystyle\int \frac{(x - 2)}{x^2 - 4x + 6} \, dx$

(e) $\displaystyle\int \frac{\sec^2 \theta}{4 \tan \theta + 8} \, d\theta$

(f) $\displaystyle\int \frac{\sin 2\theta}{3 \cos 2\theta + 4} \, d\theta$

(g) $\displaystyle\int \frac{e^x}{5 - 2e^x} \, dx$

(h) $\displaystyle\int \frac{1}{x \ln x} \, dx$

7. Find the integral

$$\int (x + 3)^2 \, dx$$

both with and without expanding the integrand. Are the results equivalent?

8. Explain why Formula 3 cannot be used to integrate

$$\int (x^2 - 4)^{1/2} \, dx$$

9. Show that

$$\int \sin \theta \cos \theta \, d\theta = \frac{1}{2} \sin^2 \theta + C$$

and

$$\int \cos \theta \sin \theta \, d\theta = -\frac{1}{2} \cos^2 \theta + C$$

How can these results be reconciled?

10. It is given that f' is the derivative of f, g' is the derivative of g, and $g(x) = af(x)$, where a is a constant. From the integral of $g'(x)$ show that

$$\int af'(x) \, dx = a \int f'(x) \, dx$$

thereby proving Formula 2 for any function f'.

11. Prove Formula 1.

14.3 Integration of Trigonometric Functions

We continue the process of transforming differentiation rules into integration formulas, applying this process to trigonometric functions. Since, for example,

$$\frac{d}{dx}\sin x = \cos x$$

then

$$\int \cos x \, dx = \sin x + C$$

In this manner, the following integration formulas can be written down directly from the corresponding rules for differentiation.

5. $\displaystyle\int \cos v \, dv = \sin v + C$

6. $\displaystyle\int \sin v \, dv = -\cos v + C$

7. $\displaystyle\int \sec^2 v \, dv = \tan v + C$

8. $\displaystyle\int \csc^2 v \, dv = -\cot v + C$

9. $\displaystyle\int \sec v \tan v \, dv = \sec v + C$

10. $\displaystyle\int \csc v \cot v \, dv = -\csc v + C$

Aside from these, there are four additional formulas for integrating the tangent, cotangent, secant, and cosecant functions.

11. $\displaystyle\int \tan v \, dv = \ln(\sec v) + C$

12. $\displaystyle\int \cot v \, dv = -\ln(\csc v) + C$

13. $\displaystyle\int \sec v \, dv = \ln(\sec v + \tan v) + C$

14. $\displaystyle\int \csc v \, dv = -\ln(\csc v + \cot v) + C$

Formulas 11 to 14 do not arise directly from familiar differentiation rules. You can derive these yourself by following the hints in the exercises. The formulas are not difficult to remember when you think about how the integrations are done.

You must exercise great care in matching a given problem to the formula to be used. In Formula 5 for instance, you must be certain that the quantity dv is exactly the differential of the function v, which is the argument of $\cos v$. When constant factors are needed, they can of course be supplied as before.

EXAMPLE A

$$\int \cos(3x - 2)\,dx$$

SOLUTION

If $(3x - 2) = v$ then $3\,dx = dv$

$$\int \cos(3x - 2)\,dx = \frac{1}{3}\int \cos(3x - 2) \cdot 3\,dx \quad \left(= \frac{1}{3}\int \cos v\,dv \right)$$

$$= \frac{1}{3}\sin(3x - 2) + C$$

❑

EXAMPLE B

$$\int \sec^2(5x^2)\,x\,dx$$

SOLUTION

If $5x^2 = v$ then $10x\,dx = dv$

$$\int \sec^2(5x^2)\,x\,dx = \frac{1}{10}\int \sec^2(5x^2) \cdot 10x\,dx \quad \left(= \frac{1}{10}\int \sec^2 v\,dv \right)$$

$$= \frac{1}{10}\tan(5x^2) + C$$

❑

EXAMPLE C

$$\int \cot \frac{x}{2}\,dx$$

SOLUTION

If $\dfrac{x}{2} = v$ then $\dfrac{1}{2}dx = dv$

$$\int \cot \frac{x}{2} dx = 2 \int \cot \frac{x}{2} \cdot \frac{1}{2} dx \quad \left(= 2 \int \cot v \, dv \right)$$

$$= -2 \ln\left(\csc \frac{x}{2} \right) + C$$

Exercise 14.3

1. Determine the following integrals using Formulas 5 and 6.

(a) $\displaystyle\int \sin 3x \, dx$

(b) $\displaystyle\int \cos(2x + 1) \, dx$

(c) $\displaystyle\int \frac{dx}{\csc(5x - 2)}$

(d) $\displaystyle\int \cot \frac{1}{2}x \sin \frac{1}{2}x \, dx$

(e) $\displaystyle\int 4 \sin \pi x \, dx$

2. Determine the following integrals using Formulas 7 to 10.

(a) $\displaystyle\int 3 \sec^2 3x \, dx$

(b) $\displaystyle\int \sec 2x \tan 2x \, dx$

(c) $\displaystyle\int \frac{dx}{2 \cos^2 4x}$

(d) $\displaystyle\int \frac{\csc 3x}{\tan 3x} dx$

(e) $\displaystyle\int \frac{1}{\sin x \cos x \tan x} dx$

3. Show that

(a) $\displaystyle\int \cos ax \, dx = \frac{1}{a} \sin ax + C$

(b) $\displaystyle\int \sin ax \, dx = -\frac{1}{a} \cos ax + C$

4. (a) Derive Formula 11 for

$$\int \tan \theta \, d\theta$$

Hint: $\tan \theta = \dfrac{\sin \theta}{\cos \theta}$ and use Formula 4

(b) Derive Formula 12 for

$$\int \cot \theta \, d\theta$$

5. (a) Derive Formula 13 for

$$\int \sec \theta \, d\theta$$

Hint: Multiply and divide the integrand by $(\sec \theta + \tan \theta)$

(b) Derive Formula 14 for

$$\int \csc \theta \, d\theta$$

6. Determine the integral

(a) $\displaystyle\int \sec \phi \sin \phi \, d\phi$

(b) $\displaystyle\int \sec \phi \cot \phi \, d\phi$

(c) $\displaystyle\int \csc \phi \tan \phi \, d\phi$

(d) $\displaystyle\int \csc \phi \cos \phi \, d\phi$

7. Determine the integral

(a) $\displaystyle\int \sec(2x - 3) \, dx$

(b) $\displaystyle\int \frac{dx}{\sin 2x}$

(c) $\int \dfrac{\tan x}{\sin x}\,dx$

(f) $\int \dfrac{\sin 2\theta}{\sin \theta \cos^2 \theta}\,dx$

(d) $\int \dfrac{\tan x}{\sin 2x}\,dx$

Hint: Use an identity for $\sin 2x$.

(g) $\int \dfrac{\tan x}{\cos x}\,dx$

8. Show that Formula 14 can be written

(e) $\int \dfrac{2 \cos x}{\sin 2x}\,dx$

$$\int \csc \theta \, d\theta = \ln(\csc \theta - \cot \theta) + C$$

Integration of Exponential and Logarithmic Functions   14.4

Formulas for the integration of exponential functions follow directly from the differentiation rules.

> 15. $\displaystyle\int e^v\,dv = e^v + C$
>
> 16. $\displaystyle\int a^v\,dv = \dfrac{a^v}{\ln a} + C$

EXAMPLE A

$$\int e^{-4x}\,dx$$

SOLUTION

$$\int e^{-4x}\,dx = \frac{1}{(-4)}\int e^{-4x}(-4)\,dx \quad \left(= -\frac{1}{4}\int e^v\,dv\right)$$

$$= -\frac{1}{4}e^{-4x} + C$$

A formula for integrating the logarithm function is more difficult to find. It is not obvious what function has the logarithm function as its derivative. It is possible, however, to write down a function that has the logarithm function as a *part* of its derivative. This will lead to the integration formula we are seeking. Take the function $x \ln x$, for instance, and apply the product rule:

$$\frac{d}{dx}(x \ln x) = x \cdot \frac{1}{x} + (\ln x) \cdot 1$$

$$= 1 + \ln x$$

Now since

$$\frac{d}{dx}(x \ln x) = 1 + \ln x$$

then

$$\int (1 + \ln x)\, dx = x \ln x + C$$

$$\int dx + \int \ln x\, dx = x \ln x + C$$

$$x + \int \ln x\, dx = x \ln x + C$$

$$\int \ln x\, dx = x \ln x - x + C$$

This, then, is the integration formula for the function $\ln x$. In a similar manner, you can derive the formula for $\log_a x$.

17. $\displaystyle\int \ln v\, dv = v \ln v - v + C$

18. $\displaystyle\int \log_a v\, dv = v \log_a v - v \log_a e + C$

EXAMPLE B

$$\int \ln(x^2)\, x\, dx$$

SOLUTION

If $v = x^2$, then $dv = 2x\, dx$

$$\int \ln(x^2)\, x\, dx = \frac{1}{2} \int \ln(x^2) \cdot 2x\, dx \quad \left(= \frac{1}{2} \int \ln v\, dv \right)$$

$$= \frac{1}{2}\left(x^2 \ln(x^2) - x^2 \right) + C$$

$$= x^2 \ln x - \frac{1}{2}x^2 + C$$

Observe that it is possible to write the integral in this example as

$$\int (2 \ln x)\, x\, dx \qquad \text{or} \qquad 2 \int (\ln x)\, x\, dx$$

When it is expressed in this way, it is not in the form $\int \ln v\, dv$ and so cannot be done using the methods described here.

Exercise 14.4

1. Determine the following integrals using Formulas 15 and 16.

(a) $\int e^{5x}\,dx$

(b) $\int 3e^{-2x}\,dx$

(c) $\int \dfrac{dx}{e^x}$

(d) $\int x\,e^{x^2}\,dx$

(e) $\int 2^{3x}\,dx$

(f) $\int 10^{-x}\,dx$

2. Determine the following integrals using Formulas 17 and 18.

(a) $\int \dfrac{\ln x}{4}\,dx$

(b) $\int \ln(x^3)\,dx$

(c) $\int \ln(3x)\,dx$

(d) $\int \log_{10}(5x)\,dx$

(e) $\int \log_2\left(\dfrac{x}{2}\right)\,dx$

3. Show that $\int e^{ax}\,dx = \dfrac{1}{a}e^{ax} + C$

4. Show that
$$\int \ln(ax)\,dx = x\ln x - (1 - \ln a)\,x + C$$

5. Derive Formula 18. Start by differentiating $v \log_a v$ with respect to v.

Additional Integration Formulas 14.5

Two more integration formulas can be written down directly from the rules for differentiating $\sin^{-1} v$. and $\tan^{-1} v$. They are

19. $\displaystyle\int \dfrac{dv}{\sqrt{a^2 - v^2}} = \sin^{-1}\dfrac{v}{a} + C$

20. $\displaystyle\int \dfrac{dv}{v^2 + a^2} = \dfrac{1}{a}\tan^{-1}\dfrac{v}{a} + C$

Similar integration formulas involving the other inverse trigonometric functions could also be written down. They are of minor importance, however, and will not be discussed.

Two other integrals closely resembling those in Formulas 19 and 20 are the following:

$$21. \quad \int \frac{dv}{\sqrt{v^2 \pm a^2}} = \ln(v + \sqrt{v^2 \pm a^2}) + C$$

$$22. \quad \int \frac{dv}{v^2 - a^2} = \frac{1}{2a} \ln \frac{v - a}{v + a} + C$$

The derivation of these formulas will be postponed until special methods of integration are discussed in later chapters. Formula 21 requires a trigonometric substitution. Formula 22 requires the use of partial fractions.

You should learn to recognize which of these four formulas applies to a given problem. It will usually be necessary to rewrite the integrand to make it fit the formula exactly.

EXAMPLE A

$$\int \frac{1}{1 + 4x^2} \, dx$$

SOLUTION

Since there is no radical and since the denominator can be written as the sum of squares, Formula 20 applies:

$$\int \frac{1}{1 + 4x^2} \, dx = \int \frac{1}{1 + (2x)^2} \, dx$$

$$= \frac{1}{2} \int \frac{2 \, dx}{(2x)^2 + 1} \quad \left(= \frac{1}{2} \int \frac{dv}{v^2 + 1^2} \right)$$

$$= \frac{1}{2} \tan^{-1}(2x) + C$$

EXAMPLE B

$$\int \frac{dx}{\sqrt{x^2 - 10x + 16}}$$

SOLUTION

In order to use one of the Formulas 19 to 22, the expression in the radical must be rewritten as the sum or difference of squares. By completing the square,

$$x^2 - 10x + 16 = x^2 - 10x + 25 - 9$$

$$= (x - 5)^2 - 3^2$$

Once written in this way, it is evident that Formula 21 with $v = x - 5$ is the correct formula to use.

$$\int \frac{dx}{\sqrt{x^2 - 10x + 16}} = \int \frac{dx}{\sqrt{(x - 5)^2 - 3^2}} \quad \left(= \int \frac{dv}{\sqrt{v^2 - 3^2}} \right)$$

$$= \ln\left(x - 5 + \sqrt{(x - 5)^2 - 3^2} \right) + C$$

$$= \ln\left(x - 5 + \sqrt{x^2 - 10x + 16} \right) + C \quad \square$$

Exercise 14.5

1. Determine the following integrals using Formulas 19 to 22:

(a) $\int \dfrac{dx}{x^2 + 25}$

(b) $\int \dfrac{dx}{\sqrt{x^2 + 25}}$

(c) $\int \dfrac{dx}{\sqrt{6 - x^2}}$

(d) $\int \dfrac{2\,dx}{16 - x^2}$

(e) $\int \dfrac{dx}{9x^2 + 4}$

(f) $\int \dfrac{dx}{4x^2 - 9}$

(g) $\int \dfrac{dx}{\sqrt{9 + 4x^2}}$

(h) $\int \dfrac{dx}{\sqrt{4 - 9x^2}}$

(i) $\int \dfrac{dx}{\sqrt{x^2 - 16}}$

(j) $\int \dfrac{dx}{\sqrt{8 - 4x^2}}$

2. Determine the following integrals:

(a) $\int \dfrac{dx}{(x - 1)^2 + 7}$

(b) $\int \dfrac{dx}{4x^2 + 4x + 13}$

Hint: $4x^2 + 4x + 13 = (2x + 1)^2 + 12$

(c) $\int \dfrac{dx}{\sqrt{x^2 - 6x - 7}}$

(d) $\int \dfrac{dx}{9x^2 + 12x}$

(e) $\int \dfrac{dx}{\sqrt{2x - x^2}}$

Hint: $2x - x^2 = 1 - (x - 1)^2$

(f) $\int \dfrac{dx}{12x - x^2 + 13}$

3. Prove Formula 21 by differentiation.

4. Prove Formula 22 by differentiation.

14.6   Summary and Review

Integration is the inverse of differentiation. This fact permits numerous formulas for the integration of elementary functions to be deduced by inverting the corresponding differentiation rules.

Success in integration depends to a great extent on your ability to recognize which integration formula is the appropriate one to use. Solving a large number of problems will increase your skill. There is no short cut.

The following are some suggestions that will help you to master the material in this chapter:

1. Memorize the integration formulas. Say them aloud: "The integral of $\cos x$ is $\sin x$," etc.

2. Use correct mathematical notation and proper form when working out problems. Write down the original problem first before you proceed with its solution. Do not forget to write down the differential dx.

3. In integration problems, look both for functions *and* their derivatives in the integrand.

4. Expect to do some algebraic manipulations, such as completing the square, using trigonometric identities, supplying constants, etc., to cast an integrand into a suitable form.

5. Do not be satisfied with merely learning *how* certain steps are done. Seek motivation for *why* they are done. Why is one approach successful, but another not?

6. Look ahead. If you take a certain approach, where will it get you? Does it look promising?

Exercise 14.6

Determine the integrals in problems 1 to 9 using the integration formulas introduced in this chapter.

1. (a) $\displaystyle\int 7x^8\,dx$

    (b) $\displaystyle\int 2x^{6/5}\,dx$

    (c) $\displaystyle\int \sqrt{5x^3}\,dx$

    (d) $\displaystyle\int \frac{6}{x^{4/3}}\,dx$

    (e) $\displaystyle\int (x^{1/3} + 4x^{-1/3})\,dx$

    (f) $\displaystyle\int (3x^3 + 2\sqrt{x})\sqrt{x}\,dx$

2. (a) $\displaystyle\int \frac{(x - 2x^3)}{3\sqrt{x}}\,dx$

    (b) $\displaystyle\int \frac{(6x^2 - 4x^{3/2})}{5x^3}\,dx$

    (c) $\displaystyle\int (\sqrt{3x} + 6x)^2\,dx$

(d) $\displaystyle\int\left(3x + \frac{1}{2x^2}\right)x^{3/2}\,dx$

(e) $\displaystyle\int(2 + x^2)(3 - x)\,dx$

3. (a) $\displaystyle\int(4\sec^2 x - \csc x\cot x)\,dx$

(b) $\displaystyle\int(1 + \sin^2\theta\csc\theta)\,d\theta$

(c) $\displaystyle\int\frac{1}{4\sin^2 x}\,dx$

(d) $\displaystyle\int\frac{d\theta}{2\sin\theta\tan\theta}$

(e) $\displaystyle\int\sec x(\sec x + \tan x)\,dx$

(f) $\displaystyle\int\sec x(\tan x + \cos x)\,dx$

4. (a) $\displaystyle\int(1 + \tan\theta)^2\,d\theta$

(b) $\displaystyle\int(2 - \tan^2\theta)^2\sec^2\theta\,d\theta$

(c) $\displaystyle\int\sin^3\theta\,d\theta$

(d) $\displaystyle\int\tan^3\theta\,d\theta$

5. (a) $\displaystyle\int(e^{x/2} + e^{-x/2})\,dx$

(b) $\displaystyle\int 4e^{\ln(3x)}\,dx$

6. (a) $\displaystyle\int x\sqrt{4x^2 + 2}\,dx$

(b) $\displaystyle\int x\sqrt{9 - x^2}\,dx$

(c) $\displaystyle\int\sqrt{2x^3 - 6x^2}\,dx$

7. (a) $\displaystyle\int\frac{(3x^2 + 5x + 8)}{x + 3}\,dx$

Hint: Use long division to write the integrand as $3x - 4 + \dfrac{20}{x + 3}$

(b) $\displaystyle\int\frac{(2x + 12)}{x - 2}\,dx$

(c) $\displaystyle\int\frac{(4x + 15)}{2x + 5}\,dx$

(d) $\displaystyle\int\frac{(4x^3 - 8)}{x + 4}\,dx$

(e) $\displaystyle\int\frac{x(x - 2)}{x + 1}\,dx$

(f) $\displaystyle\int\frac{(x - 5)(x + 3)}{x - 4}\,dx$

8. (a) $\displaystyle\int\frac{1}{16x^2 + 1}\,dx$

(b) $\displaystyle\int\frac{3}{5x^2 + 4}\,dx$

(c) $\displaystyle\int\frac{2}{\sqrt{4x^2 + 5}}\,dx$

(d) $\displaystyle\int\frac{15}{\sqrt{x^2 - 25}}\,dx$

(e) $\displaystyle\int\frac{3}{x^2 - 6x}\,dx$

(f) $\displaystyle\int\frac{1}{x^2 + 3x - 4}\,dx$

9. (a) $\displaystyle\int\frac{(\ln x)^3}{3x}\,dx$

(b) $\displaystyle\int\frac{1}{2x\ln 2x}\,dx$

(c) $\displaystyle\int\frac{(1 - \sin x)}{x + \cos x}\,dx$

(d) $\displaystyle\int\frac{(x + \cos x)}{2x^2 + 4\sin x}\,dx$

(e) $\displaystyle\int\frac{1}{\cos^2 x(3\tan x + 5)}\,dx$

(f) $\displaystyle\int 4\sin x\cos 2x\,dx$

Hint: Use an identity for $\cos 2x$.

10. (a) Differentiate $x \sin x$. From the result, work out a formula for the integral

$$\int x \cos x \, dx$$

(b) Find an integration formula for

$$\int x \sin x \, dx$$

11. (a) Differentiate $x \tan^{-1} x$. From the result, work out a formula for the integral

$$\int \tan^{-1} x \, dx$$

(b) Find an integration formula for

$$\int \sin^{-1} x \, dx$$

Application of the Indefinite Integral to Rate Problems

15

I n Chapter 7, when rate problems were first encountered, the relationship between the dependent variable and the independent variable was known. For example, profit was defined as a function of price, or position was defined as a function of time. Questions were asked about the rate of change of one variable with respect to the other. When the instantaneous rate of change was required, the derivative had to be found. In this chapter the instantaneous rate of change is known. Since the instantaneous rate of change is a derivative, determining the relationship between the variables requires the calculation of the indefinite integral.

The Indefinite Integral of Rates of Change

15.1

Suppose the instantaneous rate of change of y with respect to x is a given function f of x: $dy/dx = f(x)$. An equation such as this, that includes a derivative, is called a **differential equation**. A solution of this differential equation is any function $y = F(x)$, for which $dF(x)/dx = f(x)$. From the previous chapter we know $y = F(x)$ is the indefinite integral or antiderivative of $f(x)$. That is,

$$F(x) = \int f(x)\,dx$$

The function $F(x)$ includes a constant which has various interpretations depending upon the rate of change application, as the following examples show.

Figure 15.1

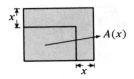

EXAMPLE A

The length and width of a given rectangle are increased by the same amount x. The instantaneous rate of change of area $A(x)$ with respect to x is given by $dA/dx = 2x + 150$. The area $A(x)$ is measured in cm^2 and the increase x is measured in cm. What is the area as a function of x? What is the significance of the constant of integration?

SOLUTION

The area $A(x)$ is the indefinite integral of dA/dx.

$$A(x) = \int (2x + 150)\,dx$$
$$= x^2 + 150x + C$$

The area of the rectangle is $x^2 + 150x + C$ cm^2.

The value of the constant C cannot be determined from the data in the example. However, the value of C is equivalent to $A(0)$ because

$$A(0) = 0^2 + (150)0 + C$$
$$= C$$

Therefore C can be thought of as the area of the rectangle when there has been no increase in the length and width. In other words, C is the initial area of the rectangle.   ❑

EXAMPLE B

The rate of change of cost $C(x)$ with respect to x, the number of units produced by a manufacturing firm, is $dC/dx = 0.15x^2 - 0.3x + 3$ in dollars per unit produced. What is the cost as a function of production? What is the significance of the constant in the answer?

SOLUTION

The cost $C(x)$ is given by the antiderivative of dC/dx.

$$C(x) = \int (0.15x^2 - 0.3x + 3)\,dx$$
$$C(x) = 0.05x^3 - 0.15x^2 + 3x + K \qquad \longleftarrow \quad \text{K is used as a constant instead of C to avoid confusion with the cost function $C(x)$}$$

The cost, as a function of production, is $0.05x^3 - 0.15x^2 + 3x + K$ dollars. The constant K cannot be determined from the information given in this example. However, the value of K is equivalent to $C(0)$ because

$$C(0) = 0.05(0)^3 - 0.15(0)^2 + 3(0) + K$$
$$= K$$

So K is the cost of not producing any units. In business and other applications this is referred to as the **fixed cost** because it is not affected by the level of production. ❏

The **change** in $F(x)$ as x changes from x_1 to x_2 is defined as $\Delta F = F(x_2) - F(x_1)$. If $y = F(x)$ is a solution to $dy/dx = f(x)$, then the change in $F(x)$ can be found without determining the precise value of the constant C that results from the indefinite integral. The constant C which is part of both $F(x_1)$ and $F(x_2)$ will disappear when $F(x_1)$ is subtracted from $F(x_2)$. This concept is demonstrated in the next example.

EXAMPLE C

The instantaneous rate of change of the consumption of luxuries is dependent on the amount of income earned above the level needed to satisfy basic needs such as food, clothing, and shelter. It has been determined that the instantaneous rate of change of consumption of luxuries is given by $dL/dx = 1.5x^{-0.35}$, $x \geq 0$. Both the level of consumption $L(x)$ and the amount of income above the basic level x are measured in dollars. What is the change in consumption when the amount of extra income changes from $100 to $200?

SOLUTION

The consumption level $L(x)$ is the indefinite integral of dL/dx.

$$L(x) = \int 1.5x^{-0.35}\, dx$$
$$= \frac{30}{13}x^{0.65} + C$$

When x changes from $100 to $200

the change in $L(x) = \Delta L$

$$= L(200) - L(100)$$

$$= \left(\frac{30}{13}(200)^{0.65} + C\right) - \left(\frac{30}{13}(100)^{0.65} + C\right)$$

$$\doteq 26.21$$

When the amount of extra income goes from \$100 to \$200, the amount spent on luxuries changes by \$26.21.   ❏

In rate problems where the instantaneous rate of change $dy/dx = f(x)$ is given, it is often necessary to find the particular function $y = F(x)$ that satisfies certain conditions called **boundary** or **initial conditions**. These conditions are used to determine a value for the constant C that results from evaluating the indefinite integral.

Figure 15.2

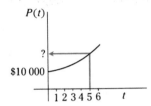

EXAMPLE D

The instantaneous rate of change of profit with respect to time of a small retail store is given by $dP/dt = 1500e^{0.15t}$, $t \geq 0$ where $P(t)$ is in dollars and t is in years. The initial condition is that at time $t = 0$ the profit is \$10 000 (see Figure 15.2). What is the profit as a function of time? What is the profit after 5 years?

SOLUTION

The profit $P(t)$ is the indefinite integral of $dP(t)/dt$.

$$P(t) = \int 1500e^{0.15t}\, dt$$
$$= 10\,000\, e^{0.15t} + C$$

Since the initial profit is \$10 000,

$$P(0) = 10\,000$$
$$10\,000\, e^{(0.15)0} + C = 10\,000$$
$$C = 0$$

The profit as a function of time is $P(t) = 10\,000e^{0.15t}$ in dollars. The profit after 5 years is

$$P(5) = 10\,000\, e^{(0.15)5}$$
$$\doteq 21\,170$$

The profit after 5 years is \$21 170.   ❏

EXAMPLE E

The rate of change of cost with respect to production for a firm manufacturing toy cars is given by $dC/dp = 100/(1 + p)$, $p \geq 0$. $C(p)$

is measured in dollars and p is the number of cars produced. What is the cost of producing 1000 cars if the fixed cost is $1500?

SOLUTION

The cost $C(p)$ is the indefinite integral of dC/dp.

$$C(p) = \int \frac{100}{1 + p} \, dp$$

$$= 100 \ln(1 + p) + K$$

The fixed cost is $1500 so the cost of producing 0 cars is $1500.

$$C(0) = 1500$$

$$100 \ln(1 + 0) + K = 1500$$

$$K = 1500$$

The cost of producing p cars is $C(p) = 100 \ln(1 + p) + 1500$ dollars. The cost of producing 1000 cars is $C(1000)$.

$$C(1000) = 100 \ln(1 + 1000) + 1500$$

$$\doteq 2190.88$$

The cost of producing 1000 cars is $2190.88. ◻

Exercise 15.1

1. Find the indefinite integral of each of the following instantaneous rates of change.

(a) $\dfrac{dP(t)}{dt} = 4t^2 + 3t^{-2} + \sqrt{t}$

(b) $\dfrac{dR(x)}{dx} = xe^{-x^2}$

(c) $\dfrac{dA(t)}{dt} = \dfrac{1}{3t + 1}$

(d) $\dfrac{dL(s)}{ds} = 4 \sin(2s)$

2. Find the indefinite integral for each of the following instantaneous rates of change. Then find

the particular function that satisfies the given boundary condition.

(a) $\dfrac{dM(w)}{dw} = w\sqrt{w^2 + 5}$ $\qquad M(2) = 50$

(b) $\dfrac{dV(r)}{dr} = 4\pi r^2$ $\qquad V(0) = 0$

(c) $\dfrac{dP(t)}{dt} = 100 \ln t$ $\qquad P(1) = 400$

(d) $\dfrac{dH(x)}{dx} = 5 \sin 3x \cos 3x$ $\qquad H\left(\dfrac{\pi}{6}\right) = 9$

3. The height of an onion plant is dependent upon the time the onion bulb is in the ground. During the first month of growth the rate of change of height $H(t)$ is $dH/dt = 0.04t$ cm/d. What is the height as a function of time? What is the significance of the constant of integration?

4. The instantaneous rate of change of the cost of production $C(x)$ with respect to x, the volume of production, is $dC/dx = 0.8x^{-0.25} + 0.20, x \geq 0$ for a company producing antifreeze. The cost is measured in dollars and the volume of production in litres. Find the cost of producing the thousandth litre.

5. An ivy plant grows so that the instantaneous rate of change of its length $L(t)$ is given by $dL/dt = 1.5$ cm/d. How long is a plant, which is 20 cm long at present, after 60 d?

6. The instantaneous rate of change of the value of an initial investment of $1000 is given by $dV(t)/dt = 100e^{0.1t}, t \geq 0$. The value $V(t)$ is measured in dollars and the time t in years. When will the investment be worth $100 000?

7. The demand for a product is dependent on the price charged. The instantaneous rate of change of demand $D(x)$ with respect to price x is $dD/dx = -1000x + 4000$ units/dollar. At a price of $1.00 the demand is 3500 units. At what price is the demand the greatest? What is the demand at this price?

8. A manufacturing firm produces 5000 hockey helmets per day using 36 employees. By adding more staff, production can be increased. It has been empirically determined that the instantaneous rate of change of daily production $P(s)$ with respect to the number of additional workers s is $dP/ds = 100 - 9\sqrt{s}$ helmets per additional worker. What is the production if 9 workers are added?

9. The instantaneous rate of change of a volume of a sphere with respect to its radius is $dV/dr = 4\pi r^2$ where $V(r)$ is in cm³ and r is in centimetres. If initially the radius is 0 cm and the volume is 0 cm³, find the volume when the radius is 10 cm.

10. The total value of ticket sales to a school dance is dependent upon the length of time the tickets are on sale. The instantaneous rate of change of total sales $S(t)$ with respect to time t is $dS/dt = 30t^2, 0 \leq t \leq 5$ where $S(t)$ is in dollars and t is in days. If $10 worth of tickets are sold on the first day, what is the total sales at the end of the fifth day?

11. A speed reading course promises to increase the number of words read per minute by 25 for each day in the course. When will a person's reading speed be doubled if the person's current reading speed is 200 words per minute?

12. The instantaneous rate of change of the average length of a person's hair is given by $dL/dt = t/225$. The average length $L(t)$ is in centimetres and the time t is in days. A haircut is obtained when the average length changes by 2 cm. How long will a person go between haircuts?

13. The instantaneous rate of change of the value of a share of a particular oil stock is predicted to be $dV/dt = 20(1 + 10t)^{-0.5}$. The value is measured in dollars and the time in months. The initial value of the stock is $104 per share. When will the share be worth $120?

15.2 Motion with Given Velocity

In Chapter 7 the position of an object moving along a straight line relative to a fixed point O was given by $y = s(t)$. The instantaneous velocity $v(t)$ at any time t was defined as the derivative of position so that $v(t) = ds(t)/dt$. Therefore $y = s(t)$ can be thought of as the function satisfying the differential equation $dy/dt = v(t)$. The position $s(t)$ is the antiderivative of velocity. The position function $s(t)$ can be found by evaluating the indefinite integral of $v(t)$

$$s(t) = \int v(t) \, dt$$

EXAMPLE A

The instantaneous velocity of an object moving along a horizontal straight line relative to a fixed point O is $v(t) = 8t - 5$ cm/s $t \geq 0$. What is the displacement from $t = 2$ s to $t = 5$ s?

SOLUTION

The displacement is defined as the change in position. If position at any time t is given by $s(t)$, then the displacement is $s(5) - s(2)$. The function $s(t)$ is found by finding the indefinite integral of $v(t)$.

$$s(t) = \int (8t - 5) \, dt$$
$$= 4t^2 - 5t + C$$

the displacement $= \Delta s$

$$= s(5) - s(2)$$
$$= (4(5)^2 - 5(5) + C) - (4(2)^2 - 5(2) + C)$$
$$= 69$$

The displacement from time $t = 2$ s to $t = 5$ s is 69 cm. ◻

In the previous example, it was not necessary to determine the value of the constant C to find the displacement. The constant C is always evaluated by requiring the function $s(t)$ to satisfy some special condition. Usually this is an **initial condition**, which assigns a value to $s(0)$, the position of the object at the start of the motion. The initial positon $s(0)$ is often written s_0 (read, "s sub zero"). In the special case when $v(t)$ is a polynomial function, the constant C is exactly equal to s_0.

EXAMPLE B

A particle moves along the x-axis with the origin as a reference point so that at time t in seconds its velocity $v(t)$ in cm/s is given by $v(t) = 2t^2 - 3t + 2, t \geq 0$. Its initial position is 5 cm to the left of the origin. What is its position after 10 s?

SOLUTION

The position $s(t)$ is the indefinite integral of velocity. The constant of integration can be expressed as s_0 because $v(t)$ is a polynomial function.

$$s(t) = \int (2t^2 - 3t + 2)\, dt$$
$$= 2/3\, t^2 - 3/2\, t^2 + 2t + s_0$$

Since the initial position is $5\,\text{cm}$ to the left of the origin $s(0) = -5$. Therefore $s_0 = -5$ and $s(t) = 2/3\, t^3 - 3/2\, t^2 + 2t - 5$.

The position after $10\,\text{s}$ is $s(10)$.

$$s(10) = \frac{2}{3}(10)^3 - \frac{3}{2}(10)^2 + 2(10) - 5$$

$$= \frac{1595}{3}$$

$$\doteq 531.7$$

The position of the particle after $10\,\text{s}$ is $531.7\,\text{cm}$ to the right of the origin.   ❏

EXAMPLE C

The instantaneous velocity of an object moving along a horizontal straight line relative to a fixed point 0 is $v(t) = 5\pi\ \sin(\pi t/2)\,\text{cm/s}$, $t \geq 0$. Its initial position is $10\,\text{cm}$ to the left of 0. Find an expression for $s(t)$. Evaluate the position for $t = 0, 1, 2, 3, 4, 5,$ and $6\,\text{s}$ and use this information to describe the nature of the motion of the object.

SOLUTION

The position $s(t)$ is the indefinite integral of velocity.

$$s(t) = \int \left(5\pi \sin \frac{\pi t}{2}\right) dt$$

$$= -10\cos(\pi t/2) + C$$

The initial condition is $s(0) = -10$.

Since

$$s(0) = -10\cos(0) + C$$
$$= -10 + C$$

then

$$-10 = -10 + C$$

and

$$C = 0$$

The position in centimetres relative to 0 at any time t in seconds is $s(t) = -10 \cos(\pi t/2)$.

The positions at $t = 0, 1, 2, 3, 4, 5, 6$ are $-10, 0, 10, 0, -10, 0, 10$ respectively. The particle moves between a position 10 cm to the left of O and a position 10 cm to the right of O. The path of the particle is pictured in Figure 15.3.

Figure 15.3

It takes 4 s for the object to return to its initial position. Since the period of the position function is $(2\pi) \div (\pi/2) = 4$, the object is oscillating between points A and B and completes one oscillation, that is from A to B and back to A, every 4 s.

EXAMPLE D

Two particles moving in the same straight line relative to a fixed point O have instantaneous velocities $v_1(t) = 6t - 8$ and $v_2(t) = 4t$ at time t s. Their velocities are measured in m/s. If they are both initially located at O, determine their velocities when their positions are the same.

SOLUTION

The positions $s_1(t)$ and $s_2(t)$ are the indefinite integrals of the velocities.

$$s_1(t) = \int (6t - 8)\, dt$$

$$= 3t^2 - 8t + C$$

$$s_2(t) = \int (4t)\, dt$$

$$= 2t^2 + K$$

The initial position of both particles is 0. Therefore $C = 0$ and $K = 0$.
Therefore $s_1(t) = 3t^2 - 8t$ and $s_2(t) = 2t^2$.

For the particles to be in the same position $s_1(t) = s_2(t)$

$$3t^2 - 8t = 2t^2$$

$$t^2 - 8t = 0$$

$$t(t - 8) = 0$$

$$t = 0 \text{ or } 8$$

The particles have the same position at $t = 0$ s and $t = 8$ s.
The velocities are $v_1(0)$, $v_1(8)$, $v_2(0)$, and $v_2(8)$.

$$v_1(0) = -8, \ v_1(8) = 40, \ v_2(0) = 0, \text{ and } v_2(8) = 32.$$

At time $t = 0$ s the particles are together and the velocity of the first
particle is -8 m/s and the velocity of the second is 0 m/s. The particles are together again at $t = 8$ s and the velocity of the first particle
is 40 m/s and the velocity of the second is 32 m/s.   ❏

Exercise 15.2

1. For each of the following instantaneous velocities determine an expression for the position
$s(t)$ at any time $t \geq 0$.

   (a) $v(t) = 4t + 3\sqrt{t}$ (b) $v(t) = 5\cos(2t)$

   (c) $v(t) = 4/(3 + t)$ (d) $v(t) = 4e^{3t+1}$

2. Two objects are moving along the same straight
line relative to a fixed point O such that at time
t their respective velocities are $v_1(t) = 4t - 5$,
$t \geq 0$ and $v_2(t) = 2t + 1$, $t \geq 0$. The velocities
are measured in cm/s and the initial position
of both particles is at the same point. Determine when they next coincide.

3. The velocity of a particle moving along the
x-axis with position measured relative to the
origin is $v(t) = 3t^2 - 2t + 6$ cm/s, $t \geq 0$. The initial position of the particle is at the origin. What
is the velocity when it is 6 cm to the right of
the origin?

4. The velocity of a particle moving along the
x-axis with position measured relative to the
origin is $v(t) = 3t - 6$ cm/s, $t \geq 0$. At $t = 3$ s its
position is 5 cm to the right of the origin. What
is its position when its velocity is 0 cm/s?

5. The velocity of a particle moving along the
x-axis with position measured relative to the
origin is $v(t) = 2t - 5$ cm/s, $t \geq 0$. The initial
position of the particle is 6 cm to the right of
the origin. When is the particle moving toward
the origin?

6. The velocity of a particle moving along the
x-axis with position measured relative to the
origin is $v(t) = -6\pi \sin(\pi t/2)$ cm/s, $t \geq 0$. The
initial position of the particle is 12 cm to the
right of the origin. When is the particle located
to the right of the origin?

7. How far apart are two objects at time $t = 4$ s if
they are moving along the same straight line
with positions measured relative to the same
fixed point O? Their velocities at any time $t \geq 0$
are $v_1(t) = 9t^2 - 8t$ and $v_2(t) = 3t - 6$ measured in cm/s. The initial position of each object
is 3 cm to the right of the point O.

8. Two objects start at the same point on the same straight line and their positions are measured relative to the same fixed point O. Their velocities at any time $t \geq 0$ are $v_1(t) = 6t + 5$ and $v_2(t) = 9t - 10$ measured in cm/s. How far apart are they when their velocities are equal?

9. A rock dropped from rest over a cliff moves so that its velocity measured relative to the ground at the bottom of the cliff is $v(t) = -10t, t \geq 0$. The velocity is measured in m/s and the rock falls to the ground a distance of 20 m. What is the rock's velocity when it hits the ground?

10. A ball thrown vertically upward moves so that its velocity relative to the ground at any time $t \geq 0$ is $v(t) = 10 - 10t$ m/s. The ball is thrown from a position 2 m above the ground. What is the initial velocity of the ball? How high above the ground does the ball travel? When does the ball hit the ground?

11. A driver of a car travelling on a straight slippery road sees a stop sign 50 m ahead and starts to brake. After the brakes are applied, the velocity of the car is given by $v(t) = 11 - t^{3/2}$ m/s, $t \geq 0$. Will the car go beyond the stop sign?

12. Two bicycle riders are 105 m apart on a straight level road. At time $t \geq 0$, the one in front maintains a constant velocity of $v_1(t) = 4$ m/s, while the speed of the one behind is increasing such that $v_2(t) = 0.5t$ m/s. How long will it take for the one behind to catch up?

13. Two cars are travelling in the same direction along a straight level road at constant velocities of 60 km/h and 80 km/h respectively. The slower car passes a gas station 10 min ahead of the faster car. How far beyond the gas station will they meet?

Motion With Given Acceleration   15.3

For an object moving along a straight line relative to a fixed point O with position given by $s(t)$ and velocity given by $v(t)$, the instantaneous acceleration at any time t was defined as $a(t) = dv(t)/dt$. Therefore $y = v(t)$ can be thought of as the function satisfying the differential equation $dy/dt = a(t)$. The velocity $v(t)$ is the antiderivative of acceleration. The velocity function $v(t)$ is found by evaluating the indefinite integral of $a(t)$.

$$v(t) = \int a(t)\,dt$$

EXAMPLE A

The instantaneous acceleration of an object moving along a horizontal straight line relative to a fixed point O is $a(t) = -6$ cm/s², $t \geq 0$. What is the change in velocity from $t = 2$ s to $t = 5$ s?

SOLUTION

If velocity at any time t is given by $v(t)$ then the change in velocity is $v(5) - v(2)$. The function $v(t)$ is found by finding the indefinite integral of $a(t)$.

$$v(t) = \int -6\,dt$$
$$= -6t + C$$

the change in velocity $= \Delta v$

$$= v(5) - v(2)$$
$$= (-6(5) + C) - (-6(2) + C)$$
$$= -18$$

The change in velocity from time $t = 2$ s to $t = 5$ s is -18 cm/s.   ❏

After finding the indefinite integral of $a(t)$, the constant of integration is evaluated by requiring $v(t)$ to satisfy some special condition. Like the initial position in the preceding section, it is usually the initial velocity $v(0)$ which is specified. This is often denoted by v_0. Again, in the special case when $a(t)$ is a polynomial function, the constant of integration is v_0.

EXAMPLE B

An object moves along the x-axis with its position measured relative to the origin so that its acceleration at any time $t \geq 0$ is 4 cm/s². The initial velocity is -50 cm/s. Determine the instantaneous velocity at any time t.

SOLUTION

The velocity $v(t)$ is the indefinite integral of $g(t)$. The constant of integration can be expressed as v_0 because $a(t)$ is a polynomial function.

$$a(t) = \int 4\,dt$$
$$= 4t + v_0$$

Since the initial velocity is -50 cm/s, the instantaneous velocity at any time t is $v(t) = 4t - 50$ cm/s.   ❏

If given the acceleration $a(t)$ at any time, the velocity $v(t)$ can be found by finding the indefinite integral of $a(t)$, and the position at any time t can be found by finding the indefinite integral of $v(t)$. Each indefinite integral introduces a constant. To determine the value of the constants, two boundary conditions must be given. The velocity at a particular time and the position at a particular time need to be known.

EXAMPLE C

An object moves along a straight line with its position measured relative to a fixed point O. Its acceleration at any time t is $a(t) = 6t$ m/s^2, $t \geq 0$. The initial velocity of the object is -25 m/s, and at $t = 1$ s its position is 24 m left of the reference point. Find expressions for the object's position and velocity. Determine when the object is located left of the reference point.

SOLUTION

The instantaneous velocity is the indefinite integral of the acceleration.

$$v(t) = \int 6t \, dt$$
$$= 3t^2 + v_0$$

The initial velocity is -25 m/s so $v_0 = -25$. The velocity at time t is $v(t) = 3t^2 - 25$ m/s.

The position is the indefinite integral of the velocity.

$$s(t) = \int (3t^2 - 25) \, dt$$
$$= t^3 - 25t + s_0$$

At $t = 1$ s the object is 24 m to the left of O, that is

$$s(1) = -24$$
$$1^3 - 25(1) + s_0 = -24$$
$$s_0 = 0$$

The position at any time is $s(t) = t^3 - 25t$ m.

The object is to the left of the reference point when

$$s(t) < 0$$
$$t^3 - 25t < 0$$
$$t(t - 5)(t + 5) < 0$$
$$t < -5 \text{ or } 0 < t < 5$$

Because $t \geq 0$ the particle is to the left of the origin between time 0 s and time 5 s. ❏

EXAMPLE D

A car travelling along a straight flat road at an initial velocity of 108 km/h is brought to a halt by a steadily increasing braking force. The acceleration measured from the moment at which the brakes are applied is $a(t) = -0.6t\,\text{m/s}^2$, $t \geq 0$. How far will the car travel before it stops?

SOLUTION

The instantaneous velocity is the indefinite integral of the acceleration.

$$v(t) = \int -0.6t\,dt$$
$$= -0.3t^2 + v_0$$

The initial velocity is 108 km/h or 30 m/s. The instantaneous velocity is $v(t) = -0.3t^2 + 30\,\text{m/s}$.

The car will be stopped when $v(t) = 0$.

$$-0.3t^2 + 30 = 0$$

$$t = 10\,\text{or} -10$$

Since $t \geq 0$ the car stops in 10 s. The position is the indefinite integral of the velocity.

$$s(t) = \int (-0.3t^2 + 30)\,dt$$
$$= -0.1t^3 + 30t + s_0$$

The distance travelled by the car is the displacement $s(t) - s_0$, the difference between its final and its initial position. In 10 s

$$s(10) - s_0 = -0.1(10)^3 + 30(10)$$

$$= 200$$

The car travels 200 m before it stops.   □

Newton's law of motion states that the acceleration of an object of mass M which is acted upon by a force F is given by $a = F/M$. In situations where the force is non-zero and never changes, the acceleration is constant. Near the surface of a planet, for example, the force of gravity is constant to a good approximation. Motion under such circumstances is called motion with constant acceleration.

The symbol for the constant acceleration due to gravity on earth is the letter g. For vertical motion under gravity the usual convention is that the direction upward is positive and the direction downward is negative. Therefore the acceleration at any time t of an object acted upon by a gravitational force is $a(t) = -g$. Near the earth's surface g has the value 9.8 m/s².

EXAMPLE E

A stone is thrown vertically upward from a point 1.5 m above the earth's surface with an initial velocity of 10 m/s (see Figure 15.4). Determine when the stone hits the ground.

Figure 15.4

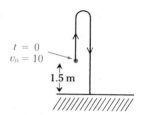

SOLUTION

The instantaneous velocity is the indefinite integral of $a(t) = -9.8$ m/s².

$$v(t) = \int -9.8 \, dt$$
$$= -9.8t + v_0$$

The initial velocity is 10 m/s so $v_0 = 10$.

$$v(t) = -9.8t + 10$$

The position at time t is the indefinite integral of $v(t)$.

$$s(t) = \int (-9.8t + 10) \, dt$$
$$= -4.9t^2 + 10t + s_0$$

The initial position is 1.5 m above the earth's surface so $s_0 = 1.5$.

$$s(t) = -4.9t^2 + 10t + 1.5$$

The stone hits the ground when $s(t) = 0$.

$$s(t) = -4.9t^2 + 10t + 1.5$$
$$= 0$$

The quadratic formula yields two answers.

$$t = -0.14 \text{ or } 2.18$$

Since time is positive, the stone hits the ground after 2.18 s.

Exercise 15.3

1. Given the instantaneous acceleration, determine the instantaneous velocity in each of the following cases.

   (a) $a(t) = \sqrt{5t + 3}$ (b) $a(t) = e^{0.5t}$

   (c) $a(t) = 4\sin(3t)$ (d) $a(t) = 4/(3 + t)$

2. In each of the following determine equations for $v(t)$ and $s(t)$.

   (a) $a(t) = 3\sqrt{t} - 9$ (b) $a(t) = 2e^{5t}$

   (c) $a(t) = 5\cos(-2t)$ (d) $a(t) = 15t^{2/3}$

3. For each of the given instantaneous accelerations below, determine the particular velocity $v(t)$ and position $s(t)$ satisfying the given boundary conditions.

   (a) $a(t) = 12t + 8$ $v(0) = -2$ $s(0) = 5$

   (b) $a(t) = 6t - 5$ $v(1) = 4$ $s(1) = -7$

   (c) $a(t) = 3\pi\sin(\pi t)$ $v(0) = -3$ $s(0) = 5$

   (d) $a(t) = e^{2t}$ $v(0) = 0.5$ $s(0) = 9$

4. A car is moving along a straight flat road at a velocity of 90 km/h. The driver applies the brakes so that it decelerates at 10 m/s². How long will it take to stop? How far will the car go from the time the brakes are applied?

5. Two particles are moving along the same straight line with positions measured relative to a fixed point O, and their accelerations at time t are $a_1(t) = 2t + 5$ and $a_2(t) = 3t - 5$. Time is measured in seconds and distance in metres and both particles have the same initial velocity. At what other time do they have the same velocity?

6. An object is moving along the x-axis relative to the origin so that its acceleration in m/s² at any time $t \geq 0$ is $a(t) = 8t - 4$. If the initial velocity is 0 m/s, when is the velocity positive?

7. An object is moving along the x-axis relative to the origin so that its acceleration in m/s² at any time t is $a(t) = 6t - 14, t \geq 0$. Initially it is at the origin with a velocity of 10 m/s. When does it return to the origin? What is its acceleration at these times?

8. An object moves vertically so that $a(t) = -10$ m/s² and $v(0) = 20$ m/s. For how long does the object move upward?

9. A ball is dropped from the 553-m tall CN Tower in Toronto. What is the velocity of the ball when it hits the ground?

10. An object is moving along a straight line with its position measured relative to a fixed point O. The velocity of the object at any time t is $v(t) = 3t^2$ cm/s. Initially the object is 8 cm to the left of the point O. Determine the object's acceleration when it is at O.

11. A ball rolls from rest down a smooth board inclined at 45° to the horizontal. The acceleration at any time t measured relative to its starting point is $a(t) = 10\sin 45°$ m/s². If the ball rolls a distance of $5\sqrt{2}$ m what will be its velocity?

12. The moon's gravitational force is one-sixth that of the earth's gravitational force. What would be the high jump record on the moon equivalent to the Olympic record of 2.36 m?

13. What constant acceleration would enable a car travelling at 30 km/h to increase its velocity to 100 km/h while travelling 2 km?

14. A ball rolls in a straight line with an initial velocity of 5 m/s. It is slowing down because of the effect of friction. If its acceleration is -1.4 m/s², how far will it roll?

Rate Problems Requiring the Separation of Variables

In the first three sections of this chapter the rate of change problems have involved the solution of a differential equation of the form $dy/dx = f(x)$. In some applications the rate of change of y with respect to x is expressed as a function of y or both y and x rather than a function of x, that is $dy/dx = f(x, y)$. Differential equations of this type are solved using a method called the **separation of variables**. This method is applicable if $dy/dx = f(x, y)$ can be separated into two relations, one purely in x and the other purely in y. This allows the differential equation to be rewritten in the differential form $h(y)\, dy = g(x)\, dx$. The solution can be found by integrating both sides.

$$\int h(y)\, dy = \int g(x)\, dx$$

The constant that results from each indefinite integral can be combined into one constant value.

EXAMPLE A

Solve the differential equation $dy/dx = -x/y$ given that y is 8 when x is 6.

SOLUTION

Using the separation of variables method the differential form is

$$y\, dy = -x\, dx$$

Integrating both sides gives

$$\int y\, dy = \int -x\, dx$$

$$\frac{y^2}{2} = \frac{-x^2}{2} + C$$

$$y^2 = -x^2 + 2C$$

When $x = 6$, $y = 8$

$$\therefore 8^2 = -6^2 + 2C$$

$$C = 50$$

$$y^2 = -x^2 + 100$$

$$y = \pm\sqrt{-x^2 + 100}$$

The solution of the differential equation is $y = \pm\sqrt{-x^2 + 100}$. ❏

In many applications the instantaneous rate of change of y with respect to x is proportional to the value of y. In these situations $dy/dx = ky$ where k is a constant. This type of differential equation can be solved by using the method of separation of variables.

$$\frac{1}{y}dy = k\,dx$$

Integrating both sides gives

$$\int\frac{1}{y}dy = \int k\,dx$$

$$\ln y = kx + S \quad S \text{ is a constant}$$

$$y = e^{kx + S}$$

$$y = e^S e^{kx}$$

Since S is a constant e^S is a constant and can be called C.

The solution to $dy/dx = ky$ is $y = Ce^{kx}$. Because the solution involves an exponential function, variables that obey the law $dy/dx = ky$ are said to follow an **exponential model**. If the value of k is negative, the relationship is called **exponential decay**, and if k is positive, the relationship is called **exponential growth**.

EXAMPLE B

The number of fruit flies in a colony at a particular time increases in proportion to the number present at that time. A colony of 500 grows at the rate of 3% per day. In how many days will the colony have 1000 flies?

SOLUTION

Let $N(t)$ represent the number of fruit flies present at time t.

$$\frac{dN}{dt} = kN$$

Since the rate of change of the number present is 3% of the number present at any time t, $k = 0.03$.

$$\frac{dN}{dt} = 0.03\,N$$

Using the separation of variables method, the differential form is

$$\frac{dN}{N} = 0.03\,dt$$

Integrating both sides gives

$$\int \frac{dN}{N} = \int 0.03\,dt$$

$$\ln N = 0.03t + S$$

$$N(t) = e^{0.03t + S}$$

$$N(t) = e^S e^{0.03t}$$

$$N(t) = Ce^{0.03t}$$

Initially there are 500 fruit flies so $N(0) = 500$.

$$Ce^{0.03(0)} = 500$$

$$C = 500$$

$$N(t) = 500e^{0.03t}$$

The colony has 1000 fruit flies when $N(t) = 1000$.

$$1000 = 500e^{0.03t}$$

$$0.03t = \ln 2$$

$$t = (\ln 2)/0.03$$

$$t \doteq 23$$

The colony will have 1000 fruit flies in 23 days.

EXAMPLE C

Newton's law of cooling states that the rate at which an object cools is proportional to the difference in the temperature between the object and the surrounding medium. The temperature of a cup of coffee is 95° C, and it is in a room held at a constant temperature of 20° C. If it cools to 50° C in 30 min, what will the temperature be after 1 h?

SOLUTION

Let $T(t)$ be the temperature in $°C$ at any time t in minutes. According to Newton's law of cooling

$$\frac{dT}{dt} = k(T - 20)$$

Using the separation of variables method the differential form is

$$\frac{dT}{T - 20} = k\,dt$$

Integrating both sides gives

$$\int \frac{dT}{T - 20} = \int k\,dt$$

$$\ln(T - 20) = kt + S$$

$$T - 20 = e^{kt + S}$$

$$T(t) = e^{S}e^{kt} + 20$$

$$T(t) = Ce^{kt} + 20$$

Initially the temperature is 95°C. Therefore, $T(0) = 95$. This initial condition is used to evaluate the constant of integration C.

$$T(0) = Ce^{K(0)} + 20$$

$$95 = C + 20$$

$$C = 75$$

Therefore, $T(t) = 75e^{kt} + 20$.

After 30 min, the temperature has fallen to 50° C. Therefore, $T(30) = 50$. This condition is used to evaluate the proportionality constant k.

$$T(30) = 75e^{k(30)} + 20$$

$$50 = 75e^{k(30)} + 20$$

$$75e^{k(30)} = 30$$

$$30k = \ln(30/75)$$

$$k \doteq -0.03$$

Therefore, $T(t) = 75e^{-0.03t} + 20$.

After 1 h the temperature is $T(60)$

$$T(60) = 75e^{-0.03(60)} + 20$$

$$\doteq 32.4$$

After 1 h the temperature of the coffee is 32.4° C.   ☐

EXAMPLE D

A tank initially contains 400 L of pure water. A salt solution of concentration 0.1 kg/L is added at the rate of 20 L/min. The tank is being stirred continually to ensure even mixing. At the same time, the mixed solution is drained from the tank at the rate of 20 L/min. How much salt is in the tank after 1 h?

SOLUTION

Let $S(t)$ kg be the amount of salt in the solution at any time t min.

$$\frac{dS}{dt} = \text{rate of change of salt}$$

$$= \text{rate of inflow of salt} - \text{the rate of outflow of salt}$$

$$= \begin{pmatrix} \text{concentration} \\ \text{of salt in} \\ \text{added solution} \end{pmatrix} \cdot \begin{pmatrix} \text{rate of} \\ \text{inflow of} \\ \text{solution} \end{pmatrix}$$

$$- \begin{pmatrix} \text{concentration} \\ \text{of salt in} \\ \text{withdrawn solution} \end{pmatrix} \cdot \begin{pmatrix} \text{rate of} \\ \text{outflow of} \\ \text{solution} \end{pmatrix}$$

$$= \left(\frac{0.1\,\text{kg}}{\text{L}}\right)\left(\frac{20\,\text{L}}{\text{min}}\right) - \left(\frac{S\,\text{kg}}{400\,\text{L}}\right)\left(\frac{20\,\text{L}}{\text{min}}\right)$$

$$= 2 - 0.05S$$

Using the separation of variables method, the differential form is

$$\frac{dS}{2 - 0.05\,S} = dt$$

Integrating both sides gives

$$\int \frac{dS}{2 - 0.05S} = \int dt$$

$$\frac{1}{-0.05}\ln(2 - 0.05S) = t + K$$

$$\ln(2 - 0.05S) = -0.05(t + K)$$

$$2 - 0.05S = e^{-0.05(t + K)}$$

$$-0.05S = e^{-0.05(t + K)} - 2$$

$$\therefore \ \ S(t) = -20e^{-0.05K}e^{-0.05t} + 40$$

Using C to stand for the constant value $-20e^{-0.05K}$

$$S(t) = Ce^{-0.05t} + 40$$

Initially there is no salt in the tank so $S(0) = 0$.

$$Ce^{-0.05(0)} + 40 = 0$$

$$C = -40$$

Therefore, $S(t) = -40e^{-0.05t} + 40$.
   After 1 h, or 60 min, the amount of salt in the tank is $S(60)$.

$$S(60) = -40e^{-0.05(60)} + 40$$

$$\doteq 38.0$$

After 1 h there are 38.0 kg of salt in the tank.   ❑

Exercise 15.4

1. For each of the following, use the method of separation of variables to solve the differential equation.

   (a) $dy/dx = xy^2$

   (b) $dy/dx = y\sqrt{x}$

   (c) $dy/dx = 1 - x - y + xy$

   (d) $dy/dx = y^{-1}e^x$

2. The rate at which a substance dissolves in a solution is proportional to the amount of the substance that is present. Thirty grams of sodium chloride are put in a litre of water, and it is found that 1 h later only 10 g are left. How much will be left after 30 min more?

3. Bacteria increase in proportion to the amount present. If there are initially 1500 bacteria present and the amount doubles in 2 h, how many bacteria will there be in 5.5 h?

4. The population of a colony of mosquitoes increases in proportion to the number present. If initially there are 1200 mosquitoes and 2000 after 24 h, when will the population be 7500 mosquitoes?

5. A glass of hot water is put in a freezer which is kept at a constant temperature of $-5°$ C. It cools according to Newton's law of cooling and changes from its original temperature of $80°$ C to $60°$ C in 1 h. How long will it take to reach $10°$C?

6. A town is growing so that at any time t in years its instantaneous rate of change of population $dP/dt = kP$ where k is a constant. If the town originally has a population of 2000 people and has a 3% growth in the first year, when will it reach a population of 10 000?

7. In radioactive decay the rate of change of the amount of a radioactive isotope present at any time is proportional to the amount present at that time. The half-life of a radioactive isotope is the time for one half of the isotope to disappear. Carbon-14 has a half-life of 5568 a. How old is bone which has lost 20% of its carbon-14?

8. In radioactive decay the rate of change of the amount of a radioactive isotope present at any time is proportional to the amount present at that time. If the half-life of the radioactive isotope polonium-218 is 3 min, determine the percentage of a sample that remains after 0.5 h.

9. A tank initially contains 2 kg of salt dissolved in 400 L of water. A salt solution of concentration 0.05 kg/L is added at the rate of 25 L/min. The tank is being stirred continuously to ensure even mixing. At the same time, the mixed solution is drained from the tank at the rate of 25 L/min. How much salt is in the tank after 30 min?

10. A tank initially contains 50 kg of salt dissolved in 1000 L of water. Pure water is added at the rate of 20 L/min. The tank is being stirred continuously to ensure even mixing. At the same time the mixed solution is drained from the tank at the rate of 20 L/min. When will the salt concentration be one-half of the initial concentration?

11. In an experiment to investigate the rate of learning it is determined that the rate of change of the number or symbols known by an individual is proportional to the difference between the number of symbols known and the number of symbols required to be learned. An individual who knows 10 symbols initially is given 1000 symbols to learn and has learned 30 after 1 h. How many symbols will the individual know in 5 h? How long will it take to learn 100 symbols?

12. If money is invested at an interest rate which compounds continuously, the instantaneous rate of change of the amount accumulated is proportional to the amount accumulated. That is, $dA/dt = kA$ where $A(t)$ is the amount accumulated at time t. If the rate of interest is 9% compounded continuously then $k = 0.09$. What is the value of an investment of $10 000 after 20 a if the interest rate is 9% compounded continuously? Show that if the interest rate is r% compounded continuously then $dA/dt = 0.01rA$.

Summary and Review   15.5

In certain applications, the instantaneous rate of change of a quantity is a known function, that is, $dy/dx = f(x)$. The solution of this differential equation is found by evaluating the antiderivative or the indefinite integral of $f(x)$. In mathematical notation this means if $\qquad \dfrac{dy}{dx} = f(x)$

then $\qquad y = \displaystyle\int f(x)\, dx$

$\qquad\qquad = F(x) + C$

A boundary condition or ititial condition is used to evaluate C, the constant of integration.

Prime examples are the relations between velocity and position, and acceleration and velocity.

Since $\qquad \dfrac{ds}{dt} = v(t) \qquad$ and $\dfrac{dv}{dt} = a(t)$

then $\quad s(t) = \displaystyle\int v(t)dt \qquad$ and $v(t) = \displaystyle\int a(t)dt$

When dy/dx is a function of both x and y of the form $g(x)/h(y)$, then the differential equation $dy/dx = g(x)/h(y)$ can be solved by separation of variables. Rewriting the equation in the form $h(y)dy = g(x)dx$ and then integrating both sides gives a relation between y and x:

$$\int h(y)dy = \int g(x)dx$$

A particularly important special case is the equation

$$\frac{dy}{dt} = kt$$

which has the solution

$$y = Ce^{kt}$$

This equation describes the exponential growth or decay of a quantity depending on whether k is positive or negative.

To master the material in this chapter, it is helpful to

1. Ask yourself what quantities are varying and how. Express the relationships in words. Use phrases such as "... is a function of ..." or "... varies as ..." or "... depends on ..." to help you decide which variable is the dependent variable and which is the independent variable.

2. Express the relation between the variables as a differential equation, then integrate.

3. Examine the resulting function. For instance, does it increase or decrease as you expect it to?

4. Remember to evaluate the constant of integration using particular values of the variables.

Exercise 15.5

1. Find the indefinite integral of each of the following instantaneous rates of change.

(a) $dR(t)/dt = 12t^3 - \sqrt{t} + 5$

(b) $dH(x)/dx = (3x + 5)^2$

(c) $dS(y)/dy = 5\pi \sin(\pi y)$

(d) $dW(z)/dz = 5e^{4z}$

2. Find the indefinite integral for each of the following instantaneous rates of change. Find the particular function that satisfies the given boundary condition.

(a) $dH(a)/da = 32a + 4$ $H(0) = 5$

(b) $dS(t)/dt = 4e^{5t}$ $S(0) = 9$

(c) $dM(x)/dx = 1/(5 + x)$ $M(-4) = 12$

(d) $dP(t)/dt = \sqrt{3t + 1}$ $P(5) = 1$

3. The instantaneous rate of change of the cost of production of x metres of rope is $dC/dx = 0.6x^{0.5}$, $x \geq 0$. The cost $C(x)$ is measured in dollars. Find the cost of producing the five-hundredth metre.

4. The instantaneous rate of change of the value $V(t)$ of a car is given by $dV/dt = -4000/(1 + t)$ in dollars per year. What is the change in value of a $20\ 000 car two years from the date of purchase?

5. The price $P(t)$ of an oil stock originally purchased at $58.00 per share has been changing at the rate $dP/dt = -3e^{0.02t}$ measured in dollars per month. The investor plans to sell the stock when the price declines to 50% of its original value. In how many months will the investor sell the stock?

6. The number of people using public transit each day is dependent upon the price charged for the service. A small city has found that the instantaneous rate of change of public transit revenue $R(x)$ with respect to the price x in dollars is $dR/dx = 12\ 000 - 4000x$. When $1.00 is charged, the number of riders per day is 10 000. What price produces the maximum revenue for the city? How many people use the service at this price?

7. The population $P(t)$ of a small city is changing with respect to time t according to $dP/dt = 1000e^{0.05t}$. The time t is measured in years. If the initial population of the city is 20 000, what will be its population in 20 years?

8. An employee is given the choice of two pay scales. $S(t)$ stands for the daily salary of the employee. The first includes an initial daily salary of $40 with $dS/dt = 2t$ $/month. The second includes an initial daily salary of $2 with $dS/dt = 1.6e^{0.8t}$ $/month. After 5 months, which pay scale yields the higher daily salary? How much higher is this salary at this time?

9. For each of the following instantaneous velocities determine an expression for the position $s(t)$ at any time t.

(a) $v(t) = 9t^2 + 3t^{1.5}$

(b) $v(t) = 5 \sin(3t + 9)$

(c) $v(t) = 6/(8 + t)$

(d) $v(t) = -6e^{9.2t}$

10. For each of the following instantaneous velocities determine the particular equation for the position $s(t)$ satisfying the given boundary condition.

(a) $v(t) = t^2 - \sqrt{t}$ $s(0) = 9$

(b) $v(t) = 4\pi \cos(\pi t)$ $s(0) = 4$

(c) $v(t) = 8t - 7$ $s(2) = -4$

(d) $v(t) = \sqrt{2t + 15}$ $s(5) = 4$

11. Two objects are moving along the same straight line relative to a fixed point O so at time $t \geq 0$ their respective velocities are $v_1(t) = 2t + 4$ and $v_2(t) = 2t - 5$. The velocities are measured in cm/s. Initially the first object is 3 cm left of O, and at $t = 1$ s the second one is 11 cm to the right of O. Determine when they are together and their velocities at this time.

12. The velocity of a particle moving along the x-axis with position measured relative to the origin is $v(t) = -8t + 12$ cm/s, $t \geq 0$. The initial position of the particle is at the origin. What is the velocity when it is 8 cm to the right of the origin?

13. A rocket is fired vertically upward from the ground and its velocity with respect to the ground at time t s after being fired is $v(t) = 196 - 9.8t$ m/s. What is the maximum height reached by the rocket?

14. A train moves along a straight section of track so that $v(t) = 15 - 0.3t$ m/s at any time $t \geq 0$. What is the initial velocity of the train? How far will the train travel before it stops?

15. Given the instantaneous acceleration, determine the instantaneous velocity in each of the following cases.

(a) $a(t) = 8\sqrt{3t - 9}$ (b) $a(t) = 6e^{-0.3t}$

(c) $a(t) = 12 \sec^2(3t)$ (d) $a(t) = (2t - 1)^{2/3}$

16. In each of the following determine expressions for $v(t)$ and $s(t)$.

(a) $a(t) = 2t^3 - 5$

(b) $a(t) = t^{0.5} - t^{-0.5}$

(c) $a(t) = 5\sin(-2t + 6)$

(d) $a(t) = -4e^{-2t}$

17. For each of the instantaneous accelerations below determine the particular velocity $v(t)$ and position $s(t)$ satisfying the given boundary conditions.

(a) $a(t) = 24t - 8$ $v(0) = 5$ $s(0) = 2$

(b) $a(t) = 12t - 4$ $v(2) = 5$ $s(2) = 0$

(c) $a(t) = 4\pi \cos(\pi t)$ $v(0) = 4$ $s(0) = 5$

(d) $a(t) = 4e^{5t}$ $v(0) = 8$ $s(0) = 9$

18. Two particles are moving along the same straight line with positions measured relative to a fixed point O, and their accelerations at time t are $a_1(t) = 6t + 4$ and $a_2(t) = 12t - 8$. Time is measured in seconds and distance in metres, and both particles have the same initial velocity of 2 m/s and start from the same position 3 m to the right of O. What are their positions when they again have the same velocity?

19. What constant deceleration would enable a car travelling at 90 km/h to stop in a distance of 500 m?

20. A ball rolls from rest down a smooth board inclined at 30° to the horizontal. The acceleration at any time t is $a(t) = 10 \cos 60°$ m/s². The ball rolls a distance of 10 m along the board and then continues rolling on a horizontal floor with an acceleration of -1.6 m/s² because of the effect of friction. How far does the ball roll along the floor before it stops?

21. A motorcycle moving at 2 m/s is 25 m in front of a car which is travelling at a rate of 72 km/h. To avoid collision the driver of the motorcycle accelerates at the rate of 4 m/s². The driver of the car decelerates at the rate of 2 m/s². Will the car hit the motorcycle?

22. A boy throws a snowball from a height of 1.5 m above the ground. The initial velocity of the snowball is 10 m/s at an angle of 60° to the ground. The motion of the snowball can be analyzed by considering two independent motions. These are the vertical motion under the influence of gravity with an initial velocity of $10 \cos 30°$ m/s upward and the horizontal motion independent of gravity with a constant velocity of $10 \cos 60°$ m/s. Analyze the vertical motion to determine how long the snowball was in the air. How far from the boy's feet did the snowball land?

23. For each of the following use the method of separation of variables to solve the differential equation.

(a) $dy/dx = x^2y^{-2}$ (b) $dy/dx = \sqrt{x}/y$

(c) $dy/dx = y - xy$ (d) $dy/dx = y^{-2}e^{3x}$

24. The slope of the tangent at any point (x, y) on a curve is $6x^2y^{-2}$. The point $(1, -1)$ is on the curve. Determine the equation of the curve.

25. In a chemical reaction a substance is used up according to $dS/dt = kS$. $S(t)$ is the amount of the substance in grams present at time t in hours. If 75% of the substance is used up in 2 h what proportion remains after 3 h?

26. Bacteria increase in proportion to the amount present. Initially there are 1000 bacteria present and after 2 h the number of bacteria has doubled. When will the number of bacteria grow to 10 000?

27. A glass of hot water is put in a freezer that is kept at a constant temperature of $0°$ C. It cools according to Newton's law of cooling and changes from its original temperature of $90°$ C to $50°$ C in 1 h. How long will it take to reach $10°$ C?

28. A town grows so that any time t in years its instantaneous rate of change of population $dP/dt = kP$ where k is a constant. The popu-lation of the town on January 1, 1980 was 12 500, and on January 1, 1985 the population was 20 000. What will be the population on January 1, 2000?

29. In radioactive decay the rate of change of the amount of a radioactive isotope present at any time is proportional to the amount present at that time. The half-life of the radioactive iso-tope bismuth-210 is 5 days. If you start with a 5 g sample of this isotope, when will you have 1 g left?

30. A tank initially contains 5 kg of salt dissolved in 500 L of water. A salt solution of concentra-tion 0.02 kg/L is added at the rate of 15 L/min. The tank is being stirred continually to ensure even mixing. At the same time the mixed solu-tion is drained from the tank at the rate of 15 L/min. How much salt is in the tank after 2 h?

31. If money is invested at an interest rate which compounds continuously, the instantaneous rate of change of the amount invested is pro-portional to the amount invested, that is, $dA/dt = kA$. If \$10 000 is invested at 8% com-pounded continuously what will the invest-ment be within 15 years?

H.   Sigma Notation

Sigma notation is a shorthand that is used to express the sum of a series of terms in a compact form. This notation and the results of several of the exercises will be helpful in the discussion of the definite integral and the area under a curve in Chapter 16.

The basic idea is illustrated by the series

$$S = 1^2 + 2^2 + 3^2 + 4^2 + 5^2$$

Each term in the series is of the form k^2, where k is one of the integers from 1 to 5. The expression k^2 represents the **general term** or **kth term** in the series. In sigma notation, this series is written

$$S = \sum_{k=1}^{5} k^2$$

The symbol Σ is the upper case Greek letter sigma. The notation indicates that the successive integers 1 to 5 are to be substituted into the expression k^2, and the resulting values added together. The expression is read: "the summation of k^2, where k runs from 1 to 5."

The numbers 1 and 5 in this case are called, respectively, the **lower** and **upper limits of summation**. The letter k is called the **index of summation**. Any letter may be used as the index of summation. Moreover, the summation index may start at some value other than 1 if the summed expression is changed accordingly. For example, all of the following denote the same sum $1^2 + 2^2 + 3^2 + 4^2 + 5^2$:

$$\sum_{j=1}^{5} j^2, \quad \sum_{m=1}^{5} m^2,$$

$$\sum_{k=0}^{4} (k+1)^2, \quad \sum_{k=3}^{7} (k-2)^2, \quad \sum_{k=-2}^{2} (k+3)^2$$

EXAMPLE A

Expand and evaluate the following sums:

(a) $\displaystyle\sum_{k=4}^{7} k^3$

(c) $\displaystyle\sum_{k=1}^{6} (2k - 1)$

(b) $\displaystyle\sum_{k=1}^{7} 2k$

(d) $\displaystyle\sum_{k=0}^{4} (-1)^k(2k + 1)$

SOLUTION

The result in each case is found by working out the numerical values of the terms and performing the addition (and/or subtraction).

(a) $\displaystyle\sum_{k=4}^{7} k^3 = 4^3 + 5^3 + 6^3 + 7^3$

$\qquad = 64 + 125 + 216 + 343$

$\qquad = 748$

(b) $\displaystyle\sum_{k=1}^{7} 2k = 2 \cdot 1 + 2 \cdot 2 + 2 \cdot 3 + \cdots + 2 \cdot 7$

$\qquad = 2 + 4 + 6 + 8 + 10 + 12 + 14$

$\qquad = 56$

(c) $\displaystyle\sum_{k=1}^{6} (2k - 1) = 1 + 3 + 5 + 7 + 9 + 11$

$\qquad = 36$

(d) $\displaystyle\sum_{k=0}^{4} (-1)^k(2k + 1) = 1 - 3 + 5 - 7 + 9$

$\qquad = 5$

In (d), observe that powers of (-1) cause the signs to alternate. ◻

EXAMPLE B

Express the series $0 + 3 + 8 + 15 + 24 + \cdots + 143$ in sigma notation.

SOLUTION

In problems of this sort, it is necessary to detect some pattern in the numbers. In other words, you must find some expression $f(k)$ such that

$$f(1) = 0, \quad f(2) = 3, \quad f(3) = 8, \quad \text{etc.}$$

In this case, you may notice that each of the numbers is one less than a perfect square, that is

$$f(1) = 1 - 1$$
$$= 0$$
$$f(2) = 4 - 1$$
$$= 3$$
$$f(3) = 9 - 1$$
$$= 8, \text{etc.}$$

Thus, all have the form $k^2 - 1$. For the last term, in particular, $k^2 - 1 = 143$, giving $k = 12$. Thus, there are 12 terms in the series, and using sigma notation, it is written:

$$0 + 3 + 8 + \cdots + 143 = \sum_{k=1}^{12} (k^2 - 1)$$

Alternatively, you might factor the numbers and see a different pattern:

$$0 + 3 + 8 + 15 + 24 + \cdots + 143$$
$$= 0 \cdot 2 + 1 \cdot 3 + 2 \cdot 4 + 3 \cdot 5 + 4 \cdot 6 + \cdots + 11 \cdot 13$$

Here, each term has the form $(k - 1)(k + 1)$ so the series is expressed as

$$0 + 3 + 8 + \cdots + 143 = \sum_{k=1}^{12} (k - 1)(k + 1)$$

The two expressions for the series are equivalent.   ❏

Take note of the following properties of sigma notation:

(1) When the summation index is not present in the expression to the right of the sigma sign, all terms in the sum are the same:

$$\sum_{k=1}^{5} 6 = 6 + 6 + 6 + 6 + 6$$

$$= 5 \cdot 6$$

(2) Factors that are independent of the summation index can be factored out of the sum:

$$\sum_{k=1}^{5} 2k = 2 \sum_{k=1}^{5} k$$

$$= 2(1 + 2 + 3 + 4 + 5)$$

(3) The summation of a sum or difference of terms can be separated into two different summations:

$$\sum_{k=1}^{5} (a_k + b_k) = \sum_{k=1}^{5} a_k + \sum_{k=1}^{5} b_k$$

Exercise H

1. Express in expanded form. Do not evaluate.

(a) $\sum_{k=1}^{5} k^4$

(b) $\sum_{k=1}^{4} (k^2 + 3)$

(c) $\sum_{k=1}^{6} (2k - 3)$

(d) $\sum_{k=-3}^{3} 2k^2$

(e) $\sum_{k=1}^{6} \dfrac{k + 1}{k + 2}$

(f) $\sum_{k=0}^{4} (-1)^k 2^k$

(g) $\sum_{k=0}^{8} \dfrac{(-1)^k}{(2k + 1)}$

2. Express the sums in sigma notation. Do not evaluate.

(a) $1 + 2 + 3 + \cdots + 23$

(b) $2 + 4 + 6 + \cdots + 24$

(c) $1 + 4 + 7 + \cdots + 25$

(d) $4 + 8 + 16 + \cdots + 256$

(e) $1 + 3 + 9 + \cdots + 729$

(f) $1 + 3 + 5 + \cdots + (2n - 1)$

(g) $1 + \dfrac{1}{2} + \dfrac{1}{6} + \cdots + \dfrac{1}{n!}$

3. Evaluate:

(a) $\displaystyle\sum_{k=1}^{7} k$

(b) $\displaystyle\sum_{i=1}^{7} 7$

(c) $\displaystyle\sum_{m=3}^{3} 5m$

(d) $\displaystyle\sum_{k=0}^{5} (10 - k^2)$

(e) $\displaystyle\sum_{k=1}^{6} \dfrac{12}{k}$

(f) $\displaystyle\sum_{j=0}^{7} \sin\dfrac{j\pi}{2}$

(g) $\displaystyle\sum_{k=0}^{10} (1 - \cos k\pi)$

4. What is the sum of the series

$$\sum_{k=1}^{n} (-1)^k$$

if n is odd? If n is even?

Formulas for the sums of powers of the integers are needed for some of the work on finding areas in Chapter 16. The following questions are designed to help you derive such formulas in different ways and, in doing so, increase your skill in the use of sigma notation.

5. The sum of the integers from 1 to n is an arithmetic series (see A. Preparation for Calculus, pages 16–29)

$$S_n = 1 + 2 + 3 + \cdots + n$$

(a) What is the first term? The general term? The common difference? The last term?

(b) Express this series in sigma notation.

(c) Find an expression for the sum of n terms.

6. Certain formulas for sums of integers can be visualized as patterns of dots:

- Squares

$$\sum_{k=1}^{n} n = n + n + n + \cdots + n = n^2$$

$\qquad = n$ rows of n dots.

$\qquad$ 1 $\qquad$ 4 $\qquad$ 9 $\qquad$ 16 $\qquad$ etc.

- Sums of integers (sometimes called triangular numbers):

$$\sum_{k=1}^{n} k = 1 + 2 + 3 + \cdots + n$$
$$= T_n$$
$$= n \text{ rows, increasing in length from 1 to } n.$$

$\qquad$ 1 $\qquad$ 3 $\qquad$ 6 $\qquad$ 10 $\qquad$ etc.

(a) Using these dot patterns, show that the sum of the integers from 1 to n is

$$T_n = \sum_{k=1}^{n} k$$

$$= \frac{n(n+1)}{2}$$

Hint: $= 2T_3$

(b) Using dot patterns, show that

$$n^2 = 2T_n - n$$

(c) Write down the series of the numbers suggested by each of the dot patterns:

Express these series in sigma notation. Show that the sum of the first n odd integers is n^2:

$$\sum_{k=1}^{n} (2k - 1) = n^2$$

What is the connection between this formula and the one in part (b)?

7. (a) By direct expansion of the sum and cancellation of terms show that

$$\sum_{k=1}^{n} [(k + 1)^2 - k^2] = (n + 1)^2 - 1$$

(b) Show that

$$\sum_{k=1}^{n} [(k + 1)^2 - k^2] = \sum_{k=1}^{n} (2k + 1)$$

$$= 2T_n + n$$

(c) Use the results of parts (a) and (b) to show that

$$2T_n = n(n + 1)$$

or

$$(1 + 2 + 3 + \cdots + n) = \frac{n(n + 1)}{2}$$

8. Find a formula for the sum of the first n even integers.

9. Show that

$$\sum_{k=1}^{n} 2k + \sum_{k=1}^{n} (2k - 1) = \sum_{k=1}^{2n} k$$

Write a sentence in English that expresses the content of this equation. Draw a dot pattern to illustrate this equation.

10. (a) By direct expansion of the sum and cancellation of terms show that

$$\sum_{k=1}^{n} [(k + 1)^3 - k^3] = (n + 1)^3 - 1$$

(b) Show that

$$\sum_{k=1}^{n} [(k + 1)^3 - k^3] = \sum_{k=1}^{n} (3k^2 + 3k + 1)$$

$$= 3 \sum_{k=1}^{n} k^2 + 3T_n + n$$

(c) Use the results in parts (a) and (b) to show that

$$3 \sum_{k=1}^{n} k^2 = \frac{2n^3 + 3n^2 + n}{2}$$

or

$$(1^2 + 2^2 + 3^2 + \cdots + n^2) = \frac{n(n + 1)(2n + 1)}{6}$$

11. Starting with the sum

$$\sum_{k=1}^{n} [(k + 1)^4 - k^4]$$

use the methods and results of the previous problem to show that

$$\sum_{k=1}^{n} k^3 = \left(\sum_{k=1}^{n} k \right)^2$$

or

$$(1^3 + 2^3 + 3^3 + \cdots + n^3) = \left[\frac{n(n + 1)}{2} \right]^2$$

12. Show that

(a) $\displaystyle\sum_{k=1}^{n} (k - 1) = \frac{(n - 1)n}{2}$

(b) $\displaystyle\sum_{k=1}^{n} (k - 1)^2 = \frac{(n - 1)n(2n - 1)}{6}$

13. Show that

$$\sum_{k=a}^{b} k = \frac{(a + b)(b - a + 1)}{2}$$

Isaac Newton

From its beginnings until the nineteenth century the calculus was closely associated with problems in mechanics and theoretical astronomy. The great mathematicians of the eighteenth century—Euler, d'Alembert, Clairaut and Lagrange—were also leading mathematical physicists. The physical tradition in analysis originated with Isaac Newton, the greatest scientist of the seventeenth century and the co-inventor with Leibniz of the differential and integral calculus. The son of an English farmer, Newton became in 1669 the Lucasian professor of mathematics at Cambridge University. According to the statutes of the University, the Lucasian professor was to read and expound "some part of Geometry, Astronomy, Geography, Optics, Statics, or some other mathematical discipline."

The celebrated Newtonian synthesis was presented in 1687 in Newton's *Principia mathematica philosophiae naturalis* (Mathematical Principles of Natural Philosophy). Essentially a treatise on the mathematical dynamics of particle motion, the *Principia* combined the concept of universal gravitation with Galileo's laws of terrestial motion and Kepler's laws of planetary motion. At the beginning of the treatise Newton introduced in the form of eleven lemmas his calculus or method of first and last ratios, a geometrical theory of limits that provided the mathematical basis of his dynamics.

Newton's use of the calculus in the *Principia* is illustrated by Proposition 11 of Book One: If the trajectory of a particle moving under the action of a central force is an ellipse with the centre of force at one focus, then the force is inversely proportional to the square of the distance of the particle from the centre. Because the planets were known by Kepler's laws to move in ellipses with the Sun at one focus, this proposition provided support for his inverse-square law of gravitation. To establish the proposition Newton derived an approximate measure for the force using small lines defined in terms of the radius (the line from the force centre to the particle) and the tangent to the curve at a point. This result expressed geometrically the proportionality of force to acceleration. Using properties of the ellipse known from classical geometry Newton calculated the limit of this measure and showed that it was equal to a constant times one over the square of the radius.

The *Principia mathematica* was Newton's most important contribution to science. Although he devoted several other treatises to the calculus, delays in publication meant that his results were duplicated and published independently by other researchers. The *Principia* was written in an austere style that discouraged most readers. Indeed, his mathematical dynamics became influential only after it was translated into the idiom of the Leibnizian calculus in the early 1700s. The great achievements of classical mechanics in the eighteenth century owed their physical basis to Newton but were developed mathematically using the calculus of Leibniz.

Areas and the Definite Integral

16

The indefinite integral, studied in the preceding chapters, represents only one aspect of the subject of integration. In this chapter, the concept of the **definite integral** is introduced. It results from taking the limit of a sum of terms in such a way that the number of terms added becomes infinitely large while simultaneously the size of each term becomes infinitesimally small.

In this chapter, the focus is on the interpretation of the definite integral as the area under a curve. In this context, the relationship between differentiation and integration will be further explored. There are many other applications, however. In later chapters, the definite integral will be used to determine such things as the volume of a solid, the length of an arc, and the work done by a force.

Computation of Area

16.1

Advances in mathematics are often prompted by practical problems. The problem of finding the area of a region is a case in point. If a region is bounded by straight line segments, it is possible to find its area by methods of plane geometry. Divide the region, for instance, into triangles, rectangles, or other simple geometrical figures, and take the total area to be the sum of these elementary areas:

$$A = A_1 + A_2 + A_3 \cdots + A_n$$

If, however, the region in question is bounded by segments of curves rather than straight lines, finding its area is more difficult. It is possible to approximate such an area using rectangles, triangles, and the like. Even though the result would not be exact, this scheme may nevertheless produce a numerically acceptable answer if the subdivisions are made fine enough.

It is natural to want to pursue this strategy to its limit—taking the area to be the sum of an infinite number of infinitesimally small pieces. This step is crucial, for it opens up the possibility of finding exact answers for the areas of regions having arbitrary curvilinear boundaries.

Approximation to the Area Under a Curve

Let $y = f(x)$ be a continuous, non-negative function defined over the closed interval $[a, b]$. Consider the region enclosed by this curve, the lines $x = a$ and $x = b$, and the x-axis, (Figure 16.1). To find the area of this region:

Figure 16.1

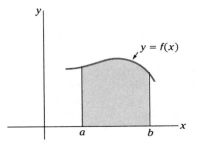

1. Divide the interval $[a, b]$ into n equal subintervals. The width of each subinterval is

$$\Delta x = \frac{(b - a)}{n}$$

2. Denote the x-coordinates at the right end of each subinterval by

$$x_1, x_2, x_3, \ldots x_k, \ldots, x_n$$

where

$$x_k = a + k \cdot \Delta x$$

The corresponding values of the function f are then

$$f(x_1), f(x_2), f(x_3), \ldots, f(x_k), \ldots, f(x_n)$$

3. Construct a rectangle of width Δx and height $f(x_k)$ on each sub-interval (Figure 16.2). The area A_k of a typical rectangle is

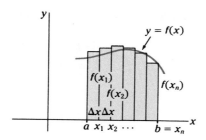

Figure 16.2

$$A_k = f(x_k) \cdot \Delta x$$

4. Find the sum of the areas of all the rectangles. This sum is approximately equal to the area under the curve:

$$A \doteq f(x_1) \cdot \Delta x + f(x_2) \cdot \Delta x + f(x_3) \cdot \Delta x + \ldots + f(x_n) \cdot \Delta x$$

$$= \sum_{k=1}^{n} f(x_k) \cdot \Delta x$$

(Refer to H. Sigma Notation, page 400–405, for a discussion of this notation for sums.)

EXAMPLE A

By summing a set of rectangles, calculate approximately the area of a circle of radius 4.

SOLUTION

To simplify matters, take the centre of the circle to be at the origin, and calculate the area of the circle in the first quadrant (Figure 16.3). Multiplying the result by 4 will give the area of the full circle. In the first quadrant, the function describing the circle is

$$y = \sqrt{4^2 - x^2}$$

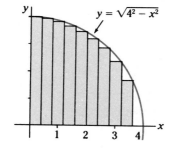

Figure 16.3

The accuracy of the calculation will clearly depend on the number of subintervals into which the interval $x = 0$ to $x = 4$ is divided. Choose, arbitrarily, $n = 10$ for a first approximation. This makes Δx equal to

$$\Delta x = \frac{(4 - 0)}{10}$$

$$= 0.4$$

The area of a typical rectangle is

$$A_k = f(x_k) \cdot \Delta x$$

$$= \sqrt{4^2 - x_k} \cdot (0.4)$$

The sum that gives the area is thus

$$A \doteq \sum_{k=1}^{n} \sqrt{4^2 - x_k^2} \cdot (0.4)$$

where

$$x_k = a + k \cdot \Delta x$$

$$= 0.4k$$

Table 16.1

$\Delta x = 0.4$			
$x_k = 0.4k$			
$f(x_k) = \sqrt{4^2 - x_k^2},$			
k	x_k	$f(x_k)$	$f(x_k) \cdot \Delta x$
1	0.4	3.98	1.59
2	0.8	3.92	1.57
3	1.2	3.82	1.53
4	1.6	3.67	1.47
5	2.0	3.46	1.38
6	2.4	3.20	1.28
7	2.8	2.86	1.14
8	3.2	2.40	0.96
9	3.6	1.74	0.70
10	4.0	0	0
Total:			11.62

The calculations for ten rectangles are summarized in Table 16.1. The area of the full circle is therefore determined to be

$$A \doteq 4(11.62) = 46.48$$

Compare this to the exact value found from the formula $A = \pi r^2$, which is $A = 16\pi \doteq 50.27$. Considering that all the rectangles in this example fall below the edge of the circle, one should expect that this calculation would underestimate the area, as indeed has happened. The error is about 8%. ◻

The repetitive nature of the calculation in Example A suggests the use of a computer, especially if more than two or three digit accuracy is required. A program to compute an area by summing rectangles is given in Figure 16.4. As you can see in Table 16.2, the answer to Example A improves as n increases, at the price of an increase in the computation time. A numerical solution of this sort is sometimes the only practical solution to many scientific and engineering problems. More sophisticated numerical methods will be discussed in Section 16.5.

```
10  REM AREA UNDER A CURVE
100 :
110 DEF FNF(X) = SQR(16 − X ↑ 2)
120 :
130 INPUT "FROM X = ";A
140 INPUT "   TO X = ";B
150 INPUT "NUMBER OF INTERVALS = ";N
160 :
170 DX = (B − A)/N
200 :
210 AREA = 0
220 FOR K = 1 TO N
230    XK = A + K*DX
240    Y = FNF(XK)
250    AREA = AREA + Y*DX
260 NEXT K
300 :
310 PRINT "AREA = ";AREA
390 END
```

Figure 16.4

Remarks: DX stands for Δx. The rectangular areas are computed and added one at a time in line 250 inside the FOR-NEXT loop. This loop structure is the counterpart of the summation sign in the area formula.

Table 16.2
Sum of Rectangles

Number of Subintervals	Area of Quarter Circle	Computation Time (s)
10	11.6180733	1
100	12.4816681	13
1000	12.5582219	128
10000	12.5655658	1289
	$16\pi/4 = 12.5663706$	

The Limit of a Sum

In the foregoing example, it is clear that the value calculated for the area becomes more precise as the number of rectangles increases. It is logical, therefore, to define the area under a curve as the limit of the sum as $n \rightarrow \infty$:

$$A = \lim_{n \to \infty} \sum_{k=1}^{n} A_k$$

$$= \lim_{n \to \infty} \sum_{k=1}^{n} f(x_k) \cdot \Delta x$$

if this limit exists. As the number of subintervals increases, the width of each rectangle, and therefore its area, necessarily approaches zero; picture an exceedingly large number of extremely thin rectangles covering the region in question. Example B below illustrates the use of this definition of area. You will later appreciate the power and elegance of methods of calculus when you have worked through a few problems of this sort.

The following formulas for sums of powers of the integers will be needed in the next example and in some of the exercises. Refer to Exercises 5 to 13, H. Sigma Notation, pages 400–405, for a derivation of these formulas.

$$\sum_{k=1}^{n} 1 = 1 + 1 + 1 + \cdots + 1 = n$$

$$\sum_{k=1}^{n} k = 1 + 2 + 3 + \cdots + n = \frac{n(n+1)}{2}$$

$$\sum_{k=1}^{n} k^2 = 1^2 + 2^2 + 3^3 + \cdots + n^2 = \frac{n(n+1)(2n+1)}{6}$$

$$\sum_{k=1}^{n} k^3 = 1^3 + 2^3 + 3^3 + \cdots + n^3 = \left[\frac{n(n+1)}{2}\right]^2$$

EXAMPLE B

Find the area under the curve $y = x^2$ from $x = 0$ to $x = 5$ (see Figure 16.5).

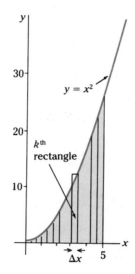

Figure 16.5

SOLUTION

First divide the interval from $x = 0$ to $x = 5$ into n subintervals, and find expressions for x_k and Δx:

$$\Delta x = \frac{5 - 0}{n}$$

$$= \frac{5}{n}$$

$$x_k = 0 + k \cdot \Delta x$$

$$= k \cdot \frac{5}{n}$$

The area of the kth rectangle is then

$$A_k = f(x_k) \cdot \Delta x$$

$$= \left(k \cdot \frac{5}{n}\right)^2 \cdot \frac{5}{n}$$

$$= \frac{125}{n^3} \cdot k^2$$

The sum of the n rectangular areas is

$$\sum_{k=1}^{n} f(x_k) \cdot \Delta x = \frac{125}{n^3} \cdot \sum_{k=1}^{n} k^2$$

$$= \frac{125}{n^3} \cdot \frac{n(n+1)(2n+1)}{6}$$

$$= \frac{125}{6} \cdot \frac{2n^3 + 3n^2 + n}{n^3}$$

The area under the curve is therefore

$$A = \lim_{n \to \infty} \sum_{k=1}^{n} A_k$$

$$= \lim_{n \to \infty} \sum_{k=1}^{n} f(x_k) \cdot \Delta x$$

$$= \lim_{n \to \infty} \frac{125}{6} \cdot \frac{2n^3 + 3n^2 + n}{n^3}$$

$$= \frac{125}{6} \lim_{n \to \infty} \left(2 + \frac{3}{n} + \frac{1}{n^2} \right)$$

$$= \frac{125}{6} \cdot 2$$

$$= \frac{125}{3} \text{ square units}$$

Exercise 16.1

1. Sketch the curve whose equation is given and shade in the area under the curve in the specified interval:

(a) $y = 6x - x^2$; $x = 1$ to $x = 6$

(b) $y = \sin x$; $x = \pi/2$ to $x = \pi$

(c) $y = 5 - \sqrt{25 - x^2}$; $x = -3$ to $x = 3$

(d) $y = 8x - x^3$; $x = 1$ to $x = 2$

(e) $y = \sec x$; $x = 0$ to $x = \pi/3$

(f) $y = \ln(x + 1)$; $x = 1$ to $x = 10$

2. Find the shaded areas in **Figure 16.6 a–d:**

Figure 16.6

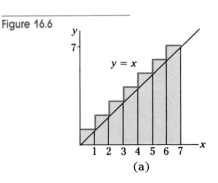

(a)

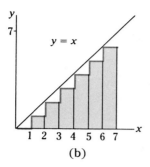

(b)

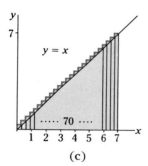

(c)

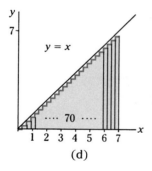

(d)

3. Estimate the shaded areas in Figure 16.7 a–d, assuming they are triangles:

Figure 16.7

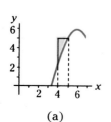

(a)

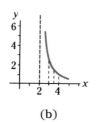

(b)

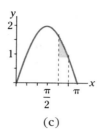

(c)

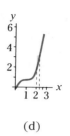

(d)

4. Find the area of the four rectangles in the diagram in Figure 16.8 that approximate the area under the curve $y = -x^2 + 6x + 16$ from $x = 1$ to $x = 7$.

Figure 16.8

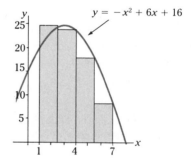

5. Obtain an approximation for the area under the curve $y = x^3/8$ in the interval $x = 3$ to $x = 4$, with the interval divided into four parts.

6. Find an approximate value for the area under the curve $y = x^2 + 2$ for $0 \leq x \leq 3$ by dividing the interval into

(a) 3 equal parts

(b) 6 equal parts.

(c) The exact area is 15 square units. With the help of a diagram, show why the approximation improves when the number of rectangles is doubled.

7. Express x_k in terms of k, if x_k is taken to be the x-coordinate at the *left* end of each subinterval. Repeat the calculation of Example A under these circumstances. Compare this result to the result of Example A and discuss the difference.

8. Sketch the graph of each of the following functions and roughly estimate the area under the curve in the given interval. Then numerically calculate the area with the help of a computer program.

(a) $y = (x + 2)(x - 1)^2$; $x = -1$ to 2

(b) $y = (x - 1)e^{x/2}$; $x = 1$ to 5

(c) $y = (\pi/2 - x)\tan x$; $x = 0$ to $\pi/4$

(d) $y = \ln\sqrt{x^2 + 16}$; $x = 4$ to 12

9. Consider the region defined by the function $f(x) = 2x + 2$, the line $x = 5$, and the coordinate axes.

(a) Find an expression for A_k, the area of the kth rectangle. Choose the height of the rectangle to be the value of f at the right side.

(b) Determine the area of the region by finding the limit of the sum

$$A = \lim_{n \to \infty} \sum A_k$$

(c) Determine the area of the region using the formula for the area of a trapezoid. Compare to (b).

10. Repeat question 9 for the region defined by the function $f(x) = -x/2 + 4$, the x-axis, and the lines $x = \pm 2$.

11. Determine the area under the curve $y = 12 - x^2$ from $x = -2$ to $x = 2$ as the limit of a sum.

12. Determine the area under the curve $y = x(x - 3)^2$ from $x = 0$ to $x = 3$ as the limit of a sum.

13. (a) Show that the area of a regular, n-sided polygon inscribed in a circle of radius R is

$$A = \frac{n}{2}R^2 \sin \frac{2\pi}{n}$$

(b) Show that the area of the polygon approaches the area of the circle in the limit as $n \to \infty$.

14. Find the area of a regular, n-sided polygon circumscribed about a circle of radius R, and show that its area approaches the area of the circle in the limit as $n \to \infty$.

16.2 The Definite Integral

You have seen in the preceding section that the area under a curve can be defined as the limit of a sum:

$$A = \lim_{n \to \infty} \sum_{k=1}^{n} f(x_k)\Delta x$$

Central to this definition is the notion that a quantity can be represented as the sum of an infinite number of infinitesimally small terms. This idea is of fundamental importance, arising in all sorts of physical and mathematical applications. Many quantities such as volume, arc length, and work, to mention only a few, for example, can be thought of in the same way.

In what follows, you will see that such sums can be calculated by a process of integration, a process that will prove more illuminating than the numerical methods and more versatile than the algebraic methods described in the previous section. In order to understand just where and how integration comes into the picture, consider again the problem of finding the area under a curve, with the question put somewhat differently.

Suppose that $y = f(x)$ is a continuous function whose graph lies above the x-axis in the interval $[a, b]$ and that x is some value in the interval (Figure 16.9). The area from a up to x clearly depends on what value is ultimately chosen for x. The area under the curve in the interval $[a, x]$ is therefore a function of x, which we shall call the **area function**, A. How does A depend on x, and what is its connection with f?

If $A(x)$ is the area under the curve from a to x, then $A(x + \Delta x)$ is a slightly larger area, and the difference $\Delta A = A(x + \Delta x) - A(x)$ is the area of a thin strip as shown in Figure 16.10b. This strip is not rectangular in shape. Its area is exactly the area under the curve on the interval of width Δx. It should be evident from the geometry of the situation that, in value, the area of the strip lies between the areas of the two rectangular regions shown in Figure 16.10a and c.

Figure 16.9

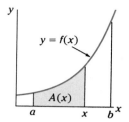

Figure 16.10

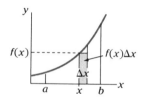

(a)

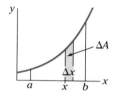

(b)

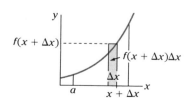

(c)

$$f(x) \cdot \Delta x < A(x + \Delta x) - A(x) < f(x + \Delta x) \cdot \Delta x$$

Dividing by Δx,

$$f(x) < \frac{A(x + \Delta x) - A(x)}{\Delta x} < f(x + \Delta x)$$

In the limit as $\Delta x \to 0$,

$$f(x) < \lim_{\Delta x \to 0} \frac{A(x + \Delta x) - A(x)}{\Delta x} < f(x)$$

This limit is by definition the derivative of the area function:

$$\lim_{\Delta x \to 0} \frac{A(x + \Delta x) - A(x)}{\Delta x} = \frac{dA}{dx}$$

The derivation shows that the value of the derivative lies between two other quantities, both of which equal $f(x)$ in the limit as $\Delta x \to 0$. Thus, the derivative must equal $f(x)$:

$$\frac{dA}{dx} = f(x)$$

Consequently, $A(x)$ must be an integral of $f(x)$:

$$A(x) = \int f(x) \, dx$$
$$= F(x) + C$$

This is the required area function. The actual value of area under the curve in the interval $[a, b]$ can now be determined. When $x = a$ at the lower end of the interval, the area function must be zero. From this, the constant of integration can be evaluated:

$$A(x) = F(x) + C$$
$$A(a) = F(a) + C$$
$$0 = F(a) + C$$

So,

and, $$C = -F(a)$$

$$A(x) = F(x) - F(a)$$

Now setting $x = b$, gives $A(b)$, which is the area under the curve over the entire interval from a to b:

$$A = A(b)$$
$$= F(b) - F(a)$$

The conclusion is that the area under the curve is given exactly by

$$A = F(b) - F(a)$$

where

$$F(x) = \int f(x)\,dx$$

The **definite integral** of a continuous function f, from a to b is denoted by the symbol

$$\int_a^b f(x)\,dx$$

and defined to be equal to the infinite sum:

$$\int_a^b f(x)\,dx \equiv \lim_{n \to \infty} \sum_{k=1}^n f(x_k) \cdot \Delta x$$

You may think of the integral sign in this context as a new summation symbol—an elongated S—that stands for this particular type of sum. The value b is called the **upper limit** of integration, and the value a, the **lower limit**. As shown above, the value of the definite integral, which is equal to $F(b) - F(a)$, can be interpreted as the area under the graph of function f from a to b:

$$A = \int_a^b f(x)\,dx$$
$$= F(b) - F(a)$$

In practice, a definite integral is evaluated in two steps:

1. Find the indefinite integral in the usual way (omitting the constant of integration).
2. Substitute the upper and lower limits into $F(x)$ and subtract.

$$A = \int_a^b f(x)\,dx$$

$$= F(x)\Big|_a^b$$

$$= F(b) - F(a)$$

Here, the symbol

$$F(x)\Big|_a^b$$

is a notational device to keep track of the limits of integration between the two steps.

EXAMPLE A

Evaluate the definite integral

$$\int_1^3 (2x + 5)\,dx$$

SOLUTION

$$\int_1^3 (2x + 5)\,dx = (x^2 + 5x)\Big|_1^3$$

$$= (3^2 + 5\cdot 3) - (1^2 + 5\cdot 1)$$

$$= 24 - 6$$

$$= 18 \qquad\qquad \square$$

EXAMPLE B

Find the area between the curve $y = x^2$ and the x-axis in the interval $0 \le x \le 5$.

SOLUTION

$$A = \int_a^b f(x)\,dx$$

$$= \int_0^5 x^2\,dx \qquad = \frac{x^3}{3}\Big|_0^5$$

$$= \frac{5^3}{3} - \frac{0^3}{3}$$

$$= \frac{125}{3} \text{ square units}$$

Compare this solution to the solution of the same problem given in Example B in the preceding section. ❏

EXAMPLE C

Find the area under the curve

$$y = (x + 1)(x - 2)^2$$

between the zeros of the function.

SOLUTION

The function is a cubic with zeros at $x = -1$ and $x = 2$. Figure 16.11 shows its graph and the required area.

Figure 16.11

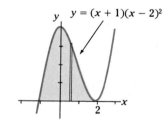

$$A = \int_a^b f(x)\, dx$$

$$= \int_{-1}^{2} (x + 1)(x - 2)^2\, dx$$

$$= \int_{-1}^{2} (x^3 - 3x^2 + 4)\, dx$$

do not try to substitute the limits until the integration is completely finished

$$= \left(\frac{x^4}{4} - x^3 + 4x \right) \Big|_{-1}^{2}$$

do not do any numerical simplification at all until the limits have been properly substituted

$$= \left(\frac{2^4}{4} - 2^3 + 4 \cdot 2 \right) - \left(\frac{(-1)^4}{4} - (-1)^3 + 4 \cdot (-1) \right)$$

$$= (4 - 8 + 8) - \left(\frac{1}{4} + 1 - 4 \right)$$

simplify carefully— improper treatment of negative signs is a common source of error

$$= \frac{27}{4} \text{ square units}$$

❏

Properties of the Definite Integral

Although the definite integral has been introduced in connection with the area under a curve, it has a much wider significance. Its most important properties are given below. They can be proven from the definition of the integral (see the exercises).

1. A definite integral is a real number, in contrast to an indefinite integral, which is a function. Its value depends on the limits of integration but not on what letter is chosen for the variable of integration:

$$\int_a^b f(x)\,dx = \int_a^b f(z)\,dz$$

For this reason, the variable of integration is sometimes called a **dummy variable**.

2. When the limits are the same, the value of the definite integral is zero:

$$\int_a^a f(x)\,dx = 0$$

Interpretation: there is zero area between a curve $y = f(x)$ and a single point on the x-axis.

3. When the limits of a definite integral are reversed, the sign of the integral changes:

$$\int_a^b f(x)\,dx = -\int_b^a f(x)\,dx$$

4. Since a definite integral is just a sum:

$$\int_a^b f(x)\,dx + \int_b^c f(x)\,dx = \int_a^c f(x)\,dx$$

The whole is the sum of its parts.

The First Fundamental Theorem of Calculus

The ideas developed in this section rest on the following theorem: If the function f is continuous on the interval $[a, b]$, then

$$\int_a^b f(x)\,dx = F(b) - F(a)$$

where F is any function such that $F'(x) = f(x)$ for all x in $[a, b]$ and the definite integral stands for the sum:

$$\int_a^b f(x)\,dx \equiv \lim_{n \to \infty} \sum_{k=1}^{n} f(x_k) \cdot \Delta x$$

This major theorem establishes the relationship between definite integrals and antiderivatives. The main concepts have been outlined in this section. For a complete proof, you must consult a more advanced calculus text.

Exercise 16.2

1. Sketch the area that corresponds to each of the following definite integrals.

(a) $\displaystyle\int_{-1}^{1} (3 - x^2)\,dx$

(b) $\displaystyle\int_{1}^{4} (4x - x^2 - 5)\,dx$

(c) $\displaystyle\int_{0}^{3} 5e^{-x}\,dx$

(d) $\displaystyle\int_{1}^{5} \frac{1}{x}\,dx$

(e) $\displaystyle\int_{0}^{\pi} (1 + \cos 2x)\,dx$

2. Express each of the areas shown in Figure 16.12 as a definite integral:

Figure 16.12

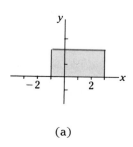

(a)

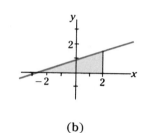

(b)

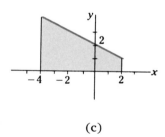

(c)

3. Evaluate the following definite integrals.

(a) $\displaystyle\int_{2}^{6} (3x + 5)\,dx$

(b) $\displaystyle\int_{-2}^{3} (x^2 + 4)\,dx$

(c) $\displaystyle\int_{0}^{2} x^3\,dx$

(d) $\displaystyle\int_{-2}^{2} x^4\, dx$

(e) $\displaystyle\int_{-4}^{1} (x^2 + 4x + 8)\, dx$

4. Make a sketch and calculate the areas of the regions under the curves:

(a) $y = 3x^2$,   from $x = -2$ to $x = 0$

(b) $y = x^2 + x$,   from $x = 2$ to $x = 5$

(c) $y = x^3 - x^2$,   from $x = 0$ to $x = 3$

(d) $y = 4x^{2/3}$,   from $x = 1$ to $x = 8$

5. In each of the following cases, determine the area of the region bounded by the given curve and the x-axis:

(a) $y = \sin x$,   from $x = 0$ to $x = \pi$

(b) $y = 2 \ln x$,   from $x = 1$ to $x = 10$

(c) $y = \sec(x/2)$,   from $x = -\pi/3$ to $x = \pi/3$

(d) $y = \dfrac{x^2 + 4}{x}$,   from $x = 1$ to $x = 4$

(e) $y = e^{2x} - e^{-2x}$, from $x = 0$ to $x = 2$

6. Compute the area under the curve for those intervals where the function lies above the x-axis.

(a) $y = 3 - 2x - x^2$

(b) $y = 16 - x^4$

(c) $y = \dfrac{2 \sin x - 1}{2}$, $0 \le x \le 2\pi$

7. Compute the area of the figure bounded by the two curves and the x-axis. Hint: Find where the curves intersect.

(a) $y = 2x$ and $y = (x - 4)^2$

(b) $y = x^2$ and $y = 6x - x^2$

8. Find the area between the curve $y = 3 + 2x - x^2$ and the line $y = -5$. Hint: Consider a vertical translation of the function.

9. Find the area enclosed by the two curves $y = x^2 - 6x + 11$ and $y = x + 1$ by subtracting the area under one from the area under the other. Hint: Where do the curves intersect?

10. In each of the figures shown in Figure 16.13, the area is bounded by an arc of a circle of radius 5. In each case, express the area as a definite integral, then evaluate it with the help of the integration formula:

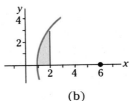

$$\int \sqrt{a^2 - v^2}\, dv = \frac{1}{2} v \sqrt{a^2 - v^2} + \frac{1}{2} a^2 \sin^{-1} \frac{v}{a}$$

Figure 16.13

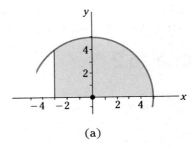

(a)

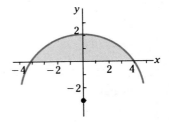

(b)

(c)

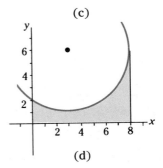

(d)

11. A function is odd if $f(-x) = -f(x)$, and even if $f(-x) = f(x)$. What can be said about

$$\int_{-a}^{a} f(x)\,dx$$

if f is odd? If f is even?

12. Prove that

(a) $\displaystyle\int_{a}^{a} f(x)\,dx = 0$

(b) $\displaystyle\int_{a}^{b} f(x)\,dx = -\int_{b}^{a} f(x)\,dx$

(c) $\displaystyle\int_{a}^{c} f(x)\,dx = \int_{a}^{b} f(x)\,dx + \int_{b}^{c} f(x)\,dx$

Areas Between Curves   16.3

As you have seen, the strategy for computing an area is to consider it to be the sum of an infinite number of rectangles with vanishingly small width. This idea of summing a large number of similar elements is an important one that arises repeatedly in problems of all sorts. The basic unit of such a sum is called the **element of integration**.

In finding areas as above, it is convenient to choose the element of integration to be a narrow rectangular strip, but you should realize that the element of integration is not always a rectangle. Depending on the application, it could be a thin triangle, a narrow band, a thin circular disk, a cylindrical shell, or something else. You will see examples of these in later chapters. The key to many problems is to find an expression for a suitable element of integration. Once this is done, it is merely a matter of working out an integral to complete the solution.

Consider now a more general area problem. This is the problem of finding the area between two curves, that is, an area whose lower boundary is not the x-axis, but rather some function of x. Take the element of integration to be a rectangle, whose upper end is on the top curve at $y_{top} = f(x)$ and whose lower end is on the bottom curve at $y_{bottom} = g(x)$ (Figure 16.14). The length of this rectangle is $(y_{top} - y_{bottom})$. Its width is Δx. Its area is therefore

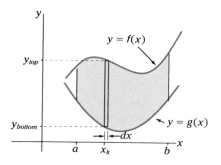

Figure 16.14

$$A_k = (y_{top} - y_{bottom}) \cdot \Delta x$$

$$= [f(x_k) - g(x_k)] \cdot \Delta x$$

The total area between the curves is therefore the limit of the sum

$$A = \lim_{\Delta x \to 0} \sum_{k=1}^{n} A_k$$

$$= \lim_{\Delta x \to 0} \sum_{k=1}^{n} \left[f(x_k) - g(x_k) \right] \Delta x$$

$$= \int_{a}^{b} \left[f(x) - g(x) \right] dx$$

In setting up an integral of this sort, it is customary to by-pass the summation step altogether and express the area of the element of integration directly as a **differential** (see Chapter 13):

$$dA = \text{Area of element}$$

$$= (\text{length}) \cdot (\text{width})$$

$$= \left[f(x) - g(x) \right] \cdot dx$$

In this context, the differential dA is viewed as a typical, infinitesimally small element in an infinite sum. The integral for the area then follows directly from

$$A = \int_{a}^{b} dA$$

$$= \int_{a}^{b} \left[f(x) - g(x) \right] dx$$

EXAMPLE A

Find the area between the curves $y = 2 - x$ and $y = 12 - (x - 2)^2$ in the interval $1 \le x \le 5$.

SOLUTION

First, sketch graphs of the functions and shade in the required area (Figure 16.15). The bottom boundary of this area is the line $y = 2 - x$. The element of integration is a rectangle reaching from the parabola on top down to the line. The area of this rectangle is

$$dA = [y_{\text{top}} - y_{\text{bottom}}] \, dx$$

$$= [12 - (x - 2)^2 - (2 - x)] \, dx$$

The area is therefore the sum of such rectangles which is the definite integral

$$A = \int_a^b [y_{\text{top}} - y_{\text{bottom}}] \, dx$$

$$= \int_1^5 [12 - (x - 2)^2 - (2 - x)] \, dx \qquad \longleftarrow \text{simplify the integrand before carrying out the integration}$$

$$= \int_1^5 [-x^2 + 5x + 6] \, dx$$

$$= -\frac{x^3}{3} + \frac{5x^2}{2} + 6x \, \Big|_1^5$$

$$= \left[-\frac{5^3}{3} + \frac{5 \cdot 5^2}{2} + 6 \cdot 5 \right] - \left[-\frac{1^3}{3} + \frac{5 \cdot 1^2}{2} + 6 \cdot 1 \right]$$

$$= \frac{128}{3} \text{ square units}$$

Observe that it does not matter if part or all of the area lies below the x-axis. Choosing the length of the rectangular element of integration to be $(y_{\text{top}} - y_{\text{bottom}})$ automatically gives the total area *between* the curves, wherever they lie.

Figure 16.15

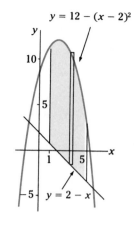

$y = 12 - (x - 2)^2$

$y = 2 - x$

EXAMPLE B

Find the area in the first quadrant between the curves
$y = 10 - x^2$ and $y = 9/x^2$.

SOLUTION

The required area is shown in Figure 16.16. The limits of integration are found by determining the points where the two curves intersect:

$$10 - x^2 = \frac{9}{x^2}$$

$$x^4 - 10x^2 + 9 = 0$$

$$(x^2 - 1)(x^2 - 9) = 0$$

$$x = \pm 1, \pm 3$$

Figure 16.16

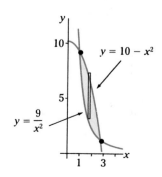

$y = 10 - x^2$

$y = \frac{9}{x^2}$

In the first quadrant, the area corresponds to values of x in the interval $[1, 3]$, therefore, 1 and 3 are the limits of integration. The element of integration is a rectangle with area

$$dA = [y_{top} - y_{bottom}] \, dx$$

$$= \left[(10 - x^2) - \frac{9}{x^2} \right] dx$$

The area is thus given by

$$A = \int_1^3 \left[(10 - x^2) - \frac{9}{x^2} \right] dx$$

$$= 10x - \frac{x^3}{3} + \frac{9}{x} \Big|_1^3$$

$$= \left[10 \cdot 3 - \frac{3^3}{3} + \frac{9}{3} \right] - \left[10 \cdot 1 - \frac{1^3}{3} + \frac{9}{1} \right]$$

$$= \frac{16}{3} \text{ square units} \qquad \square$$

EXAMPLE C

Find the area between the curves $y = 2x - 8$ and $y^2 = 8x$.

SOLUTION

A diagram is essential. Figure 16.17 shows the required area. The points at which the curves intersect can be found by eliminating y:

Figure 16.17

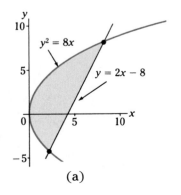
(a)

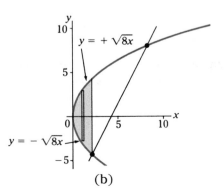
(b)

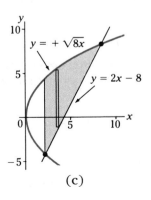
(c)

$$(2x - 8)^2 = 8x$$

$$4x^2 - 32x + 64 = 8x$$

$$4x^2 - 40x + 64 = 0$$

$$4(x - 2)(x - 8) = 0$$

$$x = 2, 8$$

The intersection points are thus $(2, -4)$ and $(8, 8)$.

This problem is complicated by the fact that you must consider two areas: the function to be used for y_{bottom} depends on whether the element of integration is to the right or to the left of $x = 2$. The area must be written, therefore, as the sum of two definite integrals:

$$A = A_1 + A_2$$

$$= \int_0^2 [\sqrt{8x} - (-\sqrt{8x})] \, dx + \int_2^8 [\sqrt{8x} - (2x - 8)] \, dx$$

Taking these areas one at a time:

$$A_1 = \int_0^2 [\sqrt{8x} - (-\sqrt{8x}] \, dx$$

$$= 2\sqrt{8} \int_0^2 \sqrt{x} \, dx$$

$$= 2\sqrt{8} \frac{x^{3/2}}{3/2} \Big|_0^2$$

$$= \frac{4\sqrt{8}}{3} \cdot (2)^{3/2}$$

$$= \frac{32}{3} \text{ square units}$$

The area in the interval $[2, 8]$ is given by the integral

$$A_2 = \int_2^8 [\sqrt{8x} - (2x - 8)] \, dx$$

$$= \sqrt{8} \frac{x^{3/2}}{3/2} - x^2 + 8x \Big|_2^8$$

$$= \left[\frac{2\sqrt{8}}{3}(8)^{3/2} - 8^2 + 8 \cdot 8 \right] - \left[\frac{2\sqrt{8}}{3}(2)^{3/2} - 2^2 + 8 \cdot 2 \right]$$

$$= \left[\frac{128}{3} \right] - \left[\frac{16}{3} - 4 + 16 \right]$$

$$= \frac{76}{3} \text{ square units}$$

The total area is therefore

$$A = A_1 + A_2$$

$$= \frac{32}{3} + \frac{76}{3}$$

$$= 36 \text{ square units} \qquad \square$$

Exercise 16.3

1. Each of the following definite integrals corresponds to the area between two curves. Sketch the curves and shade in the area.

(a) $\displaystyle\int_0^2 [4x - x^3]\, dx$

(b) $\displaystyle\int_1^4 [((x - 2)^2 + 2) - (-(x - 3)^2)]\, dx$

(c) $\displaystyle\int_2^5 [\sqrt{x - 2} - (-x + 2)]\, dx$

(d) $\displaystyle\int_0^1 [\sin(\pi x/2) - x^2]\, dx$

(e) $\displaystyle\int_1^5 [e^x - \ln x]\, dx$

2. Explain in terms of areas why the following two integrals are equal:

$$\int_{-2}^{2} [(x^2 + 8) - \sin \pi x]\, dx = \int_{-2}^{2} [(x^2 + 8)]\, dx$$

3. The areas shown in Figure 16.18 are bounded by parabolas congruent to $y = x^2$, and straight lines with slope ± 1 (and vertical lines in two cases). In each case, determine the equations of the bounding curves, and express the area as a definite integral.

Figure 16.18

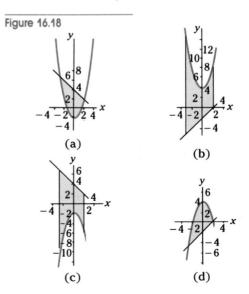

(a)

(b)

(c)

(d)

4. Find the areas between the curves over the given interval:

(a) $y = x^2 - 4x$
  $y = 2x$ $[0, 3]$

(b) $y = (x - 2)^2$
  $y = -2x + 1,\ [0, 3]$

5. Find the areas enclosed by the curves.

(a) $y = x^2 - 2x - 4$
  $y = -x^2 + 8$

(b) $y = 4x^2 - 16x + 12$
  $y = x^3 - 3x^2 - 4x + 12$

6. Sketch the curves and find the areas they enclose.

(a) $y = \sin \pi x$
  $y = x^2 - x$ $[0, 1]$

(b) $y = \cos (\pi x/2)$
  $y = -x + 1$ $[0, 1]$

7. Determine the areas enclosed by the following curves. Hint: Find their intersection points.

(a) $y = \sin x$
  $y = \sin(x - \pi/3)$

(b) $y = \sqrt{3} \sin x$
  $y = \cos x$

(c) $y = \sin x$
  $y = \cos 2x$ (two different areas)

8. Given the two curves $y = -(x - 2)^2 + 9$ and $y = -x + 9$, find the area of the region in the first quadrant that is

(a) between the curves

(b) bounded by the curves and the y-axis

(c) bounded by the curves and the x-axis.

9. (a) Find the equation of the line tangent to the parabola

$$y = x^2 + bx + c \quad \text{at} \quad x = 0.$$

(b) Find the area between the tangent and the parabola in the interval $-h \le x \le h$.

(c) Show that the chord that intersects the parabola at $x = \pm h$ is parallel to the tangent of part (a), and find its equation.

(d) Find the area between the chord of part (c) and the parabola for the interval $-h \le x \le h$.

(e) Compare the areas calculated in parts (b) and (d).

10. (a) Show that the curve $y = ax^3 + cx$ has an inflection point at the origin.

(b) Find the conditions under which the line $y = mx$ intersects the cubic of part (a) at three points and determine the x-coordinates of the intersection points.

(c) Show that the two areas enclosed by the cubic $y = ax^3 + cx$ and the line $y = mx$ are equal. Find the areas.

Improper Integrals   16.4

There are two situations in which the process of finding the definite integral of a function requires special attention. What if the interval is infinite in extent, that is, one or both of the limits of integration are infinite? Or what if the function has a vertical asymptote and is consequently unbounded in the interval?

These types of integrals are called **improper integrals**. They can be evaluated using a limit process, as illustrated in the following examples.

EXAMPLE A

Find the value of

$$\int_1^{+\infty} \frac{1}{x^2}\,dx$$

Figure 16.19

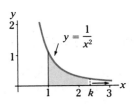

SOLUTION

If this improper integral is written with a finite value k as its upper limit, then it represents the area under the curve $y = 1/x^2$ from $x = 1$ to k, as in Figure 16.19. This area clearly depends on the value of k and increases as k increases. It is reasonable therefore to say that the value of the improper integral is the limit of this area as $k \to +\infty$ if this limit exists:

$$\int_1^{+\infty} \frac{1}{x^2}\,dx = \lim_{k \to +\infty} \int_1^k \frac{1}{x^2}\,dx$$

Now since

$$\int_1^k \frac{1}{x^2}\,dx = -\frac{1}{x}\bigg|_1^k$$

$$= -\frac{1}{k} + 1$$

therefore,

$$\int_1^{\infty} \frac{1}{x^2}\,dx = \lim_{k \to +\infty}\left(-\frac{1}{k} + 1\right)$$

$$= 1$$

Thus, in spite of the fact that the interval in question is infinite, the area under the curve is finite and equal to 1 square unit. This improper integral is said to **converge** to a finite value.   ❑

EXAMPLE B

Find the value of

$$\int_1^{+\infty} \frac{1}{\sqrt{x}}\,dx$$

SOLUTION

Figure 16.20

This integral is the area under the curve shown in Figure 16.20. Just as in Example A, first do the integral with the upper limit equal to k, and then take the limit as $k \to \infty$:

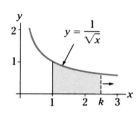

$$\int_1^k \frac{1}{\sqrt{x}}\,dx = 2\sqrt{x}\,\Big|_1^k$$

$$= 2\sqrt{k} - 2$$

Therefore,

$$\int_1^{+\infty} \frac{1}{\sqrt{x}}\,dx = \lim_{k \to +\infty} (2\sqrt{x} - 2)$$

$$= +\infty$$

The function $y = 1/\sqrt{x}$ looks superficially like the function in Example A in that both approach the x-axis asymptotically. However, in this case the area under the curve is infinite and the improper integral **diverges**. ❏

EXAMPLE C

Find the area under the curve $y = 1/\sqrt{x}$ in the interval $0 \le x \le 1$.

SOLUTION

As you can see in Figure 16.21, the y-axis is a vertical asymptote for this function. Since the function is unbounded at $x = 0$, you must handle the lower limit of integration with care: set it equal to k, and then after integration, let $k \to 0$.

Figure 16.21

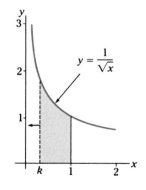

$$\int_k^1 \frac{1}{\sqrt{x}}\,dx = 2\sqrt{x}\,\Big|_k^1$$

$$= 2 - 2\sqrt{k}$$

Therefore,

$$\int_0^1 \frac{1}{\sqrt{x}}\,dx = \lim_{k \to 0} (2 - 2\sqrt{k})$$

$$= 2$$

Thus, although the function is unbounded in this interval, the definite integral converges, and the area under the curve is 2 square units. ◻

EXAMPLE D

Find the area bounded by the x- and y-axes, the line $x = 2$, and the curve $y = 1/(x - 1)^2$.

Figure 16.22

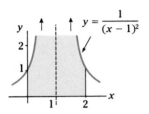

SOLUTION

This function has a vertical asymptote at $x = 1$ (see Figure 16.22). Therefore, the integral for the area must be split into two parts:

$$\int_0^2 \frac{dx}{(x - 1)^2} = \int_0^1 \frac{dx}{(x - 1)^2} + \int_1^2 \frac{dx}{(x - 1)^2}$$

Evaluate the first integral on the right on an interval $[0, k]$ that does not include the asymptote, then take the limit as $k \to 1$ from the left:

$$\int_0^k \frac{dx}{(x - 1)^2} = \frac{-1}{(x - 1)} \bigg|_0^k$$

$$= \frac{-1}{k - 1} - 1$$

Therefore,

$$\int_0^1 \frac{dx}{(x - 1)^2} = \lim_{k \to 1^-} \left(\frac{-1}{k - 1} - 1 \right)$$

$$= +\infty$$

Thus, this integral and likewise the second integral on the right are divergent, so the area of the given region is infinite. ◻

It is critical for the function in the integrand to be defined over the entire interval from the lower to the upper limit of integration. Observe in Example D that if you overlook the fact that there is a vertical asymptote in the interval $[0, 2]$, you would be tempted to write

$$\int_0^2 \frac{dx}{(x - 1)^2} = \frac{-1}{(x - 1)} \bigg|_0^2 \qquad \text{incorrect!}$$

$$= \frac{-1}{2 - 1} - \frac{-1}{0 - 1}$$

$$= -2$$

thus obtaining a result which is clearly false.

Exercise 16.4

1. Sketch the area that corresponds to each of the following improper integrals. Then evaluate the integral or show that it is divergent.

(a) $\int_1^\infty \dfrac{1}{x^3} dx$

(b) $\int_{12}^\infty \dfrac{1}{x+6} dx$

(c) $\int_0^\infty \dfrac{x}{1+x^2} dx$

(d) $\int_1^\infty \dfrac{x^2}{(x^3+1)^2} dx$

(e) $\int_{-\infty}^1 \dfrac{1}{x^3} dx$

(f) $\int_3^\infty \dfrac{1}{x(\ln x)^2} dx$

(g) $\int_0^\infty xe^{-x^2} dx$

2. Evaluate each of the following integrals or show that it is divergent. Sketch the area in each case.

(a) $\int_2^3 \dfrac{1}{\sqrt{x-2}} dx$

(b) $\int_0^3 \dfrac{6\,dx}{(3x-1)^{2/3}}$

(c) $\int_3^5 \dfrac{x\,dx}{\sqrt{x^2-9}}$

(d) $\int_1^3 \dfrac{dx}{(2x-4)^3}$

(e) $\int_0^8 \dfrac{dx}{\sqrt{8-x}}$

(f) $\int_{-1}^4 \dfrac{dx}{(x-2)^{2.5}}$

3. Find the area bounded by the x-axis, the lines $x = \pm 2$, and the curve $y = 6x/\sqrt{4-x^2}$.

4. Find the area bounded by the lines $x = 0$, $y = 0$, $x = 3$, and the curve $y = 4/(3-x)^2$.

5. Find the area between the curve $y = 12/(x^2 + 16)$ and the x-axis. Hint: Use Formula (20), Section 14.5.

6. Show that if $n > 1$,

$$\int_1^\infty \dfrac{1}{x^n} dx = \dfrac{1}{n-1}$$

but if $n \leq 1$, the integral diverges.

Numerical Integration   16.5

Numerical integration is a way of determining the value of a definite integral without doing the integration in a formal way. Although a numerical process gives an approximate rather than an exact value for an integral, it is possible to obtain results accurate to several decimal places. Numerical integration is employed when it is impractical or impossible to work out an integral any other way.

You have already seen an example of numerical integration in Example A, Section 16.1, page 409–411. We suggest you reread that example at this time. In that problem the object was to find the area of a quarter circle. The method was to divide the area into a large number of narrow rectangles and compute their sum. That calculation showed that the approximation improved as the number of rectangles increased and their width decreased. The principal drawback was that although thousands of rectangles were used, an accuracy of only four digits was achieved. A higher accuracy would require an inordinately long computation time.

The difficulty clearly lies in the fact that the top of a rectangle does not fit an arbitrary curve very well (Figure 16.23a). Using other shapes it is possible to do much better. In the **trapezoidal rule**, the curve in each small interval is approximated by a straight line, and the area under it by a trapezoid (Figure 16.23b). In **Simpson's rule**, the area under the curve in a pair of intervals is approximated by a section of a parabola (Figure 16.23c). These two methods of numerical integration are outlined below.

Figure 16.23

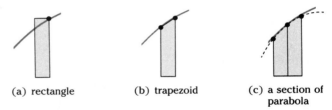

(a) rectangle   (b) trapezoid   (c) a section of parabola

The Trapezoidal Rule

Let f be a continuous function of x in the interval $[a, b]$ (Figure 16.24). Divide the interval $[a, b]$ into n subintervals, each of width h where

$$h = \frac{(b - a)}{n}$$

Figure 16.24

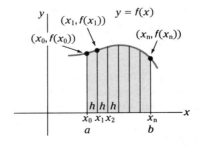

The endpoints of the subintervals are located at

$$x_0, x_1, x_2, x_3, \ldots, x_n$$

where $x_0 = a$, and $x_n = b$. At these points, the function has the values

$$f(x_0), f(x_1), f(x_2), f(x_3), \ldots, f(x_n)$$

Consider the first subinterval $[x_0, x_1]$. Join the points $(x_0, f(x_0))$ and $(x_1, f(x_1))$ by a straight line. The area under the curve in this interval is approximately equal to the area of the trapezoid, which is

$$A_1 = \frac{1}{2}[f(x_0) + f(x_1)] \cdot h$$

The total area under the curve in the interval $[a, b]$ is approximately equal to the sum of all such trapezoidal areas:

$$\text{Area} = \sum_{k=1}^{n} A_k$$

$$= \frac{1}{2}[f(x_0) + f(x_1)] \cdot h + \frac{1}{2}[f(x_1) + f(x_2)] \cdot h$$

$$+ \cdots + \frac{1}{2}[f(x_{n-1}) + f(x_n)] \cdot h$$

$$= \frac{h}{2}[f(x_0) + 2f(x_1) + 2f(x_2) + \cdots + f(x_n)]$$

EXAMPLE A

Evaluate the definite integral

$$\int_0^3 (3 + 2x - x^2)\, dx$$

using the trapezoidal rule with $n = 6$.

SOLUTION

The integral is equivalent to the area under the curve $y = 3 + 2x - x^2$ over the interval $[0, 3]$ as shown in Figure 16.25. For $n = 6$, this area is approximately equal to

$$\text{Area} = \frac{h}{2}[f(x_0) + 2f(x_1) + 2f(x_2) + 2f(x_3) + 2f(x_4)$$
$$+ 2f(x_5) + f(x_6)]$$

Since $n = 6$, the width of each subinterval is

Figure 16.25

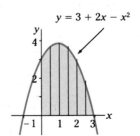

$y = 3 + 2x - x^2$

$$h = \frac{3 - 0}{6}$$

$$= 0.5$$

Values of the function are thus required at seven points:

k	x_k	$f(x_k)$
0	0	3.00
1	0.5	3.75
2	1.0	4.00
3	1.5	3.75
4	2.0	3.00
5	2.5	1.75
6	3.0	0

The total area is therefore approximately

$$\text{Area} \doteq \frac{0.5}{2}[3 + 2(3.75) + 2(4) + 2(3.75) + 2(3) + 2(1.75) + 0]$$

$$= \frac{0.5}{2}(35.5)$$

$$= 8.875$$

Compare this result to the exact value found by direct integration:

$$\int_0^3 (3 + 2x - x^2)\, dx = 3x + \frac{2x^2}{2} - \frac{x^3}{3} \Big|_0^3$$

$$= 3 \cdot 3 + 3^2 - \frac{3^2}{3} - 0$$

$$= 9 \qquad \qquad \square$$

Table 16.3
Trapezoidal Rule

Number of Subintervals	Area of Quarter Circle	Computation Time (s)
10	12.4180733	1
100	12.5616681	12
1000	12.5662219	123
10000	12.5663659	1232
$16\pi/4 = 12.5663706$		

Although it is possible to obtain a rough estimate of a definite integral using just a few subintervals, for accurate results, n must be large, which means that normally such calculations are done by computer. A program for calculating areas by the trapezoidal rule is given in Figure 16.26. For comparison, the calculation of the area of a quarter circle carried out in Section 16.1, page 411, has been repeated here. The results shown in Table 16.3 should be compared to those in Table 16.2, page 412, found by summing rectangles.

Figure 16.26

```
10  REM TRAPEZOIDAL RULE
100 :
110 DEF FNF(X)=SQR(16−X↑2)
120 :
130 INPUT "FROM X = ";A
140 INPUT "   TO X = ";B
150 INPUT "NUMBER OF SUBINTERVALS = ";N
160 :
170 H=(B−A)/N
200 :
210 SUM=FNF(A)
220 FOR K=1 TO N−1
240    XK=A + K*H
250    SUM=SUM + 2*FNF(XK)
260 NEXT K
270 SUM=SUM + FNF(B)
280 :
290 AREA=(H/2)*SUM
300 :
310 PRINT "AREA =";AREA
390 END
```

Simpson's Rule

Simpson's rule uses a parabola rather than a straight line to approximate the graph of a function. Unlike the case of a rectangle or trapezoid, however, the formula for the area under the arc of a parabola is not commonly known. It will be necessary first to derive this formula. The calculation of an area by Simpson's rule then consists of summing a large number of such areas.

The area A under the parabola $f(x) = Ax^2 + Bx + C$ over a double interval extending from $x = -h$ to $x = h$ (Figure 16.27) is given by the integral

$$A = \int_{-h}^{h} (Ax^2 + Bx + C)\,dx$$

$$= \left. \frac{Ax^3}{3} + \frac{Bx^2}{2} + Cx \right|_{-h}^{h}$$

Figure 16.27

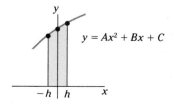

$y = Ax^2 + Bx + C$

$$= \left[\frac{Ah^3}{3} + \frac{Bh^2}{2} + Ch \right] - \left[-\frac{Ah^3}{3} + \frac{Bh^2}{2} - Ch \right]$$

$$= \frac{2Ah^3}{3} + 2Ch$$

$$= \frac{h}{3}(2Ah^2 + 6C)$$

This result can be expressed in terms of the values of $f(x)$ at $x = -h, 0,$ and h.

Since

$$f(h) = Ah^2 + Bh + C$$

$$4f(0) = \qquad\qquad 4C$$

$$f(-h) = Ah^2 - Bh + C$$

it follows by addition that

$$f(-h) + 4f(0) + f(h) = (2Ah^2 + 6C)$$

Consequently, the area under the parabola on this double interval about the origin is

$$A = \frac{h}{3}[f(-h) + 4f(0) + f(h)]$$

By translating the axes, the interval $[-h, h]$ can be made to coincide with any pair of subintervals. Thus, in general,

$$A_k = \frac{h}{3}(f(x_k) + 4f(x_{k+1}) + f(x_{k+2}))$$

where

$$x_{k+1} - x_k = x_{k+2} - x_{k+1} = h$$

According to Simpson's rule, the area under a curve whose equation is $y = f(x)$ is found by approximating the curve on each pair of subintervals by the arc of a parabola and summing the areas under these arcs:

$$\text{Area} = \frac{h}{3}[f(x_0) + 4f(x_1) + f(x_2)]$$

$$+ \frac{h}{3}[f(x_2) + 4f(x_3) + f(x_4)]$$

$$+ \frac{h}{3}[f(x_4) + 4f(x_5) + f(x_6)] + \cdots$$

$$\text{Area} = \frac{h}{3}[f(x_0) + 4f(x_1) + 2f(x_2) + 4f(x_3) + 2f(x_4) + 4f(x_5)$$

$$+ \cdots + f(x_n)]$$

The procedure for evaluating a definite integral by Simpson's rule is similar to that shown in Example A for the trapezoidal rule. A computer program for Simpson's rule is given in Figure 16.28. As a test case, the area of the quarter circle has been repeated with the results shown in Table 16.4. Observe that although the formula for area according to Simpson's rule is quite similar to that of the trapezoidal rule, there is a noticeable improvement in the accuracy of the calculation.

Figure 16.28

```
10   REM SIMPSON'S RULE
100 :
110 DEF FNF(X) = SQR(16 − X ↑ 2)
120 :
130 INPUT "FROM X  =  ";A
140 INPUT "   TO X  =  ";B
150 INPUT "NUMBER OF SUBINTERVALS (EVEN)";N
160 :
170 H = (B − A)/N
200 :
210 SUM = FNF(A) : C = 2
220 FOR K = 1 TO N − 1
230    C = 6 − C   : REM 4 2 4 2...
240    XK = A + K*H
250    SUM = SUM + C*FNF(XK)
260 NEXT K
270 SUM = SUM + FNF(B)
280 :
290 AREA = (H/3)*SUM
300 :
310 PRINT "AREA  = ";AREA
390 END
```

Table 16.4
Simpson's Rule

Number of Subintervals	Area of Quarter Circle	Computation Time (s)
10	12.5080326	1
100	12.5645328	13
1000	12.5663125	132
10000	12.5663689	1326
$16\pi/4 = 12.5663706$		

Exercise 16.5

1. Sketch the trapezoid determined by the lines $y = 2x + 7$, $x = 3$, $x = 5$, and the x-axis, and determine its area.

2. Determine the width of the subintervals and the values of x that divide the interval

   (a) $[6, 10]$ into 8 equal parts

   (b) $[-2, 1]$ into 12 equal parts

   (c) $[-3, 7]$ into 50 equal parts.

3. How many subintervals of width h will result when each of the following intervals is subdivided?

   (a) $[-1, 3]$, $h = 0.25$

   (b) $[3, 6]$, $h = 0.05$

   (c) $[-7, 0]$, $h = 0.333\ldots$

4. Sketch the area under the curve $y = (x - 4)^2$ in the interval $[0, 4]$, and find its approximate value using the trapezoidal rule

   (a) with 2 subintervals

   (b) with 4 subintervals

   (c) with 8 subintervals.

5. Sketch the area corresponding to each of the following definite integrals. Then evaluate the integral in two ways: by the trapezoidal rule with the given value of n, and by integration.

   (a) $\displaystyle\int_2^7 x^2\,dx$, $n = 5$

   (b) $\displaystyle\int_0^4 x\sqrt{x^2 + 4}\,dx$, $n = 4$

   (c) $\displaystyle\int_0^\pi \sin\theta\,d\theta$, $n = 6$

   (d) $\displaystyle\int_0^{\pi/4} \tan\theta\,d\theta$, $n = 4$

   (e) $\displaystyle\int_0^2 e^{-x}\,dx$, $n = 8$

   (f) $\displaystyle\int_1^2 \frac{dx}{x}$, $n = 5$

6. Evaluate each of the following integrals using the trapezoidal rule with the given value of n.

   (a) $\displaystyle\int_1^6 \sqrt{x^2 + 3x}\,dx$, $n = 5$

   (b) $\displaystyle\int_1^5 \frac{x^2}{\sqrt{x^2 + 8}}\,dx$, $n = 4$

   (c) $\displaystyle\int_1^4 x\sqrt[3]{x^3 + 2}\,dx$, $n = 6$

   (d) $\displaystyle\int_1^2 e^{-x^2}\,dx$, $n = 10$

   (e) $\displaystyle\int_0^{\pi/4} \sqrt{2 + \sin^2\theta}\,d\theta$, $n = 5$

   (f) $\displaystyle\int_0^{\pi/2} \frac{1}{\sqrt{2 - \cos^2\theta}}\,d\theta$, $n = 6$

7. Repeat the evaluation of each of the integrals in the preceding problem using the computer program for the trapezoidal rule with $n = 100$.

8. At $x = -1$, 0, and 1, determine the values of the parabola whose equation is $y = 6 - x^2$. Using these values, find the area under the parabola in the interval $[-1, 1]$. Sketch the area.

9. At $x = 2.5$, 3, and 3.5, determine the values of the parabola whose equation is $y = (x - 1)^2 + 1$. Using these values, find the area under the parabola in the interval $[2.5, 3.5]$. Sketch the area.

10. Sketch the area corresponding to each of the following definite integrals. Then evaluate the integral by Simpson's rule with the given value of n. Compare the approximation with the actual value found by integration.

(a) $\int_0^2 x^3 \, dx$, $\qquad\qquad n = 4$

(b) $\int_1^5 \ln x \, dx$, $\qquad\qquad n = 8$

(c) $\int_{-\pi/3}^{\pi/3} \sec^2 \theta \, d\theta$, $\qquad n = 6$

11. Evaluate each of the following integrals using Simpson's rule with the given value of n.

(a) $\int_1^3 (9 - x^2)^{2/3} \, dx$, $\qquad n = 4$

(b) $\int_0^{10} \dfrac{x}{x^2 + 1} \, dx$, $\qquad n = 10$

(c) $\int_{-1}^1 \sin(\pi x^2) \, dx$, $\qquad n = 4$

(d) $\int_{\pi/2}^{\pi} \dfrac{\sin \theta}{\theta} \, d\theta$, $\qquad\qquad n = 6$

12. Repeat the evaluation of each of the integrals in the preceding problem using the computer program for Simpson's rule with $n = 100$.

13. Use the relationship

$$\int_0^1 \frac{1}{1 + x^2} \, dx = \tan^{-1}(1) = \frac{\pi}{4}$$

and Simpson's rule to estimate the value of π.

14. The length L of the arc of the sine function from 0 to π is given by the integral

$$L = \int_0^{\pi} \sqrt{1 + \cos^2 \theta} \, d\theta$$

Determine the value of this integral correct to five significant digits.

15. Use Simpson's rule to approximate the area under the curve determined by the given experimental data:

x	y
1	2.85
2	4.72
3	7.15
4	8.89
5	6.76
6	5.08
7	4.65

Summary and Review $\qquad\qquad$ 16.6

The value of a definite integral is by definition

$$\int_a^b f(x) \, dx = \lim_{\Delta x \to 0} \sum_{k=1}^n f(x_k) \cdot \Delta x$$

$$= F(b) - F(a)$$

where F is an antiderivative of f. The definite integral can be interpreted as the area under the curve $y = f(x)$ on the interval $[a, b]$.

In applications of the definite integral, the element of integration is expressed as a differential. For instance, when finding an area, the element of integration is taken to be a narrow rectangle of height $f(x)$, width dx, and area $dA = f(x)\,dx$.

There are two types of improper integrals. One type is immediately recognized by having a limit of integration at $\pm\infty$. The other type has an integrand which is unbounded at some point in the interval $[a, b]$. Both types are evaluated by an appropriate use of limits.

Numerical integration is a way to evaluate an integral when other methods are inconvenient or cannot be found. The essential idea is to divide the area under the curve $y = f(x)$ into a large number of thin strips and find the sum of their areas. The accuracy of the approximation depends on the number of strips used and on how closely the tops of the strips match the given curve.

In working the problems in this chapter it is helpful to

(a) sketch the curve or curves pertaining to an integral, shade in the area in question, and draw the element of integration;

(b) write down the expression for the element of integration dA as the first step in setting up an integral;

(c) substitute for the limits of integration as a separate step after integrating but before doing any simplification.

Exercise 16.6

1. Find the area under the graphs of each of the following functions in the interval given.

(a) $f(x) = x^3 + 2x^2 + 3,$ $[-1, 0]$

(b) $f(x) = \dfrac{2x - 3}{\sqrt{x}},$ $[2, 4]$

(c) $f(x) = \dfrac{9}{2x + 5},$ $[2, 5]$

2. Find the area of the region between the curves

(a) $y = x^2/4$ $y = 2\sqrt{x}$

(b) $y = x^2 - 1,$ $y = x^3 - 3x^2 - x + 3$

3. Find the area of the region bounded by

(a) $y = 0,$ $y = (x^2 - 4)^2$

(b) $y = x,$ $y = 1 - x,$ $y = -x/2$

(c) $y = \cos x,$ $y = \sin x,$ $\pi/4 < x < 5\pi/4$

4. Find the areas of the three regions enclosed by the graphs of

$$y = x + 2, y = -2x + 4, y = -x^2 + 2x + 4.$$

5. Determine the value of each of the following improper integrals if it exists:

(a) $\displaystyle\int_3^\infty \dfrac{1}{(x-1)^3}\,dx$

(b) $\displaystyle\int_2^3 \dfrac{1}{\sqrt{x-2}}\,dx$

(c) $\displaystyle\int_{-\infty}^0 \dfrac{1}{(3x-4)^2}\,dx$

6. Find the total area between the following curves and the x-axis in the interval given:

(a) $y = x(x - 5),$ $0 \le x \le 10$

(b) $y = \dfrac{x - 2}{x},$ $1 \le x \le 4$

7. Using a numerical method, estimate the value of the following definite integrals:

(a) $\displaystyle\int_{\pi/4}^{\pi/2} \ln(\sin x)\,dx$

(b) $\displaystyle\int_{0}^{3} \sqrt{x^2 + 9}\,dx$

(c) $\displaystyle\int_{1}^{4} \frac{e^x}{x}\,dx$

Sofia Kovalevskaia

UNTIL THE NINETEENTH CENTURY few women had the opportunity to study science or to pursue a career as a professional scientist. It was in mathematics that the first woman was admitted as a regular member of the international research community. Sofia Kovalevskaia, the daughter of a Russian general, was the first woman to be awarded a doctorate and the first woman outside of Italy to hold a chair at a European university. At the time of her death in 1890 she participated as an equal in the scientific community and stood as a peer among the outstanding mathematicians of Europe.

Coming from a well-to-do background in Russia, Kovalevskaia travelled to Germany in 1869 to study mathematics at Heidelberg and Berlin. Barred from lectures at the University of Berlin because she was a woman, Kovalevskaia studied privately for three years with the distinguished mathematician Karl Weierstrass. In 1874 she was awarded a doctorate *in absentia* from the University of Göttingen. After several years of study and travel she was appointed lecturer in mathematics at the University of Stockholm, a position that was arranged by her friend the Swedish mathematician Gösta Mittag-Leffler. In 1889 Mittag-Leffler secured a life professorship at Stockholm for her.

Kovalevskaia wrote her doctoral dissertation on a problem in partial differential equations. These equations occur frequently in nature, whenever a changing quantity depends on more than one independent variable. The initial and most basic mathematical question in the theory of partial differential equations is to determine conditions under which a solution exists. Using techniques that involved the expansion of functions in power series, Kovalevskaia derived conditions that assure the existence of one and only one solution of the given equation. The theorem she proved remains a basic result in the modern subject.

Apart from her research in mathematics, Sofia Kovalevskaia was committed throughout her career to the support of progressive socialist movements in Europe. In several literary works she presented her views on public education, feminism, and socialism. In her beautifully written *Memories of Childhood* Kovalevskaia traced her political development and that of her sister Aniuta in a provincial gentry family. Kovalevskaia's political sympathies contributed to the opposition she encountered during her life. After her death a tsarist government minister is said to have opposed the translation of a memoir dealing with her life with the words: "People have already concerned themselves too much with a woman who, in the last analysis, was a nihilist."

Methods of Integration

There is no straightforward algorithmic procedure for integration as there is for differentiation. The process of integration depends, for the most part, on finding ways to transform integrals into simple forms to which the basic formulas can be applied.

Chapter 17 begins by showing how to use a **table of integrals** that contains many integration formulas in addition to the basic ones. From the viewpoint of a scientist or engineer, a problem is essentially solved when it has been reduced to an integral that can be looked up in a table. But what steps can be taken if a particular integral cannot be found in the table, and how do such integration formulas originate in the first place?

Several general methods of integration are sampled in this chapter. One method is to use an **algebraic substitution** to change the variable of integration. Another method uses **partial fractions** to deal with integrals, in which the integrand consists of the quotient of two polynomials. A third method called **integration by parts** is generally used when the integrand contains the product of two functions.

Methods of handling powers of trigonometric functions and the use of **trigonometric substitutions** will be discussed in Chapter 18. Skill in integration depends on insight and experience in recognizing which method to apply in a particular case.

Integration by Use of Tables   17.1

The integration formulas you have been using so far were established by inverting the corresponding differentiation rules. However, many integrals cannot be easily done using these basic formulas alone. You must either employ more sophisticated methods of integration or resort to the use of a **table of integrals**.

Look closely at I. Table of Selected Integrals on pages 469–473 following this chapter. Following the list of basic formulas, the integrals are classified by the form of the expression in the integrand. Sometimes you will find a formula in the table that has exactly the form you require. At other times, it may be necessary to change the algebraic form of an integral somewhat to make it match one of the formulas.

EXAMPLE A

$$\int x\sqrt{5 + 2x}\,dx$$

SOLUTION

The integrand has a factor $\sqrt{5 + 2x}$ which is of the form $\sqrt{a + bx}$. Integrals containing an expression of this type are found in Formulas 33 to 36 in the table on page 471. Formula 33 in particular has the form required here:

$$33. \quad \int x\sqrt{a + bx}\,dx = \frac{-2(2a - 3bx)\sqrt{(a + bx)^3}}{15b^2} + C$$

If $a = 5$ and $b = 2$, then the factor $(2a - 3bx) = (10 - 6x)$. The integral is, therefore,

$$\int x\sqrt{5 - 2x}\,dx = \frac{-2(10 - 6x)\sqrt{(5 + 2x)^3}}{60} + C$$

$$= \frac{-(5 - 3x)\sqrt{(5 + 2x)^3}}{15} + C \qquad \square$$

EXAMPLE B

$$\int \frac{\sqrt{8x^2 - 5}}{3x}\,dx$$

SOLUTION

If an 8 is removed from the radical, then this integral contains a factor of the form $\sqrt{x^2 - a^2}$. Integrals containing an expression of this sort are found in Formulas 37 to 41. Formula 41 is the one which matches the integral in this example:

$$41. \quad \int \frac{\sqrt{x^2 - a^2}}{x}\,dx = \sqrt{x^2 - a^2} - a\cos^{-1}\frac{a}{x} + C$$

Thus,

$$\int \frac{\sqrt{8x^2 - 5}}{3x} \, dx = \frac{\sqrt{8}}{3} \int \frac{\sqrt{x^2 - 5/8}}{x} \, dx, \qquad \text{where } a^2 = \frac{5}{8}$$

$$= \frac{\sqrt{8}}{3} \left[\sqrt{x^2 - 5/8} - \frac{\sqrt{5}}{\sqrt{8}} \cos^{-1} \frac{\sqrt{5}}{x\sqrt{8}} \right]$$

$$= \frac{\sqrt{8x^2 - 5}}{3} - \frac{\sqrt{5}}{3} \cos^{-1} \frac{\sqrt{5}}{x\sqrt{8}} + C \qquad \qquad \square$$

EXAMPLE C

$$\int \frac{dx}{\sqrt{2x^2 - 5x + 6}}$$

SOLUTION

Integrals containing quadratic expressions of the form $(a + bx + cx^2)$ are found in Formulas 45 to 50. Formula 47 where $a = 6$, $b = -5$, and $c = 2$ is the one that corresponds to the integral in this example. Since c is positive, the first of the two possibilities is the one to use:

$$47. \quad \int \frac{dx}{\sqrt{X}} = \frac{1}{\sqrt{c}} \ln \left| \sqrt{X} + x\sqrt{c} + \frac{b}{2\sqrt{c}} \right| + C, \qquad \text{if } c > 0$$

Therefore,

$$\int \frac{dx}{\sqrt{2x^2 - 5x + 6}}$$

$$= \frac{1}{\sqrt{2}} \ln \left| \sqrt{2x^2 - 5x + 6} + x\sqrt{2} - \frac{5}{2\sqrt{2}} \right| + C \qquad \qquad \square$$

EXAMPLE D

$$\int x^2 \sin 3x \, dx$$

SOLUTION

A few integrals containing transcendental functions are given in Formulas 51 to 60. The correct formula to use in this case is Formula 59:

$$59. \quad \int x^n \sin ax \, dx = \frac{-x^n \cos ax}{a} + \frac{n}{a} \int x^{n-1} \cos ax \, dx$$

When $n = 2$, this becomes

$$\int x^2 \sin ax \, dx = \frac{-x^2 \cos ax}{a} + \frac{2}{a} \int x \cos ax \, dx$$

Formulas such as this are called **reduction formulas** because the exponent is reduced, making the integral that remains to be done on the right simpler than the original one on the left. In this case, the second integral is done using Formula 58:

58. $\int x \cos ax \, dx = \frac{1}{a^2}(\cos ax + ax \sin ax) + C$

Putting this all together, with $a = 3$,

$$\int x^2 \sin 3x \, dx = \frac{-x^2 \cos 3x}{3} + \frac{2}{3} \int x \cos 3x \, dx$$

$$= -\frac{x^2 \cos 3x}{3} + \frac{2}{3} \cdot \frac{1}{9}(\cos 3x + 3x \sin 3x) + C$$

$$= -\frac{x^2}{3} \cos 3x + \frac{2x}{9} \sin 3x + \frac{2}{27} \cos 3x + C$$

Exercise 17.1

1. Integrate using Formulas 19, 20, 21, or 22:

(a) $\int \dfrac{dx}{x^2 - 9}$

(b) $\int \dfrac{18 \, dx}{16 - 9x^2}$

(c) $\int \dfrac{4 \, dx}{\sqrt{8 - 2x^2}}$

(d) $\int \dfrac{3 \, dx}{2x^2 + 5}$

2. Integrate using Formulas 26 to 36:

(a) $\int \dfrac{-5x}{2 + 3x} \, dx$

(b) $\int \dfrac{2x - 1}{x - 2} \, dx$

(c) $\int \dfrac{4x \, dx}{\sqrt{2x + 3}}$

(d) $\int \dfrac{2}{x^2(3x - 1)} \, dx$

3. Integrate using Formulas 37 to 44:

(a) $\int \dfrac{\sqrt{9x^2 - 6}}{3x} \, dx$

(b) $\int \dfrac{8 \, dx}{x\sqrt{4x^2 + 1}}$

(c) $\int \dfrac{\sqrt{15 - 8x^2}}{2x} \, dx$

(d) $\int \sqrt{4x^2 + 9} \, dx$

4. Integrate using Formulas 45 to 50:

(a) $\displaystyle\int \frac{dx}{x^2 + 6x + 10}$

(b) $\displaystyle\int \frac{dx}{x\sqrt{x^2 + 2x - 1}}$

(c) $\displaystyle\int \frac{dx}{(2x^2 + 5x - 3)^2}$

(d) $\displaystyle\int \sqrt{6 - 2x - x^2}\, dx$

5. Integrate using Formulas 51 to 60:

(a) $\displaystyle\int \sin^2(4x)\, dx$

(b) $\displaystyle\int 3x^2 e^{4x}\, dx$

(c) $\displaystyle\int x^3 e^{5x}\, dx$

(d) $\displaystyle\int x^2 \sin(x/2)\, dx$

6. Integrate using a suitable formula:

(a) $\displaystyle\int \frac{3dx}{2x - 5}$

(b) $\displaystyle\int \frac{x\, dx}{(2 - x)^2}$

(c) $\displaystyle\int \frac{dx}{7x^2 + 8x}$

(d) $\displaystyle\int \frac{dx}{1 - 5x^2}$

7. Integrate using a suitable formula:

(a) $\displaystyle\int \frac{dx}{x\sqrt{x + 1}}$

(b) $\displaystyle\int \sqrt{4x^2 + 5x^3}\, dx$

(c) $\displaystyle\int \frac{8\, dx}{x\sqrt{20x^2 - 16}}$

(d) $\displaystyle\int \frac{x\, dx}{\sqrt{1 + 2x - 4x^2}}$

(e) $\displaystyle\int \frac{\sqrt{9x^2 + 144}}{x}\, dx$

(f) $\displaystyle\int \frac{dx}{x\sqrt{2 + 2x - x^2}}$

8. Integrate using a suitable formula:

(a) $\displaystyle\int x^5 \ln 5x\, dx$

(b) $\displaystyle\int 4x\, e^{x/2}\, dx$

(c) $\displaystyle\int x^3 \cos x\, dx$

Algebraic Substitutions   17.2

Many integrals are too complicated to be done by inspection using one of the basic formulas and cannot be found in the tables. In such cases, you should try to change the expression in the integrand into a more suitable form. One way to do this is to use an **algebraic substitution**.

In an algebraic substitution, the variable of integration x is replaced by a different variable v that is related to x in some way. What relation you choose will depend on what expressions you find in the integrand in question. Example A illustrates the general procedure.

EXAMPLE A

$$\int \cos 5x \, dx$$

SOLUTION

This integral does not coincide with any of the basic formulas because the argument of the cosine function is $5x$, rather than just x. You can get around this obstacle by setting

$$v = 5x$$

Then $\cos 5x = \cos v$

and $dv = 5 \, dx$ or $dx = \dfrac{1}{5} dv$

The variable of integration is now changed from x to v by substituting these expressions into the integral. The new integral is one of the basic integrals in the Table of Selected Integrals (pages 469–473):

$$\int \cos 5x \, dx = \int \cos v \cdot \frac{1}{5} dv$$

$$= \frac{1}{5} \int \cos v \, dv$$

$$= \frac{1}{5} \sin v + C \qquad \text{Formula 5}$$

The integration is completed, but the result is expressed in terms of the variable v. It is now necessary to eliminate v in favour of the original variable x.

$$\int \cos 5x \, dx = \frac{1}{5} \sin v + C$$

$$= \frac{1}{5} \sin 5x + C$$

The substitution $v = ax$ illustrated by Example A is a simple one, but it is useful because it takes care of constant factors, leaving nothing to guesswork.

The key to the substitution method lies in what substitution you choose to simplify a given integral. You are, of course, free to make any change of variable you please, but you may have to experiment a bit before finding a substitution that works. Sometimes a considerable amount of ingenuity is required. Generally, the most promising substitution is one in which v replaces the argument of a function, resulting in terms like cos v, ln v, e^v, v^n, $\sqrt{v}$, etc. It is not possible to be more specific. The following examples will give you some idea of the many possibilities.

EXAMPLE B

$$\int \frac{dx}{e^{6x-3}}$$

SOLUTION

Let $v = 6x - 3$

Then $dv = 6\, dx$ 　or 　$dx = \frac{1}{6} dv$

$$\int \frac{dx}{e^{6x-3}} = \int \frac{1}{e^v} \cdot \frac{1}{6} dv$$

$$= \frac{1}{6} \int e^{-v}\, dv$$

$$= -\frac{1}{6} e^{-v} + C$$

Now change back to x:

$$\int \frac{dx}{e^{6x-3}} = -\frac{1}{6} \cdot \frac{1}{e^{6x-3}} + C$$

EXAMPLE C

$$\int \frac{5x}{(4 - x^2)^3} dx$$

SOLUTION

Let $v = 4 - x^2$

Then $dv = -2x\, dx$ 　or 　$x\, dx = -\frac{1}{2} dv$

$$\int \frac{5x}{(4 - x^2)^3} dx = 5 \int \frac{1}{v^3} \left(-\frac{1}{2} dv \right)$$

$$= -\frac{5}{2} \int v^{-3}\, dv$$

$$= -\frac{5}{2} \frac{v^{-2}}{-2} + C$$

$$= \frac{5}{4}(4 - x^2)^{-2} + C$$

$$= \frac{5}{4(4 - x^2)^2} + C$$

❏

EXAMPLE D

$$\int x\sqrt{3x + 4}\, dx$$

SOLUTION

Let $v = 3x + 4$

Then $x = \frac{1}{3}(v - 4)$ and $dx = \frac{1}{3}dv$

Substituting these expressions into the integral

$$\int x\sqrt{3x + 4}\, dx = \int \frac{1}{3}(v - 4) \cdot \sqrt{v} \cdot \frac{1}{3}dv$$

$$= \frac{1}{9}\int (v^{3/2} - 4v^{1/2})\, dv$$

$$= \frac{1}{9}\left(\frac{2}{5}v^{5/2} - 4 \cdot \frac{2}{3}v^{3/2}\right) + C$$

$$= \frac{2}{45}v^{5/2} - \frac{8}{27}v^{3/2} + C$$

Converting back to the variable x

$$\int x\sqrt{3x + 4}\, dx = \frac{2}{45}(3x + 4)^{5/2} - \frac{8}{27}(3x + 4)^{3/2} + C$$

Observe that this answer could be expressed differently as a product of two factors. If you factor the solution before replacing x, it becomes

$$\frac{2}{135}(3v - 20)v^{3/2} + C$$

Now, replacing x gives

$$\int x\sqrt{3x + 4}\, dx = \frac{2}{135}(9x - 8)(3x + 4)^{3/2} + C \quad \text{compare to Formula 33}$$

❏

EXAMPLE E

$$\int \frac{1}{\sqrt{4 - 49x^2}}\, dx$$

SOLUTION

Let $\qquad v = 7x$

Then $\qquad 49x^2 = v^2$

and $\qquad dv = 7\, dx \qquad$ or $\qquad dx = \frac{1}{7}\, dv$

Substituting these expressions into the integral

$$\int \frac{1}{\sqrt{4 - 49x^2}}\, dx = \frac{1}{7}\int \frac{1}{\sqrt{2^2 - v^2}}\, dv$$

$$= \frac{1}{7}\sin^{-1}\frac{v}{2} + C \qquad \longleftarrow \qquad \text{using Formula 19}$$

Converting back to the variable x

$$\int \frac{1}{\sqrt{4 - 49x^2}}\, dx = \frac{1}{7}\sin^{-1}\frac{7x}{2} + C$$

Exercise 17.2

1. Determine the following integrals:

(a) $\displaystyle\int \frac{dx}{(2 - 3x)^4}$

(b) $\displaystyle\int \frac{3x}{(4x^2 + 5)^2}\, dx$

(c) $\displaystyle\int (18x - 15)(3x^2 - 5x + 8)^4\, dx$

(d) $\displaystyle\int \sqrt{3x + 1}\, dx$

(e) $\displaystyle\int \sqrt{ax + b}\, dx$

(f) $\displaystyle\int \frac{x}{\sqrt{x^2 - 1}}\, dx$

2. Determine the following integrals:

(a) $\displaystyle\int x^2(1 + 2x^3)^{1/4}\, dx$

(b) $\displaystyle\int \frac{x^2 - 1}{\sqrt{x^3 - 3x}}\, dx$

(c) $\displaystyle\int \frac{1}{\sqrt{x}}(1 - 2\sqrt{x})^5\, dx$

(d) $\displaystyle\int 3x\sqrt{5 - 2x}\, dx$

(e) $\displaystyle\int x^2\sqrt{2x - 5}\, dx$

(f) $\displaystyle\int x^5(1 + x^3)^{1/4}\, dx$

3. Determine the following integrals:

(a) $\displaystyle\int \frac{dx}{1 + \sqrt{x}} \qquad$ Hint: Let $v = \sqrt{x} + 1$.

(b) $\displaystyle\int \frac{\sqrt{x}}{\sqrt{x} - 1}\, dx$

4. Determine the following integrals:

(a) $\displaystyle\int \frac{\sec^2 \theta\, d\theta}{3 + 5\tan \theta} \qquad$ Hint: Let $v = 3 + 5\tan \theta$.

(b) $\displaystyle\int (2 - \cos 2\theta)^3 \sin 2\theta\, d\theta$

(c) $\displaystyle\int e^{\cos 3x} \sin 3x \, dx$

(d) $\displaystyle\int x^2 \, e^{4x^3} \, dx$

(e) $\displaystyle\int \frac{3e^x}{2e^x - 1} \, dx$

(f) $\displaystyle\int \frac{dx}{\sqrt{1 + e^x}}$

5. Evaluate the definite integrals:

(a) $\displaystyle\int_0^3 x\sqrt{25 - x^2} \, dx$

(b) $\displaystyle\int_3^4 (6 - x)^{-3} \, dx$

(c) $\displaystyle\int_0^1 x(x^2 + 1)^3 \, dx$

6. Derive Formula 30 in the Table of Selected Integrals.

7. Derive Formula 31 in the Table of Selected Integrals.

8. Derive Formula 33 in the Table of Selected Integrals.

9. Derive Formula 34 in the Table of Selected Integrals.

10. Show that

$$\int x^{n-1} \sqrt{a + bx^n} \, dx = \frac{2}{3bn}(a + bx^n)^{3/2}$$

17.3 Integration of Rational Functions

A rational function is a function in the form of a fraction

$$f(x) = \frac{N(x)}{D(x)}$$

where $N(x)$ and $D(x)$ are polynomials in x. You will see that it is the denominator that gives the clue as to how to find the integral:

$$\int \frac{N(x)}{D(x)} \, dx$$

This section deals with the cases where the denominator is a linear expression or an irreducible quadratic, that is, one which cannot be factored. The important case of a factorable denominator and the method of partial fractions will be dealt with in the next section.

Case I. Denominator is a single linear expression of the form $(a + bx)$

EXAMPLE A

$$\int \frac{8}{(5x + 4)} \, dx$$

SOLUTION

This is an integral of the form $\int dv/v$, a fact that can be recognized upon making the substitution

$$v = 5x + 4$$

Then $dv = 5dx$ or $dx = \dfrac{1}{5}dv$

$$\int \frac{8}{(5x + 4)}\,dx = \int \frac{8}{v}\cdot\frac{1}{5}\,dv$$

$$= \frac{8}{5}\int \frac{dv}{v}$$

$$= \frac{8}{5}\ln v + C$$

$$= \frac{8}{5}\ln(5x + 4) + C \qquad \square$$

Integrals of this sort will occur frequently when working with rational functions, so it is worthwhile to remember the general formula:

$$\boxed{\int \frac{1}{(a + bx)}\,dx = \frac{1}{b}\ln(a + bx) + C}$$ Formula 26

Case II. Denominator is an irreducible quadratic expression of the
         form $(ax^2 + bx + c)$

In this case, the method is to complete the square in the denominator. When v is substituted for the squared term, the denominator becomes an expression of the form $v^2 \pm k^2$. The resulting integrals can be done using Formulas 20 or 22 from the Table of Integrals. (In the next section, you will find out how to derive Formula 22.)

EXAMPLE B

$$\int \frac{dx}{x^2 + 8x + 6}$$

SOLUTION

Start by completing the square in the denominator:

$$\int \frac{dx}{x^2 + 8x + 6} = \int \frac{dx}{(x + 4)^2 - 10}$$

Next, substitute for the squared term:

$$v = x + 4 \quad \text{or} \quad x = v - 4 \quad \text{and} \quad dx = dv$$

$$\int \frac{dx}{x^2 + 8x + 6} = \int \frac{dv}{v^2 - \sqrt{10}^2}$$

$$= \frac{1}{\sqrt{10}} \ln\left(\frac{v - \sqrt{10}}{v + \sqrt{10}}\right) + C \quad \longleftarrow \quad \text{using Formula 22}$$

$$= \frac{1}{\sqrt{10}} \ln\left(\frac{x + 4 - \sqrt{10}}{x + 4 + \sqrt{10}}\right) + C \quad \square$$

Whenever the numerator $N(x)$ of a rational function has degree greater than or equal to the degree of the denominator $D(x)$, the fraction should first be reduced by long division. Integration of such functions will be dealt with in the exercises.

Exercise 17.3

1. Determine the following integrals:

(a) $\int \frac{6}{2x - 5} dx$

(b) $\int \frac{5}{3 - 4x} dx$

2. Determine the following integrals:

(a) $\int \frac{4}{x^2 + 2x + 10} dx$

(b) $\int \frac{8}{x^2 - 8x + 14} dx$

3. (a) Using long division, show that

$$\frac{x^2 + 3x + 4}{x - 2} = x + 5 + \frac{14}{x - 2}$$

(b) Determine the integral

$$\int \frac{x^2 + 3x + 4}{x - 2} dx$$

4. With the help of long division, determine the following integrals:

(a) $\int \frac{x + 5}{x + 3} dx$

(b) $\int \frac{3 - 7x - 3x^2}{3x - 2} dx$

(c) $\int \frac{x^3 + 1}{x - 1} dx$

5. (a) After completing the square in the denominator and using the substitution $v = x - 2$, show that

$$\frac{4x - 7}{x^2 - 4x + 8} = \frac{4v}{v^2 + 4} + \frac{1}{v^2 + 4}$$

(b) Determine the integral

$$\int \frac{4x - 7}{x^2 - 4x + 8} dx$$

6. Determine the following integrals:

(a) $\int \frac{3x + 4}{x^2 - 6x + 10} dx$

(b) $\int \frac{3x}{x^2 - 2x - 4} dx$

7. Determine the following integrals:

(a) $\int \dfrac{x^2 - 3x + 11}{x^2 + 4x + 13}\,dx$

(b) $\int \dfrac{x^4}{x^2 + 1}\,dx$

8. Determine the following integrals:

(a) $\int \dfrac{x + 1}{x\sqrt{x - 3}}\,dx$ Hint: Let $v = \sqrt{x - 3}$.

(b) $\int \dfrac{5}{\sqrt{x^2 - 4x + 8}}\,dx$

Hint: Complete the square under the radical and use Formula 21.

9. Derive the two parts of Formula 45 in the Table of Selected Integrals by completing the square and then using Formulas 20 and 22.

The Method of Partial Fractions   17.4

The method of partial fractions is applicable when the denominator of a rational function can be factored into linear or quadratic factors. The idea is to break up a rational function into a sum of simple fractions, which then can be integrated immediately. For example,

$$\int \frac{11x + 7}{x^2 + 4x - 5}\,dx = \int \left[\frac{A}{(x - 1)} + \frac{B}{(x + 5)} \right] dx$$

$$= A \ln(x - 1) + B \ln(x + 5) + C$$

For this approach to be successful, it is clearly essential to be able to find the constants A and B.

To break up a rational function into its partial fractions, start by factoring the denominator. Staying with the previous example,

$$\frac{11x + 7}{x^2 + 4x - 5} = \frac{11x + 7}{(x - 1)(x + 5)}$$

Next, express the quantity on the right as a sum of fractions, having unknown constants A and B for the numerators:

$$\frac{11x + 7}{x^2 + 4x - 5} = \frac{A}{(x - 1)} + \frac{B}{(x + 5)}$$

Finding a common denominator and adding the fractions on the right gives

$$\frac{11x + 7}{x^2 + 4x - 5} = \frac{A(x + 5) + B(x - 1)}{(x - 1)(x + 5)}$$

This equation is an identity that is true for all values of x (except for $x = 1, -5$). Since the denominators are the same, the values of A and B can be determined from the numerators. Thus:

$$11x + 7 = A(x + 5) + B(x - 1)$$

or

$$11x + 7 = (A + B)x + (5A - B)$$

The polynomials on the left and right are equal for *all* values of x, if and only if the coefficients of like powers of x are equal. This yields a system of two equations in two unknowns to solve for A and B:

$$11 = A + B$$

$$7 = 5A - B$$

It follows that $A = 3$ and $B = 8$. Thus

$$\int \frac{11x + 7}{x^2 + 4x - 5} dx = \int \frac{3}{(x - 1)} + \frac{8}{(x + 5)} dx$$

$$= 3\ln(x - 1) + 8\ln(x - 5) + C$$

An alternate method of finding the constants A and B makes explicit use of the fact that the equation

$$11x + 7 = A(x + 5) + B(x - 1)$$

is an identity in x, which is true for any value of x (*including* $x = 1, -5$). By setting x to a particular value, it is possible to eliminate a term, simplifying the work somewhat. Choosing $x = 1$, for example, makes the B term vanish, so

$$18 = 6A \qquad \text{or} \qquad A = 3$$

On the other hand, choosing $x = -5$ makes the A term vanish, so

$$-48 = -6B, \qquad \text{or} \qquad B = 8$$

The method of partial fractions outlined above will be successful when the denominator of a rational function factors completely into linear factors, each occurring only once. The procedure must be modified somewhat when some factors are repeated or when irreducible quadratic factors occur. These and other cases will be dealt with in the exercises below.

EXAMPLE A

$$\int \frac{9x^2 - 6}{2x^3 - x^2 - 8x + 4} dx$$

SOLUTION

Use the factor theorem to factor the denominator if the factoring cannot be done by inspection. (Recall that $(x - a)$ is a factor of the polynomial $P(x)$ if $P(a) = 0$.) In this instance,

$$2x^3 - x^2 - 8x + 4 = (2x - 1)(x - 2)(x + 2)$$

The integrand expressed as a partial fraction sum is, therefore,

$$\frac{9x^2 - 6}{2x^3 - x^2 - 8x + 4} = \frac{A}{(2x - 1)} + \frac{B}{(x - 2)} + \frac{C}{(x + 2)}$$

$$= \frac{A(x - 2)(x + 2) + B(2x - 1)(x + 2) + C(2x - 1)(x - 2)}{(2x - 1)(x - 2)(x + 2)}$$

The denominators are the same; so the numerators must be equal.

$$9x^2 - 6 = A(x - 2)(x + 2) + B(2x - 1)(x + 2) + C(2x - 1)(x - 2)$$

Substituting values for x yields

$$\text{if } x = 2, \qquad 30 = 12B, \qquad \text{or} \qquad B = \frac{5}{2}$$

$$\text{if } x = -2, \qquad 30 = 20C, \qquad \text{or} \qquad C = \frac{3}{2}$$

$$\text{if } x = \frac{1}{2}, \qquad -\frac{15}{4} = -\frac{15}{4}A, \qquad \text{or} \qquad A = 1$$

These values for x were choosen so that two of the three unknowns would be eliminated from the equation in each case. Now returning to the integration:

$$\int \frac{9x^2 - 6}{2x^3 - x^2 - 8x + 4} dx$$

$$= \int \frac{1}{(2x - 1)} dx + \int \frac{5/2}{(x - 2)} dx + \int \frac{3/2}{(x + 2)} dx$$

$$= \frac{1}{2}\ln(2x - 1) + \frac{5}{2}\ln(x - 2) + \frac{3}{2}\ln(x + 2) + C$$

Exercise 17.4

1. Determine the following integrals:

(a) $\displaystyle\int \frac{x}{(x + 1)(2x + 1)}\,dx$

(b) $\displaystyle\int \frac{x}{x^2 - 7x + 10}\,dx$

(c) $\displaystyle\int \frac{x}{2x^2 - 3x - 2}\,dx$

(d) $\displaystyle\int \frac{(3 - 4x)\,dx}{2x^2 - 3x + 1}$

2. Derive Formula 22 in the Table of Integrals.

3. Determine the following integrals:

(a) $\displaystyle\int \frac{2x^2 + 41x - 91}{(x - 1)(x + 3)(x - 4)}\,dx$

(b) $\displaystyle\int \frac{16}{x^3 - 4x}\,dx$

(c) $\displaystyle\int \frac{32x}{(2x - 1)(4x^2 - 16x + 15)}\,dx$

(d) $\displaystyle\int \frac{x\,dx}{x^4 - 5x^2 + 4}$

4. The denominator of

$$\frac{x^2 + x - 4}{(x + 2)(x + 3)^2}$$

contains a repeated linear factor, which requires that the partial fraction sum be written

$$\frac{A}{(x + 2)} + \frac{B}{(x + 3)} + \frac{C}{(x + 3)^2}$$

(a) Find A, B, and C.

(b) Determine the integral

$$\int \frac{x^2 + x - 4}{x^3 + 8x^2 + 21x + 18}\,dx$$

5. Determine the following integrals:

(a) $\displaystyle\int \frac{(x^2 - 3x + 2)}{x(x^2 + 2x + 1)}\,dx$

(b) $\displaystyle\int \frac{(x^2 + 1)}{x^3 - x^2}\,dx$

6. The denominator of

$$\frac{-9x - 8}{(x^2 + 2x + 5)(x - 2)}$$

contains an irreducible quadratic factor, which requires that the partial fraction sum be written

$$\frac{A + Bx}{(x^2 + 2x + 5)} + \frac{C}{(x - 2)}$$

(a) Find A, B, and C.

(b) Determine the integral

$$\int \frac{-9x - 8}{x^3 + x - 10}\,dx$$

7. Determine the following integrals:

(a) $\displaystyle\int \frac{dx}{x^3 + x}$

(b) $\displaystyle\int \frac{dx}{1 + x^3}$

(c) $\displaystyle\int \frac{(2x^2 - 3x - 3)}{(x - 1)(x^2 - 2x + 5)}\,dx$

Integration by Parts   17.5

One of the most valuable tools for working out integrals is the method known as **integration by parts**. This method uses a formula derived from the derivative of a product. Suppose that u and v are functions of x. Applying the product rule to differentiate $u \cdot v$ gives

$$\frac{d}{dx}(u \cdot v) = u \cdot \frac{dv}{dx} + v \cdot \frac{du}{dx}$$

Integrating both sides results in

$$\int \frac{d}{dx}(u \cdot v) \cdot dx = \int u \cdot \frac{dv}{dx} dx + \int v \cdot \frac{du}{dx} dx$$

On the left, the integration "undoes" the differentiation leaving just $u \cdot v$. The expressions on the right can be rewritten in terms of the differentials dv and du:

$$\frac{dv}{dx} \cdot dx = dv \qquad \text{and} \qquad \frac{du}{dx} \cdot dx = du$$

Therefore,

$$u \cdot v = \int u \cdot dv + \int v \cdot du$$

Rearranging the result produces the formula for integration by parts:

$$\int u \cdot dv = u \cdot v - \int v \cdot du$$

This formula changes the problem of integrating $\int u \cdot dv$ into the problem of integrating $\int v \cdot du$. The hope is that the new integral will be easier to work out than the original one. The arbitrary constant will be added after the final integration on the right is completed.

EXAMPLE A

$$\int x \sin x \, dx$$

SOLUTION

This integral cannot be done by any of the previous methods.

However, $\sin x$ is simple to integrate. Therefore, u and dv are chosen as follows:

$$u = x \quad \text{and} \quad dv = \sin x \, dx$$

Differentiating to obtain du and integrating to obtain v gives

$$du = dx \quad \text{and} \quad v = -\cos x$$

The formula for integration by parts now gives

$$\int x \sin x \, dx = x(-\cos x) - \int (-\cos x) \, dx$$

$$= -x \cos x + \sin x + C \quad \text{compare to Formula 57} \quad \square$$

The success of the integration-by-parts process depends critically on the choice of u and dv. Consider the situation if, in this example, the choice of u and dv had been made in a different way:

$$u = \sin x \quad \text{and} \quad dv = x \, dx$$

Then,

$$du = \cos x \, dx \quad \text{and} \quad v = \frac{1}{2}x^2$$

and the formula for integration by parts gives

$$\int x \sin x \, dx = \frac{1}{2}x^2 \cdot \sin x - \int \frac{1}{2}x^2 \cos x \, dx$$

This result is certainly true, but it is not useful since the new integral is more difficult to handle than the original one.

EXAMPLE B

$$\int \ln 3x \, dx$$

SOLUTION

Choose,

$$u = \ln 3x \qquad dv = dx$$

$$du = \frac{1}{x} dx \qquad v = x$$

Therefore,

$$\int \ln 3x \, dx = x \ln 3x - \int x \cdot \frac{1}{x} \, dx$$

$$= x \ln 3x - x + C \qquad \text{compare to Formula 17} \qquad \square$$

How does one recognize when it is appropriate to apply integration by parts to a given problem? How does one determine what to choose for u and dv? Examine the functions in the integrand. If you encounter integrals of the following types, integration by parts will generally be successful.

Choose u to be a power of x when the integrand is of the form:

$$\int (\text{polynomial}) \cdot (\text{trigonometric}) \, dx$$

$$\int (\text{polynomial}) \cdot (\text{exponential}) \, dx$$

u can be chosen to be either function when the integrand contains products like

$$\int (\text{exponential}) \cdot (\text{trigonometric}) \, dx$$

Choose u to be the logarithmic or inverse trigonometric function when the integrand contains one of these functions:

$$\int (\text{polynomial}) \cdot (\text{inverse trigonometric}) \, dx$$

$$\int (\text{polynomial}) \cdot (\text{logarithmic}) \, dx$$

Exercise 17.5

1. Use integration by parts to determine the following integrals:

(a) $\int x \cos x \, dx$

(b) $\int 2x \sin(2x + 1) \, dx$

(c) $\int x \sec^2 x \, dx$

(d) $\int x^2 \sin x \, dx$ Hint: Apply integration by parts twice.

2. Determine the following integrals:

(a) $\int x e^x \, dx$

(b) $\int x^3 e^{-x} \, dx$ Hint: Apply integration by parts three times.

3. Determine the following integrals:
  Hint: In each case, apply integration by parts twice, then solve for the original integral.

(a) $\int e^{2x} \sin x \, dx$

(b) $\int \sin 3x \cos 4x \, dx$

4. Determine the following integrals:

(a) $\int \ln 7x \, dx$

(b) $\int x \ln 2x \, dx$

(c) $\int \ln^2 x \, dx$

(d) $\int x^n \ln x \, dx$

5. Derive Formula 23 in the Table of Selected Integrals.

6. Determine the following integrals using integration by parts:

(a) $\int \sin^{-1} 4x \, dx$

(b) $\int x \tan^{-1} x \, dx$

7. Determine the following integrals:

(a) $\int \frac{x}{\sqrt{3x + 1}} \, dx$ Hint: Choose $u = x$.

(b) $\int \frac{x^3}{\sqrt{1 - x^2}} \, dx$ Hint: Choose $u = x^2$.

8. Evaluate the definite integral:

(a) $\int_0^1 x e^{-4x} \, dx$

(b) $\int_0^{\pi/2} x \sin 2x \, dx$

9. Derive Formula 51 in the Table of Selected Integrals.

10. Using integration by parts once, derive Formula 59 in the Table of Selected Integrals.

11. Show that

$$\int e^x \sin nx \, dx = \frac{e^x (\sin nx - n \cos nx)}{1 + n^2}$$

12. (a) Use integration by parts to derive the reduction formula

$$\int \ln^n x \, dx = x \ln^n x - n \int \ln^{n-1} x \, dx$$

(b) Use the formula in part (a) to integrate

$$\int \ln^3 x \, dx$$

17.6 Summary and Review

Three basic methods of integration were introduced in this chapter. An algebraic substitution is usually called for when the argument of a function is an expression other than just x. To integrate rational functions, it is necessary to split the integrand into a sum of partial fractions and sometimes complete the square when an irreducible quadratic is present. Integration by parts is the method of choice when the integrand contains a product of two functions.

  Sometimes you will discover that a particular integral can be handled by more than one method. Some problems will require perseverance and ingenuity. To be successful at working out integrals of all sorts, there are a number of things to keep in mind:

1. You should be thoroughly acquainted with all of the basic integration formulas discussed in Chapter 14.

2. Try to understand the reason why a certain method works for one integral but not another.

3. Be willing to experiment. If one method is not successful, try something different.

4. Do not try to work out an integral in your head. Write down your work. It often takes several algebraic steps before you can see how the solution comes about.

5. Learn to recognize the general types of integrals that are susceptible to a particular method.

Exercise 17.6

The following is a list of integrals for practice. In each case, determine which of the methods of integration described in this chapter is the appropriate one to use, then determine the integral.

1. $\displaystyle\int \frac{5\,dx}{x^2 + 3}$

2. $\displaystyle\int \frac{x + 2}{2x - 1}\,dx$

3. $\displaystyle\int \frac{x}{2x + 1}\,dx$

4. $\displaystyle\int \frac{dx}{x\sqrt{3 + 4x}}$

5. $\displaystyle\int \frac{x\,dx}{\sqrt{2 + 4x}}$

6. $\displaystyle\int \frac{x\,dx}{3 - (x/2)}$

7. $\displaystyle\int \frac{5\,dx}{3x(5 - 2x)}$

8. $\displaystyle\int \frac{3\,dx}{x^2 - x}$

9. $\displaystyle\int \sqrt{x^2 - 2x - 1}\,dx$

10. $\displaystyle\int \sqrt{6x - x^2}\,dx$

11. $\displaystyle\int \frac{\sqrt{x^2 + 2x}}{x}\,dx$

12. $\displaystyle\int (7x - 4)^{3.4}\,dx$

13. $\displaystyle\int x\sqrt{x}(1 + x^2\sqrt{x})^2\,dx$

14. $\displaystyle\int \frac{\sqrt{x - 1} + 1}{\sqrt{x - 1} - 1}\,dx$

15. $\displaystyle\int \frac{\cos 2x}{(1 + \sin 2x)^{3/2}}\,dx$

16. $\displaystyle\int \frac{x}{e^{x^2}}\,dx$

17. $\displaystyle\int \frac{b^3 x^3}{\sqrt{1 - a^4 x^4}}\,dx$

18. $\displaystyle\int \frac{2}{3 - 7x}\,dx$

19. $\displaystyle \int \frac{x}{2x+3}\,dx$

20. $\displaystyle \int \frac{4x^2-1}{x+4}\,dx$

21. $\displaystyle \int \frac{7-2x}{3x-5}\,dx$

22. $\displaystyle \int \frac{2x+5}{x^2+4x+1}\,dx$

23. $\displaystyle \int \frac{(x-2)\,dx}{x^2-7x+12}$

24. $\displaystyle \int \frac{x^5+x^4-8}{x^3-4x}\,dx$

25. $\displaystyle \int x\tan^2 x\,dx$

26. $\displaystyle \int e^x \cos 4x\,dx$

27. $\displaystyle \int \ln(x+a)\,dx$

28. $\displaystyle \int x^{-2}\ln x\,dx$

29. $\displaystyle \int \frac{\ln x}{(x-1)^2}\,dx$

30. $\displaystyle \int \cos^{-1} x\,dx$

31. $\displaystyle \int x^3\sqrt{x^2+a^2}\,dx$

32. Show that

$$\int e^{mx} \sin nx\,dx = \frac{e^{mx}(m\sin nx - n\cos nx)}{m^2+n^2}$$

I.   Table of Selected Integrals

Basic Integration Formulas

1. $\displaystyle \int (f(x) + g(x))\,dx = \int f(x)\,dx + \int g(x)\,dx$

2. $\displaystyle \int a f(x)\,dx = a \int f(x)\,dx$

3. $\displaystyle \int x^n\,dx = \frac{x^{n-1}}{n+1} + C,\, n \neq -1$

4. $\displaystyle \int \frac{dx}{x} = \ln x + C$

5. $\displaystyle \int \cos x\,dx = \sin x + C$

6. $\displaystyle \int \sin x\,dx = -\cos x + C$

7. $\displaystyle \int \sec^2 x\,dx = \tan x + C$

8. $\displaystyle \int \csc^2 x\,dx = -\cot x + C$

9. $\displaystyle \int \tan x \sec x\,dx = \sec x + C$

10. $\displaystyle \int \csc x \cot x\,dx = -\csc x + C$

11. $\displaystyle \int \tan x\,dx = \ln |\sec x| + C$

12. $\displaystyle \int \cot x\,dx = -\ln |\csc x| + C$

13. $\displaystyle \int \sec x\,dx = \ln |\tan x + \sec x| + C$

14. $\displaystyle \int \csc x\,dx = -\ln |\cot x + \csc x| + C$

15. $\displaystyle \int e^x\,dx = e^x + C$

16. $\displaystyle \int a^x\,dx = \frac{a^x}{\ln a} + C$

17. $\displaystyle \int \ln x\,dx = x \ln x - x + C$

18. $\int \log_a x \, dx = x \log_a x - x \log_a e + C$

19. $\int \dfrac{dx}{\sqrt{a^2 - x^2}} = \sin^{-1} \dfrac{x}{a} + C$

20. $\int \dfrac{dx}{x^2 + a^2} = \dfrac{1}{a} \tan^{-1} \dfrac{x}{a} + C$

21. $\int \dfrac{dx}{\sqrt{x^2 \pm a^2}} = \ln \left| x + \sqrt{x^2 \pm a^2} \right| + C$

22. $\int \dfrac{dx}{x^2 - a^2} = \dfrac{1}{2a} \ln \left| \dfrac{x - a}{x + a} \right| + C$

23. $\int \sin^{-1} x \, dx = x \sin^{-1} x + \sqrt{1 - x^2} + C$

24. $\int \cos^{-1} x \, dx = x \cos^{-1} x - \sqrt{1 - x^2} + C$

25. $\int \tan^{-1} x \, dx = x \tan^{-1} x - \dfrac{1}{2} \ln \left| 1 + x^2 \right| + C$

Forms containing $(a + bx)$

26. $\int \dfrac{1}{(a + bx)} \, dx = \dfrac{1}{b} \ln |a + bx| + C$

27. $\int \dfrac{1}{x(a + bx)} \, dx = -\dfrac{1}{a} \ln \left| \dfrac{a + bx}{x} \right| + C$

28. $\int \dfrac{1}{x^2(a + bx)} \, dx = -\dfrac{1}{ax} + \dfrac{b}{a^2} \ln \left| \dfrac{a + bx}{x} \right| + C$

29. $\int \dfrac{1}{x(a + bx)^2} \, dx = \dfrac{1}{a(a + bx)} - \dfrac{1}{a^2} \ln \left| \dfrac{a + bx}{a} \right| + C$

30. $\int \dfrac{x}{(a + bx)} \, dx = \dfrac{a + bx}{b^2} - \dfrac{a}{b^2} \ln |a + bx| + C$

31. $\int \dfrac{x}{(a + bx)^2} \, dx = \dfrac{1}{b^2} \left[\dfrac{a}{a + bx} + \ln |a + bx| \right] + C$

32. $\int \dfrac{(a' + b'x)}{(a + bx)} \, dx = \dfrac{b'x}{b} + \dfrac{a'b - ab'}{b^2} \ln |a + bx| + C$

Forms containing $\sqrt{a + bx}$

33. $\displaystyle \int x\sqrt{a + bx}\, dx = -\frac{2(2a - 3bx)\sqrt{(a + bx)^3}}{15b^2} + C$

34. $\displaystyle \int \frac{x}{\sqrt{a + bx}}\, dx = -\frac{2(2a - bx)\sqrt{a + bx}}{3b^2} + C$

35. $\displaystyle \int \frac{1}{x\sqrt{a + bx}}\, dx = \frac{1}{\sqrt{a}} \ln \left| \frac{\sqrt{a + bx} - \sqrt{a}}{\sqrt{a + bx} + \sqrt{a}} \right| + C, \qquad \text{if } a > 0$

$\displaystyle = \frac{2}{\sqrt{-a}} \tan^{-1} \sqrt{\frac{a + bx}{-a}} + C, \qquad \text{if } a < 0$

36. $\displaystyle \int \frac{\sqrt{a + bx}}{x}\, dx = 2\sqrt{a + bx} + a \int \frac{dx}{x\sqrt{a + bx}}$

Forms containing $\sqrt{x^2 \pm a^2}$

37. $\displaystyle \int \sqrt{x^2 \pm a^2}\, dx = \frac{x}{2}\sqrt{x^2 \pm a^2} \pm \frac{a^2}{2} \ln \left| x + \sqrt{x^2 \pm a^2} \right| + C$

38. $\displaystyle \int \frac{1}{x\sqrt{x^2 + a^2}}\, dx = -\frac{1}{a} \ln \left| \frac{a + \sqrt{x^2 + a^2}}{x} \right| + C$

39. $\displaystyle \int \frac{1}{x\sqrt{x^2 - a^2}}\, dx = \frac{1}{a} \cos^{-1} \frac{a}{x} + C$

40. $\displaystyle \int \frac{\sqrt{x^2 + a^2}}{x}\, dx = \sqrt{x^2 + a^2} - a \ln \left| \frac{a + \sqrt{x^2 + a^2}}{x} \right| + C$

41. $\displaystyle \int \frac{\sqrt{x^2 - a^2}}{x}\, dx = \sqrt{x^2 - a^2} - a \cos^{-1} \frac{a}{x} + C$

Forms containing $\sqrt{a^2 - x^2}$

42. $\displaystyle \int \sqrt{a^2 - x^2}\, dx = \frac{x}{2}\sqrt{a^2 - x^2} + \frac{a^2}{2} \sin^{-1} \frac{x}{a} + C$

43. $\displaystyle \int \frac{1}{x\sqrt{a^2 - x^2}}\, dx = -\frac{1}{a} \ln \left| \frac{a + \sqrt{a^2 - x^2}}{x} \right| + C$

44. $\displaystyle \int \frac{\sqrt{a^2 - x^2}}{x}\, dx = \sqrt{a^2 - x^2} - a \ln \left| \frac{a + \sqrt{a^2 - x^2}}{x} \right| + C$

Forms containing $(a + bx + cx^2)$

$$X = a + bx + cx^2, \quad q = 4ac - b^2$$

45. $$\int \frac{dx}{X} = \frac{2}{\sqrt{q}} \tan^{-1}\left[\frac{2cx + b}{\sqrt{q}}\right] + C, \quad \text{if } q > 0$$

$$= \frac{1}{\sqrt{-q}} \ln\left|\frac{2cx + b - \sqrt{-q}}{2cx + b + \sqrt{-q}}\right| + C, \quad \text{if } q < 0$$

46. $$\int \frac{dx}{X^2} = \frac{2cx + b}{qX} + \frac{2c}{q} \int \frac{dx}{X}$$

47. $$\int \frac{dx}{\sqrt{X}} = \frac{1}{\sqrt{c}} \ln\left|\sqrt{X} + x\sqrt{c} + \frac{b}{2\sqrt{c}}\right| + C, \quad \text{if } c > 0$$

$$= \frac{1}{\sqrt{-c}} \sin^{-1}\left[\frac{-2cx - b}{\sqrt{b^2 - 4ac}}\right] + C, \quad \text{if } c < 0$$

48. $$\int \sqrt{X}\, dx = \frac{(2cx + b)\sqrt{X}}{4c} + \frac{q}{8c} \int \frac{dx}{\sqrt{X}}$$

49. $$\int \frac{x\, dx}{\sqrt{X}} = \frac{\sqrt{X}}{c} - \frac{b}{2c} \int \frac{dx}{\sqrt{X}}$$

50. $$\int \frac{dx}{x\sqrt{X}} = -\frac{1}{\sqrt{a}} \ln\left|\frac{\sqrt{X} + \sqrt{a}}{x} + \frac{b}{2\sqrt{a}}\right| + C, \quad \text{if } a > 0$$

$$= \frac{1}{\sqrt{-a}} \sin^{-1}\left[\frac{bx + 2a}{x\sqrt{b^2 - 4ac}}\right] + C, \quad \text{if } a < 0$$

$$= -\frac{2\sqrt{X}}{bx} + C, \quad \text{if } a = 0$$

Forms containing $\sin ax$, $\cos ax$, e^{ax}, and $\ln ax$

51. $$\int xe^{ax}\, dx = \frac{e^{ax}}{a^2}(ax - 1) + C$$

52. $$\int x^2 e^{ax}\, dx = \frac{e^{ax}}{a^3}(a^2x^2 - 2ax + 2) + C$$

53. $$\int x^n e^{ax}\, dx = \frac{x^n e^{ax}}{a} - \frac{n}{a} \int x^{n-1} e^{ax}\, dx$$

54. $\displaystyle\int x^n \ln ax\, dx = \frac{x^{n+1} \ln ax}{n+1} - \frac{x^{n+1}}{(n+1)^2} + C, (n \neq -1)$

55. $\displaystyle\int \sin^2 ax\, dx = \frac{1}{2a}(ax - \sin ax \cos ax) + C$

56. $\displaystyle\int \cos^2 ax\, dx = \frac{1}{2a}(ax + \sin ax \cos ax) + C$

57. $\displaystyle\int x \sin ax\, dx = \frac{1}{a^2}(\sin ax - ax \cos ax) + C$

58. $\displaystyle\int x \cos ax\, dx = \frac{1}{a^2}(\cos ax + ax \sin ax) + C$

59. $\displaystyle\int x^n \sin ax\, dx = -\frac{x^n \cos ax}{a} + \frac{n}{a}\int x^{n-1} \cos ax\, dx$

60. $\displaystyle\int x^n \cos ax\, dx = \frac{x^n \sin ax}{a} - \frac{n}{a}\int x^{n-1} \sin ax\, dx$

Trigonometric Integrals

18

The process of integration consists largely of transforming an integral into a simpler form that corresponds to one of the basic integration formulas. The previous chapter dealt with problems that could be simplified by algebraic methods. Other problems of integral calculus are simplified by making a **trigonometric substitution**. Such substitutions lead to integrals of powers of the trigonometric functions. Methods of handling such integrals are described in this chapter.

Integrals of Powers of Sin x and Cos x

18.1

The purpose of this section is to discover how to work out integrals of the form

$$\int \sin^m x \cos^n x \, dx$$

To begin with, observe that when one of the exponents is 1, the integration is readily performed. In this special case, the integral reduces to

$$\int \sin^m x \cos x \, dx \qquad \text{or} \qquad \int \cos^n x \sin x \, dx$$

Both of these have the form $\int v^n \, dv$. For instance, if $v = \sin x$, then $dv = \cos x \, dx$, and

$$\int \sin^m x \cos x \, dx = \int v^m \, dv$$

$$= \frac{v^{m+1}}{m+1}$$

$$= \frac{1}{m+1} \sin^{m+1} x + C \quad (m \neq -1)$$

In other words, to integrate a power of $\sin x$, there must be a $\cos x$ present and vice versa.

Generally, then, to integrate powers of sines and/or cosines, the strategy is to use trigonometric identities to change them into one of the two forms:

$$\int \sin^m x \cos x \, dx \qquad \text{or} \qquad \int \cos^n x \sin x \, dx$$

There are two approaches depending on whether the exponents are positive *even* or positive *odd* integers.

Case I. One of the exponents a positive odd integer

EXAMPLE A

$$\int \cos^3 x \, dx$$

SOLUTION

First factor the integrand into $\cos^2 x \cdot \cos x$. Then, using the Pythagorean identity

$$\cos^2 x = 1 - \sin^2 x$$

change the even power of $\cos x$ into powers of $\sin x$. This will result in terms of the form

$$\int \sin^n x \cos x \, dx$$

Thus,

$$\int \cos^3 x \, dx = \int \cos^2 x \cos x \, dx$$

$$= \int (1 - \sin^2 x) \cos x \, dx$$

$$= \int \cos x \, dx - \int \sin^2 x \cos x \, dx$$

$$= \sin x - \frac{1}{3}\sin^3 x + C$$

❑

EXAMPLE B

$$\int \sin^3 x \cos^4 x \, dx$$

SOLUTION

Should this integral be converted into powers of sine or powers of cosine? Since the exponent of the sine function is odd, set aside a factor of $\sin x$. The remaining even power, $\sin^2 x$ can be converted to a polynomial in $\cos x$.

$$\int \sin^3 x \cos^4 x \, dx = \int \sin^2 x \cos^4 x \sin x \, dx$$

$$= \int (1 - \cos^2 x) \cos^4 x \sin x \, dx$$

$$= \int \cos^4 x \sin x \, dx - \int \cos^6 x \sin x \, dx$$

$$= -\frac{1}{5}\cos^5 x + \frac{1}{7}\cos^7 x + C$$

❑

Note carefully that this procedure will always work when one of the exponents is a *positive odd* integer, no matter what the other exponent is: positive, negative, or zero; integer or fraction.

Case II. Both of the exponents positive even integers

In this case the integration is accomplished by making use of one or more of the double angle identities:

$$\sin^2 \theta = \frac{1}{2}(1 - \cos 2\theta)$$

$$\cos^2 \theta = \frac{1}{2}(1 + \cos 2\theta)$$

$$\sin \theta \cos \theta = \frac{1}{2}\sin 2\theta$$

These identities, in effect, change even powers of the functions into odd powers.

EXAMPLE C

$$\int \cos^2 x \, dx$$

SOLUTION

Using the second identity above changes the even power of $\cos x$ into an odd power of $\cos 2x$, which can then be handled as in Case I.

$$\int \cos^2 x \, dx = \int \frac{1}{2}(1 + \cos 2x) \, dx$$

$$= \frac{1}{2}\int dx + \frac{1}{2}\int \cos 2x \, dx$$

$$= \frac{1}{2}x + \frac{1}{4}\sin 2x + C$$

☐

Exercise 18.1

1. By means of the substitution $v = ax$, show that

(a) $\displaystyle\int \sin ax \, dx = -\frac{1}{a}\cos ax + C$

(b) $\displaystyle\int \cos ax \, dx = \frac{1}{a}\sin ax + C$

2. Express in the form $\int \sin^m x \cos x \, dx$ and integrate:

(a) $\displaystyle\int \cos^3 3x \, dx$

(b) $\displaystyle\int \cos^7 x \, dx$

(c) $\displaystyle\int \sin^2 2x \cos^3 2x \, dx$

3. Express in the form $\int \cos^n x \sin x \, dx$ and integrate:

(a) $\displaystyle\int \sin^3 x \, dx$

(b) $\displaystyle\int \sin^5 \frac{x}{2} \, dx$

(c) $\displaystyle\int \sin^3 2x \cos^3 2x \, dx$

4. Integrate with the help of the double angle identities:

(a) $\displaystyle\int \sin^2 x \, dx$

(b) $\displaystyle\int \cos^4 x \, dx$

(c) $\displaystyle\int \sin^2 4x \cos^2 4x \, dx$

5. Integrate:

(a) $\displaystyle\int \cos^5 4x \, dx$

(b) $\displaystyle\int \sin x \cos^5 x \, dx$

(c) $\displaystyle\int \tan^3 x \, dx$ Hint: Express in terms of sin x and cos x.

(d) $\displaystyle\int \tan x \sec^3 x \, dx$

(e) $\displaystyle\int \cos x \csc^4 x \, dx$

(f) $\displaystyle\int \sqrt{\cos x} \sin^3 x \, dx$

6. Determine the integral:

(a) $\int \sin^4 2x \, dx$

(b) $\int \sin^4 x \cos^2 x \, dx$

(c) $\int (\cos^2 x - \sin^2 x)^3 \, dx$

(d) $\int (\cos^2 x + \sin^2 x)^3 \, dx$

(e) $\int (\sqrt{\cos x} + 3 \sin x)^2 \, dx$

7. Determine the integral:

(a) $\int \sin x \sin 2x \, dx$

(b) $\int \sin x \cos x \cos^4 2x \, dx$

(c) $\int \sin^2 \frac{x}{2} \cos^2 x \, dx$

8. With the help of the trigonometric identity

$$\cos mx \cos nx = \frac{1}{2} \cos(m + n)x$$
$$+ \frac{1}{2} \cos(m - n)x$$

integrate

$$\int \cos 6x \cos 4x \, dx$$

9. Do the following integrals using the suitable trigonometric identities.

(a) $\int \cos 7x \sin 2x \, dx$

(b) $\int \sin 8x \sin 4x \, dx$

(c) $\int \cos 2x \cos 5x \, dx$

10. Integrate. Hint: Each integral is already in the form $\int v^n \, dv$.

(a) $\int \tan^6 x \sec^2 x \, dx$

(b) $\int \sec^3 x \tan x \sec x \, dx$

11. Express in the form $\int \tan^n x \sec^2 x \, dx$ and integrate:

(a) $\int \tan^3 x \, dx$

(b) $\int \sec^6 2x \, dx$

(c) $\int \sec^6 x \tan^2 x \, dx$

12. Express in the form $\int \sec^n x \tan x \sec x \, dx$ and integrate:

(a) $\int \sec \theta \tan^5 \theta \, d\theta$

(b) $\int \tan \theta \sec^5 \theta \, d\theta$

(c) $\int \sin 2x \sec^6 x \, dx$

Trigonometric Substitutions 18.2

In the original collection of integration formulas given in Chapter 14, there were two which involved inverse trigonometric functions. These are Formulas 19 and 20 in the Table of Integrals:

$$\int \frac{dx}{\sqrt{a^2 - x^2}} = \sin^{-1} \frac{x}{a} + C \qquad \text{Formula 19}$$

$$\int \frac{dx}{x^2 + a^2} = \frac{1}{a} \tan^{-1} \frac{x}{a} + C \qquad \text{Formula 20}$$

It was possible to state these formulas at the outset because they arose directly from the corresponding differentiation rules for $\sin^{-1} x$ and $\tan^{-1} x$.

Is it possible, however, to work out these integrals directly and arrive at the same results? The answer is yes, but not by any of the methods discussed up to this point. Instead, the way to proceed is to substitute a trigonometric function for the variable x. A suitable **trigonometric substitution** will unlock these and many otherwise intractable integrals, especially those involving radicals.

Consider first the integral

$$\int \frac{1}{\sqrt{a^2 - x^2}}\, dx$$

The radical expression is reminiscent of the Pythagorean theorem. It is the third side of a right triangle with hypotenuse a and one side x (Figure 18.1). In this triangle

Figure 18.1

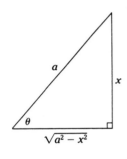

$$x = a \sin \theta$$

and this relation can be used to change the variable of integration from x to θ. Making this substitution, eliminates the radical:

$$\sqrt{a^2 - x^2} = \sqrt{a^2 - a^2 \sin^2 \theta}$$
$$= \sqrt{a^2(1 - \sin^2 \theta)}$$
$$= \sqrt{a^2 \cos^2 \theta}$$
$$= a \cos \theta$$

This result could, of course, have been written down directly from Figure 18.1. The differential, dx, must also be replaced:

$$dx = a \cos \theta\, d\theta$$

The substitution of $a \sin \theta$ for x in the integral thus leads to

$$\int \frac{1}{\sqrt{a^2 - x^2}}\, dx = \int \frac{1}{a \cos \theta} \cdot (a \cos \theta\, d\theta)$$
$$= \int d\theta$$
$$= \theta + C$$

It is necessary now to convert this result back into terms of x.

Since $a \sin \theta = x$ then $\theta = \sin^{-1}\dfrac{x}{a}$

and we have arrived at the expected result:

$$\int \frac{1}{\sqrt{a^2 - x^2}}\,dx = \sin^{-1}\frac{x}{a} + C$$

The integral leading to the inverse tangent function

$$\int \frac{1}{x^2 + a^2}\,dx$$

is handled in much the same way. This time, however, the expression $x^2 + a^2$ is the square of the hypotenuse of a right triangle with sides a and x (Figure 18.2). In this case, the substitution is

$$x = a \tan \theta$$

Figure 18.2

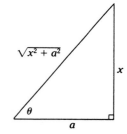

Completion of this integration will be left to the exercises.

Generally speaking, a trigonometric substitution can be applied when expressions of the form $(a^2 - x^2)$ or $(x^2 \pm a^2)$ appear in the integrand. When such an expression occurs in a radical, the radical will be eliminated. As the following examples will show, these substitutions lead to integrals of powers of trigonometric functions, which can be handled by the methods discussed in the first part of this chapter.

EXAMPLE A

$$\int \sqrt{9 - x^2}\,dx$$

SOLUTION

What is the appropriate trigonometric substitution? Draw a right triangle with hypotenuse 3 and one side x (Figure 18.3). From this diagram, it follows that

Figure 18.3

$$x = 3 \sin \theta$$

$$\sqrt{9 - x^2} = 3 \cos \theta$$

Also $dx = 3 \cos \theta\, d\theta$

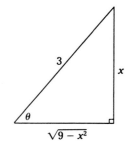

Consequently, upon substituting into the integral, we have

$$\int \sqrt{9 - x^2}\, dx = \int (3 \cos \theta) \cdot (3 \cos \theta\, d\theta)$$

$$= 9 \int \cos^2 \theta\, d\theta$$

$$= 9 \int \frac{1}{2} (1 + \cos 2\theta)\, d\theta$$

$$= \frac{9}{2}\theta + \frac{9}{4} \sin 2\theta + C$$

$$= \frac{9}{2}\theta + \frac{9}{2} \sin \theta \cos \theta + C$$

It is necessary now to convert this result back into terms of x.

Since

$$\sin \theta = \frac{x}{3}$$

then

$$\theta = \sin^{-1} \frac{x}{3}$$

and

$$\cos \theta = \frac{\sqrt{9 - x^2}}{3}$$

Therefore,

$$\int \sqrt{9 - x^2}\, dx = \frac{9}{2} \sin^{-1} \frac{x}{3} + \frac{9}{2} \cdot \frac{x}{3} \cdot \frac{\sqrt{9 - x^2}}{3}$$

$$= \frac{9}{2} \sin^{-1} \frac{x}{3} + \frac{x\sqrt{9 - x^2}}{2} + C$$

Compare Formula 42 in the Table of Integrals. ❑

EXAMPLE B

$$\int \frac{dx}{\sqrt{9x^2 + 4}}\, dx$$

SOLUTION

Figure 18.4

The denominator in this example looks like the hypotenuse of a right triangle with sides $3x$ and 2 (Figure 18.4). Thus

$$\tan \theta = \frac{3x}{2}$$

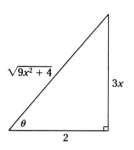

$$x = \frac{2}{3}\tan \theta$$

$$dx = \frac{2}{3}\sec^2 \theta \, d\theta$$

$$\sqrt{9x^2 + 4} = 2\sec \theta$$

Substituting

$$\int \frac{dx}{\sqrt{9x^2 + 4}}\, dx = \frac{2}{3}\int \frac{\sec^2 \theta \, d\theta}{2\sec \theta}$$

$$= \frac{1}{3}\int \sec \theta \, d\theta$$

$$= \frac{1}{3}\ln(\sec \theta + \tan \theta) + C$$

$$= \frac{1}{3}\ln\left(\frac{\sqrt{9x^2 + 4}}{2} + \frac{3x}{2}\right) + C$$

$$= \frac{1}{3}\ln(\sqrt{9x^2 + 4} + 3x) - \frac{1}{3}\ln 2 + C$$

$$= \frac{1}{3}\ln(\sqrt{9x^2 + 4} + 3x) + C'$$

In the last step, the constant $-(1/3)\ln 2$ is included as a part of the arbitrary constant C'. Compare this result to Formula 16 in the Table of Selected Integrals. ❑

A summary of the three trigonometric substitutions is shown in Table 18.1. In choosing the correct substitution, right triangle diagrams are a helpful guide. They are also valuable for making the conversion from θ back to x once the integration has been completed. Note: It is a rather common error merely to substitute $d\theta$ for dx. You should not forget to differentiate to find the proper expression for dx in terms of θ and $d\theta$.

Table 18.1 Summary of Trigonometric Substitutions

Expression	Substitution	Triangle
$(a^2 - x^2)$ or $\sqrt{a^2 - x^2}$ or $(a^2 - x^2)^{n/2}$	$x = a \sin \theta$ $\sqrt{a^2 - x^2} = a \cos \theta$ $dx = a \cos \theta \, d\theta$	
$(x^2 + a^2)$ or $\sqrt{x^2 + a^2}$ or $(x^2 + a^2)^{n/2}$	$x = a \tan \theta$ $\sqrt{x^2 + a^2} = a \sec \theta$ $dx = a \sec^2 \theta \, d\theta$	
$(x^2 - a^2)$ or $\sqrt{x^2 - a^2}$ or $(x^2 - a^2)^{n/2}$	$x = a \sec \theta$ $\sqrt{x^2 - a^2} = a \tan \theta$ $dx = a \tan \theta \sec \theta \, d\theta$	

Exercise 18.2

1. Integrate using a substitution of the form $x = a \sin \theta$.

(a) $\displaystyle \int x(9 - x^2)^{3/2} \, dx$

(b) $\displaystyle \int \frac{dx}{x^2 \sqrt{7 - 4x^2}}$

(c) $\displaystyle \int \frac{dx}{x \sqrt{16 - x^2}}$

2. Integrate using a substitution of the form $x = a \tan \theta$:

(a) $\displaystyle \int \frac{3x}{\sqrt{x^2 + 16}} \, dx$

(b) $\displaystyle \int \frac{x^3}{\sqrt{9 + x^2}} \, dx$

(c) $\displaystyle \int \frac{x^2}{6x^2 + 1} \, dx$

3. Integrate using a substitution of the form $x = a \sec \theta$:

(a) $\displaystyle \int x \sqrt{(x^2 - 16)^3} \, dx$

(b) $\displaystyle\int \frac{dx}{x\sqrt{3x^2 - 5}}$

(c) $\displaystyle\int \frac{dx}{x^2\sqrt{x^2 - 6}}$

4. Using a suitable trigonometric substitution, show that

$$\int \frac{1}{x^2 + a^2}\, dx = \frac{1}{a}\tan^{-1}\frac{x}{a} + C \qquad \text{Formula 12}$$

5. Integrate using a suitable trigonometric substitution:

(a) $\displaystyle\int \frac{x^2\, dx}{\sqrt{(1 - x^2)^3}}$

(b) $\displaystyle\int \frac{\sqrt{x^2 - 1}}{x}\, dx$

(c) $\displaystyle\int \frac{\sqrt{x^2 - 4}}{x^4}\, dx$

(d) $\displaystyle\int \frac{dx}{x^2\sqrt{9x^2 + 25}}$

(e) $\displaystyle\int x^3\sqrt{4 - x^2}\, dx$

(f) $\displaystyle\int \frac{x^3\, dx}{\sqrt{x^2 - 9}}$

6. Derive Formula 39 in the Table of Selected Integrals.

7. Derive Formula 43 in the Table of Selected Integrals.

8. Show that

$$\int \frac{dx}{x^2\sqrt{a^2 + x^2}} = -\frac{\sqrt{a^2 + x^2}}{a^2 x} + C$$

9. Show that

$$\int \frac{\sqrt{a^2 - x^2}}{x^2}\, dx = -\frac{\sqrt{a^2 - x^2}}{x} - \sin^{-1}\frac{x}{a} + C$$

10. $\displaystyle\int \frac{dx}{\sqrt{(x + 1)^2 + 4}}$ Hint: Let $x + 1 = v$, and $v = 2\tan\theta$.

11. Integrate using a suitable trigonometric substitution:

(a) $\displaystyle\int \frac{x^2\, dx}{\sqrt{(x^2 + 49)^3}}$

(b) $\displaystyle\int \frac{x^3\, dx}{\sqrt{a^2 - x^2}}$

(c) $\displaystyle\int x\sqrt{4x^2 - 25}\, dx$

(d) $\displaystyle\int x\sqrt{4 + x^2}\, dx$

(e) $\displaystyle\int x(9 - 4x^2)^{3/2}\, dx$

(f) $\displaystyle\int \frac{dx}{x^3\sqrt{x^2 - 1}}$

Summary and Review   18.3

The principal method for solving integrals of powers of trigonometric functions is to express them in the form $\int v^n\, dv$. To accomplish this, various trigonometric identities are required. On the few occasions when this method does not work, integration by parts is the best alternative.

Integrals which contain factors of the form $(x^2 \pm a^2)$ or $(a^2 - x^2)$ can be done using a trigonometric substitution. This method is especially useful when there are radical expressions in the integrand.

To master the material in this chapter:

1. Memorize the eight basic trigonometric identities, the double angle formulas for sine and cosine, and the derivatives of the trigonometric functions if you have not done so already.

2. Try to recognize the pattern $\int v^n dv$ in each trigonometric integral. Be flexible in switching your attention from one possibility to another.

3. Find out how to tell *when* a trigonometric substitution is called for and *which* substitution is the one to use.

Exercise 18.3

The following is a list of integrals for practice. In each case, determine which of the methods of integration described in this chapter is the appropriate one to use, then work out the integral.

1. $\int \sin^5 x \cos^2 x \, dx$

2. $\int \sin^3 x \cos^2 x \, dx$

3. $\int \frac{\cos^3 x}{\sqrt{\sin^3 x}} \, dx$

4. $\int (\sin x + \cos x)^2 \, dx$

5. $\int \tan^2 x \cos^3 x \sin x \, dx$

6. $\int \sin^2 nx \, dx$

7. $\int \tan^5 3x \, dx$

8. $\int \sec^4 \frac{x}{2} \, dx$

9. $\int \tan^5 x \sec^4 x \, dx$

10. $\int \frac{\sin^2 x}{\cos^4 x} \, dx$

11. $\int (\tan 2x + \cot 2x)^2 \, dx$

12. $\int \csc^3 \frac{x}{3} \cot \frac{x}{3} \, dx$

13. $\int \frac{dx}{1 + \sin x}$

14. $\int \frac{\cos x}{1 - \cos x} \, dx$

15. $\int \frac{1 + \cos x}{1 - \sin x} \, dx$

16. $\int \frac{dx}{(1 + x^2)^2}$

17. $\int \frac{\sqrt{4 - 9x^2}}{x^2} \, dx$

18. $\int \frac{dx}{\sqrt{5x^2 - x^4}}$

19. $\int \frac{\sqrt{x^2 + 16}}{x^4} \, dx$

20. $\int \frac{\sqrt{3x^2 - 16}}{x^4} \, dx$

21. $\int \frac{\sqrt{4x^2 - 9}}{x} \, dx$

22. $\int \frac{dx}{x^2 \sqrt{4x^2 - 9}}$

23. $\int \frac{x^7}{(1 + x^4)^2} \, dx$ Hint: Let $v = x^2$.

24. $\int e^x \sqrt{1 - e^{2x}} \, dx$ Hint: Let $v = e^x$.

25. $\int \frac{\cos x}{\sqrt{2 - \sin^2 x}} \, dx$ Hint: Let $v = \sin x$.

26. $\int \dfrac{1}{\sqrt{x}(x+1)}\,dx$ Hint: Let $v = \sqrt{x}$.

27. $\int \dfrac{\sqrt{1-x}}{\sqrt{x}}\,dx$

28. Show that if

$$\tan\frac{x}{2} = v, \quad \text{that is,} \quad x = 2\tan^{-1}v$$

then

$$dx = \frac{2}{1+v^2}\,dv$$

$$\sin x = \frac{2v}{1+v^2} \qquad \text{Hint:}\sin x = 2\sin\frac{x}{2}\cos\frac{x}{2}$$

$$\cos x = \frac{1-v^2}{1+v^2} \qquad \text{Hint:}\cos x = 2\cos^2\frac{x}{2} - 1$$

29. Use the substitution described in Problem 28 to work out the following integrals.

(a) $\displaystyle\int \frac{1}{1 + \sin x + \cos x}\,dx$

(b) $\displaystyle\int \frac{1}{\sin x - 2\cos x - 2}\,dx$

(c) $\displaystyle\int \frac{1}{5 + 4\cos x}\,dx$

(d) $\displaystyle\int \frac{1}{\sin x + \cos x}\,dx$

J. Three-Dimensional Solids

An important application of the definite integral, that of finding volumes and surface areas of three-dimensional solids, is the subject of Chapter 19, which follows. The object here is to describe how to draw such solids with reference to a three-dimensional coordinate system. A realistic drawing is a considerable aid in the solution of a problem. This section also contains formulas for certain volumes and surface areas.

The xy-coordinate system, to which you are accustomed, can be extended to three dimensions by adding a third axis, the **z-axis**, perpendicular to the other two. The positive z-axis may be represented in a diagram by drawing it as though it were pointing out of the page, toward you, as shown in Figure J.1.

By following a few simple practices, you can make drawings that have an acceptable three-dimensional appearance.

Figure J.1

1. Represent planes parallel to the coordinate planes by rectangles, or parallelograms whose sides are parallel to the coordinate axes. In Figure J.2, for instance, rectangles represent the front and the back of a rectangular box, while parallelograms are used for the other four sides.

Figure J.2

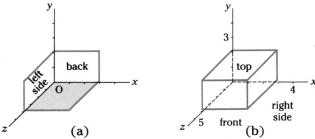

(a)   (b)

2. Use ellipses to represent circular areas that extend in the z direction as in Figure J.3a. Circular cylinders then look like those in Figure J.3b and c. In drawing a sphere, as in Figure J.4, use

Figure J.3

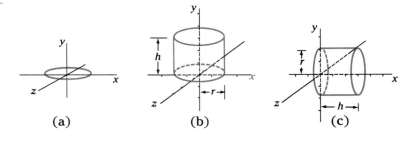

(a)   (b)   (c)

one ellipse for the "equator" and a second one for the "prime meridian." These represent circles in the xz-plane and the yz-plane respectively.

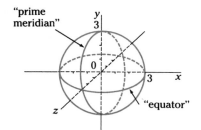

Figure J.4

3. Make the scale on the z-axis somewhat smaller than the x and y scales. Lengths in the z direction are foreshortened in a three-dimensional perspective.

4. Draw hidden lines of a figure and hidden parts of the axes using dotted lines.

Certain solids, known as **solids of revolution**, are generated by rotating an area in the xy-plane about the x-axis or the y-axis. To depict the solid formed by rotating the area under the curve shown in Figure J.5a about the x-axis, for instance, first draw the mirror image of the curve using the x-axis as the line of symmetry. Then trace the paths followed by selected points during rotation. These are circles, which appear as ellipses in the drawing. The result is the bugle-shaped solid shown in Figure J.5b. Rotating the area to the left of the curve about the y-axis generates the bowl shape shown in Figure J.5c.

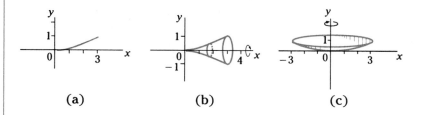

(a)   (b)   (c)

Figure J.5

Table J.1 contains formulas for the volume and surface area of some solids. In Chapter 19, you will learn how to derive formulas of this sort using integration.

Table J.1

Name		Volume	Surface Area
Rectangular box	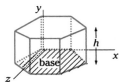	$V = lwh$	$S = 2(lh + lw + hw)$
Prism with polygonal base		$V = $ (area of base) $\times$ height	$S = 2 \times$ (area of base) $+$ sum of areas of rectangular sides
Cylinder	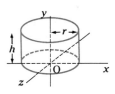	$V = \pi r^2 h$	$S = 2\pi rh + 2\pi r^2$
Cylinder with irregular base		$V = $ (area of base) $\times$ height	$S = 2 \times$ (area of base) $+$ sum of areas of sides
Cone	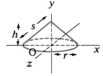	$V = \dfrac{1}{3}\pi r^2 h$	$S = \pi rs + \pi r^2$

Table J.1 continued

Name		Volume	Surface Area
Pyramid with square base	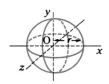	$V = \dfrac{4a^2h}{3}$	$S = 4as + 4a^2$
Sphere		$V = \dfrac{4}{3}\pi r^3$	$S = 4\pi r^2$
Spherical cap	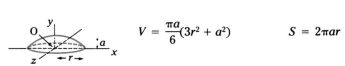	$V = \dfrac{\pi a}{6}(3r^2 + a^2)$	$S = 2\pi ar$
Torus (doughnut)	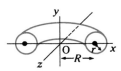	$V = 2\pi^2 R r^2$	$S = 4\pi^2 R r$

Exercise J

1. Draw a rectangular solid with its centre at the origin.

2. Draw a rectangular solid having a square cross section with its length
   (a) much greater than its width
   (b) much less than its width.

3. Draw (a) a cylinder and (b) a cone showing a cross section (i) perpendicular to and (ii) parallel to its axis.

4. Draw a hemispherical bowl of radius 4 with its base at the origin.

5. Cylinders, spheres, and cones are all solids of revolution. What areas will generate these solids by rotation?

6. Sketch the solids formed by rotating the following curves about the x-axis and about the y-axis.
   (a) $y = x^2$, $0 \le x \le 3$
   (b) $y = \sin x$, $0 \le x \le \pi$
   (c) $(x - 4)^2 + y^2 = 4$

Applications of the Definite integral

19

The definite integral was introduced in Chapter 16 in connection with the problem of finding the area under a curve. In this chapter, the definite integral is used to find volumes of three-dimensional solids (in particular, solids of revolution) and also lengths of arcs and areas of surfaces.

These examples can only begin to suggest the varied applications of the definite integral. What is important is the idea that a quantity can be thought of as the sum of an infinite number of infinitesimally small parts.

Volumes by Integration

19.1

The use of integration to find the volume of a three-dimensional solid, such as that shown in Figure 19.1, parallels to a great extent the process used to find areas in Chapter 16. Start by dividing the solid into n paper-thin slices of width Δx. A typical slice is known as a **volume element**.

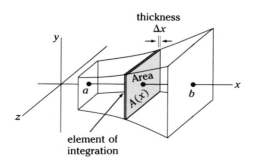

Figure 19.1

The volume ΔV of the kth slice located at x_k is the product of the surface area $A(x_k)$, which generally varies with its position x_k, and its thickness Δx:

$$\Delta V = (\text{area}) \cdot (\text{thickness})$$

$$= A(x_k) \cdot \Delta x$$

The sum of the volumes of all the slices is approximately equal to the volume of the solid:

$$V \doteq \sum_{k=1}^{n} A(x_k) \cdot \Delta x$$

In the limit as Δx approaches zero, the volume of each slice approaches zero, while the number of slices in the interval from $x = a$ to $x = b$ increases without bound. The exact value of the volume is given by the limit of the sum

$$V = \lim_{n \to \infty} \sum_{k=1}^{n} A(x_k) \cdot \Delta x$$

which is by definition the integral

$$V = \int_{a}^{b} A(x)dx$$

In practice, this integral for the volume is usually set up in terms of differentials. The element of integration is the volume dV of a typical slice, an expression which can be written down directly from the figure:

$$dV = (\text{area}) \cdot (\text{thickness})$$

$$= A(x) \cdot dx$$

It follows immediately that

$$V = \int_{a}^{b} A(x)dx$$

The choice of the element of integration is critical. For this process to be successful, you must be able to find an expression for the surface area of the element in terms of its position x. A second concern has to do with the choice of a coordinate system. You must decide where to place the origin of coordinates and how the axes are to be oriented with respect to the volume. A careful choice here can simplify the integral for the volume.

EXAMPLE A

Determine the volume of a rectangular container that has sides 2 cm, 3 cm, and 5 cm in length.

SOLUTION

A diagram of the container is shown in Figure 19.2. A coordinate system has been attached with the origin at one corner and axes lying along the sides of the container. A suitable element of integration is a narrow slice perpendicular to the x-axis. Because of the way it is oriented with respect to the coordinate system, this volume element has a width dx and a constant cross-sectional area of 6 cm². The volume dV of the element of integration is therefore

$$dV = (\text{area}) \cdot (\text{thickness})$$

$$= 6\,dx$$

Figure 19.2

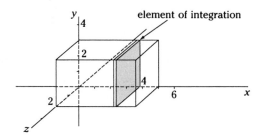

The total volume of the container is the sum of all such elements for $x = 0$ to $x = 5$. It is found by integrating

$$V = \int_0^5 6\,dx$$

$$= 6x \Big|_0^5$$

$$= 6 \cdot 5 - 6 \cdot 0$$

$$= 30$$

The volume of the container is 30 cm³. This result could, of course, have been obtained by multiplying length × width × height. The point of this example is to illustrate how a volume can be found by summing an infinite number of infinitesimally small volume elements using integration. ❑

EXAMPLE B

Find the volume of a right circular cone with radius 6 cm and height 10 cm.

SOLUTION

A diagram of the cone is shown in Figure 19.3a. It is oriented with the centre of its base at the origin and its axis of symmetry along the y-axis. Every cross section of the cone perpendicular to its axis is a circle. Therefore, a suitable element of integration is a thin circular disk perpendicular to the y-axis. This disk has a radius r, a cross sectional area πr^2, and a thickness dy. Its volume is therefore

$$dV = (\text{area}) \cdot (\text{thickness})$$

$$= \pi r^2 \, dy$$

Figure 19.3

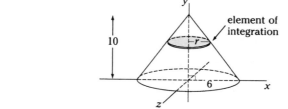

(a)   (b)

The total volume can be pictured as a stack of an infinite number of infinitesimally thin disks, with radii varying from $r = 6$ at $y = 0$ to $r = 0$ at $y = 10$. This volume is given by the integral

$$V = \int_0^{10} \pi r^2 \, dy$$

In this instance, the integration cannot yet be done. The radius r is a function of y and must be expressed in terms of y in the integrand. The relationship between the two can be found with the help of the similar triangles shown in Figure 19.3b:

$$\frac{r}{6} = \frac{10 - y}{10}$$

$$\therefore \quad r = \frac{3(10 - y)}{5}$$

$$r^2 = \frac{9(10 - y)^2}{25}$$

It follows that

$$V = \int_0^{10} \pi r^2 \, dy$$

$$= \int_0^{10} \pi \cdot \frac{9(10 - y)^2}{25} \, dy$$

$$= \frac{9\pi}{25} \cdot (-1) \cdot \frac{(10 - y)^3}{3} \Big|_0^{10}$$

$$= -\frac{3\pi}{25}(10 - y)^3 \Big|_0^{10}$$

$$= \left[-\frac{3\pi}{25}(10 - 10)^3 \right] - \left[-\frac{3\pi}{25}(10 - 0)^3 \right]$$

$$= \frac{3\pi}{25} \cdot 10^3$$

$$= 120\pi$$

The volume of the cone is 120π cm^3. This result can be checked using the formula for the volume of a cone:

$$V = \frac{\pi r^2 h}{3}$$

$$= \frac{\pi \cdot 6^2 \cdot 10}{3}$$

$$= 120\pi$$

EXAMPLE C

Find the volume inside the pagoda-shaped roof shown in Figure 19.4a. Horizontal cross sections of the roof are square. The curve of the right side of the roof in the xy plane is $y = 4/x$, for $x = 1$ to $x = 4$.

Figure 19.4

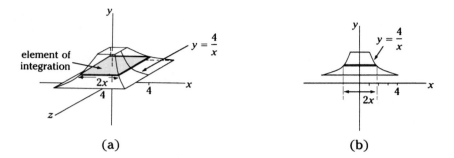

(a)   (b)

SOLUTION

In this example, the coordinate system is oriented so that the x-axis and z-axis are parallel to the edges of the roof, and the y-axis goes through its centre. A suitable element of integration is a square, horizontal slice. Since x is the horizontal distance from the y-axis to the roof surface (see Figure 19.4b), the square volume element has a side of length $2x$. Its area is then $(2x)^2$, and its thickness is dy. Its volume dV is

$$dV = (\text{area}) \cdot (\text{thickness})$$

$$= (2x^2)\, dy$$

The volume of the roof is found by integrating

$$V = \int_1^4 (2x)^2\, dy$$

As in Example B, the integration cannot be carried out until the integrand is expressed in terms of the independent variable y.

Since $y = \dfrac{4}{x}$

then $x = \dfrac{4}{y}$

and $(2x)^2 = \dfrac{64}{y^2}$

Therefore,

$$V = \int_1^4 (2x)^2 \, dy$$

$$= \int_1^4 \frac{64}{y^2} \, dy$$

$$= -\frac{64}{y} \Big|_1^4$$

$$= \left[-\frac{64}{4} \right] - \left[-\frac{64}{1} \right]$$

$$= 48$$

The volume of the pagoda roof is 48 m³.

Exercise 19.1

1. A cylindrical pillar has a height of 5 m and a circular base of radius 12 cm.

   (a) Draw the solid.

   (b) Draw an element of integration as a horizontal slice perpendicular to the axis of the pillar.

   (c) Find an expression for the volume element dV.

   (d) Find the volume of the pillar by integration.

2. A right pyramid has an altitude of 6 cm and a square base of side 5 cm.

   (a) Draw the solid.

   (b) Draw the element of integration as a slice perpendicular to the y-axis.

   (c) Find an expression for the volume element dV.

   (d) Find the volume of the pyramid by integration.

3. By integration, show that the volume of a right circular cylinder of radius R and height H is given by the formula $V = \pi R^2 H$. Choose the element of integration to be a slice parallel to the axis of the cylinder.

4. Find the volume of a cube of side 4 cm by integration. Choose the element of integration to be a slice of thickness dx, parallel to a diagonal of a face of the cube.

5. The base of a solid is bounded by the curves $x = z^2$ and $x = 4$ in the xz-plane. Cross sections of the solid perpendicular to the x-axis are rectangles for which the height is four times the base. Draw the solid and find its volume.

6. The base of a solid is a circle of radius 4 cm. Determine the volume of the solid if the cross sections perpendicular to a given diameter are (a) squares and (b) equilateral triangles. Draw the solid in each case.

7. A solid has a base which is a right-angled isosceles triangle, formed by the x-axis, the z-axis, and the line $x + z = 3$. If cross sections perpendicular to the x-axis are squares, draw the solid and find its volume.

8. A lumberjack is preparing to cut down a large maple tree. With a chain saw, he makes a horizontal cut exactly halfway through the trunk, and then makes a second cut at 45°, meeting the first cut along the diameter. Determine the volume of the wedge of wood cut out if the diameter is 0.9 m. Hint: Take the element of integration to be a vertical slice parallel to the thin edge of the wedge.

19.2 Volumes of Revolution: The Circular Disk Method

Consider a region bounded by a continuous, non-negative function $y = f(x)$ and the lines $x = a$, $x = b$, and the x-axis, as shown in **Figure 19.5a**. If this region is rotated about the x-axis, a solid like that shown in **Figure 19.5b** is generated. A solid formed in this way is called a **solid of revolution**. Its volume is a **volume of revolution**.

Figure 19.5

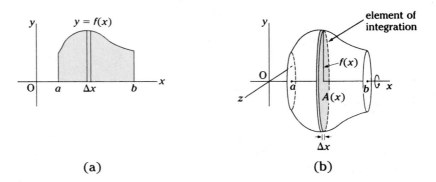

(a)   (b)

A cross section of the solid perpendicular to the x-axis is a circle of radius $y = f(x)$. Therefore, to find the volume of the solid, it is appropriate to choose for the element of integration a thin circular disk having an area πy^2 and a thickness dx. In terms of differentials, the volume element is then

$$dV = (\text{area}) \cdot (\text{thickness})$$

$$= \pi y^2\, dx$$

The volume of the solid is the integral

$$V = \int_a^b \pi y^2\, dy$$

$$= \pi \int_a^b [f(x)]^2\, dx$$

EXAMPLE A

Find the volume of the paraboloid formed by rotating the parabola $y^2 = x$, $0 \leqslant x \leqslant 4$, about the x-axis.

SOLUTION

A diagram of the solid is shown in Figure 19.6. The element of integration is a disk perpendicular to the x-axis with area πy^2 and thickness dx, the volume of which is

$$dV = \pi y^2\, dx$$

Figure 19.6

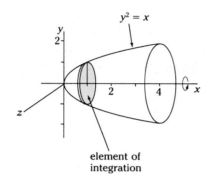

element of
integration

The total volume is given by the integral

$$V = \int_0^4 \pi y^2\, dx$$

$$= \pi \int_0^4 x\, dx$$

$$= \frac{\pi x^2}{2}\bigg|_0^4$$

$$= \frac{\pi(4)^2}{2} - \frac{\pi(0)^2}{2}$$

$$= 8\pi$$

The volume of the paraboloid is 8π cubic units.

The circular disk method can be generalized to cases in which the region rotated lies between two curves. When the region shown in Figure 19.7a is rotated, for instance, it produces a solid of revolution containing a hole as shown in Figure 19.7b. The cross section of this solid is in the shape of a ring. Its area is found by subtracting the area of the inner circle πy_1^2 from that of the outer circle πy_2^2:

$$\text{area of ring} = \pi y_2^2 - \pi y_1^2$$

Figure 19.7

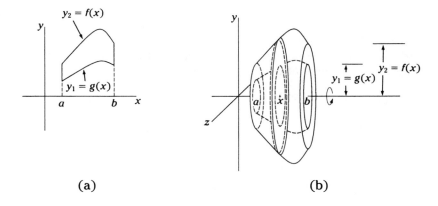

(a)   (b)

The volume of the ring shaped element of integration is

$$dV = (\text{area}) \cdot (\text{thickness})$$

$$= (\pi y_2{}^2 - \pi y_1{}^2)dx$$

The volume of the solid is therefore

$$V = \int_a^b (\pi y_2{}^2 - \pi y_1{}^2)dx$$

Figure 19.8

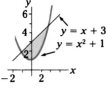

(a)

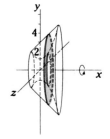

(b)

EXAMPLE B

Find the volume of the solid obtained by rotating the region between $y = x^2 + 1$ and $y = x + 3$ about the x-axis.

SOLUTION

The graphs of the curves are shown in Figure 19.8a. To find the points of intersection solve the equation

$$x^2 + 1 = x + 3$$
$$x^2 - x - 2 = 0$$
$$(x - 2)(x + 1) = 0$$
$$\therefore \qquad x = 2 \text{ or } -1$$

Figure 19.8b shows the solid generated when the region between the curves from $x = -1$ to $x = 2$ is rotated about the x-axis. Choosing the element of integration to be a thin circular ring, the volume of the solid is

$$V = \int_{-1}^{2} (\pi y_2{}^2 - \pi y_1{}^2)dx$$

$$= \pi \int_{-1}^{2} \left((x + 3)^2 - (x^2 + 1)^2 \right) dx$$

$$= \pi \int_{-1}^{2} (-x^4 - x^2 + 6x + 8)dx$$

$$= \pi \left(-\frac{x^5}{5} - \frac{x^3}{3} + \frac{6x^2}{2} + 8x \right) \Bigg|_{-1}^{2}$$

$$= \pi \left(-\frac{(2)^5}{5} - \frac{(2)^3}{3} + \frac{6(2)^2}{2} + 8(2) \right)$$

$$- \pi \left(-\frac{(-1)^5}{5} - \frac{(-1)^3}{3} + \frac{6(-1)^2}{2} + 8(-1) \right)$$

$$= \frac{117\pi}{5}$$

The volume of the solid is $\dfrac{117\pi}{5}$ cubic units.

Exercise 19.2

1. Find the volume of the solid of revolution formed by rotating the given region about the x-axis. In each case, draw the solid.

(a) $y = x^2$, $x = 0$, $x = 2, y = 0$

(b) $y = \dfrac{x}{3}$, $x = 1$, $x = 5, y = 0$

(c) $y = \dfrac{1}{x}$, $x = 1$, $x = 5, y = 0$

(d) $y = \sqrt{x}$, $y = x$

(e) $x^2 - y^2 = 4$, $x = 5$

2. Find the volume of the solid of revolution formed by rotating the given region about the y-axis. In each case, draw the solid.

(a) $y = x^3$, $x = 0, y = 1$

(b) $y^2 = x$, $y = x - 2$

(c) $y^2 = x - 5$, $y = -2, y = 1$

3. Find the volume of the solid of revolution formed by rotating the given region about the specified axis.

(a) $y = \cos x$, $x = 0, x = \pi/2, y = 0$, about the x-axis

(b) $y = \sin x^2$, $x = 0, y = 1$, about the y-axis

(c) $y = \sin x$, $y = \cos x, x = 0$, about the x-axis

(d) $y = \tan x$, $x = \pi/4, y = 0$, about the x-axis

(e) $y = \sin^{-1}x$, $x = 0, y = \pi/4, y = \pi/2$, about the y-axis

4. Show that the volume of the paraboloid from $x = 0$ to $x = a$ formed by rotating the parabola $y^2 = x$ about the x-axis is $V = \pi a^2/2$.

5. A sphere is a volume of revolution formed by rotating the circle $x^2 + y^2 = r^2$ about a diameter. Show by integration that the volume of a sphere is given by the formula $V = 4\pi r^3/3$.

6. Find the volume of the solid formed by rotating the ellipse $4x^2 + 9y^2 = 36$ about the x-axis. This solid is called a **prolate spheroid**.

7. Find the volume generated by rotating the region formed by $y^2 = a^2$ and $x^2 - y^2 = a^2$

   (a) about the y-axis

   (b) about the x-axis.

8. Find the volume of a spherical cap, given that the radius of the sphere is 8 cm and the maximum height of the cap is 5 cm.

9. Show that the formula for the volume of a spherical cap of radius r and height a is $V = \pi a(3r^2 + a^2)/6$.

10. Find the volume of a torus (doughnut) which measures 9 cm in radius and has a hole of radius 3 cm. The torus is formed by rotating the circle $(x - 6)^2 + y^2 = 3^2$ about the y-axis. Hint: The outer and inner radii of the volume element are $6 + \sqrt{9 - y^2}$ and $6 - \sqrt{9 - y^2}$ respectively (see Figure 19.9).

Figure 19.9

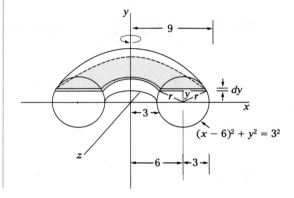

19.3 Arc Length by Integration

The now familiar method of summing a large number of small elements can be applied to finding the length of a curve. Consider the graph of the function $y = f(x)$ shown in Figure 19.10, where the curve has been cut into n arcs by dividing the interval $a \le x \le b$ into n equal parts, Δx. When Δx is small, each arc PQ is approximately equal in length to its secant. According to the Pythagorean theorem, the length Δs of the secant of the kth arc is

$$(\Delta s_k)^2 = (\Delta x_k)^2 + (\Delta y_k)^2$$

$$= (\Delta x_k)^2 + \frac{(\Delta y_k)^2}{(\Delta x_k)^2} \cdot (\Delta x_k)^2$$

$$= \left[1 + \left(\frac{\Delta y}{\Delta x} \right)^2_k \right] (\Delta x_k)^2 \qquad \longleftarrow \quad \left(\frac{\Delta y}{\Delta x} \right)_k \text{ is the slope of the secant of the } k\text{th arc}$$

$$\Delta s_k = \sqrt{1 + \left(\frac{\Delta y}{\Delta x} \right)^2_k} \cdot \Delta x_k$$

Figure 19.10

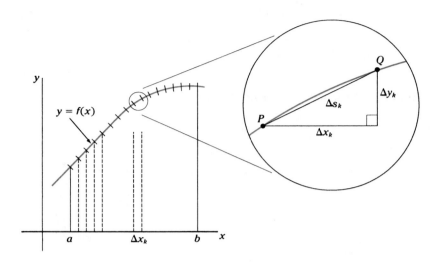

The required arc length L is found by summing the lengths of all the secants from $x = a$ to $x = b$ and then taking the limit of this sum as $\Delta x \to 0$:

$$L = \lim_{\Delta x \to 0} \sum_{k=1}^{n} \sqrt{1 + \left(\frac{\Delta y}{\Delta x}\right)_k^2} \cdot \Delta x_k$$

This is the integral

$$L = \int_a^b \sqrt{1 + \left(\frac{dy}{dx}\right)^2} \, dx$$

In taking the limit, the slope $\Delta y / \Delta x$ of the secant has become the slope dy/dx of the tangent.

In terms of differentials, this integral formula for arc length can be understood as follows. Take the element of integration to be ds, an infinitesimally short segment of the *tangent* to the curve at P (see Figure 19.11). An expression for ds is

$$(ds)^2 = (dx)^2 + (dy)^2$$

$$= (dx)^2 + \left(\frac{dy}{dx}\right)^2 \cdot (dx)^2 \qquad \longleftarrow \qquad \text{since } dy = \left(\frac{dy}{dx}\right) dx$$

$$= \left[1 + \left(\frac{dy}{dx} \right)^2 \right] (dx)^2$$

$$ds = \sqrt{1 + \left(\frac{dy}{dx} \right)^2} \cdot dx$$

from which the formula for arc length follows immediately.

Figure 19.11

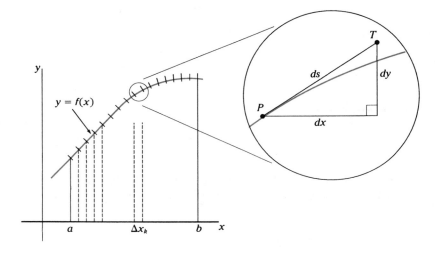

EXAMPLE A

Find the length of the curve $y = x^{2/3}$ from the point $A(1, 1)$ to the point $B(8, 4)$.

SOLUTION

Figure 19.12 shows the part of the curve whose length is required. It is best to proceed in a step-by-step fashion to set up and work out the integral for arc length:

Figure 19.12

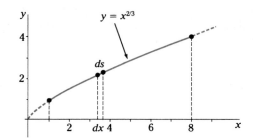

$$y = x^{2/3}$$

$$\frac{dy}{dx} = \frac{2}{3x^{1/3}}$$

$$1 + \left(\frac{dy}{dx}\right)^2 = 1 + \frac{4}{9x^{2/3}}$$

$$= \frac{9x^{2/3} + 4}{9x^{-1/3}}$$

The arc length is therefore

$$L = \int_1^8 \sqrt{\frac{9x^{2/3} + 4}{9x^{2/3}}}\, dx$$

$$= \int_1^8 \frac{1}{3} \cdot x^{-1/3} \sqrt{9x^{2/3} + 4}\, dx$$

$$= \frac{1}{18} \int_1^8 6x^{-1/3} \sqrt{9x^{2/3} + 4}\, dx \quad \longleftarrow \quad \begin{array}{c} \text{has the form} \\ \int \sqrt{v}\, dv \end{array}$$

$$= \frac{1}{18} \cdot \frac{(9x^{2/3} + 4)^{3/2}}{3/2} \Bigg|_1^8$$

$$= \frac{1}{27} \left[9(8)^{2/3} + 4 \right]^{3/2} - \frac{1}{27} \left[9(1)^{2/3} + 4 \right]^{3/2}$$

$$= \frac{1}{27} \cdot 40^{3/2} - \frac{1}{27} \cdot 13^{3/2}$$

$$\doteq 7.634$$

The length of the curve from A to B is approximately 7.634 units. $\square$

Exercise 19.3

1. Find the length of the upper branch of the semi-cubical parabola $y^2 = x^3$ from the origin to the point $(1, 1)$. Compare this length with that of the straight line between the two points.

2. Find the arc length from $x = 1$ to $x = 3$ of

$$y = \frac{x^3}{3} + \frac{1}{4x}$$

Hint: show that $1 + \left(\dfrac{dy}{dx}\right)^2$ is a perfect square.

3. Find the arc length of the curve $24xy = x^4 + 48$ from $x = 2$ to $x = 4$.

4. Find the length of the graph of $y = (x - 1)^{3/2}$ from $(1, 0)$ to $(5, 8)$.

5. Find the circumference of the loop of $3y^2 = x(x - 1)^2$.

6. Find the arc length of the astroid $x^{2/3} + y^{2/3} = 1$ in the first quadrant (see Figure 19.13). Then find the *total* arc length. Compare your answer with the perimeter of the square with the same vertices and also with the circumference of the unit circle.

Figure 19.13

7. Find the arc length of the curve $y = \ln(x + \sqrt{x^2 - 1})$ from $x = 1$ to $x = 10$.

8. Find the length of the graph of $y = (e^x + e^{-x})/2$ on the interval $[0, \ln 2]$.

9. Find the length of the arc of the parabola $y = x^2 - 4x$ that lies below the x-axis. Hint: In the integral, substitute $v = 2x - 4$ and use Formula 37 from Section I. Table of Integrals, page 471.

19.4   Areas of Surfaces of Revolution

When a region in the xy-plane is rotated about one of the coordinate axes, it forms a solid of revolution. The surface of the solid generated by the rotating boundary of the region is a **surface of revolution**. An integral formula for the area of such a surface can be found using differentials.

Consider the curved surface shown in Figure 19.14 formed by rotating the graph of the function $y = f(x)$ in the interval $a \le x \le b$ about the x-axis. To find the area of the surface, take the element of integration to be the narrow circular band shown in Figure 19.14. If the band is cut and laid out flat, it is approximately rectangular in shape, having a width ds and a length equal to the circumference $2\pi y$. This approximation gets better as the width of the band becomes infinitesimally small.

Figure 19.14

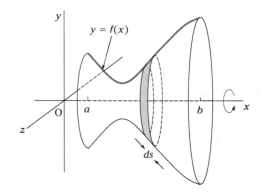

The area dA of the band is therefore

$$dA = (\text{circumference}) \cdot (\text{width})$$

$$= 2\pi y \cdot ds$$

$$= 2\pi y \cdot \sqrt{1 + \left(\frac{dy}{dx}\right)^2} \cdot dx$$

(using the formula for ds from the previous section). The total area of the surface is the sum of the areas of all the bands that cover the surface from $x = a$ to $x = b$. It is given by the integral

$$A = \int_a^b 2\pi y \cdot \sqrt{1 + \left(\frac{dy}{dx}\right)^2}\, dx$$

EXAMPLE A

Find the area of the surface formed by rotating the cubic function $y = x^3/3$ in the interval $0 \le x \le 2$ about the x-axis.

SOLUTION

The surface in question is shown in Figure 19.15. It is best to proceed in a step-by-step manner to set up and work out the integral for the surface area:

Figure 19.15

$$y = \frac{x^3}{3}$$

$$\frac{dy}{dx} = x^2$$

$$1 + \left(\frac{dy}{dx}\right)^2 = 1 + x^4$$

The surface area is therefore

$$A = \int_b^a 2\pi y \cdot \sqrt{1 + \left(\frac{dy}{dx}\right)^2}\, dx$$

$$= \int_0^2 2\pi \left(\frac{x^3}{3}\right) \sqrt{1 + x^4}\, dx$$

$$= \frac{2\pi}{3} \cdot \frac{1}{4} \int_0^2 4x^3 \sqrt{1 + x^4}\, dx \quad \longleftarrow \quad \text{has the form} \quad \int \sqrt{v}\, dv$$

$$= \frac{\pi}{6} \frac{(1 + x^4)^{3/2}}{3/2} \Bigg|_0^2$$

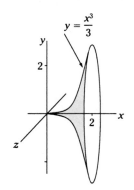

$y = \dfrac{x^3}{3}$

$$= \frac{\pi}{9}\left(1 + (2)^4\right)^{3/2} - \frac{\pi}{9}\left(1 + (0)^4\right)^{3/2}$$

$$= \frac{\pi}{9}(17^{3/2} - 1)$$

$$\doteq 24.1$$

The surface has an area of approximately 24.1 square units.   ❑

Exercise 19.4

1. What is the area of the surface generated by rotating the curve $y = 2\sqrt{x}$ from $x = 0$ to $x = 8$ about the x-axis?

2. Prove that the surface area of a sphere of radius R is $4\pi R^2$.

3. Find the area of that part of the surface of a sphere obtained by rotating the function $y = \sqrt{4 - x^2}, 0 \le x \le 1$, about the x-axis. What fraction of the spherical surface is this?

4. Show that the lateral surface area of a right circular cone of radius R and height H is $\pi R\sqrt{R^2 + H^2}$, or πRS where S is the slant height. Hint: orient the cone with its axis along the positive x-axis and its apex at the origin.

5. Find the area of the surface generated by rotat-

ing the loop of the curve $9y^2 = x(3 - x)^2$ about the x-axis.

Hint: Show that $1 + \left(\frac{dy}{dx}\right)^2 = \frac{(1 + x)^2}{4x}$

6. Find the area of the surface produced by rotating the curve $y = (e^x + e^{-x})/2$ from $x = -1$ to $x = 1$ about the x-axis.

Hint: Show that $1 + \left(\frac{dy}{dx}\right)^2$ is a perfect square.

7. Find the area of the surface generated by rotating one arch of the sine curve $y = \sin x$ about the x-axis. Hint: Evaluate the integral from 0 to $\pi/2$, then double it. In the integral, set $v = \cos x$, and then use Formula 37 from Section I. Table of Integrals, page 471.

19.5 Summary and Review

In using the definite integral to find volume, arc lengths, and surface areas, the key is to choose a suitable element of integration. To find the volume of a solid with a given cross-sectional area, use a thin slice for the element of integration:

$$V = \int_a^b A(x)\,dx$$

To find the volume of a solid of revolution, use a thin disk for the element of integration:

$$V = \int_a^b \pi y^2 \, dx$$

To find the length of the arc of a curve, use a tiny segment of a tangent for the element of integration:

$$L = \int_a^b \sqrt{1 + \left(\frac{dy}{dx}\right)^2} \, dx$$

To find the area of a surface of revolution, use a narrow circular band for the element of integration:

$$A = \int_a^b 2\pi y \sqrt{1 + \left(\frac{dy}{dx}\right)^2} \, dx$$

It is helpful in setting up integrals such as these to recall the geometrical interpretation of each part of the expression. Before integrating, remember to express the integrand in terms of x alone.

Exercise 19.5

1. Find the volume of a cone of height 3 and radius 2.

2. Prove that the volume of a cone of height H and radius R is given by the formula $V = \pi R^2 H/3$.

3. Find the volume of the solid of revolution formed when the curve $e^{-x}, 0 \le x \le 2$ is rotated about the x-axis.

4. A bowl is in the shape of a hemisphere of radius 20 cm. Calculate the volume of water it contains when the maximum depth is 2.5 cm.

5. Make a sketch and find the volume of the solid having the circle $x^2 + y^2 = 4$ as a base and cross sections perpendicular to the x-axis in the shape of

(a) a square

(b) an equilateral triangle

(c) an isoceles right triangle with its hypotenuse on a chord of the circle.

6. The arc of $3y = \sqrt{x}(x - 3)$ from $(3, 0)$ to $(9, 6)$ is rotated about the x-axis. Find the area of the surface generated.

7. Rotate the curve $y = \cos x, 0 \le x \le \pi/4$ about the x-axis and find the volume generated.

8. An ant walks along the curve $4y^2 = 9x^3$ from the origin to the point $(1, 1.5)$. Find the distance it has travelled. How much shorter is the shortest distance between the two points?

9. The base of a solid is the region enclosed by the x-axis and the curve $y = \sin x, \pi/4 \le x \le 3\pi/4$. Every cross section perpendicular to the x-axis is a square. Sketch the solid and find its volume.

10. Find the volume of the solid generated by rotating the region in the first quadrant between $y = x$ and $y = x^2$ about the y-axis.

11. Two right circular cylinders of radius 2 cm intersect at right angles. Find the volume that is common to both cylinders. Hint: One-eighth of the required volume is shown in Figure 19.16. Consider the shape of the cross section in the plane of the axes.

Figure 19.16

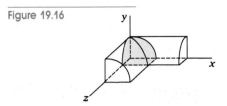

12. A craftsperson fashions a bowl from a mass of clay spinning on a potter's wheel. The axis of rotation is the y-axis. The bottom of the bowl is flat from $x = 0$ to $x = 2$. From $x = 2$ to $x = 4$ the sides of the bowl follow the curve $y = (x - 2)^2$. Sketch the bowl and find its volume.

I N THE FIRST PART of the eighteenth century advanced mathematics on the Continent was devoted to the organization and extension of calculus to problems in the geometry of curves. Maria Gaetana Agnesi, the first woman in modern times who can be accurately called a mathematician, published in 1748 a two-volume treatise presenting a systematic exposition of work in the subject. The *Istituzioni analitiche ad uso della gioventù italiana* (Analytical Lectures for the Use of Italian Youth) covered the range from elementary algebra to coordinate geometry, and then on to differential calculus, integral calculus, infinite series, and the solution of differential equations. The work won immediate acclaim from the European mathematical community and was translated into both French and English.

Calculus as it was then practised was quite intricate in its application to the study of curves, requiring the careful calculation of infinitesimals of different orders. Its complexity was evident in a problem as basic as the determination of the radius of curvature of a curve at a point. In many memoirs published during the period the mathematical exposition was very informal; it was assumed that the necessary details could be supplied by the reader. In contrast, Agnesi's book provided a thorough and sophisticated treatment of difficult technical problems that avoided the weaknesses of other works.

A committee of the Paris Academy of Sciences, which authorized the French translation of Agnesi's book, reported in 1749: "This work is characterized by its careful organization, its clarity, and its precision. There is no other book, in any language, which would enable a reader to penetrate as deeply, or as rapidly, into the fundamental concepts of analysis. We consider this treatise the most complete and best written work of its kind." The editor of the English edition wrote "He [John Colson, the English translator] found her work to be so excellent that he was at the pains of learning the Italian language at an advanced age for the sole purpose of translating her book into English, that the British Youth might have the benefit of it as well as the Youth of Italy."

Maria Agnesi was the daughter of a wealthy professor of mathematics at the University of Bologna. Her father, encouraged by her interest in scientific matters, secured a series of distinguished tutors and arranged soirées attended by scholars and local celebrities. At the age of eleven she was fluent in or thoroughly familiar with French, Latin, Greek, German, Spanish, and Hebrew. In adulthood she became retiring, preferring to work alone in mathematics or devoting herself to religious studies and social work. After the publication of the *Istituzioni analitiche*, however, she gradually withdrew from mathematics. In 1762, she declared that she was no longer concerned with the subject of calculus.

Maria Gaetana Agnesi

Polar Coordinates

20

I t is convenient in certain applications of mathematics, especially in science and engineering, to use coordinate systems other than the familiar Cartesian coordinate system. This chapter gives a brief introduction to the **polar coordinate system**. This system is useful in navigation and for describing such things as antenna radiation patterns, where direction and range are the important variables. It is instructive to study applications of calculus, such as finding area in this context.

The Polar Coordinate System   20.1

To locate a point P in a plane, two numbers are required. In polar coordinates, these two numbers are the distance r of the point from the origin or **pole** O, and the angle θ that OP makes with the **polar axis**, a line that in the Cartesian system would be the positive x-axis (see Figure 20.1). These two numbers (r, θ) are the **polar coordinates** of the point.

Points expressed in polar coordinates can be plotted directly on **polar coordinate graph paper** as shown in Figure 20.2. Rays extending from the origin represent different values of θ. Concentric circles about the origin represent different values of r. The point $P(4, 120°)$, for instance, is located on the ray making an angle of $120°$ with the polar axis (measured counterclockwise), a distance of 4 units from the origin.

Figure 20.1

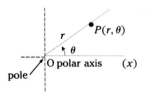

Figure 20.2

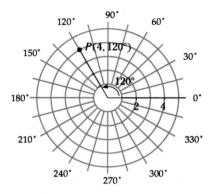

The polar coordinates of a point are not unique. One reason for this is that the angle θ may wind around the origin for more than one revolution in either direction. This means, for instance, that the point $P(4, 120°)$ may also be described by the coordinates $P(4, 480°)$, or $P(4, -240°)$ as in Figure 20.3. In general, any ordered pair of the form $(r, \theta \pm 2n\pi)$, for $n = 1, 2, \ldots$, is equivalent to (r, θ).

Figure 20.3

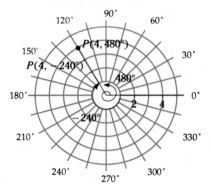

It is also possible for the radial coordinate r to be negative. When r is negative, it indicates that the point is on the extension of the terminal arm backward through the origin. For example, $P(-4, 300°)$ is still another way to denote point P in Figure 20.3 (see Figure 20.4). In general, $(-r, \theta)$ is equivalent to $(r, \theta \pm \pi)$.

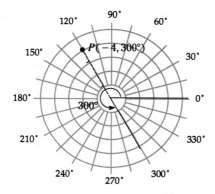

Figure 20.4

EXAMPLE A

Plot the following points:
(a) $(2, 60°)$ (b) $(5, 0°)$ (c) $(3, -150°)$ (d) $(-3, 150°)$

SOLUTION

In each case, determine the ray with the given direction θ as shown in Figure 20.5, then move r units in that direction (or in the *opposite* direction, if r is negative). ❑

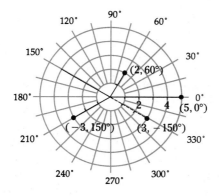

Figure 20.5

Figure 20.6

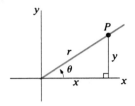

The relationship between the Cartesian Coordinates (x, y) and the polar coordinates (r, θ) can be written down directly from the diagram in Figure 20.6. The polar coordinates are given in terms of the Cartesian coordinates by

$$r = \sqrt{x^2 + y^2} \qquad \text{and} \qquad \theta = \tan^{-1}\frac{y}{x}$$

The Cartesian coordinates are given in terms of the polar coordinates by

$$x = r\cos\theta \qquad \text{and} \qquad y = r\sin\theta$$

EXAMPLE B

Determine the polar coordinates of the point P, whose Cartesian coordinates are $(2, 2\sqrt{3})$.

SOLUTION

$$r = \sqrt{x^2 + y^2} \qquad\qquad \theta = \tan^{-1}\frac{y}{x}$$

$$= \sqrt{(2)^2 + (2\sqrt{3})^2} \qquad\qquad = \tan^{-1}\frac{2\sqrt{3}}{2}$$

$$= 4 \qquad\qquad = 60° \text{ or } \frac{\pi}{3}$$

Thus, the polar coordinates of P are $(4, \pi/3)$.   ❑

EXAMPLE C

Find the Cartesian coordinates of the point Q, whose polar coordinates are $(6, 3\pi/4)$.

SOLUTION

$$x = r\cos\theta \qquad\qquad y = r\sin\theta$$

$$= 6\cos\left(\frac{3\pi}{4}\right) \qquad\qquad = 6\sin\left(\frac{3\pi}{4}\right)$$

$$= 6\left(-\frac{1}{\sqrt{2}}\right) \qquad\qquad = 6\left(\frac{1}{\sqrt{2}}\right)$$

$$= -3\sqrt{2} \qquad\qquad = 3\sqrt{2}$$

Thus, the Cartesian coordinates of Q are $(-3\sqrt{2}, 3\sqrt{2})$.   ☐

EXAMPLE D

Express the relation $x^2 + y^2 = a^2$ in polar form.

SOLUTION

The Cartesian coordinates x and y are given in terms of polar coordinates by

$$x = r\cos\theta \quad \text{and} \quad y = r\sin\theta$$

Substituting for x and y:

$$x^2 + y^2 = a^2$$
$$(r\cos\theta)^2 + (r\sin\theta)^2 = a^2$$
$$r^2(\sin^2\theta + \cos^2\theta) = a^2$$
$$r^2 = a^2$$
$$r = a$$

The original equation in Cartesian coordinates is recognized as that of a circle of radius a, centered at the origin (see Figure 20.7). In polar coordinates, this relation has a much simpler form.   ☐

$r = a$

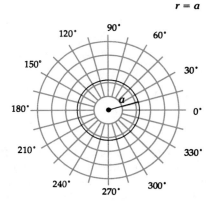

Figure 20.7

EXAMPLE E

Convert the polar equation $r = 2a\cos\theta$ into Cartesian coordinates and identify the curve.

SOLUTION

Using the fact that $\cos\theta = \dfrac{x}{r}$

$$r = 2a\cos\theta$$

$$r = 2a\left(\frac{x}{r}\right)$$

$$r^2 = 2ax$$

$$x^2 + y^2 = 2ax$$

$$x^2 - 2ax + y^2 = 0 \qquad\longleftarrow\qquad \text{complete the square}$$

$$x^2 - 2ax + a^2 + y^2 = a^2$$

$$(x - a)^2 + y^2 = a^2$$

The relation $r = 2a\cos\theta$, therefore, is a circle of radius a and centre at $(a, 0)$ (see Figure 20.8). ☐

Figure 20.8

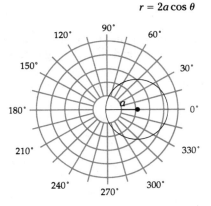

$r = 2a\cos\theta$

Examples D and E show how to convert a relation from one system of coordinates to the other. You should be aware that a relation having a simple polar form may be quite complicated when expressed in Cartesian coordinates, which is why polar coordinates are used in the first place.

Exercise 20.1

1. Plot the following points.

(a) $P(5, 0°)$ (b) $Q(0, 45°)$

(c) $R(6, 15°)$ (d) $S(4, 4\pi/3)$

(e) $T(2, 11\pi/6)$ (f) $U(3\sqrt{2}, 2\pi/3)$

2. Express the following points in polar coordinates with $r \geq 0$ and $0 \leq \theta \leq 2\pi$, and plot the points.

(a) $P(-2, 30°)$ (b) $Q(4, -225°)$

(c) $R(-3, 0)$ (d) $S(1, 13\pi/6)$

(e) $T(-1, -\pi/3)$ (f) $U(0, -\pi)$

3. Plot the following points:

(a) $P(2, -45°)$ (b) $Q(-2, 45°)$

(c) $R(4, 4\pi/3)$ (d) $S(-4, -4\pi/3)$

4. Express the following Cartesian coordinates as polar coordinates with $r \geq 0$ and $0 \leq \theta \leq 2\pi$.

(a) $(7, 0)$ (b) $(0, -6)$

(c) $(-4, 4)$ (d) $(-4, 4\sqrt{3})$

(e) $(-4\sqrt{3}, 4)$ (f) $(-3, -3\sqrt{3})$

5. Change the following polar coordinates to Cartesian coordinates.

(a) $(5, 180°)$ (b) $(6, \pi/6)$

(c) $(-2, 3\pi/2)$ (d) $(7, \pi/4)$

(e) $(3\sqrt{2}, 3\pi/4)$ (f) $(3, 5\pi/3)$

6. What is the relationship between $P(r, \theta)$ and the following:

(a) $Q(-r, \theta)$

(b) $R(r, -\theta)$

(c) $S(r, \pi - \theta)$

(d) $T(-r, \pi + \theta)$

7. Express the following equations in polar form.

(a) $x^2 + y^2 = 25$

(b) $y = 13$

(c) $5xy = 11$

(d) $x^2 - y^2 = 6$

(e) $x^2 = 13y$

(f) $(x^2 + y^2)^2 = 25(x^2 - y^2)$

8. Transform the following to Cartesian coordinates and identify the curve.

(a) $r = 6$

(b) $r \sin \theta = 5$

(c) $r = \dfrac{6}{\cos \theta}$

(d) $r = 4 \sin \theta$

(e) $r + 5 \cos \theta = 0$

(f) $r = \dfrac{6}{2 - \cos \theta}$

20.2 Graphing Relations in Polar Coordinates

In polar coordinates, θ is considered to be the independent variable, and r, the dependent variable. A relation is usually expressed in the form $r = f(\theta)$. The graph of f is the locus of points (r, θ) satisfying the relation $r = f(\theta)$.

To draw a graph, start by making a table of values. Choose a sequence of values of θ, and calculate the corresponding value of r. As the following examples show, it is best to plot the points *consecutively*. In addition, you should look for and take advantage of symmetries and other qualitative features.

Figure 20.9

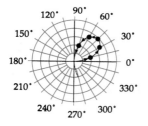

EXAMPLE A

Sketch the graph of $r = 4 \sin 2\theta$.

SOLUTION

In making a table of values for this function, take θ to be multiples of 15°. Then, values of 2θ will be multiples of 30°, which simplifies the calculation of values of r. Note that the value of the sine function never exceeds 1 so the value of r can never be greater than 4. Table 20.1 shows the results for $0° \leq \theta \leq 90°$. Plotting consecutive points as θ increases produces the graph shown in Figure 20.9.

Table 20.1

θ	2θ	$\sin 2\theta$	$4 \sin 2\theta$
\multicolumn{4}{c}{$r = 4 \sin 2\theta$}
0°	0°	0	0
15°	30°	$\dfrac{1}{2}$	2
30°	60°	$\dfrac{\sqrt{3}}{2}$	$2\sqrt{3} \doteq 3.5$
45°	90°	1	4
60°	120°	$\dfrac{\sqrt{3}}{2}$	$2\sqrt{3} \doteq 3.5$
75°	150°	$\dfrac{1}{2}$	2
90°	180°	0	0

For $90° \leq \theta \leq 180°$, the values of $\sin 2\theta$ are the negatives of those in Table 20.1, so these next points are in the fourth rather than the second quadrant. Continuing in this way produces the graph shown in Figure 20.10, which is known as a four-leaf rose. ❏

Figure 20.10

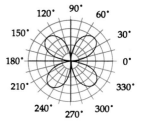

The graph of $r = 4 \sin 2\theta$ found in Example A is symmetric about the pole, the polar axis, and the line $\theta = \pi/2$. These symmetries occur when a function $r = f(\theta)$ satisfies certain conditions.

Consider, for example, the case where f is an even function. If f is even, then the values of $f(\theta)$ and $f(-\theta)$ are always the same, so every point $P(r, \theta)$ on the graph will have a mirror image $Q(r, -\theta)$, as shown in Figure 20.11a. This means that an even function is symmetric about the polar axis.

Figure 20.11

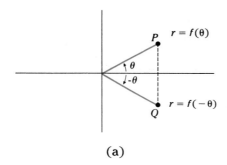

(a)

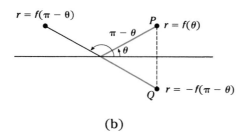

(b)

Symmetry about the polar axis also occurs if $-f(\pi - \theta) = f(\theta)$ as shown in Figure 20.11b. You must test both conditions. The function $f(\theta) = 4 \sin 2\theta$ in Example A is not even:

$$f(-\theta) = 4 \sin 2(-\theta)$$

$$= -4 \sin 2\theta$$

$$\neq f(\theta)$$

Nevertheless, it *is* symmetric about the polar axis because

$$-f(\pi - \theta) = -4 \sin 2(\pi - \theta)$$

$$= -4 \sin(-2\theta)$$

$$= 4 \sin 2\theta$$

$$= f(\theta)$$

In a similar way, Figure 20.12 shows that symmetry about the line $\theta = \pi/2$ occurs when $-f(-\theta) = f(\theta)$, which means that f is odd, or when $f(\pi - \theta) = f(\theta)$. Symmetry about the pole occurs if $f(\pi + \theta) = f(\theta)$ (see Figure 20.13). These results are summarized in Table 20.2.

Figure 20.12

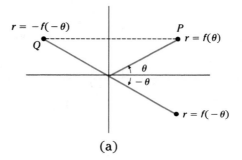

(a)

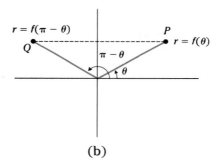

(b)

Figure 20.13

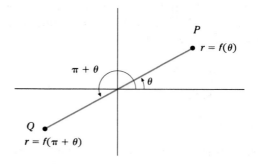

Table 20.2

condition		symmetry
$f(-\theta) = f(\theta)$	(even)	polar axis
$-f(\pi - \theta) = f(\theta)$		polar axis
$-f(-\theta) = f(\theta)$	(odd)	$\theta = \pi/2$
$f(\pi - \theta) = f(\theta)$		$\theta = \pi/2$
$f(\pi + \theta) = f(\theta)$		pole

EXAMPLE B

Determine the symmetry of $r = 1 + 2\cos\theta$ and sketch its graph.

SOLUTION

By inspection, the function $f(\theta) = 1 + 2\cos\theta$ is an even function:

$$f(-\theta) = 1 + 2\cos(-\theta)$$
$$= 1 + 2\cos\theta$$
$$= f(\theta)$$

Therefore, the graph is symmetric about the polar axis. On the other hand,

$$f(\pi \pm \theta) = 1 + 2\cos(\pi \pm \theta)$$
$$= 1 + 2[\cos\pi\cos\theta \mp \sin\pi\sin\theta]$$
$$= 1 - 2\cos\theta$$
$$\neq f(\theta)$$

This means that the graph is not symmetric about the pole or about the line $\theta = \pi/2$. A table of values for $0 \leq \theta \leq 180°$ is given in Table

Table 20.3

$r = 1 + 2\cos\theta$		
$0°$	$\cos\theta$	$1 + 2\cos\theta$
$0°$	1	3
$30°$	$\dfrac{\sqrt{3}}{2}$	$1 + \sqrt{3} \doteq 2.7$
$60°$	$\dfrac{1}{2}$	2
$90°$	0	1
$120°$	$-\dfrac{1}{2}$	0
$150°$	$-\dfrac{\sqrt{3}}{2}$	$1 - \sqrt{3} \doteq -0.7$
$180°$	-1	-1

20.3. Plotting these values produces the curve shown in Figure 20.14. Symmetry about the polar axis gives the rest of the graph shown in Figure 20.15.   ❑

Figure 20.14

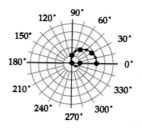

Figure 20.15

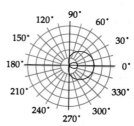

Exercise 20.2

1. Sketch the graphs of the following polar equations:

(a) $r = 4 \sin \theta$

(b) $r = -3 \cos \theta$

(c) $r = 3(1 - \sin \theta)$

(d) $r = -5 + 5 \cos \theta$

2. Determine whether the following relations are symmetric about the pole, the polar axis, or the line $\theta = \pi/2$.

(a) $r = 5 \sin \theta$

(b) $r = 2 \cos 2\theta$

(c) $r = 3 \cos 3\theta$

(d) $r = 3 \sin 4\theta$

(e) $r = 4(1 + \cos \theta)$

(f) $r = 2(\sin \theta - 2)$

3. Sketch the graphs of the following equations with the help of symmetry.

(a) $r = 6 \sin 2\theta$

(b) $r = 5 \cos 3\theta$

(c) $r = 4(1 - \cos \theta)$

(d) $r = 3(2 - 3 \sin \theta)$

4. The relations $r = \sin n\theta$ and $r = \cos n\theta$ produce roses of n leaves if n is odd and $2n$ leaves if n is even. Sketch graphs of the following functions:

(a) $r = 5 \sin 3\theta$

(b) $r = 2 \cos 2\theta$

(c) $r = 3 \cos 5\theta$

(d) $r = 3 \sin 4\theta$

5. Compare the graphs of the following two functions by plotting them on the same polar coordinate axes.

(a) $r = 4 \sin 3\theta$

(b) $r = 4 \cos 3\theta$

6. Compare the graphs of the following two polar equations:

(a) $r = \sin \theta + \cos \theta$

(b) $r^2 = \sin \theta + \cos \theta$

7. Compare the graphs of the following three functions:

(a) $r = 4 \cos \theta + 3$

(b) $r = 4 \cos \theta + 4$

(c) $r = 4 \cos \theta + 5$

8. (a) Compare the graphs of

$$r = \sin 4\theta \quad \text{and} \quad r = \sqrt{\sin 4\theta}$$

(b) Find a function that produces a rose-shaped graph having 6 leaves. Sketch the graph.

Area in Polar Coordinates

20.3

Using integration to find the area of a region in polar coordinates parallels the process used in Cartesian coordinates. Consider a region such as that shown in Figure 20.16 between a polar curve $r = f(\theta)$ and two rays $\theta = \alpha$ and $\theta = \beta$. First the region is divided into a large number of similar elements. In polar coordinates, these elements are chosen to be narrow sectors of a circle as in Figure 20.17. The area of a sector of a circle is

$$A = \frac{1}{2}r^2\theta$$

where r is the radius of the sector and θ is the measure of the central angle (see F. The Trigonometric Functions, pages 223–229). The area of a typical one of these narrow sectors is then

$$\Delta A_k = \frac{1}{2}r_k^2 \cdot \Delta\theta$$

where $\Delta\theta = \dfrac{\beta - \alpha}{n}$

and n is the number of sectors.

The total area is approximately equal to the sum of the areas of the sectors:

$$A \doteq \sum_{k=1}^{n} \Delta A_k$$

$$= \sum_{k=1}^{n} \frac{1}{2}r_k^2 \cdot \Delta\theta$$

Figure 20.16

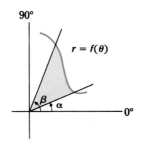

Figure 20.17

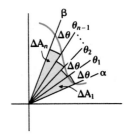

The exact value of the area is found by taking the limit of this sum as $\Delta\theta \to 0$:

$$A = \lim_{\Delta\theta \to 0} \sum_{k=1}^{n} \frac{1}{2} r_k^2 \cdot \Delta\theta$$

This limit is by definition the definite integral

$$A = \int_{\alpha}^{\beta} \frac{1}{2} r^2 \, d\theta$$

In terms of differentials, the area of a pie-shaped element of integration can be expressed as

$$dA = \frac{1}{2} r^2 \cdot d\theta$$

from which the above integral can be written down directly. dA can be visualized as the beam of a searchlight that illuminates the entire region as it sweeps from $\theta = \alpha$ to $\theta = \beta$.

EXAMPLE A

Find the area of a circle of radius R.

SOLUTION

Figure 20.18

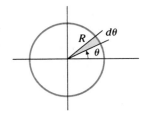

Take the centre of the circle to be at the pole. The equation of the circle is then $r = R$, where R is a constant. Let the element of integration be a narrow circular sector of angle $d\theta$ and radius r, as in Figure 20.18. The area is then

$$A = \int_0^{2\pi} \frac{1}{2} r^2 \, d\theta$$

$$= \frac{1}{2} R^2 \int_0^{2\pi} d\theta$$

$$= \frac{1}{2} R^2 \, \theta \Big|_0^{2\pi}$$

$$= \frac{1}{2} R^2 (2\pi - 0)$$

$$= \pi R^2$$

The area of the circle is $A = \pi R^2$, as expected.

EXAMPLE B

Find the area of the region enclosed by the cardioid $r = 2 + 2\cos\theta$

Figure 20.19

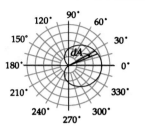

$r = 2 + 2\cos\theta$

SOLUTION

A sketch of the cardioid is shown in Figure 20.19. The element of integration is a narrow sector with area $dA = (r^2/2)d\theta$, where θ sweeps through a full revolution from 0 to 2π.

$$A = \int_0^{2\pi} \frac{1}{2} r^2 \, d\theta$$

$$= \int_0^{2\pi} \frac{1}{2} (2 + 2\cos\theta)^2 \, d\theta$$

$$= \int_0^{2\pi} \frac{1}{2} (4 + 8\cos\theta + 4\cos^2\theta) \, d\theta$$

$$= \int_0^{2\pi} (2 + 4\cos\theta + 2\cos^2\theta) \, d\theta$$

Since $2\cos^2\theta = 1 + \cos 2\theta$

$$A = \int_0^{2\pi} (2 + 4\cos\theta + 1 + \cos 2\theta) \, d\theta$$

$$= \int_0^{2\pi} (3 + 4\cos\theta + \cos 2\theta) d\theta$$

$$= (3\theta + 4\sin\theta + \frac{1}{2}\sin 2\theta) \Big|_0^{2\pi}$$

$$= 6\pi$$

The area enclosed by the cardioid is 6π square units. ◻

Exercise 20.3

1. Find the area of the region bounded by the graphs of the following functions:

   (a) $r = 3\sin\theta$

   (b) $r = 3 + 2\cos\theta$

   (c) $r = 5 - 5\sin\theta$

   (d) $r = 6\sin^2\dfrac{\theta}{2}$

2. In each of the following cases, find the area enclosed by one loop of the graph.

   (a) $r = 3\sin 2\theta$ Hint: $0 \le \theta \le \pi/2$

   (b) $r = 5\cos 3\theta$

(c) $r = 2\sqrt{\cos 2\theta}$

(d) $r = \cos(3\theta/2)$

3. Find the area of the inner loop and the area of the outer loop of the graph of $r = 1 + 2\cos\theta$.

4. Find the area of the region which is common to $r = 4\cos\theta$, and $r = 4\sin\theta$.

20.4 Summary and Review

In polar coordinates, the point (r, θ) lies at a distance r from the pole on the ray that makes an angle θ with the polar axis. The relationship between the polar coordinates (r, θ) and the Cartesian coordinates (x, y) of a point is expressed by the following equations:

$$r = \sqrt{x^2 + y^2}, \qquad \theta = \tan^{-1}\frac{y}{x}$$
$$x^2 = r\cos\theta, \qquad y = r\sin\theta$$

In graphing functions on polar coordinate axes, it is helpful to

(a) choose values of θ which make the arguments of the trigonometric functions multiples of 30°, 45°, or 60°

(b) plot points in consecutive order starting at $\theta = 0$

(c) take advantage of symmetries in the graph.

The application of the definite integral to determine area has been extended here to curves in polar coordinates. The area of a region in polar coordinates bounded by the arc of the curve $r = f(\theta)$ and the rays $\theta = \alpha$ and $\theta = \beta$ is given by

$$A = \int_{\alpha}^{\beta} \frac{1}{2}r^2 \, d\theta$$

Exercise 20.4

1. Plot the following points:

(a) $(4, 0°)$ (b) $(-6, \pi/2)$

(c) $(2, -\pi/3)$ (d) $(-7, 45°)$

(e) $(-3, 390°)$ (f) $(-2, \pi/6)$

2. Give three other sets of polar coordinates for each ordered pair in question 1.

3. Transform the following into Cartesian form.

(a) $r = 1 - 2\sin\theta$ (b) $r = 3\csc\theta$

(c) $r^2 = 4\sin^2 2\theta$ (d) $r = 2\theta$

(e) $r + 4\cos\theta = 0$ (f) $r = -4\sin 6\theta$

4. Transform the following into polar form.

(a) $4x + 7y - 2 = 0$ (b) $y^2 - 8x - 16 = 0$

(c) $x^2 - y^2 = 4$ (d) $x^2 = 9y$

(e) $y = x^2$ (f) $-7xy = 13$

5. Sketch the following curves. Use symmetry properties when possible.

(a) $r = -5\cos\theta$ (b) $r = 6 + \sin 4\theta$

(c) $r = 3\cos 2\theta$ (d) $r = 2\cos 3\theta$

(e) $r = a\theta$ (f) $r = 2\sin 3\theta$

(g) $r^2 = 4\sin 2\theta$ (h) $r = 2\sec\theta$

(i) $r = \csc\theta$ (j) $r = 4\sin\theta\tan\theta$

6. Some polar coordinate curves have been named after ancient and modern mathematicans who studied them. Sketch the graphs of

(a) the lemniscate of Bernoulli,

$$r^2 = a^2\cos 2\theta \qquad \text{(use } a = 3)$$

(b) Tschirnhausen's cubic,

$$r = \frac{1}{a\cos^3(\theta/3)} \qquad \text{(use } a = 2)$$

(c) the nephroid of Freeth,

$$r = a(1 + 2\sin(\theta/2)) \qquad \text{(use } a = 3)$$

(d) the conchoid of Nicomedes,

$$r = 4 + 2\sec\theta$$

7. Find the area of the region bounded by

(a) $r = 3\cos 3\theta$

(b) $r^2 = 2\cos 2\theta$

(c) $r^2 = 4\sin\theta$

8. Find the area of the region swept out by one revolution of the spiral $r = \theta$.

9. Find the area of the region inside $r = 3 + 2\sin\theta$ and outside the region $r = 3$.

10. Find the area of the region bounded by one loop of the graph of $r^2 = 4\sin 2\theta$.

K.   Polar Coordinate Graphing by Computer

The computer program shown in Figure K.1 for producing polar coordinate graphs is similar in many respects to the program given in D. Graphing by Computer, pages 124-127, for producing graphs in the *xy*-plane. It is recommended that you review that program at this time.

Figure K.1

```
10   REM POLAR COORDINATE PLOTTER
100 :
110 DEF FNF(TH) = 1 + 2*COS(TH)
120 :
130 NC = 40 : REM NUMBER OF COLUMNS
140 NR = 25 : REM NUMBER OF ROWS
200 :
210 (clear screen)
220 INPUT "X RANGES FROM";XA
230 INPUT "              TO";XB
240 INPUT "Y RANGES FROM";YA
250 INPUT "              TO";YB
260 INPUT "NUMBER OF POINTS";NP
270 :
280 DX = (XB – XA)/(NC – 1)
290 DY = (YB – YA)/(NR – 1)
300 :
310 (clear screen)
320 FOR TH = 0 TO 2*π STEP 2*π/NP
330    RA = FNF(TH)
340    X = RA*COS*(TH)
350    Y = RA*SIN(TH)
360    C = (X – XA)/DX + 1
370    R = (YB – Y)/DY + 1
380    IF 1< =R AND R< =NR AND 1< =C AND C< =NC THEN plot(C,R)
390 NEXT TH
400 :
999 END
```

As in the previous graphing program, it is necessary to set the number of rows and columns in lines 130 and 140 to correspond to the resolution of the screen, and put in the proper statements for clearing the screen and plotting a point. The part of the graph which is to appear on the screen is still expressed in terms of the ranges on the *x*-axis and *y*-axis. It is recommended that you choose

$x_a = -x_b$ and $y_a = -y_b$ so that the origin is in the centre of the screen.

The principal differences are

1. The function to be graphed, which appears in line 110 should express the radial distance r in terms of the polar angle θ.

2. The number of points to be plotted must be input. If the number of points is 24, for example, the angle θ ranges from 0 to 360° in steps of 15°. (Note that computer calculations are in radians.) The program computes r for each value of θ and then in lines 340 and 350 determines the corresponding xy-coordinates of the point to be plotted. Conversion from Cartesian to screen coordinates is done in the same manner as in the previous graphing program.

It is very satisfying to run this program and watch the computer trace out beautiful polar graphs on the screen. The usual caution holds, however: check and double check that the graphs you see truly represent the function of interest.

Exercise K

1. If $r = 1 + \cos \theta$

   (a) find the value of r at $\theta = 0°, 90°, 180°, 270°$.

   (b) What are suitable values for x_a, x_b, y_a, and y_b?

   (c) Draw a rectangle to represent the computer screen and make a rough sketch of this graph in the rectangle.

2. A low-resolution screen has 25 rows and 40 columns. For the scale to be the same horizontally and vertically, what should the ratio of $(y_b - y_a)/(x_b - x_a)$ be?

3. Using the program in Figure K.1, display the graph of the function $r = 1 + 2\cos \theta$ in the intervals:

   (a) $x_a = -5, x_b = 5, y_a = -3.125, y_b = 3.125$

   (b) $x_a = -4, x_b = 4, y_a = -2.5, \quad y_b = 2.5$

   (c) $x_a = -3, x_b = 3, y_a = -1.875, y_b = 1.875$

   (d) Describe what effect decreasing the scale has on the display.

4. Display the graph of $r = 2 \sin 2\theta$ in the region defined by:

$$x_a = -4, x_b = 4, y_a = -2.5, y_b = 2.5$$

   (a) either on a low resolution screen with 40 columns and 25 rows, using 12 points, 24 points, 48 points, and 96 points;

   (b) or on a high resolution screen with 320 rows and 200 rows, using 96 points, 192 points, 384 points, and 768 points.

   (c) How does increasing the number of plotted points affect the appearance of the graph? How does it affect the plotting time?

5. Display graphs of the following curves:

   (a) $r = \sin 3\theta$,   (three-leafed rose)

   (b) $r = \cos 5\theta$,   (five-leafed rose)

   (c) $r = \sin 4\theta$,   (eight-leafed rose)

   (d) $r = 1 - \sin \theta$,   (cardioid)

   (e) $r = 4 - 3\cos \theta$,   (limacon)

6. In plotting certain curves, it is necessary to avoid values of θ at which the value of a function becomes infinite or complex. Plot the lemniscate $r^2 = 9 \sin 2\theta$ by writing first $r = \pm 3\sqrt{\sin 2\theta}$ and then changing lines 110 and 330 to read as follows:

110 DEF FNF(TH)=3*SQR(SIN(2*TH))
330 IF SIN(2*TH)>=0 THEN RA=FNF(TH)

7. Display graphs of the following curves:

(a) $r^2 = 9 \sin 3\theta$

(b) $r^2 = 9 \cos 6\theta$

(c) $r = 3 \sin \theta \tan \theta$, (cissoid)

(d) $r = \sec \theta - 2$, (conchoid of Nicomedes)

Hint: Write $\sec \theta = 1/\cos \theta$.

8. Find a function that will produce a six-leafed rose. Display its graph.

9. Display graphs of the following curves. (These are special!)

(a) $r = \cos(3\theta/2)$ $0 < \theta < 4\pi$

(b) $r = \sin(4\theta/3)$ $0 < \theta < 6\pi$

(c) $r = \sin(5\theta/2)$ $0 < \theta < 4\pi$

(d) $r = \cos(5\theta/4)$ $0 < \theta < 8\pi$

Answers to Exercises

Exercise 1.1, Page 12

1. $(3, 1), 2$
2. 0
3. $y = 2x - 5$
4. (a) 11   (b) 20, between 11 and 20, 15.5
5. (a) 28.4   (b) 28.9
6. 105 km/h
   no, actual speed fluctuates about the average
7. (a) $\dfrac{\sqrt{2}}{2\sqrt{x}}$ (b) $\dfrac{1}{\sqrt{x}}$
8. (a) $5\sec^2(5x)$ (b) $\frac{1}{2}\sec^2\left(\frac{1}{2}x\right)$
9. (a) $x^{-1} + C$ (b) $\frac{2}{3}x^{3/2}$ (c) $\dfrac{1}{3.5}x^{3.5}$
10. (a) $\frac{1}{4}\sin(4x) + C$ (b) $-3\sin\left(\frac{1}{3}x\right) + C$
    (c) $2\sin(0.5x) + C$
11. rate of change of height at age 6
    yes, because at different times in one's life the rate of growth is different.
12. length decreases
    area increases then decreases
    the 5×5 square
13. 5.66, 6.00, 6.21, 6.26, 6.28
    circle of radius 1, 2π
14. (a) 1382.3 cm³   (b) overestimates the volume
    (c) use more disks
    (d) the exact volume is 1047.2 cm³
15. (a) 11.1   (b) underestimate
    (c) use more line segments
16. (a) 1.25 m, 0.625 m, 0.3125 m
    (b) 8.75 m, 9.375 m, 9.6875 m
    (c) There is no last jump.

Exercise A, Page 27

1. (a) $32x^{-4}$ (b) $-4x^{-2}$ (c) $1/2$ (d) $12x^{-2}$
2. (a) $(x - 12)(x + 2)$
   (b) $(x - 8)(x + 8)$
   (c) $(x - 2)(x + 2)(x - 3)(x + 3)$
   (d) $(6x - 5)(x - 6)$
   (e) $(x - 8)(2x + 1)(2x - 1)$
   (f) $(x + 1)(x - 5)(x - 6)$
   (g) $7(2x - 1)(4x^2 + 2x + 1)$
3. (a) $2x\sqrt{2x}$ (b) $(x + 2)\sqrt{x - 2}$
   (c) $2(\sqrt{4 + x} + \sqrt{x})$
4. (a) $-5, 2$ (b) $-\frac{5}{2}, \frac{2}{3}$ (c) $-2, -\frac{3}{4}, 2$

(d) $-2, -\frac{1}{2}, 3$

5. (a) $-3 \pm \sqrt{21}$ (b) $\dfrac{5 \pm \sqrt{5}}{5}$

6. (a) $\frac{1}{2}x^{1/2}(5x - 9)$
   (b) $(x^2 - 4)^2(5x^2 + 4)$
   (c) $(x - 6)^{3/2}(7x - 12)$

7. (a) $\dfrac{7x + 18}{(x - 2)(x + 6)}$ (b) $\dfrac{-2x + 3}{(2x + 3)^3}$
   (c) $\dfrac{5x^2 - 20x}{2\sqrt{x - 5}}$

8. (a) $3 + \sqrt{x + 4}$ (b) $3 - x + 5x^2$
   (c) $\sqrt{4 + x^{-1}}$ (d) $x + \sqrt{3}$
   (e) $x^{2/3} - 2x^{1/3} + 4$ (f) $8x - 3$

9. (a) $-3 \le x \le \frac{2}{3}$ (b) $-4 < x \le -\frac{1}{3}$
   (c) $x \ge 7$ or $-2 < x < 1$

10. (a) $x > 4$ or $x < 3$ (b) $x > 4$ or $-5 < x < 1$
    (c) $|x| > 2$

11. (a) $-\frac{1}{3}, 3$ (b) $-\frac{2}{3}, 4$

12. (a) $y = -\frac{2}{3}x + 4$ (b) $y = -\frac{2}{3}x + \frac{8}{3}$
    (c) $y = \frac{3}{4}x - 8$ (d) $y = \frac{2}{3}x + 6$

13. (a) $(x + 4)^2 + 4$ (b) $\left(x - \frac{5}{2}\right)^2 + \frac{7}{4}$
    (c) $-2(x - 3)^2 + 9$

14. (a) $-1, -1 - 3h$ (b) $7, 2h^2 + 4h + 7$
    (c) $-2, \dfrac{6}{h - 3}$

15. (a) arithmetic, $-66, 6 - 9n$
    (b) geometric, $49152, 3(4)^{n-1}$
    (c) geometric, $\dfrac{2}{243}, 2(3)^{3-n}$

16. (a) geometric, -1365 (b) arithmetic, 625
    (c) arithmetic, 240

17. (a) $\frac{15}{8}$ (b) $\frac{255}{128}$ (d) 2

Exercise B, Page 36

3. (a) zeros: $3, 4$
   positive: $x < 3, x > 4$ negative: $3 < x < 4$
   (b) zeros: 1
   positive: $x > 1$ negative: $x < 1$
   (c) zeros: $-3, 2, 4$
   positive: $x < -3, -3 < x < 2, x > 4$
   negative: $2 < x < 4$

4. (a) undefined: $x < 5/2$ (b) undefined: $|x| > 1$
5. (a) $\frac{2}{3}(x + 2)(x - 1)(x - 3)$

   (b) $\frac{1}{3}(x + 2)^2(x - 1)(x - 3)$
6. (a) even   (b) odd   (c) odd   (d) even
7. (a) Translate $y = 1/x$ right 2 units.
   (b) Translate $y = 1/x$ down 2 units.
8. (a) asymptotes: $x = 1/2, y = 0$
   (b) asymptotes: $x = -1, x = 2, y = 0$
   (c) asymptotes: $x = -2, x = 3, y = 0$
   (d) asymptotes: $x = 0, y = 2$

Exercise C, Page 38

1. (a) $25, 0, 9$ (b) $-4, -4, 14$
   (c) $-0.6, 0.8, 0.1980198$
2. (a) $1, 2, 3.162278, 7$
   (b) $0, 0.5, 0.7071068, 0.8660254$ (c) $4, 2, 0, 2$
3. (a) division by zero error
   (b) illegal quantity error
   (c) overflow error (the computer cannot hold a
   number this large)
4. (a) in steps of 1:   (b) in steps of 2:

x	y		x	y
-3	-24		-4	-64
-2	0		-2	-8
-1	8		0	0
0	6		2	0.5
1	0		4	0.25
2	-4			
3	0			
4	18			

(c) from $x = 0$
   to $x = 1.5708$
in steps of 0.2

x	y
0	0
0.2	$-2.710036E-03$
0.4	-0.0227932
0.6	-0.0841368
0.8	-0.2296386
1.0	-0.5574077
1.2	-1.372152
1.4	-4.397884

Note: $E-03$ means $\times 10^{-3}$

Exercise 2.1, Page 47

1. (a)

x	$f(x)$
5.1	-12.99
5.01	-12.9999
5.001	-12.999999
5.0001	-12.99999999

   (b) -13 (c) -13
2. the limit exists:
   $$\lim_{x \to 2^-} f(x) = \lim_{x \to 2^+} f(x) = 8$$
3. (a) 0   (b) 9   (c) not defined   (d) 4   (e) 0
   (f) 4

4. (b) 3, 7
   (c) the limit does not exist because left and right
   limits are different.
5. the limit exists:
   $$\lim_{x \to 2^-} g(x) = \lim_{x \to 2^+} g(x) = 8$$
6. (a) 3   (b) 12   (c) 463   (d) $3 - 4k + 5k^2$
7. (a) 0   (b) 9   (c) 1   (d) 4
8. (a) -1 (b) 0   (c) -9 (d) 0
   (e) 6   (f) 20/3
9. (a) $16\sqrt{2} - 1$ (b) 1/2   (c) 1/4   (d) 1
10. (a) $90.576 \le f(x) \le 107.973$
    (b) $4.994 \le f(x) \le 5.006$
    (c) $-0.25 \le f(x) \le -0.249997$
11. (a) $x = 8$ (b) $7.7143 \le x \le 8.3158$
    (c) $7.9701 \le x \le 8.0302$

Exercise 2.2, Page 51

1. (a) 1/3   (b) 8/3   (c) $-25/3$ (d) $-1/3$
2. (a) 3   (b) 1   (c) 5   (d) -5 (e) 1
3. (a) 4   (b) 8   (c) -64 (d) 0
4. (a) $-1/4$ (b) -1 (c) $1/a^2$
5. (a) 1   (b) 0   (c) 1/6   (d) $1/(2\sqrt{3})$
   (e) 1   (f) 1   (g) $2\sqrt{5}$ (h) $\dfrac{3 - \sqrt{2}}{21\sqrt{2}}$
6. (a) -2 (b) 4   (c) 4
7. (a) 6   (b) -1 (c) 42   (d) $\dfrac{-1}{2\sqrt{3}}$

Exercise 2.3, Page 55

1. step function, not continuous
2. continuous
3. discontinuous
4. (a) condition 2: limit does not exist.
   (b) condition 1: value of function is indeterminate.
   (c) condition 3: value of limit $\ne$ value of function.
   (d) conditions 1 and 2: asymptote.
   (e) condition 2: limit does not exist.
   (f) condition 1: function not defined.
   (g) condition 2: limit does not exist.
5. (a) continuous
   (b) discontinuous at $x = 2$
   (c) discontinuous at $x = 0$
   (d) discontinuous at $x = 0$ and $-1/3$
   (e) discontinuous at $x = -1$
   (f) discontinuous at $x = \pm 1$
   (g) discontinuous at $x = 0$
   (h) discontinuous at $x = -3$ and 1
6. (a) $f(a) = \dfrac{a + 1}{a^2 + 1}$

   $\lim_{x \to a} f(x) = \dfrac{a + 1}{a^2 + 1}$
7. define $f(-3) = -8$

Exercise 2.4, Page 57

1. $x \doteq 6.0156$
2. (a) $x = 2/3, -5$
   (b) $x = 2, \pm\sqrt{6}$
4. $x \doteq 2.914214, 0.0857864$

$$x = \frac{3 \pm 2\sqrt{2}}{2}$$

5. (a) $x \doteq -6.386698$
   (b) $x \doteq 0.739084$
   (c) $x \doteq 0.235456, \, 4.880011$
6. $x \doteq 1.165561$

Exercise 2.5, Page 58

1. (a) 0   (b) 3/10   (c) $-7/5$ (d) 11
   (e) $-20/3$ (f) $-1/5$
2. (a) $-17/2$ (b) 2   (c) 3   (d) 0
   (e) 8/5   (f) 1/2
3. (a) 0   (b) 1/2   (c) 5/6   (d) -3 (e) 0
4. (a) 1/3   (b) 1/27   (c) 2/3
5. $-4/3$
6. (b) $1 \leq x \leq 3/2$
7. $3.995 < x < 4.005$
8. (a) yes   (b) yes
9. (a) $x \in R, \, x \neq 4$ (b) $x \in R, \, x \neq \pm 3$ (c) $x > 0$
10. (a) \$9050   (b) the value decreases
    (c) \$2309   (d) \$62
11. $\lim\limits_{a \to 0} \dfrac{-b + \sqrt{b^2 - 4ac}}{2a} = -\dfrac{c}{b}$
    the other root receeds to infinity.
13. 4

Exercise 3.1, Page 69

1. $10, \, 10 + 3(\Delta x), \, -2, \, -2 + 3(\Delta x)$
2. (a) $5, \, 5 - 2(\Delta x), \, -2(\Delta x)$
   (b) $-1, \, -1 + (\Delta x)^2, \, (\Delta x)^2$
   (c) $0, \, \dfrac{2(\Delta x)}{3(\Delta x) + 5}, \, \dfrac{2(\Delta x)}{3(\Delta x) + 5}$
   (d) $2\sqrt{2}, \, \sqrt{8 - \Delta x}, \, \sqrt{8 - \Delta x} - 2\sqrt{2}$
3. (a) 1   (b) -5 (c) 14   (d) $-2/25$
4. (a) 3   (b) $-42 - 7(\Delta x)$
   (c) $4 + \Delta x$ (d) $\dfrac{3}{4(4 + \Delta x)}$
5. (a) no   (b) no   (c) yes, at $x = 2$
   (d) yes, at $x = 2$ (e) yes, at $x = 2$
7. (a) $2 + \Delta x, \, 2$
   (b) $\dfrac{1}{4(-2 + \Delta x)}, \, -\dfrac{1}{8}$
   (c) $\dfrac{2 + \Delta x}{3((1 + \Delta x)^2 - 4)}, \, -\dfrac{2}{9}$
   (d) $\dfrac{\sqrt{(5 + \Delta x)^2 - 9} - 4}{\Delta x}, \, \dfrac{5}{4}$
8. (a) 12   (b) -3 (c) 7   (d) -12
9. (a) $-1/18$ (b) 3/4   (c) 3/4   (d) -1
   (e) $-1/4$

Exercise 3.2, Page 77

1. (a) $y = 2x^3$ (b) $y = 5\sqrt{x}$
2. (a) $3(x + \Delta x)^2$
       $6x + 3(\Delta x)$
   (b) $\frac{1}{2}(x + \Delta x)^3 - 2$
       $\frac{3}{2}x^2 + \frac{3}{2}x(\Delta x) + \frac{1}{2}(\Delta x)^2$

(c) $\dfrac{-2}{x + \Delta x + 4}$
   $$\dfrac{2}{(x + 4)(x + \Delta x + 4)}$$
(d) $\sqrt{2(x + \Delta x) - 1}$
   $$\dfrac{2}{\sqrt{2(x + \Delta x) - 1} + \sqrt{2x - 1}}$$
(e) $\cos\left(\dfrac{\pi(x + \Delta x)}{2}\right)$
   $$\dfrac{\cos\left(\dfrac{\pi(x + \Delta x)}{2}\right) - \cos\left(\dfrac{\pi x}{2}\right)}{\Delta x}$$
(f) $7^{3(x + \Delta x)}$
   $$\dfrac{7^{3(x + \Delta x)} - 7^{3x}}{\Delta x}$$
(g) $\log_a(x + \Delta x)$
   $$\dfrac{1}{\Delta x} \log_a \dfrac{x + \Delta x}{x}$$

3. (a) $8x$ (b) $-\dfrac{3}{x^2}$
   (c) $\dfrac{1}{2\sqrt{x - 1}}$ (d) $\dfrac{3}{(2 + x)^2}$
   (e) $\dfrac{3}{2(2 - 3x)^{3/2}}$ (f) $2 + \dfrac{3}{x^2}$
4. 1/2
5. $y = x + 2$
6. $y = 2x - 1$
7. $x = 0$
8. (a) $f'(x) = 4x - 7$ (b) 2
9. (a) $\left(-4, \frac{32}{3}\right), \left(4, -\frac{32}{3}\right)$ (b) $y = \pm\frac{32}{3}$
10. $y = 5x - 3$
12. (a) $x = 0, 1$ (b) $x = \pm 2$
    (c) $1 < x < 5$
13. $(x_t, \, ax_t^2 + bx_t + c)$
    where $x_t = \dfrac{x_1 + x_2}{2}$

Exercise 3.3, Page 90

1. (a) $-3x^2$ (b) $6x - 7$ (c) $2ax + b$
   (d) $4x^{-1/3}$ (e) $\dfrac{-4}{x^2}$ (f) $\dfrac{\sqrt{6}}{2\sqrt{x}}$
2. (a) $6 - 6x^2$ (b) $4x - 20$ (c) $\frac{2}{3}\sqrt[3]{3x}^{-2/3}$
   (d) $\dfrac{18}{x^4} - 1$ (e) $-\dfrac{6}{x^3} + \dfrac{2}{x^2} + 7$
   (f) $1 - 3x^2 - 4x^3$ (g) $4 + 2x - 9x^2$
   (h) $1 - \dfrac{1}{4x^2}$
3. (a) $\dfrac{3\sqrt{2}}{4\sqrt{x}}$ (b) $x^{-1/2} + \dfrac{\sqrt[4]{6}}{4}x^{-3/4}$
   (c) $0.4x^{-3/4} - \dfrac{3}{0.9}x^2 - \dfrac{2}{5x^3}$ (d) $-x^{-3/2} - x^{-1/2}$
   (e) $\dfrac{1}{\sqrt{8}}\left(\frac{3}{2}\sqrt[3]{x} + 4x^{-3/2}\right)$

4. (a) $9(2 - u)^2$ (b) $3w^2 - 2w + 1$

 (c) $\frac{2}{5}r^{-5/3} + \frac{26}{15}r^{-4/3} - \frac{1}{2}r^{-6/5}$

5. (a) $\frac{3m}{2}t^{-1/2} + \frac{7n}{6}t^{1/6} + \frac{p}{2}t^{-3/2}$

 (b) $\frac{1}{a - b} \cdot \left(2at + b - \frac{c}{t^2} \right)$

6. (a) $-5, 3a^2 - 5$ (b) $2, 5/2$ (c) $9/2, 13$

7. $y = -x + 1$

8. $y = 10x + 18$

9. $y = 3x - 6, y = -3x - 6$

10. (a) $y = 3x - 2$ (b) $(-2, -8)$

 (c) $y = 12x + 16$

11. $\left(\frac{1}{5}, \frac{6}{5} \right)$

12. $a = 4, b = -1$

13. $x^2 - 2x - 1$

14. $x = \pm 1$

15. $x = \pm 1$

16. $\left(\frac{1}{6}, \frac{1}{4} \right)$

17. $y = 6x - 10, y = -2x - 2$

18. (a) $y = -x + 6$

 (b) $y = -4x + 14$

Exercise 3.4, Page 97

1. (a) 70 (b) -12 (c) $1/2$

2. (a) $15x^2(3 + x^3)^4$ (b) $-\frac{7}{6}\left(1 - \frac{x}{6} \right)^6$

 (c) $(36x - 12)(3x^2 - 2x + 8)^{23}$

 (d) $(9 - 54x)(2 - 3x + 9x^2)^{-4}$

 (e) $\frac{8}{(3 - 4x)^3}$ (f) $\frac{2x - 4}{(4x - x^2)^2}$

 (g) $\frac{-3}{2\sqrt{2 - 3x}}$ (h) $\frac{12(7x - 3)}{\sqrt{14x^2 - 12x + 9}}$

 (i) $\frac{2x}{3(x^2 - m^2)^{2/3}}$

3. (a) $m(2ax + b)(ax^2 + bx + c)^{m-1}$

 (b) $\frac{-4(7x - 2)}{(7x^2 - 4x + 1)^{3/2}}$

 (c) $-3x^3(6 - 0.5x^4)^{0.5}$

 (d) $\frac{-9 - 16x}{2(1 - 9x - 8x^2)^{5/4}}$

4. (a) $-2(2x^{-2} - 3x^{-1})^{-3}(-4x^{-3} + 3x^{-2})$

 (b) $(-1)(\sqrt{9 - u^2} - u)^{-2} \cdot [\frac{1}{2}(9 - u^2)^{-1/2}(-2u) - 1]$

 (c) $\frac{1}{2}\left[\sqrt{t^6 - 1} - \frac{1}{t^2} \right]^{-1/2} \cdot \left[\frac{1}{2}(t^6 - 1)^{-1/2} \cdot 6t^5 + \frac{2}{t^3} \right]$

 (d) $\frac{1}{2\sqrt{s - \sqrt{s + \sqrt{s}}}} \cdot \left[1 - \frac{1}{2\sqrt{s + \sqrt{s}}} \cdot \left(1 + \frac{1}{2\sqrt{s}} \right) \right]$

Exercise 3.5, Page 99

1. $\frac{577}{408}$

2. $\frac{301027}{110889}$

3. (a) 2.171573, 7.828427

 (b) $1, -1.25$

 (c) no real roots

4. (a) -0.483752 (b) $-2, 2.2$

 (c) $-0.703753, 1.492640$ (d) 1.100642

Exercise 3.6, Page 101

1. (a) 10

  $2(\Delta x)^2 + 9(\Delta x) + 10$

  $2(\Delta x) + 9$

 (b) $\sqrt{2}$

  $\sqrt{2 - 2(\Delta x)}$

  $\frac{\sqrt{2 - 2(\Delta x)} - \sqrt{2}}{\Delta x}$

 (c) 1

  $\frac{3 - \Delta x}{3 + \Delta x}$

  $\frac{-2}{3 + \Delta x}$

2. yes

3. (a) -4 (b) $6x$ (c) $\frac{1}{\sqrt{2x - 1}}$

 (d) $\frac{-3}{(3x + 1)^2}$ (e) $3 + \frac{8}{x^3}$ (f) $\frac{x^2 - 8x}{(x - 4)^2}$

4. (a) $6x - 4$ (b) $24x^2 + 60x^4$

 (c) $12(4x - 1)^2$ (d) $2x + \frac{7}{2x^2}$

 (e) $\frac{4}{3}x^{1/3} + \frac{2}{3}x^{-5/3}$ (f) $-\frac{18}{5}x^{-5/2}$

5. (a) $\frac{x - 2}{\sqrt{x^2 - 4x}}$ (b) $\frac{6}{(1 - 3x)^{3/2}}$

 (c) $\frac{3(3x^2 - 10x)}{\sqrt{x^3 - 5x^2}}$ (d) $\frac{12x^5 - 7x^6}{4\sqrt{2x^6 - x^7}}$

 (e) $\frac{-1}{2(2 - \sqrt{2 - x})^2} \cdot \frac{1}{\sqrt{2 - x}}$

 (f) $-2\left(\frac{x}{3} - \frac{x^2}{4} + \frac{x^3}{6} \right)^{-3} \cdot \left(\frac{1}{3} - \frac{x}{2} + \frac{x^2}{2} \right)$

 (g) $6\left(3x - \frac{1}{3x} \right)^5 \cdot \left(3 + \frac{1}{3x^2} \right)$

 (h) $8\left(\frac{1}{x - 2} + \frac{1}{x + 2} \right)^7 \cdot \left(\frac{-1}{(x - 2)^2} + \frac{-1}{(x + 2)^2} \right)$

 (i) $\frac{1}{2}\left(\frac{1}{x^3} + \frac{1}{x^2} \right)^{-1/2} \cdot \left(\frac{-3}{x^4} + \frac{-2}{x^3} \right)$

6. $2 - \frac{19}{(x - 3)^2}$

7. (a) $1 - \frac{88}{(x + 5)^2}$ (b) $2x + 2 + \frac{4}{(x - 2)^2}$

 (c) $\frac{-11}{(x - 4)^2}$

8. (a) $y = 12x + 24$ (b) $y = \frac{9}{4}x - 1$

 (c) $y = -\frac{1}{3}x - \frac{2}{3}$ (d) $y = 2x + 4$

9. $\frac{1}{2\sqrt{a + 3}}, a > -3$

10. $y = -\frac{2}{3}x + \frac{5}{3}$ $y = \frac{3}{2}x + 6$

11. $y = 0$

12. $\left(\frac{1}{6}, \frac{107}{54}\right), \left(-\frac{1}{6}, \frac{109}{54}\right)$

13. $y = 6x - 18, y = 6x + 18$

14. $y = -\frac{1}{9}x + 2, y = -\frac{1}{9}x - 2$

15. $y = -4x, y = 8x - 16, y = 8x + 16$

16. $1/2$

17. $\left(\sqrt{3}, \frac{1 - 2\sqrt{3}}{3}\right), k = \frac{1 - 4\sqrt{3}}{3}$

$\left(-\sqrt{3}, \frac{1 + 2\sqrt{3}}{3}\right), k = \frac{1 + 4\sqrt{3}}{3}$

19. $y = 0, y = \frac{27}{4}x$

20. $(1, 2), (-1, 0), y = x + 1$

Exercise 4.1, Page 106

1. (a) $4x - 5$

(b) $(6x^7 - 4x^5) \cdot (-2x) + (1 - x^2) \cdot (42x^6 - 20x^4)$

(c) $(3x^2 - 1) \cdot (2x - 2) + (x^2 - 2x + 2) \cdot (6x)$

(d) $(4 + 5x)^2 \cdot 6(8 - x)^5(-1)$
$+ (8 - x)^6 \cdot 2(4 + 5x)(5)$

(e) $(x^3 - 5x^2)^4 \cdot 2(2 - 3x^2)(-6x)$
$+ (2 - 3x^2)^2 \cdot 4(x^3 - 5x^2)^3(3x^2 - 10x)$

(f) $6x^8 \cdot 3(3x^2 - 4x + 9)^2(6x - 4)$
$+ (3x^2 - 4x + 9)^3 \cdot (48x^7)$

2. (a) $5t \cdot 3(t^2 - 4)^2(2t) + (t^2 - 4)^3 \cdot (5)$

(b) $(r + 5)^2 \cdot 7(r - 5)^6 + (r - 5)^7 \cdot 2(r + 5)$

(c) $(z^2 - 3z + 6)^2 \cdot 2(z^2 + 4z - 2)(2z + 4)$
$+ (z^2 + 4z - 2)^2 \cdot 2(z^2 - 3z + 6)(2z - 3)$

3. (a) $(x - 2) \cdot \frac{1}{2} \cdot \frac{1}{\sqrt{1 - x^2}}(-2x) + \sqrt{1 - x^2}$

(b) $4\sqrt{x} \cdot 2(9 - 4x)(-4) + (9 - 4x)^2 \cdot \frac{4}{2\sqrt{x}}$

(c) $\frac{-2x^3}{\sqrt{1 - x^4}}$

(d) $\frac{(5 - 2x) \cdot 3(3x - 1)^2(3) + (3x - 1)^3 \cdot (-2)}{2\sqrt{(5 - 2x)(3x - 1)^3}}$

(e) $(x + 1)^2 \cdot \frac{1}{2} \cdot \frac{1}{\sqrt{x^2 + 1}}(2x) + \sqrt{x^2 + 1} \cdot 2(x + 1)$

(f) $(x^2 + 1)^2 \cdot \frac{1}{2} \cdot \frac{1}{\sqrt{x + 1}} + \sqrt{x + 1} \cdot 2(x^2 + 1)(2x)$

4. (a) $\frac{9t + 5}{2\sqrt{t + 2}}$

(b) $6(r^2 + 8)^2(5r^2 + 8)r^{1/2}$

(c) $\frac{2 - 4z}{\sqrt{(3 - 2z)(1 + 2z)}}$

5. (a) $4x^3 + 3x^2 - 4x$

(b) $8x^7 - 6x^5 - 4x^3 + 2x$

(c) $4x^3 - 10x$

6. (a) $5x^{10} \cdot \frac{2}{3}(6x^2 - 1)^{-1/3}(12x) + (6x^2 - 1)^{2/3} \cdot (50x^9)$

(b) $(x - 1) \cdot \frac{1}{3}(2 - 9x)^{-2/3}(-9) + (2 - 9x)^{1/3}$

(c) $(3x + 6)^{-3} \cdot (-2)(2x + 8)^{-3}(2)$
$+ (2x + 8)^{-2} \cdot (-3)(3x + 6)^{-4}(3)$

7. (a) $(x - 3)^2 \cdot (-1)(x + 9)^{-2} + (x + 9)^{-1} \cdot 2(x - 3)$

(b) $(2x) \cdot (-\frac{1}{2})(2 - x)^{-3/2}(-1) + (2 - x)^{-1/2} \cdot (2)$

8. $y = 5x - 10, y = -5x - 15$

9. $-3.7, 0.4$

10. $\pm\sqrt{6}$

Exercise 4.2, Page 110

1. (a) $\frac{-3x^2 + 24x}{(4 - x)^2}$ (b) $\frac{10x^5 - 200x^3}{(x^2 - 10)^2}$

(c) $\frac{2}{(x + 1)^2}$ (d) $\frac{2x^2 + 8x + 1}{(x + 2)^2}$

(e) $2 + \frac{3}{2}x^{-2}$ (f) $\frac{-4x^2 + 2x - 4}{(x^2 - 4x)^2}$

2. (a) $\frac{-3x^4 - 2x^2 + 1}{(x^2 + 1)^4}$

(b) $\frac{(1 - t)^3 \cdot (6t) - (3t^2 + 1) \cdot 3(1 - t)^2(-1)}{(1 - t)^6}$

(c) $\frac{(u^2 + u + 1)(3u^2 - 2) - (u^3 - 2)(2u + 1)}{(u^2 + u + 1)^2}$

(d) $\frac{(v^2 - 4)(2v + 5) - (v^2 + 5v - 6)(2v)}{(v^2 - 4)^2}$

(e) $\frac{(r^2 - 16)^2 \cdot 2(r + 2) - (r + 2)^2 \cdot 2(r^2 - 16)(2r)}{(r^2 - 16)^4}$

(f) $\frac{(3z^2 + 6z)(2z) - (z^2 - 9)(6z + 6)}{(3z^2 + 6z)^2}$

3. (a) $\frac{(2 - x)^{1/2} - (x + 5) \cdot (\frac{1}{2})(2 - x)^{-1/2}(-1)}{(2 - x)}$

(b) $\frac{4(2x - 1)^{1/2} \cdot (9x^2) - (3x^3) \cdot 4(\frac{1}{2})(2x - 1)^{1/2}(2)}{16(2x - 1)}$

(c) $\frac{1}{2}\left(\frac{1 + x^2}{1 - x^2}\right)^{-1/2} \cdot \frac{(1 - x^2)(2x) - (1 + x^2)(-2x)}{(1 - x^2)^2}$

(d) $\frac{1}{3}\left(\frac{x}{x - 1}\right)^{-2/3} \cdot \frac{-1}{(x - 1)^2}$

4. (a) $\frac{(cx + d)a - (ax + b)c}{(cx + d)^2}$

(b) $\frac{(a^{1/2} - x^{1/2}) \cdot (\frac{1}{2})x^{-1/2} - (a^{1/2} + x^{1/2}) \cdot \left(-\frac{1}{2}\right)x^{-1/2}}{(a^{1/2} - x^{1/2})^2}$

(c) $\frac{(ax + b)(2ax + b) - (ax^2 + bx + c)a}{(ax + b)^2}$

(d) $\frac{(a^2 - x^2)^{1/2} - x \cdot (\frac{1}{2})(a^2 - x^2)^{-1/2}(-2x)}{a^2 - x^2}$

5. (a) $\frac{(x + 2) \cdot [(x + 1) + (x - 2)] - (x + 1)(x - 2)}{(x + 2)^2}$

(b) $\frac{(2x + 1)(2x - 3) - (x - 5) \cdot [(2x + 1)(2) + (2x - 3)(2)]}{(2x + 1)^2(2x - 3)^2}$

6. $y = \frac{1}{8}x - \frac{3}{8}, y = \frac{8}{25}x - \frac{3}{5}$

7. $y = -\frac{1}{4}x, y = -\frac{1}{16}x$

Exercise 4.3, Page 115

1. (a) $\dfrac{-2xy}{x^2 + 2y}$ (b) $\dfrac{2y - 1}{1 - 2x}$ (c) $-\frac{3}{2}x^{-1}y$

   (d) $\dfrac{-12x - 2y + y^3}{2x - 3xy^2}$ (e) $\dfrac{(1 - x^5)^{71}x^4}{(1 - y^{72})^4 y^{71}}$

   (f) $\dfrac{2(2x - 3)^3}{3y^3}$

2. (a) $-\dfrac{x^{-1/2}}{y^{-1/2}}$ (b) $\dfrac{-\frac{1}{2}x^{-1/2} - y}{x}$

3. (a) $\dfrac{8x - x^2}{2y(4 - x)^2}$ (b) $\dfrac{x^{-2}}{y^{-2}}$

4. (a) $-\frac{12}{121}$ (b) $\frac{3}{4}, -\frac{3}{4}$ (c) $\frac{1}{4}$ (d) $-\frac{1}{4}$ (e) $-\frac{2}{5}$

5. (a) $\dfrac{-4 - 6x^2y}{2x^3 - 5}$ (b) $\dfrac{16x^3 - 42x^2 + 20}{(2x^3 - 5)^2}$

6. $y = x - 3$

7. $(2, -1)$

8. $y = \frac{3}{2}x - 6$

12. $y = 2x - 1,\ y = 2x + 1$

Exercise 4.4, Page 120

1. (a) $y' = 5x^4 - 9x^2 + 4$ (b) $y' = 4x + 7$
    $y'' = 20x^3 - 18x$ $y'' = 4$

   (c) $y' = \frac{5}{2}x^{3/2} - 7$ (d) $y' = \frac{1}{2}x^{-1/2} + \frac{3}{2}x^{-3/2}$

    $y'' = \frac{15}{4}x^{1/2}$ $y'' = -\frac{1}{4}x^{-3/2} - \frac{3}{4}x^{-5/2}$

   (e) $y' = \dfrac{10x^2 + 15}{25x^2}$ (f) $y' = -4x^{-5} + 4x^3$
                   $y'' = 20x^{-6} + 12x^2$

    $y'' = -\dfrac{6}{5x^3}$

2. (a) $f'(x) = 3(x + 2)^2(x + 3)^2 + 2(x + 2)^3(x + 3)$
    $f''(x) = 6(x + 2)(x + 3)^2 + 12(x + 2)^2(x + 3)$
     $+ 2(x + 2)^3$

   (b) $f'(x) = 6x^5 - 12x^{11}$
     $f''(x) = 30x^4 - 132x^{10}$

   (c) $f'(x) = 10(5x - 4)(3x + 1) + 3(5x - 4)^2$
     $f''(x) = 450x - 190$

   (d) $f'(x) = \dfrac{-16x}{(x^2 - 4)^2}$

     $f''(x) = \dfrac{48x^2 + 64}{(x^2 - 4)^3}$

   (e) $f'(x) = \dfrac{-2x - 4}{(x - 4)^4}$

     $f''(x) = \dfrac{6x + 24}{(x - 4)^5}$

   (f) $f'(x) = \dfrac{-10x - 18}{(5x - 4)^3}$

     $f''(x) = \dfrac{100x + 310}{(5x - 4)^4}$

3. (a) $y'' = -x^2(4 - x^2)^{-3/2} - (4 - x^2)^{-1/2}$
   (b) $y'' = -x^3(4 - x^2)^{-3/2} - 3x(4 - x^2)^{-1/2}$
   (c) $y'' = -8(2x^2 + 6)^{-3/2} + 48x^2(2x^2 + 6)^{-5/2}$
   (d) $y'' = -144x(2x^2 + 6)^{-5/2}$

4. (a) $y'' = -\dfrac{25}{y^3}$

   (b) $y'' = -\dfrac{81}{16(y - 4)^3}$

   (c) $y'' = -\frac{1}{2}x^{-3/2}a^{1/2}$

   (d) $y'' = -\dfrac{2a^4}{9y^5}$

5. (a) $-\frac{3}{4}, -\frac{25}{64}$ (b) $\frac{4}{9}, -\frac{25}{81}$ (c) $1, -1$

   (d) $\frac{4}{5}, -\frac{16}{25}$

6. $-3, 2$

7. $f'''(x) = -18(x - 1)^{-4}$

8. $y'' = uv'' + 2u'v' + u''v$

9. $(-1)^n(n + 1)! x^{-n-2}$

Exercise 4.5, Page 122

1. (a) $8x - 20x^4$
   (b) $9x^2 + 8x - 15$
   (c) $48x^2 - 28x + 5$
   (d) $5x^4 - 4x^3 - 12x^2 + 8x - 5$
   (e) $5x^4 - 12x^3 + 9x^2 - 30x + 5$

2. (a) $(1 - 3x) \cdot 3(x^2 - 2)^2 + (x^2 - 2)^3 \cdot (-3)$
   (b) $(5 - 2x)^2 \cdot 3(x - 1)^2 + (x - 1)^3 \cdot 2(5 - 2x)(-2)$
   (c) $(x - x^3) \cdot 3(2 + x)^2 + (2 + x)^3 \cdot (1 - 3x^2)$

   (d) $(3x^2 - 4x)^2 \cdot \left(-\dfrac{1}{2\sqrt{x}}\right)$
     $+ (7 - \sqrt{x}) \cdot 2(3x^2 - 4x)(6x - 4)$

   (e) $\dfrac{x}{2\sqrt{x - 1}} + \sqrt{x - 1}$

3. (a) $x \cdot \frac{1}{3}(1 - 3x)^{-2/3}(-3) + (1 - 3x)^{1/3}$

   (b) $(x - 1) \cdot \dfrac{1}{2\sqrt{x + 1}} + \sqrt{x + 1}$

   (c) $(x + 1)^2 \cdot \dfrac{1}{2\sqrt{x^2 + 1}}(2x) + \sqrt{x^2 + 1} \cdot 2(x + 1)$

   (d) $(1 - 3x) \cdot \dfrac{1}{2\sqrt{x^2 + 3x - 2}}(2x + 3)$
     $+ \sqrt{x^2 + 3x - 2}(-3)$

   (e) $-5x^{-2} + 4x^{-3}$

   (f) $(2x)^{1/3} \cdot \left(-\frac{2}{3}\right)(x - 4)^{-5/3}$
     $+ (x - 4)^{-2/3} \cdot \left(\frac{1}{3}\right)(2x)^{-2/3}(2)$

4. (a) $xp' + p$ (b) $x^2q' + 2xq$

   (c) $6q^2q'$ (d) $\dfrac{2pqp' - p^2q'}{q^2}$

5. (a) $\dfrac{50}{(10 - x)^2}$ (b) $\frac{5}{2}x^{3/2} - \frac{3}{2}x^{-1/2}$

   (c) $\dfrac{x^2 - 6x - 6}{(6 + x^2)^2}$

   (d) $\dfrac{(3x^2 + 4)(6x^2 - 14x) - (2x^3 - 7x^2)(6x)}{(3x^2 + 4)^2}$

   (e) $\dfrac{4c^2 - x}{(c^2 - x^2)^2}$ (f) $10x + \dfrac{10}{(10 - x)^2}$

6. (a) $4x(3 - 4x^2)^{-3/2}$ (b) $\dfrac{-1}{(x^2 - 1)^{3/2}}$

(c) $\dfrac{1}{\sqrt{x}(\sqrt{x} + 1)^2}$ (d) $\dfrac{-3}{x^2\sqrt{3 - x^2}}$

(e) $\dfrac{x(3a^2 - x^2)}{(a^2 - x^2)^{3/2}}$

(f) $\dfrac{-a^2b^2c^2[(x - b)(x - c) + (x - a)(x - c) + (x - b)(x - c)]}{(x - a)^2(x - b)^2(x - c)^2}$

(g) $\dfrac{-18x}{(9 - x^2)^{1/2}(9 + x^2)^{3/2}}$

(h) $-\dfrac{(\sqrt{x^2 - 4} + x)^2}{2\sqrt{x^2 - 4}}$

7. (a) $\dfrac{-2xy}{x^2 + 2y}$ (b) $\dfrac{x^2 - 2y}{2x}$

(c) $\dfrac{ay - x^2}{y^2 - ax}$ (d) $-\dfrac{4x(1 - x^2)}{3y^2}$

(e) $-\dfrac{x^3 + y}{x + y^3}$ (f) $\dfrac{4}{38y + 4y^3}$

(g) $-\dfrac{\sqrt{y}}{2\sqrt{x}}$ (h) $\dfrac{1 - 2(6x)^{-2/3}}{(2y)^{-1/2}}$

(i) $\dfrac{y\sqrt{x^2 + y^2} + x^3}{x\sqrt{x^2 + y^2} - x^2y}$

8. (a) -2 (b) $-30 - 36x$ (c) $\dfrac{-36}{(3x + 1)^3}$

(d) $-\dfrac{4y^2 + (3 - 2x)^2}{4y^3}$ (e) $-\dfrac{2xy^3 + 2x^4}{y^5}$

9. $\dfrac{6}{(1 - x)^4}$, 6

10. 0

11. (a) $y = \frac18 x + \frac18$ (b) $y = x$ (c) $y = -\frac25 x + \frac85$

(d) $y = 8x - 6$ (e) $y = -\frac{4}{13}x + \frac{11}{13}$

(f) $y = -3x + 4$

15. $-\dfrac{Cg'(x)}{[g(x)]^2}$

16. $y = \frac34 x - \frac52$

18. $-2 - \dfrac{h}{r}$

Exercise D, Page 127

2. (a) $y_b = 2.8$
   (b) x-axis is in row 15
   y-axis is in column 16
3. (d) DX = 0.6, DY = 0.6
   DX = 0.45, DY = 0.45
   DX = 0.3, DY = 0.3
   (e) you see a smaller part of the graph in more detail.
4. (d) DX = 0.3, DY = 0.3
   DX = 0.3, DY = 0.6
   DX = 0.3, DY = 0.15
   (e) graph is compressed vertically
   graph is stretched vertically

Exercise 5.1, Page 137

1. (a) $-\frac34$ (b) $+\infty$ (c) $+\infty$
   (d) $-\infty$ (e) $+\infty$ (f) $-\frac72$
2. (a) $\frac32$ (b) 0 (c) ∞
   (d) ∞ (e) a (f) ∞
3. (a) $-\frac85$ (b) ∞ (c) 1 (d) $\frac14$
4. (a) $+\infty$ (b) $+\infty$ (c) $\sqrt{2}$
   (d) $2 - \sqrt{2}$ (e) 0 (f) $-\infty$
5. (a) $x = 4$ (b) $x = -4, y = 2$ (c) $x = -2$
   (d) $x = 1, y = 1$ (e) $x = -3, y = 1$ (f) $x = 2$
   (g) $x = \pm 1, y = 0$
6. (a) $y = x + 1$ (b) none (c) $y = 3x - 1$
   (d) none (e) none (f) none (g) none
7. (a) 0 (b) 0 (c) 0 (d) 0.693 (e) $-\infty$

Exercise 5.2, Page 144

1. (a) increasing, concave down
   (b) decreasing, concave up
   (c) increasing, concave up
   (d) decreasing, concave down
2. (a) H (b) E (c) B (d) I (e) G
5. (a) $y' = -5$
   dec. for all x
   (b) $y' = 2 - 2x$
   inc. $x < 1$
   dec. $x > 1$
   (c) $y = -\dfrac{2}{x^3}$
   inc. $x < 0$
   dec. $x > 0$
   (d) $y' = \dfrac{5}{x^2}$
   inc. for $x < 0$
   and $x > 0$
   (e) $y' = \dfrac{x}{\sqrt{x^2 + 4}}$
   dec. $x < 0$
   inc. $x > 0$
   (f) $y' = 3x^2 + 6x - 24$
   inc. $x < -4$
   dec. $-4 < x < 2$
   inc. $x > 2$
6. (a) max. value $= 0$
   min. value $= 24$
   (c) inc. $x < 0$
   dec. $0 < x < 6$
   dec. $6 < x < 12$
   inc. $x > 12$
7. (a) min. value $= -1$
   inf. values $= -\frac12$
   (c) down $x < -\dfrac{2}{\sqrt{3}}$

   up $\dfrac{-2}{\sqrt{3}} < x < \dfrac{2}{\sqrt{3}}$

down $\quad x > \dfrac{2}{\sqrt{3}}$

8. (a) dec. $\quad x < -2$
   inc. $\quad -2 < x < 5$
   dec. $\quad x > 5$
   up $\quad x < 1$
   down $\quad x > 1$
   (b) dec. $\quad x < 0$ inc. $\quad x < 0$
   dec. $\quad x > 0$ inc. $\quad x > 0$
   down $\quad x < 0$ Or   up $\quad x < 0$
   up $\quad x > 0$ down $\quad x > 0$
   (c) inc. $\quad x < -2$
   dec. $\quad -2 < x < 0$
   inc. $\quad 0 < x < 2$
   dec. $\quad x > 2$
   up $\quad x < -4$
   down $\quad -4 < x < -1$
   up $\quad -1 < x < 1$
   down $\quad 1 < x < 4$
   up $\quad x > 4$
   (d) dec. $\quad x < -1$
   inc. $\quad -1 < x < 0$
   dec. $\quad x > 0$
   down $\quad x < -2$
   up $\quad -2 < x < 0$
   up $\quad x > 0$
9. (a) no
   (b) It may have 0, 1, 2, or 3
   x intercepts
10. yes

Exercise 5.3, Page 151

1. (a) $(0, 4)$ (b) $(0, -3)$ (c) $(2, 1)$
   (d) $\left(6, \dfrac{\sqrt{3}}{6}\right)$

2. (a) $\left(\dfrac{5}{2}, -\dfrac{25}{2}\right)$ (b) $(4, 36)$
   (c) $(2, 12)$ (d) $\left(\dfrac{2}{3}, -\dfrac{4\sqrt{2}}{3\sqrt{3}}\right)$

3. (a) $\left(-\dfrac{5}{2}, \dfrac{147}{4}\right)$, max.   (b) $\left(-\dfrac{7}{2}, -\dfrac{49}{4}\right)$, min.
   (c) $\left(-\dfrac{3}{2}, -\dfrac{49}{4}\right)$, min.

5. (a) $(4, 5)$, max.   (b) $(2, 16)$, min.
   (c) $\left(6, \dfrac{1}{12}\right)$, max.   (d) $\left(-\dfrac{3}{4}, -2\right)$, min.

6. (a) none   (b) $(0, 0)$, max, $\left(\pm\dfrac{1}{3}, -\dfrac{1}{9}\right)$, min.
   (c) $(\pm 1, 2)$, min.   (d) $(0, 4)$, min.
   (e) $\left(3, \dfrac{2}{9}\right)$, max. $\left(-3, -\dfrac{2}{9}\right)$, min.
   (f) $(0, 0)$, max. $(6, 12)$, min.

7. (a) $(2, 0)$ (b) $\left(\dfrac{1}{3}, 0\right), (1, 0)$
   (c) $(-1, 0), (0, 0), (2, 0)$ (d) $(2, 0)$

8. (a) $(-1, 2)$, min.   (b) $(1, 2)$, inf.
   (c) $(0, 0)$, inf. $(3, -27)$, min.
   (d) $(3, 5)$, max.   (e) $(1, 2)$, min. $(-5, -10)$, max.
   (f) $(0, 0)$, inf. $(-3, -81)$, inf.

9. (a) $(1, 3)$, abs. max. $(4, -6)$, abs. min.

(b) $(2, 5)$, abs. max. $(5, -4)$, abs. min.
(c) $\left(0, -\dfrac{1}{4}\right)$, abs. max. $(3, -1)$, abs. min.
(d) $(3, 54)$ and $(6, 54)$, abs. max.
   $(2, 50)$ and $(5, 50)$, abs. min.
(e) $(0, 0)$, abs. min. $(3, \sqrt{15})$, abs. max.

Exercise 5.4, Page 157

1. (a) $-$, $-$ (b) $+$, $+$ (c) $+$, $-$ (d) $-$, $+$
2. (a) C   (b) E   (c) I   (d) H   (e) B
5. (a) $(0, 2)$ (b) $\left(\dfrac{1}{9}, \dfrac{646}{243}\right)$ (c) none
   (d) $\left(\pm\dfrac{1}{\sqrt{3}}, \dfrac{4}{9}\right)$ (e) none   (f) $\left(\pm\dfrac{1}{\sqrt{3}}, \dfrac{3}{4}\right)$
6. (a) $(3, 5)$, min.   (b) none
   (c) $(0, 0)$, min.
   $(1, 1)$, inf.
7. (a) $(1, 2,)$, inf.
   (b) $(0, 0)$, inf. $(2, -16)$, inf. $(3, -27)$, min.
   (c) $\left(-\sqrt{6}, -\dfrac{\sqrt{6}}{8}\right)$, inf. $\left(-\sqrt{2}, \dfrac{\sqrt{2}}{4}\right)$, min.
   $(0, 0)$, inf. $\left(\sqrt{2}, \dfrac{\sqrt{2}}{4}\right)$, max. $\left(\sqrt{6}, \dfrac{\sqrt{6}}{8}\right)$, inf.
   (d) $\left(-\sqrt{6}, -\dfrac{5}{6\sqrt{6}}\right)$, inf. $\left(-\sqrt{3}, -\dfrac{2}{3\sqrt{3}}\right)$, min.
   $\left(\sqrt{3}, \dfrac{2}{3\sqrt{3}}\right)$, max. $\left(\sqrt{6}, \dfrac{5}{6\sqrt{6}}\right)$, inf.
   (e) $\left(-\sqrt{2}, \dfrac{2}{\sqrt{6}}\right)$, inf. $(0, 1)$, max.
   $\left(\sqrt{2}, \dfrac{2}{\sqrt{6}}\right)$, inf.
9. $a = 1, b = -3, c = 7$

Exercise 5.5, Page 159

1. (a) 2   (b) 3   (c) 0   (d) 0
   (e) $\dfrac{1}{2}$ (f) 0   (g) ∞
2. (b) yes
3. (c) one possibility is $f(x) = \dfrac{x - 1}{x^2}$
4. (b) one possibility is $f(x) = \dfrac{x}{x^2 + 1}$
6. $f(3) > f(3 \pm \Delta x)$
   $f'(3) = 0$
   $f''(3) < 0$
   inc. $\quad x < 3$
   dec. $\quad x > 3$
   concave down
9. must have one minimum
   $-2 < x_{min} < 2$
   and two inflection points
   $x_{min} < x_{inf} < 2$
   and $2 < x_{inf}$
10. $(0, 0)$, max. $(3, -162)$, inf. $(4, -256)$, min.
11. $(0, 1)$, max. $\left(\pm\dfrac{1}{3}, \dfrac{3}{4}\right)$, inf.
12. (a) $\left(-\sqrt{3}, -\dfrac{3\sqrt{3}}{4}\right)$, inf. $\left(-1, -\dfrac{3}{2}\right)$, min.

$(0, 0)$, inf. $\left(1, \frac{3}{2}\right)$, max. $\left(\sqrt{3}, \frac{3\sqrt{3}}{4}\right)$, inf.

(b) $(-4, -8)$, max. $(0, 0)$, min.

(c) $\left(-\frac{1}{\sqrt{3}}, -\frac{1}{2}\right)$, inf. $(0, -1)$, min.

$\left(\frac{1}{\sqrt{3}}, -\frac{1}{2}\right)$, inf. (d) none

13. (a) $(0. 1)$, min   (b) $\left(\frac{1}{4}, -\frac{1}{2}\right)$, min

(c) $\left(\frac{4}{3}, \frac{64\sqrt{3}}{9}\right)$, min.

14. (a) $(0, -48)$, abs. min.

$\left(\frac{12 - 2\sqrt{3}}{3}, \frac{16\sqrt{3}}{9}\right)$, abs. max.

(b) no abs. min. $\left(8, \frac{8}{3}\right)$, abs. max.

Exercise E, Page 164

1. (b) $w = 0.8 + \frac{2h}{\sqrt{3}}$ (c) $V = 4h + \frac{5h^2}{\sqrt{3}}$

2. (b) $y = \sqrt{l^2 - x^2}$
   (c) x and y change l is constant

3. (b) height decreases as base increases and vice versa.
   (c) $V = hx^2$ (d) $3 = x^2 + 4hx$
   (e) $V = \frac{(3x - x^3)}{4}$
   (f) Volume decreases when base becomes very large or very small.

4. (a) \$320 000   (b) \$321 750
   (c) $(64 + n)(5000 - 50n)$

Exercise 6.1, Page 169

1. (a) $l + w = 150$ (b) area, $A = 150l - l^2$
   (c) 75 m × 75 m
2. 50 m × 75 m
3. 4 m × 4 m × 4 m
4. 2.1 m²
5. 9 m × 9 m, 9 m × 4.5 m
6. 4.6 × 10.7, 49.3
7. (a) $x^2 + 4xh = 3600$
   (b) Volume, $V = \frac{1}{4}(3600x - x^3)$
   (c) 34.6 cm × 34.6 cm × 17.3 cm
   (d) 20 760 cm³
8. 4 cm × 4 cm
9. $r = 10\sqrt{2}$ cm, $h = 20$ cm
10. $r = 2.9$ cm, $h = 4.1$ cm
12. $y = -2\sqrt{\frac{10}{3}}x + \frac{40}{3}$

Exercise 6.2, Page 175

1. 14:04
2. 186 cm
3. 45 km at 11:05
   The 747 jet has crossed in front of the DC-8.
4. $\left(3, \frac{9}{\sqrt{2}}\right)$, $9\sqrt{\frac{19}{2}}$

5. $(2, 4)$
6. (a) $(1.2, \pm 1.3)$ (b) $(1, 0)$ (c) $(1, 0)$
7. 16.5 km from main road
8. max. value $= 3$
   min. value $= 0$
9. (a) $\frac{16}{\sqrt{5}}$ km down the shore, 3.6 h
   (b) row all the way, 3.2 h

Exercise 6.3, Page 181

1. \$542.50, \$117 722.50
2. 83
3. 100 tonnes
4. 80 km/h, \$2/km
5. 16
6. 384.5 m from power station
7. width $= 0.478$ m
   length $= 0.956$ m
   height $= 0.546$ m

Exercise 6.4, Page 182

1. (a) Minimize $x^2 + y^2$ (b) $x + y = 12$ (c) 6, 6
2. 16, 8
3. 1
4. $\frac{1}{2}$
6. 9680 cm at $t = 5s$
7. $r = 0.57$ m
   $h = 1.02$ m
9. 18 000 cm³
10. 18 cm²
11. $(1, 0)$
12. 12 km/h
13. $w = 23.1$ cm
    $t = 32.7$ cm
14. height $= 5.58$ m
    width $= 10.95$ m
15. (a) 30 km/h
    (b) 91.7 km/h
16. 18, \$598.89

Exercise 7.1, Page 191

1. 126, 3
2. 252π cm³
3. \$8/a
4. -2000 L/min
5. (a) $-\frac{5}{36}$ (b) $\frac{3}{10}$ (c) 348
6. 120π cm³/cm
7. 20π cm²/cm
8. \$72/a, \$57.75/a
9. 0 people/¢, a 40¢ fare produces the maximum number of riders.
10. $\frac{1}{2}$
11. 0.62 s/m
12. 70¢
13. 2 s
14. d
15. $\frac{\sqrt{3}}{2}$ s

Exercise 7.2, Page 197

1. 5, 6
3. 3 cm/s, 0 cm/s,
4. (a) $v(t) = \frac{2}{3}t(t^2 + 2)^{-2/3}$
   (b) $v(t) = -8(8t + 1)^{-3/2}$
   (c) $v(t) = \dfrac{4 - 4t^2}{(1 + t^2)^2}$
5. (a) -51 (b) $-2/25$ (c) 1.2
6. 5.7 s
7. 6 m/s
8. 3 m/s, 2 s
9. 1 s,
   $v_a(0) = -3$ m/s
   $v_a(2) = 5$ m/s
   $v_b(0) = 3$ m/s
   $v_b(2) = -1$ m/s
10. $v_a(9) = -310$ m/s
    $v_b(9) = -382$ m/s

t	0	1	2	3	4
$s(t)$	8	-1	-4	-1	8
$v(t)$	-12	-6	0	6	12

   moving toward 0 when
   $t = 0, 3$
   $s(t) \cdot v(t) < 0$
12. 6 m, 11.9 m/s, 1.2 s, 13.2 m, 2.9 s, -16.1 m/s

Exercise 7.3, Page 202

3. (a) -9.8 (b) -15 (c) 1
4. -24 km/h^2
5. (a) 0.025 m/s^2 (b) -0.32 m/s^2 (c) 5400 m/s^2
6. (a) $\dfrac{d^2s}{dt^2} = 8$ (b) $s(t) = -5$ (c) $\dfrac{ds}{dt} = 9$
   (d) $s(t) \cdot \dfrac{ds}{dt} < 0$ (e) $\dfrac{ds}{dt} \cdot \dfrac{d^2s}{dt^2} < 0$
7. $\dfrac{ds}{dt} > 0, \dfrac{d^2s}{dt^2} < 0$
8. speeding up
9. -384 m/s
10. -1.2 m/s, 6.8 s
11. yes.
12. 0 cm/s^2
13. $x'(2) = -12$
    $y'(2) = 18$
14. $2 < t < 2.5, t > 3$

Exercise 7.4, Page 205

1. 3, 63
2. 32π cm^2
3. 43.3 m, 0.866 m/m
4. 18
5. (a) $y' = -8x^{-3} + \frac{3}{2}x^{-1/2}$ (b) $y' = \frac{4}{3}x(2x^2 + 5)^{-2/3}$
   (c) $y' = 12x(3x^2 + 4)$ (d) $y' = \dfrac{33}{(3x + 5)^2}$
   (e) $y' = \frac{16}{3}x(2x^2 - 3)^{1/3}$
6. (a) $\frac{21}{2}$ (b) $\frac{2}{3}$ (c) -3600

10. (a) $\frac{1}{2}$ m/s   (b) 4 m/s   (c) 36 m/s
11. (a) 12 m/s^2 (b) -12 m/s^2 (c) -24 m/s^2
12. (a) $v(t) = \dfrac{-30}{(9 - 8t)^2}$ (b) $v(t) = 6t(t^2 - 4)^2$
    (c) $v(t) = \frac{3}{2}(3t + 4)^{-1/2}$
13. (a) $v(2) = 16$ (b) $v(2) = -\frac{5}{36}$ (c) $v(2) = \frac{3}{2}$
14. (a) $v(1) = 10$ m/s, $a(1) = 10$ m/s^2
    (b) $v(1) = 0.25$ m/s, $a(1) = -0.25$ m/s^2
    (c) $v(1) = \frac{3}{2}$ m/s, $a(1) = -\frac{3}{4}$ m/s^2
15. (a) $\dfrac{ds}{dt} = 8$ (b) $s(t) = 0$ (c) $\dfrac{d^2s}{dt^2} < 0$
    (d) $s(t) \cdot \dfrac{ds}{dt} > 0$ (e) $\dfrac{ds}{dt} \dfrac{d^2s}{dt^2} > 0$ (f) $\dfrac{d^2s}{dt^2} = 4$
16. 8×10^5 bacteria/h
17. 15 kg/kg, yes.
19. no
20. $\dfrac{dR}{dp} = 100 - 4p$, yes
21. 0.001 \$/L
22. 340 cm^2/cm
23. -250 cd/m
24. \$10 000, $-$\$400/a, \$300/a
25. -9 people/a
26. 3.5 s
27. 45.9 m
28. 17.9 m/s
29. $\frac{3}{4}$ s,
    $x'(0) = -8$ m/s
    $y'(0) = -2$ m/s
    $x'\left(\frac{3}{2}\right) = 1$ m/s
    $y'\left(\frac{3}{2}\right) = -5$ m/s
30. $0 < t < 2, 5 < t < 8$
31. 1.4 m/s
32. $2 < t < \frac{7}{2}, t > 5$
33. decreasing
34. 14 m/s
35. 1.5 s

Exercise 8.1, Page 215

2. 100 cm^2/s
3. (a) 9 cm^3/s, 900 cm^3/s
   (b) 36 cm^2/s, 360 cm^2/s
4. 180π m^2/s
5. 0.24 cm/s
6. $\dfrac{1}{3\pi}$ m/h
7. 0.8 cm/s
8. 98 cm^2/s
9. $\dfrac{12}{25\pi}$ m/min
10. (a) 0.0044 cm/s   (b) 0.0126 cm/s
    (c) 0.0305 cm/s
11. 0.64 cm/min

12. $\frac{1}{15}$ cm²/min
13. 24 cm/min
14. 0.77 m/min
15. (a) 0.04 m/min
    (b) $h = 16.17$ m, 0.02 m/min
16. 0.48 cm/min
17. 4 cm/min

Exercise 8.2, Page 220

1. 375 km/h
2. 368.9 km/h, no
3. 64 km/h
4. (a) -0.1875 m/s   (b) -1.225 m/s
5. 1600 km/h
6. -10.12 knots
7. -1.24 m/s
8. 1.9 km/h
9. 6.55 m/min
10. 40.25 km/h
    decreasing

Exercise 8.3, Page 221

1. (a) 750 km/h   (b) 20 min
2. 0.716 cm/h
3. 0.65 m/s
4. 0.6 ohms/s
5. 14.3 m/s
6. 2.5 m/s, 1 m/s
7. (a) 30 km   (b) 18 km/h, receeding
8. 13.6 m/s
9. -0.33 L/min

Exercise F, Page 228

1. $\dfrac{2\pi}{3}, \dfrac{3\pi}{4}, \dfrac{5\pi}{4}, \dfrac{4\pi}{3}, \dfrac{11\pi}{6}, \dfrac{5\pi}{2}$
2. 270°, 240°, 150°, 216°, 255°, 414°
4. $-\dfrac{5\pi}{6}, \dfrac{19\pi}{6}$
5. $\theta = 135°$ or $\dfrac{3\pi}{4}, \phi = 315°$ or $\dfrac{7\pi}{4}$

   $\theta = -210°$ or $-\dfrac{7\pi}{6}, \phi = 390°$ or $\dfrac{13\pi}{6}$
6. $\dfrac{10\pi}{3}$
7. 24 cm, 0.625
8. (a) $-\dfrac{\sqrt{3}}{2}$ (b) -1 (c) -1 (d) $-\frac{1}{2}$

   (e) $\dfrac{1}{\sqrt{3}}$ (f) $-\dfrac{2}{\sqrt{3}}$
9. (a) $\cos\theta = \frac{5}{13}, \tan\theta = \frac{12}{5}, \cot\theta = \frac{5}{12},$

   $\sec\theta = \frac{13}{5}, \csc\theta = \frac{13}{12}$

   (b) $\sin\theta = \frac{5}{7}, \tan\theta = -\dfrac{5\sqrt{6}}{12},$

   $\cot\theta = -\dfrac{12}{5\sqrt{6}}, \sec\theta = -\dfrac{7}{2\sqrt{6}},$

   $\csc\theta = \frac{7}{5}$

(c) $\sin\theta = -\frac{24}{25}, \cos\theta = -\frac{7}{25}, \cot\theta = \frac{7}{24},$

   $\sec\theta = -\frac{25}{7}, \csc\theta = -\frac{25}{24}$
10. (a) $5, \pi, -\dfrac{\pi}{4}$ (b) $3, \dfrac{2\pi}{3}, \dfrac{2\pi}{9}$ (c) $2, 6\pi, -\dfrac{\pi}{2}$
12. (a) $0, \pi, 2\pi, \dfrac{\pi}{2}$ (b) $\dfrac{\pi}{6}, \dfrac{11\pi}{6}$ (c) $\dfrac{3\pi}{4}, \dfrac{7\pi}{4}$

    (d) $\dfrac{7\pi}{6}, \dfrac{11\pi}{6}$

Exercise 9.1, Page 238

3. $\dfrac{5 + 12\sqrt{3}}{26}$
4. (a) $\dfrac{\sqrt{6} + \sqrt{2}}{4}$ (b) $\dfrac{\sqrt{6} - \sqrt{2}}{4}$

   (c) $-\left(\dfrac{\sqrt{6} + \sqrt{2}}{4}\right)$ (d) $\dfrac{\sqrt{6} - \sqrt{2}}{4}$

   (e) $-\left(\dfrac{\sqrt{6} - \sqrt{2}}{4}\right)$
6. (b) $\tan\theta - \tan\phi = \dfrac{\sin(\theta - \phi)}{\cos\theta\cos\phi}$

   (c) $\tan\theta - \cot\phi = -\dfrac{\cos(\theta + \phi)}{\cos\theta\sin\phi)}$
7. (b) $\cos(A + B + C) = \cos A \cos B \cos C$
   $- \sin A \sin B \cos C$
   $- \sin A \cos B \sin C$
   $- \cos A \sin B \sin C$
10. (a) $A = \sqrt{5}, \phi = 231°$ (b) $A = \sqrt{17}, \phi = 14°$
    (c) $A = 6\sqrt{5}, \phi = -42°$
11. $y = \sqrt{13}\sin(x - \phi), \phi = 34°$
    $y = \sqrt{13}\cos(x - \phi), \phi = 124°$
12. $y = 6\sin\left(x + \dfrac{3\pi}{4}\right)$ or $y = 6\cos\left(x + \dfrac{\pi}{4}\right)$

    $A = -3\sqrt{2}, B = 3\sqrt{2}$
13. (a) $\dfrac{19\pi}{12}, \dfrac{23\pi}{12}$ (b) $\dfrac{11\pi}{12}, \dfrac{17\pi}{12}$
    (c) 310° or 104°
14. 149°, 329°

Exercise 9.2, Page 245

1. (a) $2 - \sqrt{3}$ (b) $2 + \sqrt{3}$
2. (a) $-2 + \sqrt{3}$ (b) $2 - \sqrt{3}$
4. 3
5. 2.68 radians, 5.82 radians $(0 \le B \le 2\pi)$
6. $\tan\theta = \dfrac{\tan(\theta + \phi) - \tan\phi}{1 + \tan(\theta + \phi)\tan\phi}$
7. (a) $\cot(\theta + \phi) = \dfrac{1 - \tan\theta\cdot\tan\phi}{\tan\theta + \tan\phi}$

   (b) $\cot(\theta + \phi) = \dfrac{\cot\theta\cot\phi - 1}{\cot\theta + \cot\phi}$
8. $\tan(A + B + C)$
   $= \dfrac{\tan A + \tan B + \tan C - \tan A \tan B \tan C}{1 - \tan A \tan B - \tan A \tan C - \tan B \tan C}$

10. 0.28 and 0.24 radians

12. $y = 3x - 4$

14. 41° or 0.71 radians

Exercise 9.3, Page 248

1. $\dfrac{\sqrt{3}}{2}, -\dfrac{1}{2}, -\sqrt{3}$

2. (a) $\dfrac{4\sqrt{2}}{9}$ (b) $\dfrac{7}{9}$ (c) $\dfrac{56\sqrt{2}}{81}$ (d) $\dfrac{17}{81}$

3. (a) $\sin 2\theta = \dfrac{24}{25}$, $\cos 2\theta = -\dfrac{7}{25}$, $\tan 2\theta = -\dfrac{24}{7}$

   $\csc 2\theta = \dfrac{25}{24}$, $\sec 2\theta = -\dfrac{25}{7}$, $\cot 2\theta = -\dfrac{7}{24}$

   (b) $\sin 2\theta = -\dfrac{24}{25}$, $\cos 2\theta = -\dfrac{7}{25}$, $\tan 2\theta = \dfrac{24}{7}$

   $\csc 2\theta = -\dfrac{25}{24}$, $\cos 2\theta = -\dfrac{25}{7}$, $\cot 2\theta = \dfrac{7}{24}$

4. $-2\sqrt{2}, \dfrac{4\sqrt{2}}{7}$

5. $\dfrac{1}{7}$

8. (c) $\sin^2\theta = \dfrac{1}{2} - \dfrac{1}{2}\cos 2\theta$

   $\sin^4\theta = \dfrac{3}{8} - \dfrac{1}{2}\cos 2\theta + \dfrac{1}{8}\cos 4\theta$

10. (a) $\dfrac{\pi}{12}, \dfrac{5\pi}{12}, \dfrac{13\pi}{12}, \dfrac{17\pi}{12}$ (b) $\dfrac{\pi}{3}, \dfrac{2\pi}{3}, \dfrac{4\pi}{3}, \dfrac{5\pi}{3}$

   (c) $\dfrac{\pi}{3}, \dfrac{2\pi}{3}, \dfrac{4\pi}{3}, \dfrac{5\pi}{3}$

   (d) $\dfrac{\pi}{3}, \dfrac{5\pi}{3}$, 2.30 radians, 3.98 radians

   (e) $\dfrac{7\pi}{6}, \dfrac{11\pi}{6}$, 0.42 radians, 2.73 radians

12. $-\dfrac{5}{13}, \dfrac{12}{13}, \dfrac{-12}{5}$

13. $\dfrac{\sqrt{2 + \sqrt{3}}}{2}$

Exercise 9.4, Page 254

1. (a) $-\dfrac{1}{2}\cos 9\theta + \dfrac{1}{2}\cos 3\theta$ (b) $\dfrac{1}{2}\sin 6\theta + \dfrac{1}{2}\sin 2\theta$

2. (a) $2\cos 5\theta \sin 3\theta$ (b) $2\cos 4\theta \cos \theta$

4. (b) $(\sin\theta - \sin\phi)^2 + (\cos\theta - \cos\phi)^2$

   $= 4\sin^2\left(\dfrac{\theta - \phi}{2}\right)$

7. $\dfrac{\pi}{12}, \dfrac{5\pi}{12}$

8. (a) $\sqrt{2}, \dfrac{2\pi}{3}, \dfrac{\pi}{12}$ to the right

   (b) $2\cos\left(\dfrac{3\pi}{8}\right), \dfrac{2\pi}{5}, \dfrac{3\pi}{40}$ to the left

9. $\sqrt{3}, 2\pi, \dfrac{\pi}{6}$ to the left

10. $\left(2\cos\dfrac{\theta}{2}\right)\sin\dfrac{9\theta}{2}$

Exercise 9.5, Page 256

1. -1

2. (a) $\cos A$ (b) $\cos A$

   (c) $\cos\left(A - \dfrac{\pi}{6}\right) - \cos\left(A + \dfrac{\pi}{6}\right)$

3. $\dfrac{4\sqrt{3} + 3}{10}$

4. $\dfrac{3\sqrt{7} + 2}{10}, \dfrac{-\sqrt{21} - 2\sqrt{3}}{10}$

6. $\sin(A + B + C)$
   $= \sin A \cos B \cos C + \cos A \sin B \cos C$
   $+ \cos A \cos B \sin C - \sin A \sin B \sin C$
   $\cos(A + B + C)$
   $= \cos A \cos B \cos C - \cos A \sin B \sin C$
   $- \sin A \cos B \sin C - \sin A \sin B \cos C$

7. (a) $\dfrac{56}{65}$ (b) $\dfrac{-33}{65}$

8. (a) $2 - \sqrt{3}$ (b) $\dfrac{\sqrt{6} + \sqrt{2}}{4}$ (c) $\dfrac{\sqrt{2 + \sqrt{2}}}{2}$

   (d) $-\dfrac{\sqrt{2 + \sqrt{2}}}{2}$ (e) $\dfrac{\sqrt{4 + 2\sqrt{2}}}{2}$

9. (a) $\dfrac{\sqrt{2}}{2}$ (b) $\dfrac{\sqrt{6}}{2}$ (c) $-\dfrac{\sqrt{6}}{2}$ (d) $\dfrac{\sqrt{2}}{2}$

10. (a) $2\sin 3A \cos A$

    (b) $2\cos 3A \cos A$

    (c) $-2\cos\dfrac{3A}{2}\sin\dfrac{A}{2}$

    (d) $2\sin\left(\dfrac{A}{2} - \dfrac{\pi}{4}\right)\sin\left(\dfrac{3A}{2} - \dfrac{\pi}{4}\right)$

    (e) $2\cos\left(A + \dfrac{B}{2}\right)\sin\left(\dfrac{B}{2}\right)$

    (f) $-2\sin\left(\dfrac{B}{2} + \dfrac{\pi}{4}\right)\sin\left(A + \dfrac{B}{2} - \dfrac{\pi}{4}\right)$

12. $\theta = \dfrac{\pi}{2}, \dfrac{3\pi}{2}, \dfrac{2\pi}{3}, \dfrac{5\pi}{3}, \dfrac{\pi}{6}, \dfrac{7\pi}{6},$

    $\dfrac{\pi}{12}, \dfrac{7\pi}{12}, \dfrac{13\pi}{12}, \dfrac{19\pi}{12}$

13. (a) $2\cos\dfrac{\theta}{2}\cos\dfrac{\phi}{2}$ (b) $\sin(\theta + \phi)\cdot\sin(\theta - \phi)$

    (c) $\dfrac{\sin(\theta + \phi)\cdot\sin(\theta - \phi)}{\cos^2\theta \cos^2\phi}$

    (d) $\sin(\theta + \phi)\cos(\theta - \phi)$

    (e) $2\cos\left(\dfrac{\theta}{2} - \dfrac{\pi}{4}\right)$

    (f) $4\sin\left(\dfrac{\theta + \phi}{2}\right)\cos\dfrac{\theta}{2}\cos\dfrac{\phi}{2}$

16. (b) $4\cos^3 A - 3\cos A$

17. $\sqrt{\dfrac{5 - \sqrt{5}}{8}}$

Exercise 10.1, Page 264

1. (a) $\dfrac{1}{2}$ (b) 0 (c) ∞ (d) 1 (e) 0 (f) ∞

   (g) $\dfrac{1}{2}$ (h) 1

2. (a) 0   (b) $\frac{1}{2}$ (c) $\frac{1}{2}$ (d) 1

3. (a) $\sqrt{2}$ (b) $\dfrac{\sqrt{2}}{2}$

4. (a) 1   (b) $-\frac{3}{2}$ (c) -1

5. (a) $\sqrt{2 - 2\cos\theta}$, θ

6. $-\sin x$

7. $2\cos 2x$

8. $\sec^2 x$

9. (a) k   (b) $\frac{a}{b}$ (c) $\frac{1}{2}$ (d) -1 (e) $\dfrac{\pi}{2}$

   (f) $-2\sin a$ (g) $\dfrac{\sqrt{2}}{8}$ (h) 1   (i) 4

   (j) $-\dfrac{a}{\pi}$

Exercise 10.2, Page 270

1. (a) $2\cos 2x$ (b) $-2\sin x - 3\cos x$
   (c) $2\tan x \sec^2 x$ (d) $-2\csc^2(2x + 1)$
   (e) 0   (f) $3\tan 3x \sec 3x$

2. (a) $6\sec^2 3x \tan 3x$ (b) $\dfrac{\cos\sqrt{x}}{2\sqrt{x}}$

   (c) $\sec^3 x + \tan^2 x \sec x$

   (d) $-2\cos^2 2x \cot^2 x \csc x - 4\cot^2 x \cos 2x \sin 2x$

   (e) $\dfrac{\sec x \tan x - \sec x}{(1 + \tan x)^2}$ (f) $\dfrac{1 - \cos x}{\sin^2 x}$

4. $\sec^2 x$

5. (a) $-\csc^2 x$ (b) $\tan x \sec x$ (c) $-\cot x \csc x$

6. (a) $\tan x \sec x$ (b) $-\cot x \csc x$

7. (a) $2\cos 2x$ (b) $-\cot x \csc x$

   (c) $\sec x \tan x + \sec^2 x$ (d) $\dfrac{2}{1 - \sin 2x}$

8. (a) 0   (b) $-\dfrac{\sqrt{6}}{4}$ (c) $8 - 6\sqrt{2}$

9. (a) $x\cos x + \sin x$ (b) $-4x^2 \sin 4x + 2x\cos 4x$
   (c) $4x^3 \tan x + x^4 \sec^2 x$ (d) $x\cos x$
   (e) $\dfrac{3x\cos 3x - \sin 3x}{x^2}$ (f) $\dfrac{x\sec^2 x - \tan x}{x^2}$

10. (a) $\dfrac{-\tan y}{x}$ (b) $\cos x \cos^2 y$

    (c) $\dfrac{\cos x}{\sin y}$ (d) $\dfrac{1 + \csc^2(x - y)}{\cot^2(x - y)}$

11. (a) $y - \left(2 - \dfrac{\sqrt{3}}{2}\right) = -\frac{1}{2}\left(x - \dfrac{5\pi}{6}\right)$

    (b) $y - \frac{1}{2} = \dfrac{\sqrt{3}}{4}\left(x - \dfrac{\pi}{3}\right)$

    (c) $y - 3\sqrt{2} = -3\sqrt{2}\left(x - \dfrac{\pi}{4}\right)$

12. (a) $21\cos(3\theta + 5)$ (b) $4\sin^2 4\theta \cos 4\theta$
    (c) $2\sec^2 x \tan x - 2\csc^2 x \cot x$
    (d) $\tan^3 \theta \sec^2 \theta$ (e) $-\sin\theta + \frac{2}{3}\cos\theta \sin\theta$

    (f) $8\sin x \cos x(1 + \sin^2 x)^3$ (g) $\dfrac{1}{1 + \cos x}$

(h) $2x\sec^2 x \tan x$ (i) $\dfrac{1 - \cos x - x\sin x}{(1 - \cos x)^2}$

(j) $\dfrac{\theta\sec^2\theta - \tan\theta}{\theta^2}$

(k) $\dfrac{\theta\cos\theta - \sin\theta}{\theta^2} + \dfrac{\sin\theta - \theta\cos\theta}{\sin^2\theta}$

(l) $\dfrac{1}{x}\sin\dfrac{1}{x} + \cos\dfrac{1}{x}$

(m) $\dfrac{(1 + \tan\theta)(\sin\theta + \theta\cos\theta) - \theta\sin\theta\sec^2\theta}{(1 + \tan\theta)^2}$

(n) $\dfrac{\sec^2\dfrac{x}{2}}{4\sqrt{\tan\dfrac{x}{2}}}$ (o) $\dfrac{\sec^2\theta}{\sqrt{3 + 2\tan\theta}}$

(p) $\dfrac{x\cos\sqrt{1 + x^2}}{\sqrt{1 + x^2}}$ (q) $\dfrac{-2\theta\csc^2\sqrt[3]{1 + \theta^2}}{3(1 + \theta^2)^{2/3}}$

Exercise 10.3, Page 275

1. (a) minima at $\dfrac{\pi}{4}, \dfrac{5\pi}{4}$

      maxima at $\dfrac{3\pi}{4}, \dfrac{7\pi}{4}$

      infl. at $0, \dfrac{\pi}{2}, \pi, \dfrac{3\pi}{2}, 2\pi$

   (b) infl. at $0, \pi, 2\pi$

   (c) minima at $\dfrac{\pi}{2}, \dfrac{3\pi}{2}$

      maxima at 0.253, 2.889 radians
      infl. at 0.8825, 2.259, 3.846, 5.579 radians

2. max. s at $\left(\dfrac{\pi}{20}, 10\right)$

   min. s at $\left(\dfrac{\pi}{4}, -10\right)$

   max. v at $\left(\dfrac{7\pi}{20}, 50\right)$

   min. v at $\left(\dfrac{3\pi}{20}, -50\right)$

   max. a at $\left(\dfrac{\pi}{4}, 250\right)$

   min. a at $\left(\dfrac{\pi}{20}, -250\right)$

3. -0.05 rad/s

4. 1024π km/min

5. 3.9 m

6. 2

7. (a) 53 m, 53 m   (b) $\dfrac{\pi}{4}$ or 45°

8. 1

10. 4.2 m

11. $1024\pi \sqrt{3}$ cm³

Exercise 10.4, Page 276

1. (a) min. at $\dfrac{\pi}{12}, \dfrac{13\pi}{12}$

max. at $\dfrac{5\pi}{12}, \dfrac{17\pi}{12}$

infl. at $\dfrac{\pi}{4}, \dfrac{3\pi}{4}, \dfrac{5\pi}{4}, \dfrac{7\pi}{4}$

(b) max. at $0, 2\pi$
  min. at π
  $\left.\begin{array}{l}\text{max. at } 1.99, 4.29 \\ \text{min. at } 1.15, 5.13\end{array}\right\}$ from $\sin x = \pm\sqrt{\tfrac{5}{6}}$
  infl. at $\dfrac{\pi}{2}, \dfrac{3\pi}{2}$

  infl. at $0.56, 2.59, 3.70, 5.73$ radians
  $\qquad$ from $\sin x = \pm\sqrt{\tfrac{5}{18}}$

3. 8 cm
4. $\sqrt{2}$

6. max. at $x = \cos^{-1}\sqrt[3]{\dfrac{-1}{2}}$

7. 0.615 radians, from $\cos\theta = \sqrt{\tfrac{2}{3}}$
8. $5/\sqrt{2}$ m
9. $\dfrac{25}{\pi}$
10. $\tfrac{3}{4}a$

Exercise 11.1, Page 286

1. (a) $\sin^{-1}\left(\tfrac{1}{2}\right) = \dfrac{\pi}{6}$ (b) $\tan^{-1}(-1) = -\dfrac{\pi}{4}$
   (c) $\cos^{-1}(0) = \dfrac{\pi}{2}$

2. (a) $\tan\left(\dfrac{\pi}{3}\right) = \sqrt{3}$ (b) $\cos\left(\dfrac{5\pi}{6}\right) = -\dfrac{\sqrt{3}}{2}$
   (c) $\sin\left(-\dfrac{\pi}{6}\right) = -\tfrac{1}{2}$

3. (a) $\dfrac{\pi}{2}$ (b) $\dfrac{2\pi}{3}$ (c) $\dfrac{\pi}{6}$ (c) $\dfrac{\pi}{4}$ (e) 0

4. (a) 0 (b) $\tfrac{3}{5}$ (c) $-\sqrt{15}$

5. (a) 2 (b) $\dfrac{1}{\sqrt{2}}$ (c) $-\dfrac{1}{\sqrt{3}}$

6. (a) $1.57, 1.57, 1.57$
   (c) acute angles of a right triangle are complementary
9. (b) no

Exercise 11.2, Page 293

2. (a) $\dfrac{4}{1 + 16x^2}$ (b) $\dfrac{-1}{\sqrt{4 - x^2}}$
   (c) $\dfrac{-2}{\sqrt{2x - x^2}}$ (d) $\dfrac{-2x}{\sqrt{1 - x^4}}$

3. (a) $\dfrac{1}{x\sqrt{x^2 - 1}}$ (b) $\sin^{-1} 2x + \dfrac{2x}{\sqrt{1 - 4x^2}}$
   (c) $2x\cos^{-1} x - \dfrac{x^2}{\sqrt{1 - x^2}}$

(d) $\dfrac{1}{(1 + 4x^2)\sqrt{\tan^{-1} 2x}}$

4. (a) $\dfrac{-a}{x\sqrt{x^2 - a^2}}$ (b) $\dfrac{a}{a^2 + x^2}$
   (c) $\dfrac{a - x}{\sqrt{a^2 - x^2}}$ (d) $\dfrac{x}{\sqrt{2ax - x^2}}$

5. (a) $\dfrac{2}{(1 + x^2)^2}$ (b) $\dfrac{2 - x^2}{(1 - x^2)^{3/2}}$

6. (a) $\dfrac{\cos x}{1 + \sin^2 x}$ (b) $\dfrac{-\sec^2 x}{\sqrt{1 - \tan^2 x}}$

7. (a) $y = \tfrac{1}{2}x + \tfrac{1}{2} - \dfrac{\pi}{4}$
   (b) $y = -x + \sqrt{3} + \dfrac{\pi}{6}$
   (c) $y = -\dfrac{2\pi}{3\sqrt{3}}x - \dfrac{\pi}{3\sqrt{3}} + \dfrac{\pi^2}{36}$

8. (a) $\left(0, \dfrac{\pi}{2}\right)$ max. (b) $(0, 0)$ min.
   (c) no extrema
     $(0, 0)$ infl.
   (d) no extrema
9. 87π km/min
10. slope of graph of $y = \sin^{-1} x$ is greater than the slope of the graph of $y = x$, except at $x = 0$
11. $18°$
12. 2.4 m/s
13. $\sqrt{10}$ m
14. 0.014 radians/s

Exercise 11.3, Page 295

1. (a) $\dfrac{\pi}{4}$ (b) $-\dfrac{\pi}{3}$ (c) $\dfrac{5\pi}{4}$
   (d) $\sqrt{3}$ (e) $\dfrac{\sqrt{3}}{\sqrt{7}}$ (f) $\dfrac{\sqrt{3}}{2}$

3. (a) $\dfrac{-2}{\sqrt{1 - (2x + 1)^2}}$ (b) $\dfrac{1}{x(1 + x^2)} - \dfrac{\tan^{-1} x}{x^2}$
   (c) $\dfrac{2\sin^{-1} x}{\sqrt{1 - x^2}}$ (d) $\dfrac{x^2}{(1 - x^2)^{3/2}}$
   (e) $\dfrac{-1}{(1 + x)\sqrt{x}}$

5. (a) min. at $x = 0$
     infl. at $x = 1$
   (b) infl. at $x = 0$
   (c) min. at $x = 0$
     infl. at $x = 0.765$

6. $(\pm 2\sqrt{3}, 0)$
7. 0.029 radians/s
8. 0.04 radians/s

Exercise G, Page 298

1. (a) $16 = 4^2$ (b) $2 = 16^{1/4}$ (c) $\tfrac{1}{16} = 2^{-4}$
2. (a) $5 = \log_2 32$ (b) $\tfrac{1}{2} = \log_{36} 6$ $-4 = \log_3\left(\tfrac{1}{81}\right)$

3. (a) 4   (b) -6 (c) $\frac{3}{2}$ (d) -2 (e) -1
   (f) 3

4. (a) $\frac{128}{5}$ (b) 10^6 (c) 4   (d) 625   (e) 16

5. (a) $2\log 5 < 3\log 3$ (b) $\frac{1}{2}\log_2 16 > \frac{1}{3}\log_3 27$

7. (a)

x	1	2	4	8	2^n
$f(x)$	0	1	2	3	n

   (b) $f\!\left(\frac{1}{2}\right) = -1$

   $f\!\left(\frac{1}{4}\right) = -2$

   $f\!\left(\frac{1}{8}\right) = -3$

8. (a) $\frac{1}{5}\log 25 \doteq 0.279588$

   (b) $\dfrac{3}{\log 2} \doteq 9.965784$

   (c) $\dfrac{\log 200}{\log 16} \doteq 1.910964$

9. (b) 100 times more intense
   (c) $1.26 \times 10^8 I_0$

10. (b) 32 000 times louder

Exercise 12.1, Page 305

1. (a) 20.085537...   (b) 22 026.47...
   (c) 2.117...   (d) 0.049787...

2. (b) Domain: $x \in R$ Range: $y > 0$
   (c) $\lim\limits_{x\to\infty} e^x = \infty$ $\lim\limits_{x\to -\infty} e^x = 0$
   (d) slope positive
       concave up

3. (a) 1.0986123...   (b) $-1.0986123...$
   (c) 2.3025851...   (d) 4.6051702...

4. (b) Domain: $x > 0$ Range: $y \in R$
   (c) $\lim\limits_{x\to\infty} \ln x = \infty$ (d) $\lim\limits_{x\to 0} \ln x = -\infty$
   (d) slope positive
       concave down

5. 2.7182540
   2.7182788
   2.7182815
   10 terms

6. 28 terms

11. (b) stretched in y-direction by a factor of e^a, $a > 0$
    (or compressed, if $a < 0$).

Exercise 12.2, Page 310

1. (a) $\dfrac{18}{x}$ (b) $\dfrac{3x;}{x^3 - 6}$ (c) $\dfrac{2}{x}$ (d) $\dfrac{x-1}{x^2 - 2x}$

2. (a) $\ln x$ (b) $\dfrac{-x}{4 - x^2}$ (c) $\dfrac{8\ln 3x}{x}$

   (d) $\dfrac{4 + 9x}{2x(2 + 3x)}$ (e) $\dfrac{2}{x(x^2 + 1)}$

5. (a) $\dfrac{1}{\sqrt{x^2 + a^2}}$ (b) $\dfrac{-2a}{x^2 - a^2}$

6. (a) $\tan^3 x$ (b) $-2\cot x$ (c) $8\csc(4x)$
   (d) $\sec x$ (e) $\sec x$

7. (a) $3 + 2\ln x$ (b) $\dfrac{2}{x}(1 + \ln x)$

(c) $\dfrac{1}{x^3}(-6 + 2\ln x)$

8. (a) $y\left[\dfrac{2}{x} + \dfrac{x}{x^2 - 4}\right]$

   (b) $y\left[\dfrac{1}{x} + \dfrac{3}{2(3x + 1)} + \dfrac{1}{2(x - 5)}\right]$

   (c) $y\left[\dfrac{1}{2x} + \dfrac{3}{9x + 4}\right]$

   (d) $y\left[\dfrac{3}{x} + \dfrac{3}{2(3x - 7)} - \dfrac{3}{(6x + 2)}\right]$

   (e) $\dfrac{y}{2}\left[\dfrac{1}{x + 3} + \dfrac{1}{x + 5} - \dfrac{2x}{x^2 - 4} - \dfrac{1}{x + 1}\right]$

9. $\dfrac{4a^2 x}{(x^2 + a^2)^2}$

10. (a) $y = x - 1$ (b) $y = 3x - 3$
    (c) $y = -x + \dfrac{\pi}{4} - \ln\sqrt{2}$ (d) $y = x - 1$

11. $\tan^{-1}\frac{1}{2}$ or $27°$

Exercise 12.3, Page 314

1. (a) $-20e^{-4x}$ (b) xe^{x^2}
   (c) $4x^3 e^x + x^4 e^x$ (d) $-\dfrac{(x + 1)e^{-x}}{x^2}$
   (e) $\dfrac{e^x}{2\sqrt{1 + e^x}}$ (f) $1 + \dfrac{e^{\sqrt{x}}}{2\sqrt{x}}$

2. (a) $3 \cdot 3^{3x} \cdot \ln 3$ (b) $-10^{-x+2} \cdot \ln 10$
   (c) $-2^{-x} \cdot \ln 2 - 2x^{-3}$
   (d) $2^x \cdot 2x + x^2 \cdot 2^x \cdot \ln 2$

3. (a) $y = 2e^2 x - e^2$ (b) $y = x$
   (c) $y = (4\ln 2)x + 2 - 4\ln 2$
   (d) $y = \dfrac{\sqrt{3e}}{2}x - \dfrac{\pi\sqrt{3e}}{12} + \sqrt{e}$

4. (a) $(x^2 - 4x + 2)e^{-x}$ (b) $(16x^3 + 24x)e^{x^2}$
   (c) $-2e^{-x}\cos x$ (d) $-\dfrac{1}{x^2}$

6. $-3, 2$

7. (a) $\dfrac{e^{x+y} - e^x}{e^y - e^{x+y}}$ (b) $-\dfrac{e^{2y} + 2ye^{2x}}{e^{2x} + 2xe^{2y}}$

8. (a) $\dfrac{2y}{x}\ln x$ (b) $y\left[\dfrac{1}{\sqrt{x}} + \dfrac{1}{2\sqrt{x}}\ln x\right]$
   (c) $y[6x + 3\ln 6]$ (d) $y[x\cot x + \ln(\sin x)]$

Exercise 12.4, Page 323

1. (a) 0   (b) ∞ (c) $-\infty$ (c) 0

4. (a) $\left(\dfrac{1}{\sqrt{3}}, \dfrac{1}{3\sqrt{3}}\right), \left(-\dfrac{1}{\sqrt{3}}, -\dfrac{1}{3\sqrt{3}}\right)$
   (b) $\left(\frac{1}{3}\ln\!\left(\frac{1}{3}\right), \frac{1}{3}\right)$ (c) $(1, \ln 3)$

5. (a) increasing, concave up
   (b) decreasing, concave up
   (c) increasing, concave down
   (d) decreasing, concave up

6. (a) max. $(0, -1)$
   (b) min. $(1, 1)$

asymp. positive y-axis

(c) max. $\left(2, \dfrac{2}{e}\right)$

infl. $\left(4, \dfrac{4}{e^2}\right)$

asymp. positive x-axis

(d) min. $\left(2, \dfrac{e^2}{4}\right)$

asymp. positive y-axis
and negative x-axis

(e) min. $\left(\dfrac{1}{e}, -\dfrac{1}{e}\right)$

(f) max. $\left(\sqrt{e}, \dfrac{1}{2e}\right)$

infl. $\left(e^{5/6}, \dfrac{5}{6e^{5/3}}\right)$

asymp. positive x-axis
and negative y-axis

(g) min. (1, 0)
infl. (e, 1)
asymp. positive y-axis

(h) min. (1, 0)

max. $\left(-1, \dfrac{4}{e}\right)$

infl. $\left(-1 + \sqrt{2}, (-2 + \sqrt{2})^2 e^{-1+\sqrt{2}}\right)$

and $\left(-1 - \sqrt{2}, (-2 - \sqrt{2})^2 e^{-1-\sqrt{2}}\right)$

asymp. negative x-axis.

7. (b) $\lim\limits_{x \to \infty} \dfrac{x^n}{e^x} = 0$

8. (b) $\lim\limits_{x \to \infty} \dfrac{\ln x}{x^n} = 0$

Exercise 12.5, Page 329

2. $68.4a$
3. $62s$
4. (a) 4% (b) 34%
6. $20.09
   $100 427
7. (a) 20 000 cans (b) $2000 (c) 10¢/can
8. 1.5 h, 34 008 bacteria
9. 10 d, 37%
10. (b) 4 min more
12. (a) $1568.31 (b) $4065.70
13. (b) 43.8 min

Exercise 12.6, Page 332

1. (a) $e^{x \ln x}(1 + \ln x)$ (b) $\dfrac{e^x(\ln x - \frac{1}{x})}{(\ln x)^2}$

(c) $\dfrac{-1}{4 - x}$ (d) $-\tan x$

(e) $\dfrac{xe^{-x} - e^{-x}}{1 - xe^{-x}}$ (f) $\dfrac{xe^{\sqrt{1 + x^2}}}{\sqrt{1 + x^2}}$

(g) $-2x^{-3}e^{1/x^2}$

2. $y = 3x - 2 + \ln 4$
3. $e, (1, e)$
5. (a) min. (0, 0)
   (b) min. $(-1, -e^{-1})$ infl. $(-2, -2e^{-2})$
   (c) min. (1, e)

(d) max. $\left(\dfrac{\sqrt{2}}{2}, \dfrac{\sqrt{2}}{2\sqrt{e}}\right)$

min. $\left(-\dfrac{\sqrt{2}}{2}, -\dfrac{\sqrt{2}}{2\sqrt{e}}\right)$

infl. (0, 0)

$\left(\sqrt{\tfrac{3}{2}}, \sqrt{\tfrac{3}{2}}e^{-3/2}\right)$

$\left(-\sqrt{\tfrac{3}{2}}, -\sqrt{\tfrac{3}{2}}e^{-3/2}\right)$

(e) min. (0, 0)
   max. $(2, 4e^{-1}), (-2, 4e^{-1})$
   infl. at $x = \pm 3.02, \pm 0.94$

(f) infl. $\left(-\dfrac{1}{2}, e^{-2}\right)$

6. (a) min. (1, 0) max. $(e^{-2}, 4e^{-2})$ infl. (e^{-1}, e^{-1})

(b) min. $\left(e, -\dfrac{e^2}{4}\right)$ infl. $\left(1, -\dfrac{3}{4}\right)$

(c) min. (2, ln 4) infl. at $x = 2 + \sqrt{2}$

(d) max. $\left(1, \ln(\tfrac{1}{2})\right)$ infl. at $x = \sqrt{2 + \sqrt{5}}$

7. The graph of the first has a maximum at (0, ln 4). The graph of the second is the graph of the first translated two units to the right.

8. extrema at $x = \tan^{-1} 4 + n\pi$
9. 12.5 a
10. $200, $5414.41
11. 5 days, 50 bacteria/mL

13. (a) max. $\left(\bar{x}, \dfrac{1}{\sqrt{2\pi}\sigma}\right)$

infl. $\left(\bar{x} + \sigma, \dfrac{1}{\sqrt{2\pi}\sigma}e^{-1/2}\right)$

(b) 0

Exercise 13.1, Page 340

1. (a) $dy = \left[1 + \frac{1}{2}(x - 5)^{-1/2}\right]dx$
   $\Delta y = \Delta x + \sqrt{x + \Delta x - 5} - \sqrt{x - 5}$
   (b) $dy = 3x^2 dx$
   $\Delta y = 3x^2(\Delta x) + 3x(\Delta x)^2 + (\Delta x)^3$
   (c) $dy = -\dfrac{3}{(x + 2)^2}dx$
   $\Delta y = -\dfrac{3(\Delta x)}{(x + \Delta x + 2)(x + 2)}$
   (d) $dy = 2e^{2x}dx$
   $\Delta y = e^{2x}(e^{2(\Delta x)} - 1)$

2. (a) $dA = 2xdx$
   $A + dA = x^2 + 2xdx$
   $A(x + \Delta x) = (x + \Delta x)^2$
   $\Delta A = 2x(\Delta x) + (\Delta x)^2$
   $\Delta A - dA = (\Delta x)^2$

3. (a) $dy = 6x(x^2 - a^2)^2 dx$
   (b) $dy = 16 \sin 2x \cos 2x dx$

(c) $dy = \dfrac{5}{(3-4x)^2}dx$

(d) $dy = (2 + \ln 3x^2)dx$

4. (a) $f(x + \Delta x) \doteq 4 - 2x^3 - 6x^2dx$

(b) $f(x + \Delta x) \doteq \dfrac{x}{\sqrt{x^2-1}} - \dfrac{1}{(x^2-1)^{3/2}}dx$

(c) $f(x + \Delta x) \doteq \cos 2x - 2\sin 2xdx$

(d) $f(x + \Delta x) \doteq 5xe^{-x} + 5(1 - x)e^{-x}dx$

5. (a) $9.95, (x = 100, dx = -1)$

(b) $283.5, (x = 3, dx = 0.1)$

(c) $3.019, (x = 81, dx = 2)$

(d) $0.203, (x = 125, dx = -5)$

(e) $0.515, \left(x = \dfrac{\pi}{3}, dx = -\dfrac{\pi}{180}\right)$

(f) $1.94, \left(x = \dfrac{\pi}{3}, dx = \dfrac{3\pi}{180}\right)$

6. 2.916

7. -0.04

8. $dy = 0.375$
   $\Delta y = 0.391$

Exercise 13.2, Page 345

1. 4π cm^2

2. 720 m^2

3. $500, 16.7\%$

4. 189 L

5. (a) $4\pi r^2 dr$

(b) 2.06×10^{10} km^3

6. diameter < 12.616 cm

8. $|x| \geq 192.5$

9. $\frac{2}{3}\%$

Exercise 13.3, Page 346

1. (a) $dy = (1 - 2x)^2(1 - 8x)dx$

(b) $dy = (3\cos^2 x - 6x\cos x \sin x)dx$

(c) $dy = \dfrac{4 + 3x^2}{(1 + x)^{3/2}}dx$

2. (a) $dy = 0.1 \quad \Delta y = 0.1$

(b) $dy = 0.067 \quad \Delta y = 0.066$

(c) $dy = -0.0030 \quad \Delta y = -0.0029$

3. 7.85 km^2

4. $\frac{1}{3}\%$

5. ± 0.00003

6. -0.087 m

7. shortened by 0.28%

8. $\dfrac{\Delta v}{v} = 3\dfrac{\Delta l}{l}$

9. (b) ± 0.314 cm^2

Exercise 14.1, Page 355

1. (a) linear   (b) cubic   (c) constant
   (d) quadratic

2. $-\dfrac{x^4}{2} + $ any constant

3. (a) $x^4 + C$ (b) $e^x + C$ (c) $\sin x + C$
   (d) $\ln x + C$

4. $f'(3) = g'(3) = h'(3) = 4$
   The functions differ only by a vertical translation.

5. $g(x) = x^2 - 2x - 7$

6. (a) $x^3 + C$ (b) $-\cos x + C$
   (c) $\sec x + C$ (d) $x^2 + 3x + C$
   (e) $4x - \dfrac{x^5}{5} + C$ (f) $\dfrac{x^4}{2} - \dfrac{4x^3}{3} + C$

7. (a) $f(x) = x^5 + 3$ (b) $f(x) = -\cos x + 1$
   (c) $g(x) = \dfrac{x^2}{2} - 4x + 3$ (d) $g(x) = e^x + 1$

Exercise 14.2, Page 360

1. (a) $\frac{1}{2}x^6 + C$ (b) $16x^4 + C$

(c) $\frac{1}{5}(x + 3)^5 + C$ (d) $9x - 6x^2 + \frac{4}{3}x^3 + C$

(e) $-\dfrac{1}{8x^2} + C$ (f) $\dfrac{2\sqrt{3}}{3}x^{3/2} + C$

(g) $-\dfrac{8}{\sqrt{x}} + C$ (h) $-\frac{9}{4}x^{-4/3} + C$

2. (a) $\dfrac{4x^5}{5} - x^3 - 2x + C$

(b) $\frac{2}{7}x^{7/2} + \frac{4}{5}x^{5/2} + C$

(c) $\frac{4}{3}x^{3/2} - \frac{6}{5}x^{5/2} + C$

(d) $3x + \dfrac{4\sqrt{3}}{3}x^{3/2} + \frac{1}{2}x^2 + C$

3. (a) $\frac{1}{5}x^5 + \frac{3}{2}x^4 + \frac{1}{3}x^3 - 12x^2 + 16x + C$

(b) $\frac{1}{3}(x^2 + 3x - 4)^3 + C$

(c) $\dfrac{-1}{(x^2 + 3x - 4)^2} + C$

(d) $-\dfrac{4}{(x - 4)} + C$

(e) $\frac{3}{8}(2x - 5)^{4/3} + C$

(f) $\frac{1}{3}(x^2 - 9)^{3/2} + C$

4. (a) $\frac{1}{6}\sin^6 \theta + C$ (b) $\frac{1}{4}\sec^4 \theta + C$

(c) $\theta + \sin^2 \theta + C$ (d) $\frac{1}{5}\tan^5 \theta + C$

(e) $\frac{1}{4}(\ln x)^4 + C$ (f) $-\frac{1}{5}(2 - e^x)^5 + C$

(g) $\frac{1}{2}(e^x + 3)^2 + C$ (h) $2\sqrt{e^x + 5} + C$

5. $\ln x + C$

6. (a) $\ln(2x + 3) + C$ (b) $-\frac{1}{3}\ln(2 - 3x) + C$

(c) $\frac{1}{4}\ln(2x^2 + 1) + C$ (d) $\frac{1}{2}\ln(x^2 - 4x + 6) + C$

(e) $\frac{1}{4}\ln(\tan \theta + 2) + C$

(f) $-\frac{1}{6}\ln(3\cos 2\theta + 4) + C$

(g) $-\frac{1}{2}\ln(5 - 2e^x) + C$ (h) $\ln(\ln x) + C$

7. $\frac{1}{3}(x + 3)^3 + C, \frac{1}{3}x^3 + 3x^2 + 9x + C'$

Exercise 14.3, Page 364

1. (a) $-\frac{1}{3}\cos 3x + C$ (b) $\frac{1}{2}\sin(2x + 1) + C$

   (c) $-\frac{1}{5}\cos(5x - 2) + C$ (d) $2\sin\frac{1}{2}x + C$

   (e) $-\frac{4}{\pi}\cos \pi x + C$

2. (a) $\tan 3x + C$ (c) $\frac{1}{2}\sec 2x + C$

   (c) $\frac{1}{8}\tan 4x + C$ (d) $-\frac{1}{3}\csc 3x + C$

   (e) $-\cot x + C$

6. (a) $\ln(\sec \phi) + C$ (b) $-\ln(\csc \phi + \cot \phi) + C$
   (c) $\ln(\sec \phi + \tan \phi) + C$ (d) $-\ln(\csc \phi) + C$

7. (a) $\frac{1}{2}\ln\left[\sec(2x - 3) + \tan(2x - 3)\right] + C$

   (b) $-\frac{1}{2}\ln(\csc 2x + \cot 2x) + C$

   (c) $\ln(\sec x + \tan x) + C$

   (d) $\frac{1}{2}\tan x + C$

   (e) $-\ln(\csc x + \cot x) + C$
   (f) $2\ln(\sec \theta + \tan \theta) + C$
   (g) $\sec x + C$

Exercise 14.4, Page 367

1. (a) $\frac{1}{5}e^{5x} + C$ (b) $-\frac{3}{2}e^{-2x} + C$

   (c) $-\frac{1}{e^x} + C$ (d) $\frac{1}{2}e^{x^2} + C$

   (e) $\frac{2^{3x}}{3\ln 2} + C$ (f) $-\frac{10^{-x}}{\ln 10} + C$

2. (a) $\frac{1}{4}(x\ln x - x) + C$ (b) $3(x\ln x - x) + C$

   (c) $x\ln 3x - x + C$ (d) $x\log_{10} 5x - x\log_{10} e + C$

   (e) $x\log_2\left(\frac{x}{2}\right) - x\log_2 e + C$

Exercise 14.5, Page 369 (see page 560)

Exercise 14.6, Page 370

1. (a) $\frac{7}{9}x^9 + C$ (b) $\frac{10}{11}x^{11/5} + C$

   (c) $\frac{2\sqrt{5}}{5}x^{5/2} + C$ (d) $-18x^{-1/3} + C$

   (e) $\frac{3}{4}x^{4/3} + 6x^{2/3} + C$ (f) $\frac{2}{3}x^{9/2} + x^2 + C$

2. (a) $\frac{2}{9}x^{3/2} - \frac{4}{21}x^{7/2} + C$ (b) $\frac{6}{5}\ln x + \frac{8}{5}x^{-1/2} + C$

   (c) $\frac{3}{2}x^2 + \frac{24\sqrt{3}}{5}x^{5/2} + 12x^3 + C$

   (d) $\frac{6}{7}x^{7/2} + x^{1/2} + C$

   (e) $6x - x^2 + x^3 - \frac{1}{4}x^4 + C$

3. (a) $4\tan x + \csc x + C$ (b) $\theta - \cos \theta + C$

   (c) $-\frac{1}{4}\cot x + C$ (d) $-\frac{1}{2\sin \theta} + C$

   (e) $\tan x + \sec x + C$ (f) $x + \sec x + C$

4. (a) $\tan \theta + 2\ln(\sec \theta) + C$

   (b) $4\tan \theta - \frac{4}{3}\tan^3 \theta + \frac{1}{5}\tan^5 \theta + C$

   (c) $-\cos \theta + \frac{1}{3}\cos^3 \theta + C$

(d) $\frac{1}{2}\tan^2 \theta - \ln(\sec \theta) + C$

5. (a) $2(e^{x/2} - e^{x/2}) + C$ (b) $6x^2 + C$

6. (a) $\frac{1}{12}(4x^2 + 2)^{3/2} + C$ (b) $-\frac{1}{3}(9 - x^2)^{3/2} + C$

   (c) $\frac{1}{9}(2x^3 - 6)^{3/2} + C$

7. (a) $\frac{3}{2}x^2 - 4x + 20\ln(x + 3) + C$

   (b) $2x + 16\ln(x - 2) + C$

   (c) $2x + \frac{5}{2}\ln(2x + 5) + C$

   (d) $\frac{4}{3}x^3 - 8x^2 + 64x - 264\ln(x + 4) + C$

   (e) $\frac{1}{2}x^2 - 3x + 3\ln(x + 1) + C$

   (f) $\frac{1}{2}x^2 + 2x - 7\ln(x - 4) + C$

8. (a) $\frac{1}{4}\tan^{-1}4x + C$ (b) $\frac{3}{2\sqrt{5}}\tan^{-1}\left(\frac{\sqrt{5}x}{2}\right) + C$

   (c) $\ln(2x + \sqrt{4x^2 + 5}) + C$
   (d) $15\ln(x + \sqrt{x^2 - 25}) + C$

   (e) $\frac{1}{2}\ln\frac{x - 6}{x} + C$ (f) $\frac{1}{5}\ln\frac{x - 1}{x + 4} + C$

9. (a) $\frac{1}{12}(\ln x)^4 + C$ (b) $\frac{1}{2}\ln(\ln 2x) + C$

   (c) $\ln(x + \cos x) + C$

   (d) $\frac{1}{4}\ln(2x^2 + 4\sin x) + C$

   (e) $\frac{1}{3}\ln(3\tan x + 5) + C$

   (f) $-\frac{8}{3}\cos^3 x + 4\cos x + C$

10. (a) $x\sin x + \cos x + C$
   (b) $-x\cos x + \sin x + C$

11. (a) $x\tan^{-1}x - \frac{1}{2}\ln(1 + x^2) + C$

   (b) $x\sin^{-1}x + \sqrt{1 - x^2} + C$

Exercise 15.1, Page 377

1. (a) $\frac{4}{3}t^3 - 3t^{-1} + \frac{2}{3}t^{3/2} + C$

   (b) $-\frac{1}{2}e^{-x^2} + C$ (c) $\frac{1}{3}\ln(3t + 1) + C$

   (d) $-2\cos(2s) + C$

2. (a) $\frac{1}{3}(w^2 + 5)^{3/2} + 41$ (b) $\frac{4}{3}\pi r^3$

   (c) $100t\ln t - 100t + 500$

   (d) $\frac{5}{6}\sin^2 3x + \frac{49}{6}$

3. $0.02t^2 + C$, the initial height of the plant
4. 34¢
5. 110 cm
6. 46 a
7. \$4.00, 8000 units
8. 5385 helmets
9. 4189 cm³
10. \$1250
11. 8 days
12. 30 days
13. 2.4 months

Exercise 15.2, Page 382

1. (a) $2t^2 + 2t^{3/2} + C$ (b) $\frac{5}{2}\sin 2t + C$

(c) $4\ln(3 + t) + C$　(d) $\frac{4}{3}e^{3t+1} + C$

2. 6 s
3. 7 cm/s
4. 3.5 cm
5. $0 \le t < 2,\ 2.5 < t < 3$
6. $0 < t < 1,\ 3 + 4k < t < 5 + 4k,\ k = 0,\ 1,\ 2\ldots$
7. 128 cm
8. 37.5 cm
9. -20 m/s
10. 10 m/s, 7 m, 2.18 s
11. no
12. 30 s
13. 40 km

Exercise 15.3, Page 388

1. (a) $\frac{2}{15}(5t + 3)^{3/2} + C$　(b) $2e^{0.5t} + C$

   (c) $-\frac{4}{3}\cos(3t) + C$　(d) $4\ln(3 + t) + C$

2. (a) $2t^{3/2} - 9t + C_1,\ \frac{4}{5}t^{5/2} - \frac{9}{2}t^2 + C_1 t + C_2$

   (b) $\frac{2}{5}e^{5t} + C_1,\ \frac{2}{25}e^{5t} + C_1 t + C_2$

   (c) $-\frac{5}{2}\sin(-2t) + C_1,\ -\frac{5}{4}\cos(-2t) + C_1 t + C_2$

   (d) $9t^{5/3} + C_1,\ \frac{27}{8}t^{8/3} + C_1 t + C_2$

3. (a) $6t^2 + 8t - 2,\ 2t^3 + 4t^2 - 2t + 5$

   (b) $3t^2 - 5t + 6,\ t^3 - \frac{5}{2}t^2 + 6t - \frac{23}{2}$

   (e) $-3\cos \pi t,\ -\frac{3}{\pi}\sin \pi t + 5$

   (d) $\frac{1}{2}e^{2t},\ \frac{1}{4}e^{2t} + \frac{35}{4}$

4. 2.5 s, 31.25 m
5. 20 s
6. $t > 1$ s
7. 2 s, 5 s, -2 m/s^2, 16 m/s^2
8. 2 s
9. 104 m/s
10. 12 cm/s^2
11. 10 m/s
12. 14.2 m
13. 2275 km/h^2
15. 8.9 m

Exercise 15.4, Page 394

1. (a) $2y^3 = 3x^2 + C$　(b) $y = C\exp\left(\frac{2}{3}x^{3/2}\right)$

   (c) $|y - 1| = C\exp\left(\frac{(x - 1)^2}{2}\right)$

   (d) $y^2 = 2e^x + C$

2. 5.8 g
3. 10 091
4. 86 hours
5. 6.5 h
6. 54.4 a
7. 1792 a
8. 0.1%
9. 17.2 kg
10. 35 min
11. 104, 4.7 h
12. $60 496.48

Exercise 15.5, Page 397

1. (a) $3t^4 - \frac{2}{3}t^{3/2} + 5t + C$　(b) $\frac{(3x + 5)^3}{9} + C$

   (c) $-5\cos(\pi y) + C$　(d) $\frac{5}{4}e^{4z} + C$

2. (a) $16a^2 + 4a + 5$　(b) $\frac{4}{5}e^{5t} + \frac{41}{5}$

   (c) $\ln(5 + x) + 12$　(d) $\frac{2}{9}(3t + 1)^{3/2} - \frac{119}{9}$

3. 3¢
4. $4394.45
5. 11 months
6. $3.00, 6000
7. 54 366
8. the second, $44.20
9. (a) $3t^3 + 1.2t^{2.5} + C$　(b) $-\frac{5}{3}\cos(3t + 9) + C$

   (c) $6\ln(8 + t) + C$　(d) $3e^{9-2t} + C$

10. (a) $\frac{1}{3}t^3 - \frac{2}{3}t^{3/2} + 9$　(b) $4\sin(\pi t) + 4$

    (c) $4t^2 - 7t - 6$　(d) $\frac{1}{3}(2t + 15)^{3/2} - \frac{113}{3}$

11. 2 s, 8 cm/s, -1 cm/s
12. 4 cm/s, -4 cm/s
13. 1960 m
14. 15 m/s, 375 m
15. (a) $\frac{16}{9}(3t - 9)^{3/2} + C$　(b) $-20e^{-0.3t} + C$

    (c) $4\tan(3t) + C$　(d) $\frac{3}{10}(2t - 1)^{5/3} + C$

16. (a) $\frac{1}{2}t^4 - 5t + C_1,\ \frac{1}{10}t^5 - \frac{5}{2}t^2 + C_1 t + C_2$

    (b) $\frac{2}{3}t^{1.5} - 2t^{0.5} + C_1,\ \frac{4}{15}t^{2.5} - \frac{4}{3}t^{1.5} + C_1 t + C_2$

    (c) $+\frac{5}{2}\cos(-2t + 6) + C_1,$

    　　$-\frac{5}{4}\sin(-2t + 6) + C_1 t + C_2$

    (d) $2e^{-2t} + C_1,\ -e^{-2t} + C_1 t + C_2$

17. (a) $12t^2 - 8t + 5,\ 4t^3 - 4t^2 + 5t + 2$

    (b) $6t^2 - 4t - 11,\ 2t^3 - 2t^2 - 11t + 14$

    (c) $4\sin(\pi t) + 4,\ -\frac{4}{\pi}\cos(\pi t) + 4t + 5 + \frac{4}{\pi}$

    (d) $\frac{4}{5}e^{5t} + \frac{36}{5},\ \frac{4}{25}e^{5t} + \frac{36}{5}t + \frac{221}{25}$

18. 107 m, 75 m
19. 0.625 m/s^2
20. 31.25 m
21. yes
22. 1.9 s, 9.6 m
23. (a) $y^3 = x^3 + C$　(b) $3y^2 = 4x^{3/2} + C$

    (c) $y = C\exp\left(x - \frac{x^2}{2}\right)$　(d) $y^3 = e^{3x} + C$

24. $y^3 = 6x^3 - 7$
25. 12.5%
26. 6.6 h
27. 3.7 h
28. 81 920
29. 11.6 days
30. 5.3 kg
31. $33 201.17

Exercise H, Page 403

1. (a) $1^4 + 2^4 + 3^4 + 4^4 + 5^4$

   (b) $(1^2 + 3) + (2^2 + 3) + (3^2 + 3) + (4^2 + 3)$

   (c) $(2 \cdot 1 - 3) + (2 \cdot 2 - 3) + (2 \cdot 3 - 3)$
      $+ (2 \cdot 4 - 3) + (2 \cdot 5 - 3) + (2 \cdot 6 - 3)$

   (d) $2(-3)^2 + 2(-2)^2 + 2(-1)^2 + 2(0)^2$
      $+ 2(1)^2 + 2(2)^2 + 2(3)^2$

   (e) $\dfrac{1+1}{1+2} + \dfrac{2+1}{2+2} + \dfrac{3+1}{3+2} + \dfrac{4+1}{4+2} + \dfrac{5+1}{5+2} + \dfrac{6+1}{6+2}$

   (f) $(-1)^0 2^0 + (-1)^1 2^1 + (-1)^2 2^2 + (-1)^3 2^3$
      $+ (-1)^4 2^4$

   (g) $\dfrac{(-1)^0}{2 \cdot 0 + 1} + \dfrac{(-1)^1}{2 \cdot 1 + 1} + \dfrac{(-1)^2}{2 \cdot 2 + 1} + \dfrac{(-1)^3}{2 \cdot 3 + 1}$

      $+ \dfrac{(-1)^4}{2 \cdot 4 + 1} + \dfrac{(-1)^5}{2 \cdot 5 + 1} + \dfrac{(-1)^6}{2 \cdot 6 + 1} + \dfrac{(-1)^7}{2 \cdot 7 + 1}$

      $+ \dfrac{(-1)^8}{2 \cdot 8 + 1}$

2. (a) $\displaystyle\sum_{k=1}^{23} k$ (b) $\displaystyle\sum_{k=1}^{12} 2k$ (c) $\displaystyle\sum_{k=0}^{8} (3k + 1)$

   (d) $\displaystyle\sum_{k=2}^{8} 2^k$ (e) $\displaystyle\sum_{k=0}^{6} 3^k$ (f) $\displaystyle\sum_{k=1}^{n} (2k - 1)$

   (g) $\displaystyle\sum_{k=1}^{n} \dfrac{1}{k!}$

3. (a) 28   (b) 49   (c) 15   (d) 5

   (e) $29\frac{2}{5}$ (f) 0   (g) 10

4. $-1, 0$

5. (a) $1, k, 1, n$ (b) $\displaystyle\sum_{k=1}^{n} k$

   (c) $S_n = \dfrac{n(n+1)}{2}$

6. (c) $1, 1 + 3, 1 + 3 + 5, 1 + 3 + 5 + 7$

      $1 + 3 + 5 + 7 = \displaystyle\sum_{k=1}^{4} (2k - 1)$

8. $\displaystyle\sum_{k=1}^{n} 2k = n(n + 1)$

9. The sum of the first n even integers and the first n odd integers equals the sum of the first $2n$ integers.

Exercise 16.1, Page 414

2. (a) 28   (b) 21   (c) 24.85   (d) 24.15

3. (a) 1.5   (b) 0.375   (c) 0.18   (d) 0.21

4. 114.75

5. 4.413

6. 20, 17.375

7. 13.22

8. (a) 5.25   (b) 55.42   (c) 0.3589   (d) 17.38

9. (a) $A_k = \left[2\left(\dfrac{5k}{n}\right) + 2 \right] \cdot \dfrac{5}{n}$

   (b) 35   (c) 35

10. (a) $A_k = \left[-\dfrac{1}{2}\left(-2 + \dfrac{4k}{n} \right) + 4 \right] \cdot \dfrac{4}{n}$

   (b) 16   (c) 16

11. $\dfrac{128}{3}$

12. 6.75

14. $A_n = R^2 n \tan\left(\dfrac{\pi}{n}\right)$

Exercise 16.2, Page 423

2. (a) $\displaystyle\int_{-1}^{3} 2\,dx$ (b) $\displaystyle\int_{-3}^{2} \left(\tfrac{1}{3}x + 1\right)dx$

   (c) $\displaystyle\int_{-4}^{2} \left(-\tfrac{1}{2}x + 2\right)dx$

3. (a) 68   (b) $31\frac{2}{3}$ (c) 4

   (d) 12.8   (e) $31\frac{2}{3}$

4. (a) 8   (b) 49.5   (c) $11\frac{1}{3}$ (d) 74.4

5. (a) 2   (b) 28.05   (c) 2.197   (d) 13.045

   (e) 26.308

6. (a) $10\frac{2}{3}$ (b) $51\frac{1}{3}$ (c) 0.6849

7. (a) $6\frac{2}{3}$ (b) 27

8. 36

9. 4.5

10. (a) $\displaystyle\int_{-3}^{5} \sqrt{25 - x^2}\,dx = 33.67$

   (b) $\displaystyle\int_{1}^{2} \sqrt{25 - (x - 6)^2}\,dx = 2.04$

   (c) $2\displaystyle\int_{0}^{4} (\sqrt{25 - x^2} - 3)dx = 11.8$

   (d) $\displaystyle\int_{0}^{8} (6 - \sqrt{25 - (x - 3)^2})dx = 14.33$

11. f odd, $\displaystyle\int_{-a}^{a} f(x)dx = 0$

   f even, $\displaystyle\int_{-a}^{a} f(x)dx = 2\displaystyle\int_{0}^{a} f(x)dx$

Exercise 16.3, Page 430

3. (a) $y = -x + 4$

      $y = x^2 - 2$

      $A = \displaystyle\int_{-3}^{1} \left[(-x + 4) - (x^2 - 2)\right]dx$

   (b) $y = x - 2$

      $y = x^2 + 4$

      $A = \displaystyle\int_{-3}^{2} \left[(x^2 + 4) - (x - 2)\right]dx$

   (c) $y = -x + 4$

      $y = -x^2 - 2$

      $A = \displaystyle\int_{-3}^{2} \left[(-x + 4) - (-x^2 - 2)\right]dx$

   (d) $y = x - 2$

      $y = -x^2 + 4$

$$A = \int_{-3}^{2}\left[(-x^2 + 4) - (x - 2)\right]dx$$

4. (a) 18   (b) 9
5. (a) $41\frac{2}{3}$ (b) 11.83
6. (a) 0.80   (b) 0.14
7. (a) 2   (b) 4   (c) $\dfrac{3\sqrt{3}}{2}, \dfrac{3\sqrt{3}}{4}$
8. (a) $\frac{9}{2}$ (b) $\frac{11}{6}$ (c) $\frac{59}{6}$
9. (a) $y = bx + c$ (b) $\dfrac{2h^3}{3}$

   (c) $y = bx + c + h^2$ (d) $\dfrac{4h^3}{3}$

10. (b) $m > c$ $x = 0, \pm\sqrt{\dfrac{m-c}{a}}$ (c) $A = \dfrac{(m-c)^2}{4a}$

Exercise 16.4, Page 435

1. (a) $\frac{1}{2}$ (b) divergent   (c) divergent
   (d) $\frac{1}{6}$ (e) $\frac{1}{2}$ (f) $\dfrac{1}{\ln 3}$ (g) $\frac{1}{2}$
2. (a) 2   (b) 54   (c) 4   (d) divergent
   (e) $4\sqrt{2}$ (f) $\frac{5}{3}(3^{3/5} + 2^{3/5})$
3. 24
4. divergent
5. 3π

Exercise 16.5, Page 442

1. 30
2. (a) $0.5, x_k = 6 + 0.5k$
   (b) $0.25, x_k = -2 + 0.25k$
   (c) $0.2, x_k = -3 + 0.2k$
3. (a) 16   (b) 60   (c) 21
4. (a) 24   (b) 22   (c) 21.5
5. (a) $112.5, 111\frac{2}{3}$ (b) 27.7, 27.1   (c) 1.95, 2
   (d) 0.350, 0.347   (e) 0.869, 0.865
   (f) 0.696, 0.693
6. (a) 23.695   (b) 8.9461   (c) 21.950
   (d) 0.13581   (e) 1.1600   (f) 1.3110
7. (a) 23.71367   (b) 8.909044   (c) 21.82485
   (d) 0.1352628   (e) 1.159336   (f) 1.311029
8. $f(-1) = 5$
   $f(0) = 6$
   $f(1) = 5$
   Area $= \frac{34}{3}$
9. $f(2.5) = 3.25$
   $f(3) = 5$
   $f(3.5) = 7.25$
   Area $= 5.0833$
10. (a) 4, 4   (b) 4.046655, 4.047189
   (c) 3.509578, 3.464102
11. (a) 5.331012   (b) 2.302777
   (c) 0.9428093   (d) 0.4811722
12. (a) 5.375771   (b) 2.307564
   (c) 1.00971   (d) 0.4811749

13. 3.141592 with $n = 100$
14. 3.820198 with $n = 100$
15. 36.7

Exercise 16.6, Page 444

1. (a) $\frac{41}{12}$ (b) $\dfrac{10\sqrt{2} - 4}{3}$ (c) 2.3
2. (a) $\frac{16}{3}$ (b) 21.1
3. (a) 34.13   (b) 0.75   (c) 2.828
4. $\frac{8}{3}, \frac{11}{6}, 8$
5. (a) $\frac{1}{8}$ (b) 2   (c) $\frac{1}{12}$
6. (a) 125   (b) 1
7. (a) -0.0864137 (b) 10.33014   (c) 17.73576

Exercise 17.1, Page 450

1. (a) $\frac{1}{6}\ln\left|\dfrac{x-3}{x+3}\right| + C$ (b) $\frac{3}{8}\ln\left|\dfrac{3x-4}{3x+4}\right| + C$

   (c) $2\sqrt{2}\sin^{-1}\dfrac{x}{2} + C$ (d) $\dfrac{3\sqrt{2}}{2\sqrt{5}}\tan^{-1}\dfrac{\sqrt{2}x}{\sqrt{5}} + C$

2. (a) $-5\left[\dfrac{2+3x}{9} - \frac{2}{9}\ln|2 + 3x|\right] + C$

   (b) $2x + 3\ln|x - 2| + C$

   (c) $-\frac{4}{3}(3 - x)\sqrt{3 + 2x} + C$

   (d) $\dfrac{2}{x} + 6\ln\left|\dfrac{3x-1}{x}\right| + C$

3. (a) $\sqrt{x^2 - \frac{2}{3}} - \dfrac{\sqrt{2}}{\sqrt{3}}\cos^{-1}\left(\dfrac{\sqrt{2}}{\sqrt{3}x}\right) + C$

   (b) $-2\ln\left|\dfrac{1 + \sqrt{4x^2 + 1}}{2x}\right| + C$

   (c) $\dfrac{1}{\sqrt{8}}\sqrt{15 - 8x^2} - \dfrac{\sqrt{15}}{\sqrt{8}}\ln\left|\dfrac{\sqrt{15} + \sqrt{15 - 8x^2}}{\sqrt{8}x}\right| + C$

   (d) $\dfrac{x}{2}\sqrt{4x^2 + 9} + \frac{9}{4}\ln|2x + \sqrt{4x^2 + 9}| + C$

4. (a) $\tan^{-1}(x + 3) + C$ (b) $\sin^{-1}\left(\dfrac{x-1}{\sqrt{2}x}\right) + C$

   (c) $-\frac{1}{49}\left[\dfrac{4x+5}{2x^2 + 5x - 3} + \frac{4}{7}\ln\left|\dfrac{2x+1}{2x+6}\right|\right] + C$

   (d) $\frac{1}{2}(x + 1)\sqrt{6 - 2x - x^2} + \frac{7}{2}\sin^{-1}\left(\dfrac{x+1}{\sqrt{7}}\right) + C$

5. (a) $\frac{1}{8}(4x + \sin(4x)\cos(4x)) + C$

   (b) $\frac{1}{32}(8x^2 - 4x + 1)e^{4x} + C$

   (c) $\frac{1}{625}(125x^3 - 75x^2 + 30x - 6)e^{5x} + C$

   (d) $-2x^2\cos\left(\dfrac{x}{2}\right) + 8x\sin\left(\dfrac{x}{2}\right) + 16\cos\left(\dfrac{x}{2}\right) + C$

6. (a) $\frac{3}{2}\ln(2x - 5) + C$

   (b) $\dfrac{2}{2-x} + \ln(2 - x) + C$

(c) $-\frac{1}{8}\ln\left|\frac{7x+8}{x}\right| + C$

(d) $-\frac{1}{2\sqrt{5}}\ln\left|\frac{\sqrt{5}x-1}{\sqrt{5}x+1}\right| + C$

7. (a) $\ln\left|\frac{\sqrt{x+1}-1}{\sqrt{x+1}+1}\right| + C$

(b) $-\frac{2}{375}(8-15x)\sqrt{(4+5x)^3} + C$

(c) $2\cos^{-1}\left(\frac{2}{\sqrt{5}x}\right) + C$

(d) $-\frac{1}{4}\sqrt{1+2x-4x^2} + \frac{1}{8}\sin^{-1}\left(\frac{4x-1}{\sqrt{5}}\right) + C$

(e) $3\sqrt{x^2+16} - 12\ln\left|\frac{4+\sqrt{x^2+16}}{x}\right| + C$

(f) $-\frac{1}{\sqrt{2}}\ln\left(\frac{\sqrt{2+2x-x^2}+\sqrt{2}}{x} + \frac{1}{\sqrt{2}}\right) + C$

8. (a) $\frac{x^6\ln 5x}{6} - \frac{x^6}{36} + C$

(b) $(8x-16)e^{x/2} + C$

(c) $x^3\sin x + 3x^2\cos x - 6x\sin x - 6\cos x + C$

Exercise 17.2, Page 455

1. (a) $\frac{1}{9(2-3x)^3} + C$ (b) $-\frac{3}{8(4x^2+5)} + C$

(c) $\frac{3}{5}(3x^2-5x+8)^4 + C$ (d) $\frac{2}{9}(3x+1)^{3/2} + C$

(e) $\frac{2}{3a}(ax+b)^{3/2} + C$ (f) $(x^2-1)^{1/2} + C$

2. (a) $\frac{2}{15}(1+2x^3)^{5/4} + C$ (b) $\frac{2}{3}(x^3-3x)^{1/2} + C$

(c) $-\frac{1}{6}(1-2\sqrt{x})^6 + C$

(d) $-\frac{5}{2}(5-2x)^{3/2} + \frac{3}{10}(5-2x)^{5/2} + C$

(e) $\frac{1}{28}(2x-5)^{7/2} + \frac{1}{2}(2x-5)^{5/2} + \frac{25}{12}(2x-5)^{3/2} + C$

(f) $\frac{4}{27}(1+x^3)^{9/4} - \frac{4}{15}(1+x^3)^{5/4} + C$

3. (a) $2(\sqrt{x}+1) - 2\ln(\sqrt{x}+1) + C$
(b) $(\sqrt{x}-1)^2 + 4(\sqrt{x}-1) + \ln(\sqrt{x}-1) + C$

4. (a) $\frac{1}{5}\ln(3+5\tan\theta) + C$ (b) $\frac{1}{8}(2-\cos 2\theta)^4 + C$

(c) $-\frac{1}{3}e^{\cos 3x} + C$ (d) $\frac{1}{12}e^{4x^3} + C$

(e) $\frac{3}{2}\ln(2e^x-1) + C$ (f) $\ln\left|\frac{\sqrt{1+e^x}-1}{\sqrt{1+e^x}+1}\right| + C$

5. (a) $\frac{61}{3}$ (b) $\frac{5}{72}$ (c) $\frac{15}{8}$

Exercise 17.3, Page 458

1. (a) $3\ln(2x-5) + C$ (b) $-\frac{5}{4}\ln(3-4x) + C$

2. (a) $\frac{4}{3}\tan^{-1}\left(\frac{x+1}{3}\right) + C$

(b) $\frac{4}{\sqrt{2}}\ln\left|\frac{x-4-\sqrt{2}}{x-4+\sqrt{2}}\right| + C$

3. (b) $\frac{x^2}{2} + 5x + 14\ln(x-2) + C$

4. (a) $x + 2\ln(x+3) + C$

(b) $-\frac{x^2}{2} - 3x - \ln(3x-2) + C$

(c) $\frac{x^3}{3} + \frac{x^2}{2} + \ln(x-1) + C$

5. (b) $2\ln(x^2-4x+8) + \frac{1}{2}\tan^{-1}\left(\frac{x-2}{2}\right) + C$

6. (a) $\frac{3}{2}\ln(x^2-6x+10) + 13\tan^{-1}(x-3) + C$

(b) $\frac{3}{2}\ln(x^2-2x-4) + \frac{3}{2\sqrt{3}}\ln\left|\frac{x-1-\sqrt{3}}{x-1+\sqrt{3}}\right| + C$

7. (a) $x - \frac{7}{2}\ln|x^2+4x+13| + 4\tan^{-1}\left(\frac{x+2}{3}\right) + C$

(b) $\frac{x^3}{3} - x + \tan^{-1}x + C$

8. (a) $2\sqrt{x-3} + \frac{2}{\sqrt{3}}\tan^{-1}\left(\frac{\sqrt{x-3}}{\sqrt{3}}\right) + C$

(b) $5\ln|(x-2) + \sqrt{x^2-4x+8}| + C$

Exercise 17.4, Page 462

1. (a) $\ln(x+1) - \frac{1}{2}\ln(2x+1) + C$

(b) $\frac{5}{3}\ln(x-5) - \frac{2}{3}\ln(x-2) + C$

(c) $\frac{1}{10}\ln(2x+1) + \frac{2}{5}\ln(x-2) + C$

(d) $-\ln(2x-1) - \ln(x-1) + C$

3. (a) $4\ln(x-1) - 7\ln(x+3) + 5\ln(x-4) + C$
(b) $-4\ln x + 2\ln(x-2) + 2\ln(x+2) + C$
(c) $\ln(2x-1) - 6\ln(2x-3) + 5\ln(2x-5) + C$
(d) $\frac{1}{6}\ln(x-2) + \frac{1}{6}\ln(x+2) - \frac{1}{6}\ln(x-1)$

$\qquad - \frac{1}{6}\ln(x+1) + C$

4. (a) $-2, 3, -2$

(b) $-2\ln(x+2) + 3\ln(x+3) + \frac{2}{x+3} + C$

5. (a) $2\ln x - \ln(x+1) + \frac{6}{x+1} + C$

(b) $2\ln(x-1) - \ln x + \frac{1}{x} + C$

6. (a) $-1, 2, -2$
(b) $\ln(x^2+2x+5) - 2\ln(x-2)$

$\qquad - \frac{3}{2}\tan^{-1}\left(\frac{x+1}{2}\right) + C$

7. (a) $\ln x - \frac{1}{2}\ln(x^2+1) + C$

(b) $\frac{1}{3}\ln(x+1) - \frac{1}{6}\ln(x^2-x+1)$

$\qquad + \frac{1}{\sqrt{3}}\tan^{-1}\left(\frac{2x-1}{\sqrt{3}}\right) + C$

(c) $-\ln(x-1) + \frac{3}{2}\ln(x^2-2x+5)$

$\qquad + \frac{1}{2}\tan^{-1}\left(\frac{x-1}{2}\right) + C$

Exercise 17.5, Page 465

1. (a) $x\sin x + \cos x + C$

(b) $-x\cos(2x+1)+\frac{1}{2}\sin(2x+1)+C$

(c) $x\tan x-\ln(\sec x)+C$

(d) $-x^2\cos x+2x\sin x+2\cos x+C$

2. (a) xe^x-e^x+C

(b) $-e^{-x}[x^3+3x^2+6x+6]+C$

3. (a) $\frac{2}{5}e^{2x}\sin x-\frac{1}{5}e^{2x}\cos x+C$

(b) $\frac{4}{7}\sin 3x\sin 4x+\frac{3}{7}\cos 3x\cos 4x+C$

4. (a) $x\ln 7x-x+C$

(b) $\frac{x^2}{2}\ln 2x-\frac{x^2}{4}+C$

(c) $x\ln^2x-2x\ln x+2x+C$

(d) $\frac{x^{n+1}}{n+1}\ln x-\frac{x^{n+1}}{(n+1)^2}+C$

6. (a) $x\sin^{-1}(4x)+\frac{1}{4}\sqrt{1-16x^2}+C$

(b) $\frac{x^2}{2}\tan^{-1}x+\frac{1}{2}\tan^{-1}x-\frac{1}{2}x+C$

7. (a) $\frac{2}{3}x\sqrt{3x+1}-\frac{4}{27}(3x+1)^{3/2}+C$

(b) $-x^2\sqrt{1-x^2}-\frac{2}{3}(1-x^2)^{3/2}+C$

8. (a) 0.05678 (b) $\frac{\pi}{4}$

12. (b) $x\ln^3x-3x\ln^2x+6x\ln x-6x+C$

Exercise 17.6, Page 467

1. $\frac{5}{\sqrt{3}}\tan^{-1}\frac{x}{\sqrt{3}}+C$

2. $\frac{1}{2}x+\frac{3}{4}\ln(2x+1)+C$

3. $\frac{2x+1}{4}-\frac{1}{4}\ln(2x+1)+C$

4. $\frac{1}{\sqrt{3}}\ln\left|\frac{\sqrt{3+4x}-\sqrt{3}}{\sqrt{3+4x}+\sqrt{3}}\right|+C$

5. $\frac{1}{6}(x-1)\sqrt{2+4x}+C$

6. $12-2x-12\ln\left(3-\frac{x}{2}\right)+C$

7. $-\frac{1}{3}\ln\left|\frac{5-2x}{5}\right|+C$

8. $3\ln|1-x|+C$

9. $\frac{1}{2}(x-1)\sqrt{x^2-2x-1}$
   $-\ln|\sqrt{x^2-2x-1}+x-1|+C$

10. $\frac{1}{2}(x-3)\sqrt{6x-x^2}+\frac{9}{2}\sin^{-1}\left(\frac{x-3}{3}\right)+C$

11. $\sqrt{x^2+2x}+\ln|x+1+\sqrt{x^2+2x}|+C$

12. $\frac{4}{49}(7x-4)^{7/4}+C$

13. $\frac{2}{15}(1+x^{5/2})^3+C$

14. $x-1+4\sqrt{x-1}+4\ln|\sqrt{x-1}-1|+C$

15. $-\frac{1}{\sqrt{1+\sin 2x}}+C$

16. $-\frac{1}{2}e^{-x^2}+C$

17. $-\frac{b^3}{2a^4}\sqrt{1-a^4x^4}+C$

18. $-\frac{2}{7}\ln|3-7x|+C$

19. $\frac{2x+3}{4}-\frac{3}{4}\ln|2x+3|+C$

20. $2x^2-16x+63\ln|x+4|+C$

21. $-\frac{2}{3}x+\frac{11}{9}\ln|3x-5|+C$

22. $\ln|x^2+4x+1|+\frac{1}{2\sqrt{3}}\ln\left|\frac{x+2-\sqrt{3}}{x+2+\sqrt{3}}\right|+C$

23. $-\ln(x-3)+2\ln(x-4)+C$

24. $\frac{x^3}{3}+\frac{x^2}{2}+4x+2\ln x+5\ln(x-2)$
    $-3\ln(x+2)+C$

25. $-\frac{x^2}{2}+x\tan x-\ln(\sec x)+C$

26. $\frac{1}{17}e^x\cos 4x+\frac{4}{17}e^x\sin 4x+C$

27. $(x+a)\ln(x+a)-(x+a)+C$

28. $-\frac{1}{x}\ln x-\frac{1}{x}+C$

29. $-\frac{\ln x}{x-1}-\ln x+\ln(x-1)+C$

30. $x\cos^{-1}x-\sqrt{1-x^2}+C$

31. $\frac{1}{15}(3x^2-2a^2)(x^2+a^2)^{3/2}$

Exercise 18.1, Page 478

2. (a) $\frac{1}{3}\sin 3x-\frac{1}{9}\sin^3 3x+C$

(b) $\sin x-\sin^3 x+\frac{3}{5}\sin^5 x-\frac{1}{7}\sin^7 x+C$

(c) $\frac{1}{6}\sin^3 2x-\frac{1}{10}\sin^5 2x+C$

3. (a) $-\cos x+\frac{1}{3}\cos^3 x+C$

(b) $-2\cos\frac{x}{2}+\frac{4}{3}\cos^3\frac{x}{2}-\frac{2}{5}\cos^5\frac{x}{2}+C$

(c) $-\frac{1}{8}\cos^4 2x+\frac{1}{12}\cos^6 2x+C$

4. (a) $\frac{1}{2}x-\frac{1}{2}\sin x\cos x+C$

(b) $\frac{3}{8}x+\frac{1}{4}\sin 2x+\frac{1}{32}\sin 4x+C$

(c) $\frac{1}{8}x-\frac{1}{128}\sin 16x+C$

5. (a) $\frac{1}{4}\sin 4x-\frac{1}{6}\sin^3 4x+\frac{1}{20}\sin^5 4x+C$

(b) $-\frac{1}{6}\cos^6 x+C$

(c) $\frac{1}{2}\cos^{-2}x+\ln(\cos x)+C$

(d) $\frac{1}{3}\cos^{-3}x+C$

(e) $-\frac{1}{3}\sin^{-3}x+C$

(f) $-\frac{2}{3}\cos^{3/2}x+\frac{2}{7}\cos^{7/2}x+C$

6. (a) $\frac{3}{8}x-\frac{1}{8}\sin 4x+\frac{1}{64}\sin 8x+C$

(b) $\frac{1}{16}x-\frac{1}{64}\sin 4x-\frac{1}{48}\sin^3 2x+C$

(c) $\frac{1}{2}\sin 2x - \frac{1}{6}\sin^3 2x + C$ (d) $x + C$

(e) $\frac{9}{2}x + \sin x - \frac{9}{4}\sin 2x - 4\cos^{3/2}x + C$

7. (a) $\frac{2}{3}\sin^3 x + C$ (b) $-\frac{1}{20}\cos^5 2x + C$

(c) $\frac{1}{4}x - \frac{1}{2}\sin x + \frac{1}{8}\sin 2x + \frac{1}{6}\sin^3 x + C$

8. $\frac{1}{20}\sin 10x + \frac{1}{4}\sin 2x + C$

9. (a) $-\frac{1}{18}\cos 9x + \frac{1}{10}\cos 5x + C$

(b) $-\frac{1}{24}\sin 12x + \frac{1}{8}\sin 4x + C$

(c) $\frac{1}{14}\sin 7x + \frac{1}{6}\sin 3x + C$

10. (a) $\frac{1}{7}\tan^7 x + C$ (b) $\frac{1}{4}\sec^4 x + C$

11. (a) $\frac{1}{2}\tan^2 x - \ln(\sec x) + C$

(b) $\frac{1}{10}\tan^5 2x + \frac{1}{3}\tan^3 2x + \frac{1}{2}\tan 2x + C$

(c) $\frac{1}{7}\tan^7 x + \frac{2}{5}\tan^5 x + \frac{1}{3}\tan^3 x + C$

12. (a) $\frac{1}{5}\sec^5\theta - \frac{2}{3}\sec^3\theta + \sec\theta + C$

(b) $\frac{1}{5}\sec^5\theta + C$

(c) $\frac{1}{2}\sec^4 x + C$

Exercise 18.2, Page 484

1. (a) $-\frac{1}{5}(9 - x^2)^{5/2} + C$

(b) $-\dfrac{\sqrt{7 - 4x^2}}{7x} + C$

(c) $-\frac{1}{4}\ln\left(\dfrac{4 + \sqrt{16 - x^2}}{x}\right)$

2. (a) $3\sqrt{x^2 + 16} + C$

(b) $\frac{1}{3}(x^2 - 18)\sqrt{9 + x^2} + C$

(c) $\frac{1}{6}x - \dfrac{1}{6\sqrt{6}}\tan^{-1}(\sqrt{6}x) + C$

3. (a) $\frac{1}{5}(x^2 - 16)^{5/2} + C$ (b) $\dfrac{1}{\sqrt{5}}\cos^{-1}\left(\dfrac{\sqrt{5}}{\sqrt{3}x}\right)$

(c). $\dfrac{\sqrt{x^2 - 6}}{6x} + C$

5. (a) $\dfrac{x}{\sqrt{1 - x^2}} - \sin^{-1}x + C$

(b) $\sqrt{x^2 - 1} - \cos^{-1}\left(\dfrac{1}{x}\right) + C$

(c) $\dfrac{(x^2 - 4)^{3/2}}{12x^3} + C$

(d) $-\dfrac{\sqrt{9x^2 + 25}}{25x} + C$

(e) $-\frac{1}{15}(3x^2 + 8)(4 - x^2)^{3/2} + C$

(f) $\frac{1}{3}(x^2 + 18)(x^2 - 9)^{1/2} + C$

10. $\ln\left(\dfrac{x + 1 + \sqrt{(x + 1)^2 + 4}}{2}\right) + C$

11. (a) $\ln|x + \sqrt{x^2 + 49}| - \dfrac{x}{\sqrt{x^2 + 49}} + C$

(b) $-\frac{1}{3}(x^2 + 2a^2)\sqrt{a^2 - x^2} + C$

(c) $\frac{1}{12}(4x^2 - 25)^{3/2} + C$

(d) $\frac{1}{3}(4 + x^2)^{3/2} + C$

(e) $-\frac{1}{20}(9 - 4x^2)^{5/2} + C$

(f) $\frac{1}{2}\cos^{-1}\left(\dfrac{1}{x}\right) + \dfrac{\sqrt{x^2 - 1}}{2x^2} + C$

Exercise 18.3, Page 486

1. $-\frac{1}{3}\cos^3 x + \frac{2}{5}\cos^5 x - \frac{1}{7}\cos^7 x + C$

2. $-\frac{1}{3}\cos^3 x + \frac{1}{5}\cos^5 x + C$

3. $\frac{2}{5}\sin^{5/2}x - \frac{2}{9}\sin^{9/2}x + C$

4. $x + \sin^2 x + C$

5. $\frac{1}{4}\sin^4 x + C$

6. $\dfrac{x}{2} - \dfrac{1}{4n}\sin 2nx + C$

7. $\frac{1}{12}\tan^4 3x - \frac{1}{6}\tan^2 3x + \frac{1}{3}\ln(\sec 3x) + C$

8. $\frac{2}{3}\tan^3\dfrac{x}{2} + 2\tan\dfrac{x}{2} + C$

9. $\frac{1}{8}\tan^8 x + \frac{1}{6}\tan^6 x + C$

10. $\frac{1}{3}\tan^3 x + C$

11. $\frac{1}{2}(\tan 2x - \cot 2x) + C$

12. $-\csc^3\dfrac{x}{3} + C$

13. $\tan x - \sec x + C$

14. $-\csc x - \cot x - x + C$

15. $\tan x + \sec x + \ln(\sec x) + \ln(\sec x + \tan x) + C$

16. $\frac{1}{2}\tan^{-1}x + \dfrac{x}{2(1 + x^2)} + C$

17. $-\dfrac{\sqrt{4 - 9x^2}}{x} - 3\sin^{-1}\left(\dfrac{3x}{2}\right) + C$

18. $-\dfrac{1}{\sqrt{5}}\ln\left|\dfrac{\sqrt{5} + \sqrt{5 - x^2}}{x}\right| + C$

19. $-\dfrac{(x^2 + 16)^{3/2}}{48x^3} + C$

20. $\dfrac{(3x^2 - 16)^{3/2}}{48x^3} + C$

21. $\sqrt{4x^2 - 9} - 3\cos^{-1}\left(\dfrac{3}{2x}\right) + C$

22. $\dfrac{\sqrt{4x^2 - 9}}{9x} + C$

23. $\frac{1}{4}\ln(1 + x^4) + \dfrac{1}{4(1 + x^4)} + C$

24. $\frac{1}{2}\sin^{-1}(e^x) + \frac{1}{2}e^x\sqrt{1 - e^{2x}} + C$

25. $\sqrt{2}\sin^{-1}\left(\dfrac{\sin x}{\sqrt{2}}\right) + C$

26. $2\tan^{-1}(\sqrt{x}) + C$

27. $\sin^{-1}(\sqrt{x}) + \sqrt{x}\cdot\sqrt{1-x} + C$

29. (a) $\ln\left(\tan\dfrac{x}{2} + 1\right) + C$

    (b) $\ln\left(\tan\dfrac{x}{2} - 2\right) + C$

    (c) $\frac{2}{3}\tan^{-1}\left(\frac{1}{3}\tan\dfrac{x}{2}\right) + C$

    (d) $-\sqrt{2}\ln\left|\dfrac{\tan\dfrac{x}{2} - 1 - \sqrt{2}}{\tan\dfrac{x}{2} - 1 + \sqrt{2}}\right| + C$

Exercise J, Page 492

5. rectangle, semicircle, right triangle

Exercise 19.1, Page 499

1. (c) $dV = 144\,\pi\,dy$ cm³ (d) $72\,000\,\pi$ cm³

2. (c) $dV = \left(\dfrac{30 - 5y}{6}\right)^2 dy$ (d) 50 cm³

4. 64 cm³

5. 128 cubic units

6. (a) $\dfrac{1024}{3}$ cm³ (b) $\dfrac{256\sqrt{3}}{3}$ cm³

7. 9 cubic units

8. 0.06075 m³

Exercise 19.2, Page 503

1. (a) $\dfrac{35\pi}{5}$ (b) $\dfrac{124\pi}{27}$ (c) $\dfrac{4\pi}{5}$

    (d) $\dfrac{\pi}{6}$ (e) 27π

2. (a) $\dfrac{3\pi}{5}$ (b) $\dfrac{72\pi}{5}$ (c) $\dfrac{558\pi}{5}$

3. (a) $\dfrac{\pi^2}{4}$ (b) $\dfrac{\pi^2}{2} - \pi$ (c) $\dfrac{\pi}{2}$

    (d) $\pi - \dfrac{\pi^2}{4}$ (e) $\dfrac{\pi^2}{8} + \dfrac{\pi}{4}$

6. 16π

7. (a) $\dfrac{8\pi a^3}{3}$ (b) $\dfrac{4\pi a^3}{3}(2\sqrt{2} - 1)$

8. $\dfrac{475\pi}{3}$ cm³

10. $108\pi^2$

Exercise 19.3, Page 507

1. 1.44, 1.41

2. $\frac{53}{6}$

3. $\frac{17}{6}$

4. 9.073

5. $\dfrac{4}{\sqrt{3}}$

6. 6, 5.66, 6.28

7. $\sqrt{99}$

8. $\frac{3}{4}$

9. 9.293

Exercise 19.4, Page 510

1. $\dfrac{208\pi}{3}$

3. $4\pi, \frac{1}{4}$

5. 3π

6. 17.68

7. 14.42

Exercise 19.5, Page 511

1. 4π

3. 1.54

4. 376.3 cm³

5. (a) $\frac{128}{3}$ (b) $\dfrac{32\sqrt{3}}{3}$ (c) $\frac{32}{3}$

6. 48π

7. $\dfrac{\pi^2}{8} + \dfrac{\pi}{4}$

8. 1.834, 0.031

9. $\dfrac{\pi}{4} + \dfrac{1}{2}$

10. $\dfrac{\pi}{6}$

11. $\frac{128}{3}$ cm³

12. $\dfrac{136\pi}{3}$

Exercise 20.1, Page 521

2. (a) $P(2, 210°)$ (b) $Q(4, 135°)$ (c) $R(3, 180°)$

    (d) $S\left(1, \dfrac{\pi}{6}\right)$ (e) $T\left(1, \dfrac{2\pi}{3}\right)$ (f) $U(0, \pi)$

4. (a) $(7, 0)$ (b) $\left(6, \dfrac{3\pi}{2}\right)$ (c) $\left(4\sqrt{2}, \dfrac{3\pi}{4}\right)$

    (d) $\left(8, \dfrac{2\pi}{3}\right)$ (e) $\left(8, \dfrac{5\pi}{6}\right)$ (f) $\left(6, \dfrac{4\pi}{3}\right)$

5. (a) $(-5, 0)$ (b) $(3\sqrt{3}, 3)$ (c) $(0, 2)$

    (d) $\left(\dfrac{7\sqrt{2}}{2}, \dfrac{7\sqrt{2}}{2}\right)$ (e) $(-3, 3)$ (f) $\left(\dfrac{3}{2}, -\dfrac{3\sqrt{3}}{2}\right)$

6. (a) reflected through origin
    (b) reflected in polar axis
    (c) reflected in y-axis (d) same point

7. (a) $r = 5$ (b) $r = 13\csc\theta$ (c) $r = \frac{22}{5}\csc 2\theta$
    (d) $r^2 = 6\sec 2\theta$ (e) $r = 13\tan\theta\sec\theta$
    (f) $r^2 = 25\cos 2\theta$

8. (a) $x^2 + y^2 = 6$, circle (b) $y = 5$, horizontal line
    (c) $x = 6$, vertical line
    (d) $x^2 + (y - 2)^2 = 4$, circle

(e) $\left(x + \frac{5}{2}\right)^2 + y^2 = \left(\frac{5}{2}\right)^2$, circle

(f) $\dfrac{(x - 2)^2}{16} + \dfrac{y^2}{12} = 1$, ellipse

Exercise 20.2, Page 526

2. (a) $\theta = \dfrac{\pi}{2}$ (b) Polar axis, pole, and $\theta = \dfrac{\pi}{2}$

(c) Polar axis   (d) Polar axis, pole, and $\theta = \dfrac{\pi}{2}$

(e) Polar axis   (f) $\theta = \dfrac{\pi}{2}$

8. (b) $y = \sqrt{\sin 6x}$

Exercise 20.3, Page 529

1. (a) $\dfrac{9\pi}{4}$ (b) 11π (c) $\dfrac{75\pi}{2}$ (d) $\dfrac{27\pi}{2}$

2. (a) $\dfrac{9\pi}{8}$ (b) $\dfrac{25\pi}{12}$ (c) 2   (d) $\dfrac{\pi}{6}$

3. $\pi - \dfrac{3\sqrt{3}}{2}, 2\pi + \dfrac{3\sqrt{3}}{2}$

4. $2\pi - 4$

Exercise 20.4, Page 530

2. (a) $(4, 360°), (-4, 180°), (-4, -180°)$

(b) $\left(6, \dfrac{3\pi}{2}\right), \left(6, -\dfrac{\pi}{2}\right), \left(-6, -\dfrac{3\pi}{2}\right)$

(c) $\left(2, \dfrac{5\pi}{3}\right), \left(-2, \dfrac{2\pi}{3}\right), \left(-2, -\dfrac{4\pi}{3}\right)$

(d) $(7, 225°), (7, -135°), (-7, -315°)$

(e) $(-3, 30°), (3, 210°), (3, -150°)$

(f) $\left(2, \dfrac{7\pi}{6}\right), \left(2, -\dfrac{5\pi}{6}\right), \left(-2, -\dfrac{11\pi}{6}\right)$

3. (a) $x^2 + y^2 = (x^2 + y^2 + 2y)^2$ (b) $y = 3$

(c) $(x^2 + y^2)^3 = 16x^2y^2$ (d) $\sqrt{x^2 + y^2} = 2\tan^{-1}\dfrac{y}{x}$

(e) $x^2 + y^2 = -4x$ (f) nasty!

4. (a) $r = \dfrac{2}{4\cos\theta + 7\sin\theta}$

(b) $r^2\sin^2\theta - 8r\cos\theta - 16 = 0$

(c) $r^2 = \dfrac{4}{\cos 2\theta}$ (d) $r = 9\tan\theta\sec\theta$

(e) $r = \tan\theta\sec\theta$ (f) $r^2 = -\dfrac{26}{7\sin 2\theta}$

7. (a) $\dfrac{9\pi}{4}$ (b) 2   (c) 8

8. $\frac{4}{3}\pi^3$

9. $12 + \pi$

10. 2

Exercise K, Page 533

1. (a) 2, 1, 0, 1   (b) $-2, 2, -1.25, 1.25$

2. $\frac{25}{40}$

3. (d) graph becomes larger.

4. (c) smoother graph, longer plotting time.

8. $r = \sqrt{\sin 6x}$
   or $r = \sqrt{\cos 6x}$

Exercise 14.5, Page 369

1. (a) $\frac{1}{5}\tan^{-1}\dfrac{x}{5} + C$

(b) $\ln\left(x + \sqrt{x^2 + 25}\right) + C$

(c) $\sin^{-1}\dfrac{x}{\sqrt{6}} + C$

(d) $-\frac{1}{4}\ln\dfrac{x - 4}{x + 4} + C$

(e) $\frac{1}{6}\tan^{-1}\dfrac{3x}{2} + C$

(f) $\frac{1}{12}\ln\dfrac{2x - 3}{2x + 3} + C$

(g) $\frac{1}{2}\ln\left(2x + \sqrt{4x^2 + 9}\right) + C$

(h) $\frac{1}{3}\sin^{-1}\dfrac{3x}{2} + C$

(i) $\ln\left(x + \sqrt{x^2 - 16}\right) + C$

(j) $\frac{1}{2}\sin^{-1}\dfrac{x}{\sqrt{2}} + C$

2. (a) $\dfrac{1}{\sqrt{7}}\tan^{-1}\dfrac{x - 1}{\sqrt{7}} + C$

(b) $\dfrac{1}{4\sqrt{3}}\tan^{-1}\dfrac{2x + 1}{2\sqrt{3}} + C$

(c) $\ln\left(x - 3 + \sqrt{x^2 - 6x - 7}\right) + C$

(d) $\frac{1}{12}\ln\dfrac{3x}{3x + 4} + C$

(e) $\sin^{-1}(x - 1) + C$

(f) $-\frac{1}{14}\ln\dfrac{x - 13}{x + 1} + C$

Photo Credits

Pages 39, 128, 258, 348, 406, 446, 513
The Bettmann Archive

Index